Printed Antennas

Printed Antennas

Theory and Design

Edited by
Prof. Binod Kumar Kanaujia,
Dr. Surendra Kumar Gupta,
Dr. Jugul Kishor, and Dr. Deepak Gangwar

CRC Press
Taylor & Francis Group
Boca Raton London New York

CRC Press is an imprint of the
Taylor & Francis Group, an informa business

First edition published 2021
by CRC Press
6000 Broken Sound Parkway NW, Suite 300, Boca Raton, FL 33487-2742

and by CRC Press
2 Park Square, Milton Park, Abingdon, Oxon, OX14 4RN

CRC Press is an imprint of Taylor & Francis Group, LLC

Library of Congress Cataloging-in-Publication Data
Names: Kanaujia, Binod Kumar, editor.
Title: Printed antennas : theory and design / edited by Prof. Binod Kumar Kanaujia, Dr. Surendra Kumar Gupta, Dr. Jugul Kishor and Dr. Deepak Gangwar.
Description: First edition. | Boca Raton, FL : CRC Press, 2021. | Includes bibliographical references and index. | Summary: "Printed antennas have become an integral part of next generation wireless communication and found to be commonly used to improve system capacity, data rate, reliability, etc. This book covers theory, design techniques, and the chronological regression of the printed antennas for various applications"—Provided by publisher.
Identifiers: LCCN 2020024665 (print) | LCCN 2020024666 (ebook) | ISBN 9780367420413 (hardback) | ISBN 9780367420451 (ebook)
Subjects: LCSH: Microstrip antennas. | Printed circuits.
Classification: LCC TK7871.67.M5 P756 2021 (print) | LCC TK7871.67.M5 (ebook) | DDC 621.382/4—dc23
LC record available at https://lccn.loc.gov/2020024665
LC ebook record available at https://lccn.loc.gov/2020024666

ISBN: 978-0-367-42041-3 (hbk)
ISBN: 978-0-367-42045-1 (ebk)

Typeset in Times
by codeMantra

Contents

Preface

The demand for high data throughput and channel capacity in modern wireless communication has been the primary motivation behind the development of printed antennas. The printed antennas have become integral part of the next generation wireless communication and are found to be commonly used to improve system capacity, data rate, reliability, etc. In this book, theory, design techniques, and a chronological regression of the printed antennas for various applications are covered. The primary objective is to focus on the different techniques used by the researchers for the design of printed antennas.

The printed antennas are highly in demand in communication engineering, biomedical sciences, crop harvesting, energy harvesting, etc., which makes it necessary for the students of colleges and universities worldwide to attain good knowledge of theory and design of printed antennas. Also, this has caused a huge interest among researchers and scientists to work on printed antennas. Although the printed antennas are in practice from 1970 and a lot of research work has been done in this field, the application areas of printed antennas are still increasing day by day. The availability of books and other reference materials to the students and researchers who wish to gain good knowledge about printed antennas and related theories is limited.

This book will help the worldwide learners of antennas to understand the theories of printed antennas and their applications. It will be useful in designing antennas and making out future scope of work in this field. It presents most of the issues that are directly related to the new age technologies in printed antennas. Even the researchers in this field will be benefitted from the ready reference and techniques of research in printed antennas included in this book.

The main objective of this book is to introduce the basic theory, analysis, design, and developing techniques of printed antennas. Each chapter contains simple and easy to understand analysis concepts of various types of printed antennas and their characteristics. In addition, this book consists of a number of graphical representations, values of design parameters, experimental results, and references. An adequate number of topics are covered, so that the book will help the students who wish to pursue bachelor/master degree, research, and career in the field of antennas. Also, this book will provide the reader with the basic conceptual knowledge about antennas along with the advanced techniques of antenna design.

We, the editors, wish to acknowledge all the contributors and the publisher for shaping this book. Many thanks go to all authors who provided their invaluable time and suggestions during the completion of this book.

Editors

Dr. Binod Kumar Kanaujia is a Professor in the School of Computational and Integrative Sciences, Jawaharlal Nehru University, New Delhi, where he has been from August 2016. Before joining Jawaharlal Nehru University, he was a Professor in the Department of Electronics & Communication Engineering in Ambedkar Institute of Advanced Communication Technologies and Research (formerly Ambedkar Institute of Technology), Delhi, from February 2011, and an Associate Professor from 2008 to 2011. Earlier, Dr. Kanaujia held the positions of Lecturer from 1996 to 2005, Reader from 2005 to 2008, and Head of Department in Department of Electronics & Communication Engineering, M.J.P. Rohilkhand University, Bareilly, India. Prior to his career in academics, Dr. Kanaujia was working as an Executive Engineer in the R&D division of M/s UPTRON India Ltd.

Dr. Kanaujia did his B. Tech. in Electronics Engineering in KNIT Sultanpur, India, in 1994. He did his M. Tech. and Ph.D. in 1998 and 2004, respectively, in Department of Electronics Engineering, Indian Institute of Technology, Banaras Hindu University, Varanasi, India. He was awarded Junior Research Fellowship by UGC, Delhi, in the year 2001–2002 for his outstanding work in the electronics field. He has a keen research interest in designing and modelling of microstrip antennas, dielectric resonator antennas, left-handed metamaterial microstrip antennas, shorted microstrip antennas, ultra-wideband antennas, reconfigurable and circularly polarized antennas for wireless communication.

He is credited to publish more than 325 research papers with more than 2,600 citations with an h-index of 25 in several peer-reviewed journals and conferences. He has supervised 50 M. Tech. and 25 Ph.D. research scholars in the field of microwave engineering. He is a reviewer for several journals of international repute, i.e. *IET Microwaves, Antennas & Propagation, IEEE Antennas and Wireless Propagation Letters, Wireless Personal Communications, Journal of Electromagnetic Wave and Application, Indian Journal of Radio and Space Physics, IETE Technical Review, International Journal of Electronics, International Journal of Engineering Science, IEEE Transactions on Antennas and Propagation, AEU-International Journal of Electronics and Communication, International Journal of Microwave and Wireless Technologies*, etc. Dr. Kanaujia has successfully executed 08 research projects sponsored by several agencies of Government of India, i.e. DRDO, DST, AICTE, and ISRO. He is a member of several academic and professional bodies, i.e. IEEE, Institution of Engineers (India), Indian Society for Technical Education, and the Institution of Electronics and Telecommunication Engineers of India. Also, he serves as an

Associate Editor in AEU-International Journal of Electronics and Communications of Elsevier, and IETE Technical Review of Taylor and Francis publishers.

Dr. Surendra Kumar Gupta did his Bachelor degree in Electronics & Telecommunication Engineering from Institution of Engineers (India), Kolkatta in 1994. He did his M.E. in Digital Systems in Motilal Nehru National Institute of Technology Allahabad (formerly MLNREC), India, in 1999. He did his Ph.D. in Uttarakhand Technical University, Dehradun, India, in 2015. He was associated as Quality Assurance Services (Avionics) – Inspector with Indian Air Force from July 1995 to January 2000. He worked as Lecturer - Electronics & Communication Engineering at Moradabad Institute of Technology, Moradabad, India, from February 2000 to June 2002. Presently, he is a faculty with Department of Electronics Engineering, Ambedkar Institute of Technology, Government of Delhi, India, from July 2002. He has published two books for undergraduate courses. He has authored/co-authored around 25 research papers in international journals/conferences. His research interest includes computer-aided design, wireless communication, and microstrip antennas. He is a reviewer for several journals of national/international repute, such as Taylor & Francis, and an Editorial Board Member of Insight-Communication Journal. He is a life-time member of Indian Society of Technical Education, India, and an Associate Member of Institute of Engineers, India.

Having lived in a village in his childhood, he loves nature and like plantation and its nurturing. He believes, for humanism, nature is the greatest asset and he wish to strengthen nature in one or other form. He feels resources are to be shared among the needy rather than utilized by an individual or a small group at advantage. He believes, let others be happy to be happy. He works for PEACE.

Dr. Jugul Kishor is presently working as an Associate Professor, JIMS Engineering Management Technical Campus, Greater Noida, UP, India. Earlier, Dr. Kishor worked as an Assistant Professor in the Department of Electronics and Communication Engineering, National Institute of Technology, Delhi, India. He also worked as an Assistant professor in the Department of Electronics and Communication Engineering, ITS Engineering College, Greater Noida, UP, India. He received B. Tech. degree in Electronics Engineering from Kamla Nehru Institute of Technology, Sultanpur, India, in 2002, M. Tech. degree in Microwave Electronics from University of Delhi, India, in 2008, and Ph. D. degree from Indian Institute of Technology (Indian school of Mines), Dhanbad, Jharkhand, India, in 2017. He has been credited to publish more than 40 papers with various reputed international journals and conferences with many citations and a high h-index. He is also a reviewer of many peer-reviewed journals such as *IEEE Access, International Journal of RF and Microwave Computer-Aided Engineering*, etc. He is also a member of academic and professional bodies such as IEEE MTT-S, URSI, etc. His research interests include designing and modelling of microstrip- and dielectric resonator-based devices, and metamaterial-based antennas including circularly polarized antennas and MIMO antennas.

Dr. Deepak Gangwar received his B. Tech. degree from Uttar Pradesh Technical University, Lucknow, India, in 2008, and his M. Tech. degree from Guru Gobind Singh Indraprastha University, Delhi, India, in 2011, and his Ph.D. degree in electronics engineering from IIT (ISM), Dhanbad, India. He is currently an Associate Professor with Bharati Vidyapeeth's College of Engineering, New Delhi. His current research interests include metamaterial-based antennas, ultra-wideband antennas, metamaterial filters, frequency-selective surfaces, metasurfaces, and RCS reduction.

Contributors

Dr Shilpee Patil received her Ph.D. degree in Electronics Engineering from AKTU (formerly UPTU), Lucknow, India, in 2019, her Master of Technology Degree in Digital Communication from GGSIP University, Delhi, India, in 2009. She received her B. Tech. degree in Electronics & Communication Engineering from AKTU, Lucknow, India. She is currently working as an Associate Professor in the Department of Electronics & Communication Engineering in Noida Institute of Engineering & Technology, Greater Noida, India. She has more than 10 years of teaching & research experience. She has keen research interest in design and modelling of microstrip antennas, circularly polarized antennas, and slot antennas for wireless communication. She is a reviewer of several journals of international repute, i.e. *Journal of Electromagnetic waves and Applications, Frequenz, International Journal of Electronics Letters.* She has published many research papers in international journals. She has attended various training programmes in the area of electronics & communication engineering.

Dr. Anil Kumar Singh received his B. Tech. degree in Electronics and Instrumentation Engineering from M.J.P. Rohilkhand University, Bareilly, in 1999, his M. Tech. degree in Instrumentation and Control Engineering from NITTTR Chandigarh, India, in 2011 and his Ph.D. degree in Electronics Engineering from IIT (ISM), Dhanbad, India, in 2016.

He worked in the department of Electronics and Instrumentation Engineering, Institute of Engineering and Technology, M. J. P. Rohilkhand University, Bareilly, as an Assistant Professor from 28 September 2002 to 01 December 2016 and after that, has joined as an Associate Professor in the same department since 02 December 2016. He has published more than 40 papers in national and international journals and conferences. His research interest includes design and analysis of microstrip antennas. He is a life-time member of the Institution of Electronics and Telecommunication Engineers.

He is a Member of the Antenna and Propagation Society and Communication Society and Institute of Electrical and Electronics Engineers (IEEE), USA. He is a reviewer of Progress in Electromagnetic Research, Microwave and Optical Technology Letter, IEEE Access, International Journal of RF and Computer Aided Engineering, International Journal of Antenna and Propagation, and several other journals of repute.

Dr. Anand Sharma was born in Agra (UP), India, in 1990. He received his B. Tech. in Electronics and Communication Engineering from Uttar Pradesh Technical University, Lucknow, India, in 2012 and his M. Tech. in ECE from Jaypee University of Engineering and Technology, Guna (MP), India. He has completed his Ph.D. in Department of Electronics Engineering from Indian Institute of Technology (Indian School of Mines), Dhanbad, India, in 2018. Currently, he is an Assistant Professor in the Department of Electronics and Communication Engineering, Motilal Nehru National Institute of Technology Allahabad, Prayagraj, UP, India. He has authored or co-authored over 80 research papers in international/national journals/conference proceedings. His research interests include dielectric resonator antennas, MIMO antennas, and microstrip antennas.

Dr Amit Bage completed B. Tech. in ECE from Punjab Technical University, Jalandhar, India, in 2012, and PhD degree from Indian Institute of Technology (Indian School of Mines), Dhanbad, India, in November 2017. He joined SRM University and worked as an Assistant Professor in the department of Electronics & Communication Engineering, from April 2018 to November 2018. He is working as an Assistant Professor in the Department of Electronics and Communication Engineering, National Institute of Technology, Hamirpur, India from December 2018. In his area of specialization, he is interested in electromagnetics and his recent research activities have focused on microwave antennas, microwave passive components, filtennas, reconfigurable filters/antennas (memristor, varactor, and PIN diode). Dr. Bage has authored more than 26 'papers in' just refereed journals and conference papers. Dr. Bage is an active member of IEEE and an associate member of IETE.

Surendra Kumar Gupta has completed B. Tech degree in Electronics and Communication Engineering from Uttar Pradesh Technical University of India, in 2013 and Master of Engineering degree in Digital Communication from B.I.E.T, Jhansi, India, in 2017. After completing M.E., he joined as a Lecturer in Department of Electronics Engineering at D.Y. Patil Ramrao Adik Institute of Technology, Navi Mumbai, India and Madan Mohan Malaviya University of Technology, Gorakhpur, India. Presently, he is pursuing Ph.D. from National Institute of Technology, Hamirpur, India. He has interest in microstrip antenna design for microwave and millimeter wave communication. He has authored 7 research papers in referred journal and conference.

Dr Ashwani Kumar received his B.Sc. (Hon.) Electronics, M.Sc. Electronics, M. Tech. Microwave Electronics, and Ph.D. degrees in 2000, 2004, 2006, and 2014, respectively, from the University of Delhi, Delhi, India. He was with the Department of Electrical and Computer Engineering, University of Central Florida, Orlando, Florida, USA, for his postdoctoral research from 2016 to 2017. Currently, he is an Assistant Professor at School of Engineering, Jawaharlal Nehru University, New Delhi, India. Earlier, he was with the Department of Electronics, Sri Aurobindo College, University of Delhi, Delhi, India. His current research interests include design and development of microwave passive components such as microstrip filters, dielectric resonator-based filters, MIMO antennas, UWB antennas, and circularly polarized antennas using metamaterials. He is a Member of IEEE Microwave Theory and Techniques Society. He has published 75 journal and conference technical papers on filters and antennas.

Prashant Chaudhary received his B.Sc. (Hon.) Electronics and M.Sc. Electronics in 2015 and 2017 from University of Delhi, Delhi, India. He is doing PhD in Dept. of Electronic Science, Delhi University. His research area of interest is planar antennas, MIMO, circularly polarized antennas, metasurfaces, magnetic substrates, and metamaterials. He is a member of IEEE. He has published nine research papers in journals and conferences.

Dr. Rakesh Nath Tiwari completed B.Sc. and M.Sc. (Electronics) degree from University of Allahabad and Deen Dayal Upadhyaya Gorakhpur University, Gorakhpur, India in 2002 and 2004 respectively. He received M.Tech. degree in Optical & Wireless Communication Technology with Gold Medal from Jaypee University of Information Technology, Waknaghat, Solan, India in 2008. He has completed Ph.D from Uttarakhand Technical University, Department of Electronics and Communication Engineering, Dehradun, India, in August 2020. He has published more than 25 papers in peer reviewed International/National journals and conferences. He is life member of Materials Research Society of India (MRSI), and International Association of Engineers (IAENG). He is reviewer of many reputed journals such as *International Journal of RF and Microwave Computer-Aided Engineering, Frequenz, PIER, etc.* His research interest includes design and modelling of slot patch antennas, UWB antennas, circularly polarized microstrip antennas, MIMO antenna, microwave/millimeter wave integrated circuits & devices.

Dr. Prabhakar Singh was born at village Semara, Chandauli (UP), India, in 1984. He received his B.Sc. and M.Sc. degrees from VBS Purvanchal University in 2004 and 2006, respectively. He received his Ph.D. degree from J.K. Institute of Applied Physics, Department of Electronics and Communication, University of Allahabad, India, in 2010. He taught B. Tech. and M. Tech. students at Delhi Technological University, Delhi, India, for 1 year from 2009 to 2010. He worked as an assistant professor at Bahra University, Shimla Hills, Himachal Pradesh, India, from 2010 to 2011. Presently, he is working as an Associate Professor at Galgotias University, Greater Noida, India. He has published more than 65 research papers in peer-reviewed international/national journals and conference proceedings with more than 500 citations. He is a reviewer of many international and national journals. He is presently working on broadband microstrip antennas, size miniaturization techniques in patch antennas, UWB, MIMO designs, and photonic band gap antennas.

Dr Ganga Prasad Pandey, a senior member IEEE, received his B. Tech. degree in Electronics and Communication Engineering from Kamla Nehru Institute of Technology, Sultanpur, UP, his M. E. from Delhi College of Engineering (now Delhi Technological University), Delhi, and his PhD from Uttarakhand Technical University, Dehradun. He worked as an Assistant Professor in Maharaja Agrasen Institute of Technology, Delhi, from January 2002 to September 2016 and then joined Pandit Deendayal Petroleum University, Gandhinagar. He has published more than 30 papers in reputed international journals and more than 15 papers in international/national conferences. He has guided two PhD students, and currently, two PhD students are working under him. He is working as a reviewer of several publishers such as Elsevier, Springer, PIER, MOTL, etc. His current research interests include machine learning in antennas, energy harvesting, ME-dipole, active, reconfigurable, frequency agile microstrip antennas, and microwave/millimetre-wave integrated circuits and devices.

Dr. Dinesh Kumar Singh received his B.E. in Electronics and Communication Engineering from Kumaon Engineering College, Dwarahat, Almora, in 2003. He has done M. Tech. in Digital Communication from RGPV University, Bhopal, India. He did his Ph.D. in Indian Institute of Technology (ISM), Dhanbad, Jharkhand, India. His area of interest is microwave engineering. He is currently working as an Associate professor in the Electronics and Communication Engineering Department, GL Bajaj Institute of Technology and Management, Greater Noida, UP, India. His research interests include designing of high-gain, compact, reconfigurable, fractal-shaped, circularly polarized microstrip antennas, substrate integrated waveguides (SIW) and magnetoelectric (ME) dipole antennas for modern communication system. He has been credited to publish more than 20 papers with various reputed international journals and conferences. He is also a reviewer of the *AEU-International Journal of Electronics and Communication, Electronics Letter.*

Dr Sachin Kumar received his B. Tech. degree in Electronics & Communication Engineering from Uttar Pradesh Technical University, Lucknow, India, in 2009. He received his M. Tech. and Ph.D. degrees in Electronics & Communication Engineering from Guru Gobind Singh Indraprastha University, Delhi, India, in 2011 and 2016, respectively. At present, he is working as a Researcher in School of Electronics Engineering, Kyungpook National University, Daegu, South Korea. His research interests include circularly polarized microstrip antennas, reconfigurable antennas, ultra-wideband antennas, defected ground structures, and microwave components.

Ghanshyam Singh received his B. Tech. in Electronics and Communication Engineering from UP Technical University (presently known as Dr. A.P.J. Abdul Kalam Technical University), Lucknow, Uttar Pradesh, India, in 2004. He received his M. E. degree in Electronics and Communication Engineering from National Institute of Technical Teachers Training and Research, Chandigarh, India, in 2013. He joined the department of Electronics and Communication Engineering, Feroze Gandhi Institute of Engineering and Technology, Raebareli, as a lecturer in 2007. Prior to joining academics, he had worked as an Electronics Engineer in the Railway Testing division of M/s Central Electronics Ltd (A Govt. of India, DSIR Public Sector Enterprise). Currently, he is doing his PhD Degree in Electronics Engineering in Dr. A.P.J. Abdul Kalam Technical University, Lucknow, India. His research interest includes patch antennas, artificial electromagnetic materials (LHMs/metamaterials), and reconfigurable and circularly polarized antennas for wireless communication. He has been credited to publish five research papers in peer-reviewed international journals and conferences.

Ankit Sharma received his B. Tech. degree in electronics and instrumentation engineering and his M. Tech. degree in signal processing from the Ambedkar Institute of Advanced Communication Technologies and Research, Delhi, India, in 2008 and 2012, respectively. He is currently pursuing his Ph.D. degree in microwave engineering with IIT (ISM), Dhanbad, India. His current research interest includes RCS reduction and gain enhancement of antennas using metasurfaces.

Dr Ravi Kumar Gangwar received his B. Tech. degree in electronics and communication engineering from Uttar Pradesh Technical University, in 2006, and his PhD degree in electronics engineering from IIT (BHU), Varanasi, India, in 2011. He is currently an associate professor with the Department of Electronics Engineering, Indian Institute of Technology (Indian School of Mines), Dhanbad, India. He has authored or co-authored over 160 research papers in international journals/conference proceedings. His research interests include dielectric resonator antennas, microstrip antennas, and bioelectromagnetics. He has guided 10 PhD and 10 M. Tech. students. He has completed four research and development projects related to dielectric resonator antennas and their applications from various funding agencies such as DRDO, SERB (DST), and ISRO. He is a member of the IEEE Antenna and Propagation Society and Communication Society, the Institution of Engineers (MIE), India, and the Institution of Electronics and Telecommunication Engineers (MIETE), India. He is a reviewer for *IEEE Antenna and Propagation Letters, IEEE Antennas and Propagation Magazine, IET Microwave and Antenna Propagation, IET Electronics Letters, Progress in Electromagnetic Research, the International Journal of RF Microwave and Computer Aided Engineering, and the International Journal of Microwave and Wireless Technology.*

Gourab Das was born in Kharagpur (WB), India, in 1990. He received his B. Tech. in Electronics and Communication Engineering from West Bengal University of Technology (WBUT), Kolkata, India, in 2012 and his M. Tech. in ECE from Indian Institute of Technology (Indian School of Mines), Dhanbad, India. He is currently pursuing his Ph.D. as Senior Research Fellow (SRF) in Department of Electronics Engineering from Indian Institute of Technology (Indian School of Mines), Dhanbad, India. He has authored or co-authored over 25 research papers in international/national journals/

conference proceedings. His research interests include dielectric resonator antennas, antenna arrays, and MIMO antennas. He is a reviewer for *IEEE Antenna and Propagation Letters, IET Microwave and Antenna Propagation, IET Electronics Letters, International Journal of RF Microwave and Computer Aided Engineering, and the Microwave and Optical Technology Letters.*

Ekta Thakur is a Research Scholar in the Department of Electronics and Communication Engineering at Jaypee University of Information Technology, Waknaghat, Solan, HP, India. She received her B. Tech. degree from the Bahra University, India, in the year 2014. She received her M. Tech. degree from Jaypee University of Information Technology, Waknaghat, HP, India, in the year of 2016. Her main research interests include ultra-wideband antennas, band-notched antennas, metamaterials, MIMO systems, electromagnetic band gap structures, and circularly polarized UWB antennas.

Dr Naveen Jaglan was born in 1989 and obtained his B. Tech. (Hons.) and M. Tech. (Hons.) degrees in Electronics and Communication Engineering from Kurukshetra University, Kurukshetra, India, in 2009 and 2011, respectively. He obtained his Ph.D. dissertation entitled "Design and Development of Microstrip Antennas integrated with Electromagnetic Band Gap structures" from Jaypee Institute of Information Technology, Sec-62, Noida, UP, India, in June 2017. He has authored/co-authored several research papers in refereed international journals and conferences. His research has included microwave communications, planar and conformal microstrip antennas including array mutual coupling, artificial materials (metamorphic, metamaterials), EBG, PBG, FSS, DGS, novel antennas, UWB antennas, MIMO systems, numerical methods in electromagnetics, composite right-/left-handed (CRLH) transmissions, and high-k dielectrics. His skills include modelling of antennas and RF circuits with Ansoft HFSS/CST Microwave Studio/ADS Momentum and taking measurements using vector network analyzer and anechoic chamber.

Prof. Samir Dev Gupta obtained his B.E. (Electronics) from Bangalore University (University Visvesvaraya College of Engineering, Bangalore), M. Tech. (Electrical Engineering) from IIT Madras, M.Sc. (Defence Studies) from Madras University, and Ph.D. from JIIT Noida. He is currently designated as Director and academic head of Jaypee University of Information Technology (JUIT), Waknaghat, Solan, HP, India. He has over three and a half decades of work experience in the areas of avionics, communication, radar systems, and teaching at UG and PG levels. His teaching career spans over 22 years and 6 months; teaching at Institute of Armament Technology under Pune University, Defence Research & Development Organization (D.R.D.O.), Air Force Technical College (A.F.T.C.), Bangalore, and Advanced Stage Trade Training Wing at Guided Weapon Training Institute, Baroda. His research interest is in the area of conformal microstrip patch antennas for aircraft systems. His publications in refereed journals and conferences are well cited. He has led and supervised highly qualified and skilled engineering officers and technicians of I.A.F. in maintenance and operation of missile, radar, and communication systems. He has been associated at JIIT, Noida, in students' welfare and discipline as Associate Dean of Students and Chairman Proctorial Board, respectively.

Kanishka Katoch is a research scholar in the Department of Electronics and Communication at Jaypee University of Information Technology, Waknaghat, Solan, HP, India. She received her M. Tech. degree from Jaypee University of Information Technology, Waknaghat, Solan, HP, India, in 2017 and her B. Tech. degree from Himachal Pradesh Technical University, India, in 2015. Dielectric resonator antennas, UWB antennas, terahertz antennas, phased antenna arrays, metamaterials, and frequency-selective surfaces are her area of research interest.

1 Basic Theory and Design of Printed Antennas

Dr. Shilpee Patil
Noida Institute of Engineering & Technology, Greater Noida

Prof. Binod Kumar Kanaujia
Jawaharlal Nehru University, New delhi

Dr. Anil Kumar Singh
M. J. P. Rohilkhand University, Bareilly

CONTENTS

1.1 EVOLUTION AND UPCOMING GROWTH OF PRINTED ANTENNAS

Wireless communication is the latest emerging pillar in the field of communication. It has reshaped the way of communication and connectivity. Its various applications such as broadcast of information, personal communication, satellite communication and mobile communication have notable merits such as simplicity in design, light weight, easy execution and estimate. The progress in wireless local area networks (WLANs), cell phone networks, satellite communication and other different wireless networks shows the demand as well as interest in the distinctive area of information technology and communication. Due to the blooming of advanced communication applications, there is a high demand of highly advanced and smart antennas, which are regarded as the key for wireless technologies. For wireless technology, the antenna with multiband operating property having high performance and features such as easy to control, easily manufactured, light weight, and simple structure etc. are preferred. Its two most interesting features are: physical simplification and flat structure, which lead the printed antennas to be the common choice of designers to assemble a printed antenna. Due to its engineering process, the simplest form of printed antenna is micro-strip antenna. Now-a-days electronic circuits not only used for a transmission line but also components such as filters, couplers, resonators, etc. frequently use an open-wave guiding structure known as micro-strip.. The use of micro-strip for constructing antennas is a new development, and an example of a micro-strip antenna is shown in Figure 1.1. The conducting ground plane supports the complete lower portion of the dielectric substrate, whereas the printed conducting strip is placed at the upper portion of the dielectric substrate.

Designers worldwide are emphasizing the use of printed antenna, especially for low profile radiators, as it is a traditional and well-established antenna. The remarkable use of printed or micro-strip antennas in many established techniques makes it more reliable. Because of the rapid growth and progress in the antenna technology in a short duration of one to two decades, the historical evolution of the present-day antenna has become significant and interesting to read. With the rapidly increasing research and development and ever-growing research publications in the field

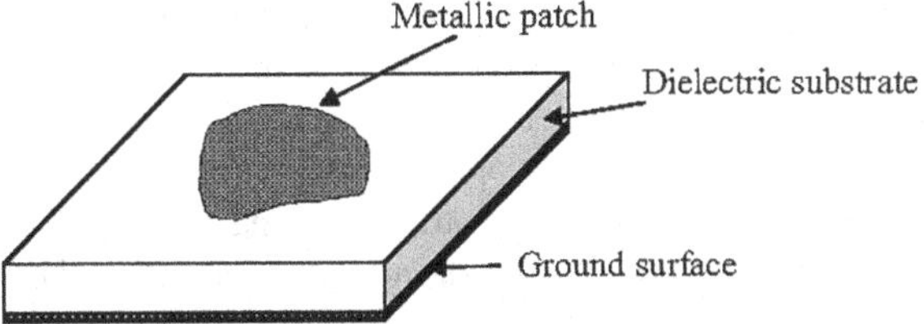

FIGURE 1.1 Basic configuration of a printed antenna.

of antenna, it would be worth saying that nowadays printed antennas have become dominant among the family of antennas. Now the demand is for more inventive printed antenna structures having reliable manufacturing approach. The main focus area for the contemporary system is lower profile antennas with low cost and less weight. However, the designer's ability to regulate the process of manufacturing in a controlled manner leads to effectively modelled structures and reduced costs of the antenna. This book brings out the foremost concern of the designers and researchers for having a detailed computer modelling of printed antennas due to the latter's challenges in the antenna domain.

Initially, Greig and Englemann [1] reported a new technique for transmission of kilo-mega range of waves and Deschamp [2] included the concept of microstrip antenna origin which is one of link from various sources. At that time the originated stripline structure emitting the undesirable radiations. To reduce the unwanted radiations from the circuit, the size of the conducting strip and substrate dimensions were reduced, which resulted in the creation of 'microstrip'. The concept that the arrival of the transistor was the turning point in the development of planar printed circuits had been questionable, but the main attention was to develop such circuits and microwave filters having low cost. Though less or no attention was given to the radiation loss, the nature of the radiation from stripline circuits was taken into consideration by Lewin [3]. Due to the development of new generation missiles, there was an urgent demand for low profile printed antennas in the early 1970s; otherwise, the concept of antenna was inactive, and there were only very few references [4–9] till that time. During this time, the progress of the microstrip antenna concept intensified, and there was a flow of numerous research publications in this field. There were many workshops, and prominent work was done during this period. In 1979, a workshop of utmost importance was held at Las Cruces, New Mexico [10]. Its proceedings were taken into account and were published in a special edition of IEEE Transactions, 1981. Two most recommended books on microstrip antenna were published by Bahl and Bhartia [11] in 1980 and James, Hall and Wood [12] in 1981. During that time, another publication based on innovative and advanced development with a different approach of flat antenna was reported as a research monograph by Dubost [13]. The early 1980s was a crucial period for printed antenna, not only for the publications but also for the practical applications and finally manufacturing. The specifications of the substrate were tightened, and a broader range of products capable of working in extreme ambient conditions were introduced by manufacturers. However, the cost of substrate was still high.

The new methods were appreciated with feeders and elements being regarded as a whole entity as they were essential, due to the problems that have arisen in connecting feeders to patch circuits in a large array. To highlight the importance of different choices that have been made in array topology, the latter elements were improved. As the printed elements cannot be freely attached using feeders, the requirement of an array structure has been made. Earlier discussion depicts that the demands of the current system are the major factors in the enhancement of printed antennas. The constantly emerging requirements for broader bandwidth in the communication systems are the basic factors for evolution of the bandwidth enhancement techniques of printed antennas, which is the field for growth. Recent awareness in this field is that of controlling the polarization characteristics of printed antennas, particularly in the domain of radar applications. The idea of 'active-array architecture' is applicable in defence systems where planar apertures have semiconductor packages integrated with radiating elements, which have a facility to oppose mechanical beam scanning. The complete concept is up to date, but the cost of these types of array is extremely high, which forces us to think about the present and the upcoming future of these printed antennas. The printed substrate technology, which has hardly been mentioned, remains to be an important source of complex electromagnetic problems, which are easily processed in university laboratories. Thus, there will be continuous flow and abundance of research publications with industrial advancement running in parallel with two main aspects: the hunt for mathematical models for prediction of realistic antennas with additional precision and improvements in manufacturing by sharpening CAD techniques for creating advanced antennas to fulfil the demands of the new systems. The emphasis is laid in the latter aspect that the conventional microwave antenna that was bulky outperformed its counterpart conformal printed antenna that was thin. However, mostly in aerospace domain many latest systems are made possible with the presence of the concept of printed antennas. And thus, it is a key for new advanced systems arising merely from the innovative antenna designs. In the future, the present trends can be infused in the direction of integrating electronically beam-scanned arrays. Thus, many radiators of conventional types placed on the surface of missiles, military aircraft, military ships, land vehicles, etc., can be replaced by conformal printed antennas. But innovative and advanced physical concepts are demanded rather than depending on the software alone, as organizing and controlling the characteristics of co-polar and cross-polar radiation patterns are difficult and complicated tasks. Therefore, the concept of printed antennas seems a doorway to compatibility of the system and optimum placement of sensors, thus implementing the several facets of conformability. It also explores the techniques of signal processing for advanced computing and facilitates lower price, semiconductor integration, and electronic control of radiation pattern. The unbeatable scenario of the popularity of printed antenna theory is indeed with its frequent growth for designing several electronic systems.

1.2 FEATURES OF PRINTED ANTENNAS

The printed antenna is a modern type of radiator. The basic configuration of printed antenna contains two metallic layers of thickness less than the free space wavelength 'λ_0', of which the first is the radiating patch on one side of the dielectric substrate and the second is the ground surface on the opposite side of the dielectric substrate, as shown in Figure 1.1. Gold and copper are usually utilized for making the metallic layers for patch or ground. In printed antenna, the patch outline is arbitrary in nature. Different shapes such as rectangular, square, circular, triangular, equi-triangular and annular ring are common in practice. As a feed, a coaxial cable or a stripline can be used for guiding the electromagnetic energy under the patch from the source to the end place. During this process, a fraction of the energy radiates into space from the patch boundary. Between the metallic patch and the ground surface, a dielectric substrate is placed. According to antenna necessity, various substrate material specifications are present. As Balanis [14] showed, between the range of $2.2 \leq \varepsilon_r \geq 12$ and $0.003\ \lambda_0 \leq h \geq 0.05\ \lambda_0$ (where λ_0 is the free space wavelength), dielectric substrates of different thicknesses having a wide range of dielectric constants (ε_r) are available. Some descriptions of printed antenna are influenced to a great extent by substrate height, such as the direct proportionality of impedance bandwidth to the substrate height of the printed antenna. With a high value of dielectric constant, the antenna attains a smaller size. For increasing the efficiency and radiation in space, the height of the substrates with low ε_r is increased, while in microwave circuits, substrates with a high value of dielectric constant and a small height are used. Some dielectric materials with their properties are listed in Table 1.1.

The ground is the metallic section created on the opposite side of the patch placed on the substrate. Some perturbed segments are introduced into the ground for better antenna performance at desired specifications. The techniques of loading different

TABLE 1.1
List of Some Dielectric Materials and Their Properties

S. No.	Dielectric Material	Relative Permittivity	Loss Tangent
1	Air	1.0006	0
2	FR4 epoxy	4.4	0.02
3	Bakelite	4.8	0.0002
4	Duroid	2.2	0.0009
5	Quartz glass	3.78	0
6	Foam	1.03	0
7	Polystyrene	2.55	0
8	Plexiglas	2.59	0.0068
9	Fused quartz	3.78	0
10	E glass	6.22	0.0023
11	RO4725JXR	2.55	0.0022
12	RO4730JXR	3	0.0023
13	Rogers RT/duroid 5870/5880	2.33/2.2	0.0012/0.0009
14	Teflon	2.1	0.001
15	Taconic CER-10	10	0.0035
16	Taconic RF-30	3	0.0014
17	Taconic RF-35	3.5	0.0018

shapes or slots into the ground plane were presented by Chen [15] and Gautam and Kanaujia [16].

1.2.1 Feeding Techniques

Figure 1.2 shows the design of a feeding network used to transmit the input energy efficiently. An appropriate feeding technique is required to match the feed and the patch, and it makes possible the transmission of the input energy from the feed line to the patch.

There are several kinds of methods to feed a printed antenna and are explained in this section. Some literature review gives details of the feeding methods. For example, coaxial probe was explained by Fan and lee [17], microstrip line feed was explained by Ramirez et al. [18], proximity coupling was explained by Tsang and Langley [19], and aperture coupling was explained by Waterhouse [20], with their own merits and demerits. To choose an appropriate feeding method for the design of a particular type of antenna, several factors must be taken into consideration. The feeding arrangement must be chosen such that it is simple and properly matches the patch. The feeding networks are divided into two types: direct contact type and indirect contact type. In the first one, the feed line is directly connected to the radiating element, whereas in the second one, they are electromagnetically coupled to input energy to the radiating element. As reported by Hsu and Wong [21], an indirect contact type feeding method has been presented to lessen the problem of enhancement of cross-polarization level that is generated from direct contact type feeding. The mainly used direct contact type feeding techniques are coaxial feed and microstrip feed. The aperture-coupled feed, proximity-coupled feed and co-planar waveguide feed will come under the indirect contact type feeding. All these types of feeding methods are explained in brief.

1.2.1.1 Coaxial Feeding

Coaxial feeding is mainly applied for transferring the input energy to the patch of a printed antenna. It is the fundamental technique that is used for feeding the microwave power. Figure 1.3 illustrates the coaxial probe feed arrangement with a microstrip patch antenna. The coaxial probe in a coaxial feed system consists of two conductors, of which the internal conductor is associated with the radiating element of the antenna and the external conductor is associated with the ground plane. The

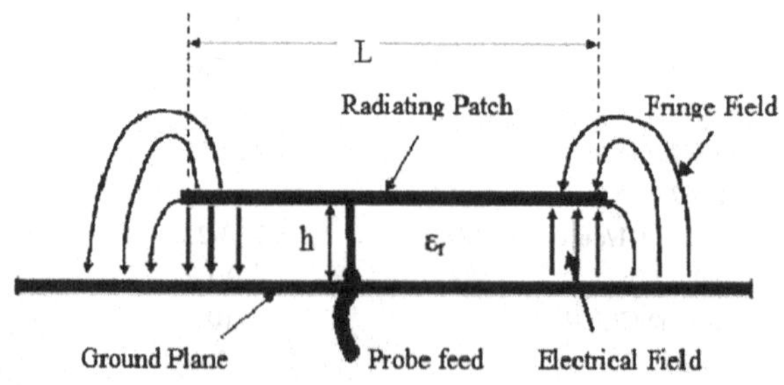

FIGURE 1.2 Variation of fringing fields.

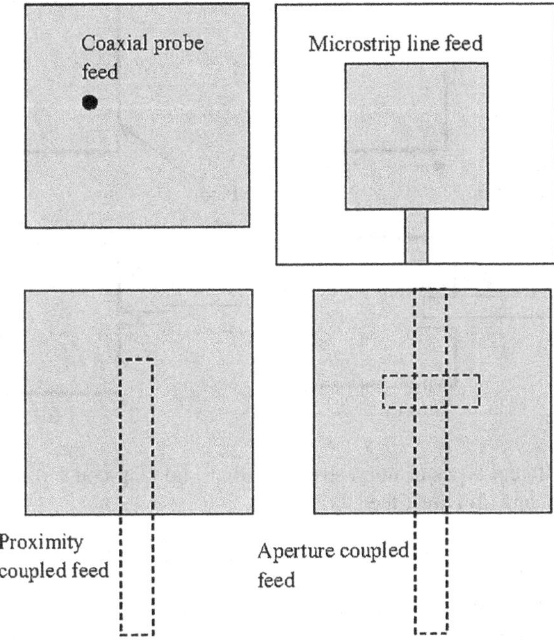

FIGURE 1.3 Various feeding techniques.

internal conductor provides the matching of input impedance to the characteristic impedance. Coaxial feeding has some benefits such as effortless design, easy fabrication, easy matching and low radiation of spurious waves. On the other hand, the coaxial feed has a drawback of the need for a high soldering precision. The problem that arises when using a coaxial feed in an array structure is the requirement of multiple solder joints. As revealed by Sharma and Vishvakarma [22], the coaxial feeding needs a longer probe when a thick substrate is utilized, which raises the surface power and the feed inductance and typically gives a narrow bandwidth.

1.2.1.2 Microstrip Feeding

In microstrip feeding method, the metallic patch of an antenna is fed via a lesser width of the microstrip line as compared to the patch in the same plane. As a result, a single structure can be found for both the feeding and the patch. Microstrip feeding has some benefits such as easy demonstration, easy matching and simple fabrication. Additionally, it is a decent decision to use this feeding in an antenna array feed network. A lower bandwidth and generation of spurious radiation in the presence of coupling between the patch and the microstrip line are the main disadvantages of this type of feeding mechanism. Microstrip feed can be of the following types:

- **Gap-coupled feed**:

  As shown in Figure 1.4a, the microstrip line of the feed is isolated from the patch; an air gap is formed between the 50 Ω microstrip feed line and

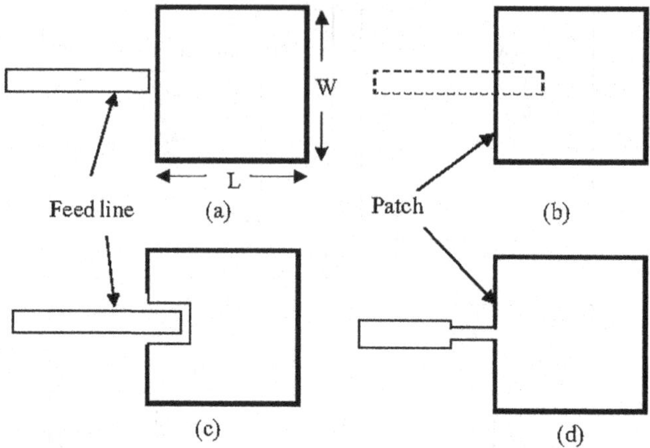

FIGURE 1.4 Different types of microstrip feed lines; (a) Gap-coupled feed, (b) Two-layer feed, (c) Inset feed, and (d) Direct feed.

the patch. The antenna has a disadvantage of coupling between the patch and the 50 Ω of microstrip line of feed.

- **Two-layer feed**:

  Figure 1.4b illustrates that the microstrip line of feed is again isolated from the patch; the location of the feed point is sandwiched between the two layers of the substrate.
- **Inset feed**:

  As shown in Figure 1.4c, a thin line of microstrip is introduced within the patch. In this type of microstrip feed method, the feed line is again isolated from the patch. The position of the microstrip line of feed is the same as that used in the coaxial feed. The 50 Ω microstrip feed line is surrounded with an air gap till the feeding point. Further, the inset feed method is more appropriate for an array of feeding networks.
- **Direct feed**:

  As shown in Figure 1.4d, the microstrip line of feed is connected with the edge of the radiating patch of the antenna. Direct feed method requires a matching network to compensate for the impedance difference between the patch and the 50 Ω microstrip feed line. A quarter-wavelength transformer is commonly used as a matching network.

1.2.1.3 Proximity-Coupled Feeding

Figure 1.3 shows the structure of the proximity-coupled feed, which is one of the recognized mechanisms of electromagnetic coupling. In this type of feed, the input energy is transferred to the patch from the microstrip line via electromagnetic coupling. The proximity-coupled feeding technique consists of a microstrip line, a patch and two different substrates. In this structure, the patch is positioned on the top side of one substrate and the microstrip feed line is placed in between the two substrates of different properties. To analyse the performance of the patch and the feed line individually, this method also provides various choices between different dielectric substrates. The parameters of the two substrates cannot be chosen to be the same to improve antenna operating features. The proximity-coupled feed decreases spurious radiation and increases operating bandwidth. In any case, it desires an exact arrangement between the two layers in multi-layer designing.

1.2.1.4 Aperture-Coupled Feeding

Figure 1.3 illustrates the structure of the aperture-coupled feeding mechanism. Its geometry contains a ground plane with an aperture slot sandwiched between two dielectric substrates, of which one has a higher-value permittivity and the other has a lower-value permittivity. The radiating patch is located at the outside of one substrate, and the microstrip line is located at the outside of the other substrate. The coupling of input energy between the patch and the feed line is achieved through a small slot cut in the ground plane. Lai et al. [23] reported that in contrast to other feeding techniques, the aperture-coupled feed does not produce spurious radiation. To minimize the cross-polarization which is produced from symmetric configuration, the coupling aperture of the slot is usually situated under the centre of the radiating patch. The drawbacks of the aperture-coupled feed due to the multi-layer fabrication are not easy to overcome, because the multi-layer geometry increases the thickness of the antenna system. The advantage of this mechanism is that it allows separate optimization of the feeding process and the radiating patch. A small area of the slot must be chosen to decrease the radiation from the ground surface. In contrast to the probe feed and microstrip line feed, a wide bandwidth is obtained from the aperture-coupled feed with improved polarization purity as reported in [23]. The features of various feed techniques are illustrated in Table 1.2.

TABLE 1.2
Features of Various Feed Techniques

Features	Coaxial Probe Feed Technique	Microstrip line Feed Technique	Proximity-Coupled Feed Technique	Aperture-Coupled Feed Technique
Spurious radiation	Large spurious radiation	Large spurious radiation	Minimum spurious radiation	Less spurious radiation
Reliability	Low reliability	High reliability	Moderate reliability	Moderate reliability
Fabrication process	Requirement of soldering and drilling process	Easily fabricated	Alignment required	Alignment required
Matching impedance	Moderate	Easily matched	Moderate	Moderate
Impedance bandwidth	2%–5%	2%–5%	13%	2%–5%

1.2.2 Performance Factors of Printed Antennas

The performance of an antenna has been evaluated by several factors that explain the antenna radiation properties, circuit properties or other properties. Such factors of the antenna are its impedance bandwidth, gain, directivity, efficiency, axial ratio and power capability. The performance enhancement of an antenna signifies the improvement in one or more than one factor, among which the following are important.

1.2.2.1 Radiation Pattern

The definition of radiation pattern as reported by IEEE Standards [24] is 'the spatial allocation of a quantity that characterizes the electromagnetic field vectors generated by antenna'. Radiation pattern is also defined by Balanis [25] that is explain in terms of the position of observers a regular arc draw by a line or surface throughout all direction of radiation of antenna. The radiation pattern is represented in terms of 2D/3D space allocation of power flux density, field strength, radiation intensity, directivity, phase or polarization. An isotropic antenna is not feasible, but may be applied in the process of comparison with practical antennas to evaluate their performance, as a reference antenna. As shown in Figure 1.5, the radiation pattern gives the information about antenna beam width, side lobes' features and antenna resolution.

The beam width of an antenna is often defined in two terms: The first is the half-power beam width (HPBW), and the second is the first null beam width (FNBW). The half-power beam width provides an angular separation of the points having the half-power from the maximum power. In decibels, this means the points at −3 dB with respect to the maximum. The first null beam width provides an angular distance between the first nulls at the side of the main lobe of the radiation pattern.

1.2.2.2 Directivity

As we know, an antenna's directivity which is expressed in dBi is defined as the radiation of power in a particular direction. The directivity of an antenna is equal to the antenna gain, if the antenna radiation efficiency is 100%. The directivity of a patch is estimated easily because its efficiency is greater than 90%. The radiating edge of the patch placed above the ground surface is working as two radiating slots. The directivity increases by 3 dB due to the overall radiation present above the patch of the antenna. By this, we achieve an excellent front-to-back ratio, where the maximum radiation takes place above the patch and the minimum at the other side of the

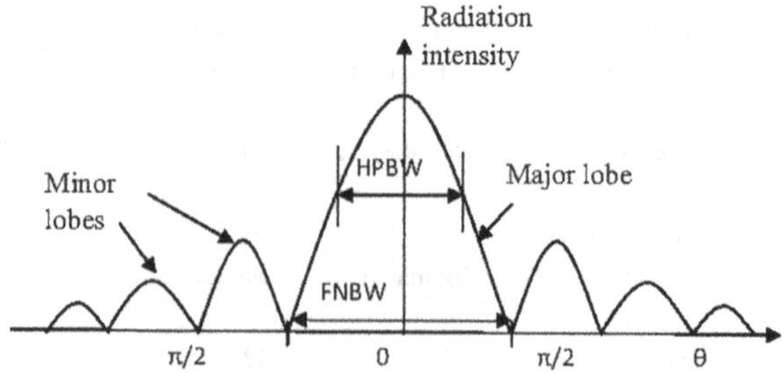

FIGURE 1.5 Antenna beam width representations.

patch. The directivity increases as we move from lower elevations to the broadside direction. The 'dBi' is used when the directivity of an antenna is measured relative to an isotropic antenna. It is also measured with respect to a dipole antenna which has 2.15 dBi of directivity, which is more than that of an isotropic element which has no directivity. The 'dBd' is used when the directivity is measured with respect to a dipole antenna.

1.2.2.3 Antenna Gain

The antenna gain is a key performance value which is measured in terms of the antenna's directivity and radiation efficiency. As the radiation efficiency never reaches the maximum, the gain of an antenna always has lower values than the directivity. Antenna efficiency is another important parameter which is most often used to measure losses in antennas and is defined as the ratio of the radiated power to the input power of the antenna. The input power is transformed into the radiated power, and it is mostly dissipated within the antenna by means of metal conduction, magnetic loss and dielectric loss. Surface wave power is also a transformed portion of the input power of the antenna. At the boundaries of the substrate, surface waves that travel within the substrate are partially radiated and reflected back. Surface waves are mostly generated when a thick substrate or a substrate with a large dielectric constant is used. When air is used as a dielectric, surface waves are not excited. Presently, various techniques are available to minimize the excitation of surface waves.

1.2.2.4 Bandwidth

Antenna bandwidth (BW) is stated as 'the range of usable frequencies within which the performance of the antenna reaches to a specified standard with respect to some characteristic'. The bandwidth is the range of frequencies from the lower side to the higher side of the operating frequency of the operating band. The bandwidth of an antenna demonstrates a standing wave ratio less than 2:1. The values of performance factors of an antenna such as input impedance, axial ratio, gain, and radiation characteristics are found to influence the desired antenna bandwidth. Various other definitions are available to realize the importance of the bandwidth, such as impedance bandwidth, directivity/gain bandwidth, electronic efficiency bandwidth, polarization bandwidth, and axial ratio bandwidth.

- **Impedance bandwidth:**
  The term impedance bandwidth is the frequency range in which the antenna gets a matching impedance relative to a given reference impedance. The quality factor 'Q', feed types and various other parameters of a printed antenna affect the antenna impedance bandwidth. The demerit of dipole and half-wave printed antennas is their typical restricted impedance bandwidth, which is between 1% and 3%.
- **Directivity/gain bandwidth**:
  The term directivity/gain bandwidth is the frequency range over which the antenna obtains the requirement of definite values of directivity/gain.
- **Efficiency bandwidth:**
  It is the range of frequency where the antenna exhibits a defined efficiency.

- **Polarization bandwidth:**
  It is the frequency range over which the printed antenna achieves the requirement of a suitable co-polarization/cross-polarization ratio.
- **Axial ratio bandwidth:**
  It is the range of frequency in which the quality of the circularly polarized waves of an antenna is explained. Polarization bandwidth and axial ratio bandwidth are related to each other.

The bandwidth of narrowband antennas in percentage [14] is expressed as 'the ratio of the frequency range over that all desired specifications are achieved to the centre frequency'. The mathematical expression is as follows:

$$BW(\%) = \frac{f_{max} - f_{min}}{f_o} \cdot 100 \tag{1.1}$$

where f_{max} represents the higher frequency, f_{min} represents the lower frequency, and f_o represents the centre frequency.

1.2.2.5 Polarization

The term polarization is stated as the orientation of electric field vectors in an electromagnetic field. If the orientation of electric field vectors is in single plane, linear polarization is achieved. The fundamental patch of an antenna mostly generates linearly polarized waves in which the orientation of the electric field falls in one direction. There are two cases obtained in linear polarization, i.e. vertical polarization and horizontal polarization. In the first case, the orientation of electric field vectors is in vertical plane, and in the second case, the orientation of electric field vectors is in horizontal plane. The earth's surface is taken the reference for the orientation of field vectors.

If the orientation of electric field vectors forms a circle in space relative to time for a specific point, then it is known as circularly polarized radiation produced by the time-harmonic waves for that point. Circular polarization is achieved if the electric field vectors produce two orthogonal linear modes. The magnitude of the two modes of electric field should have the same value, and the time phase difference of the two modes should be odd multiples of 90°. Numerous applications such as satellite communication are not suitable with the concept of linear polarization. In such cases, circular polarization is appropriate because of its sensitivity to the orientation of the antennas. The fundamental theory of the circularly polarized printed antennas is the variation of electric fields in orthogonal planes such as x- and y-directions with equal amplitude. The outcome of the circularly polarized waves is achieved by the parallel excitation of two modes in orthogonal directions. It is clear that the excitation of one mode has a phase delay of 90° with the other mode. An antenna with circular polarization can be either right-hand circularly polarized or left-hand circularly polarized. To obtain a maximum power from the surroundings, the polarization sense should be the same for both the receiver and the transmitter antennas. If the antenna polarization is 90° out of phase with the field polarization, no power has been taken out by the antenna from the surroundings.

In real world, a perfect antenna is not feasible for linear or circular polarization. If a wave is not linearly or circularly polarized, then it means that it is elliptically polarized. If the orientation of electric field vectors forms elliptical loci in space relative to time for a specific point, then it is known as elliptically polarized radiation produced by the time-harmonic waves at that point. However, an antenna is accomplished with the elliptical polarization property if the vectors of electric field produce two orthogonal linear modes of unequal amplitude.

The factor polarization sense is also used for characterization of each polarization. Polarization sense for a linearly polarized antenna is illustrated by the tilt angle of the electric field vectors. It is classified as 90° vertical, 0° horizontal, and ±45° slanted. Polarization sense for a circularly polarized antenna is illustrated by the direction of tip movement of the electric field vector. It is classified as clockwise (RHCP) and counterclockwise (LHCP).

1.2.2.6 Axial Ratio

The term axial ratio (AR) is used as an essential feature for polarization measurement of an antenna. It is generally defined as 'the ratio of electric fields in orthogonal planes'. The axial ratio of an elliptically polarized wave is explained as 'the ratio between minor and major axes of the ellipse' and it is equal to [14]

$$AR = \frac{\text{major axis}}{\text{minor axis}} \tag{1.2}$$

The value of axial ratio for an elliptically polarized wave is greater than one (> 0 dB). The value of axial ratio varies from one to infinity. An antenna radiating linearly polarized waves having one of the components of electric field with zero magnitude achieves the value of axial ratio that tends to infinity. If the values of minor and major axes are equal, then the antenna achieves ideally circularly polarized waves. That means a circularly polarized wave consists of two orthogonal electric field components (90° out of phase) of equal amplitude. In that case, the value of axial ratio is achieved as one (or 0 dB).

The axial ratio is frequently used for the characterization of the polarized waves of a circularly polarized antenna. For circularly polarized antenna, axial ratio is 3 dB or higher and axial ratio bandwidth is the frequency range over which axial ratio is less than or equal to 3 dB. Degradation in these values occurs when signal moves away from the main beam width of the antenna.

1.3 CHARACTERISTICS OF PRINTED ANTENNAS

The technology of printed circuit which is used for the development of circuit components, transmission lines and radiating elements of electronic systems is completely responsible for the development of printed antennas. Similar to the technology of printed circuit, the technology of integrated circuit presents various other advantages such as low profile, lightweight, low cost and conformability. A narrow bandwidth that arises from the concept of a resonant cavity which is the area under the patch with a high quality factor is the major drawback. In this section, some important features of printed antennas are investigated.

1.3.1 Different Shapes of Printed Antennas

A number of shapes of printed radiators and array elements are reported by designers. These are frequently available to designers with a verified checklist. In this section, details of various shapes of printed elements along with an outline sketch and source information are summarized.

Figure 1.6 illustrates some principal shapes of radiating patches for improved performance of a printed antenna. The circular sector shape and the annular ring sector shape of the patches were investigated by the cavity model in [26]. In this paper, the impedance and resonant frequency were explained in detail, but there was no explanation regarding the performance of bandwidths or radiation pattern. Star-shaped patch radiators are explained theoretically in [27] due to their good symmetry property with higher-order modes. Palanisamy and Garg [28] presented circular ring, rectangular ring and H-shaped patches and achieved performances similar to that of fundamental shapes.

Various other patch variants are investigated and suggested in [12]. Similarly, these shapes are also supposed to give the same performance as the fundamental shapes. A collection of some other patch variants are illustrated in Figure 1.7.

The fundamental shapes of patches for single-feed antennas are illustrated in Figure 1.8. A simple construction for circular polarization is achieved by a single-point feeding. A variety of geometrical distortions are presented to obtain the phase shift of 90° between two excited modes, some of which are as follows:

- Patch with rectangular shape [29].
- Patch with square notch [30].
- Patch with square slot [31].

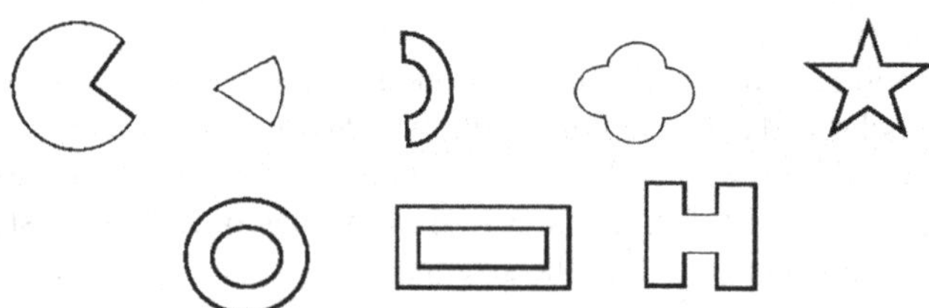

FIGURE 1.6 Variants on principal shapes.

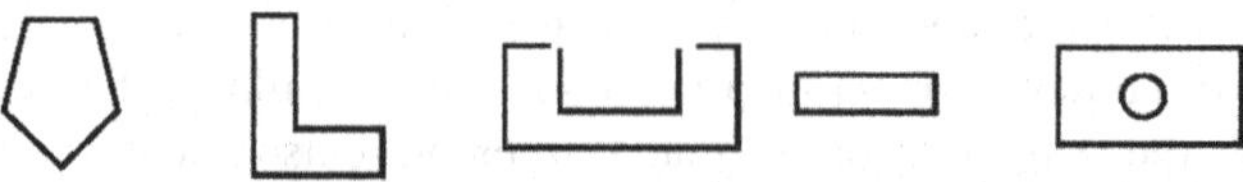

FIGURE 1.7 Various other feasible shapes.

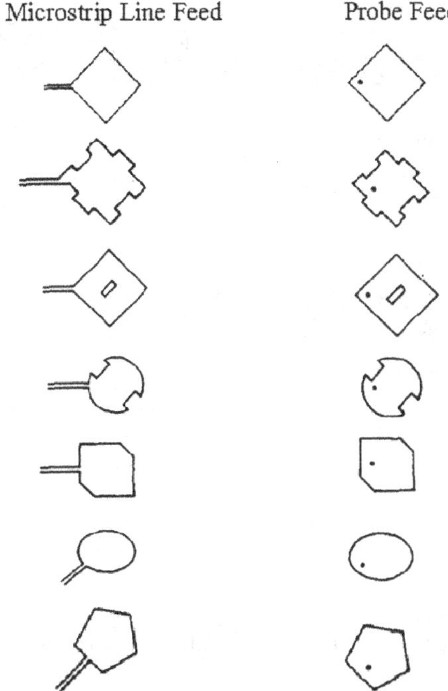

FIGURE 1.8 Different patch shapes for circular polarization.

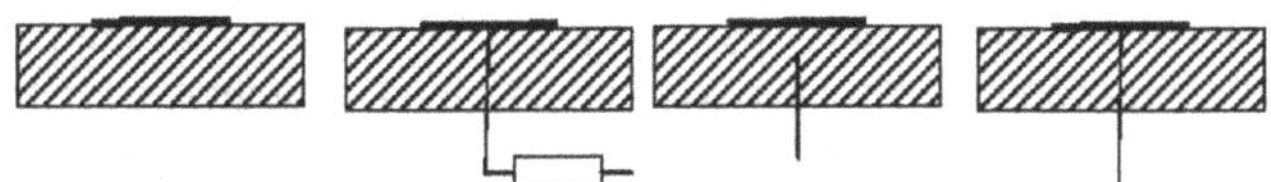

FIGURE 1.9 Thick patches for wideband operation.

- Disc-shaped patch with square notch [32].
- Square patch with truncated corner [33].
- Elliptical patch [34].
- Pentagonal patch [35].

The basic configuration of patches for wideband operation of an antenna is illustrated in Figure 1.9. As presented in [36], a lower value o dielectric constant ($\varepsilon_r = 1.0$) and a higher value of substrate's height ($h/W > 0.1$) are used to achieve bandwidths greater than 10%. Otherwise, thinner substrates also achieve a broad bandwidth of up to 30% by using external matching circuits as represented in [37]. Patches with thick substrates experience the problem of impedance matching that is overcome by using the matching gaps in the probe [12] or patch [38]. For wide bandwidth applications, patch shaping like steps, conical depression and curved depression as shown in Figure 1.10 is also a good technique.

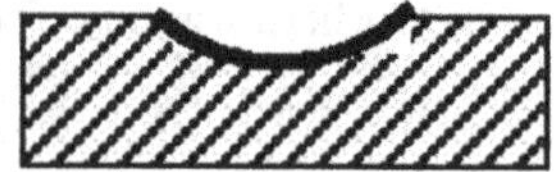

FIGURE 1.10 Different shapes of patches.

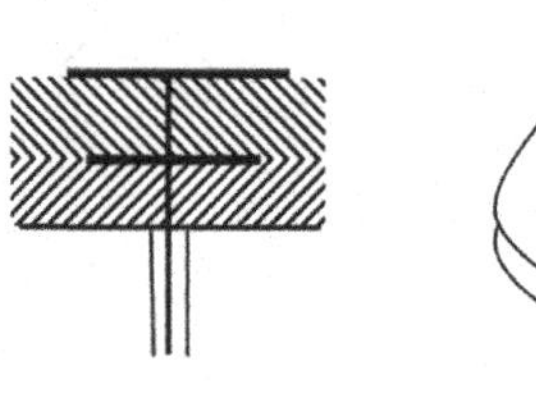
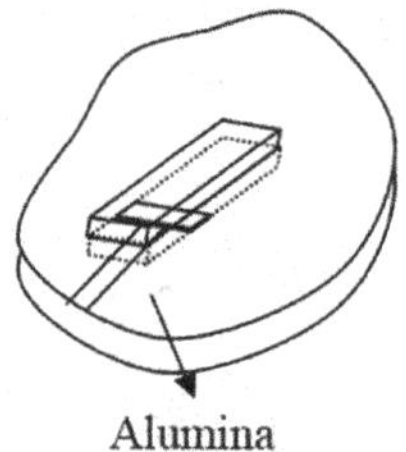

FIGURE 1.11 Parasitic patches.

Another technique to achieve wide bandwidths is the stacking of multiple patches. In the design of a stack antenna, the upper patch is electromagnetically coupled to the parasitic patches. Figure 1.11 illustrates some examples of stack antennas with feed lines of coaxial probe and microstrip line, which are designed using a base substrate of alumina ($\varepsilon_r = 10$).

Parasitic patches are also arranged in co-planar patches that are characterized as thin resonators. The additional patches in a parasitic co-planar configuration are placed either gap-coupled or line-coupled to different shapes of patches. Otherwise, to achieve a wideband patch, gap coupling is preferred for coupling thin parasitic patches. Various arrangements of parasitic co-planar patches are illustrated in Figure 1.12.

1.3.2 General Characteristics of Basic Patches

1.3.2.1 The Rectangular Patch

The basic shape of a radiating patch that is frequently used for designing a microstrip patch antenna is 'rectangular' and is shown in Figure 1.13. In this figure, the length and width of the rectangular patch are denoted by a in x-direction and b in y-direction, respectively. A cavity is assumed under the rectangular patch for the analysis of the rectangular patch. The expression of the electric field in the resonant mode is given by

$$E_r = E_o \cos(n\pi x/a) \cdot \cos(m\pi y/b) \tag{1.3}$$

where n or $m = 0, 1, 2 \ldots$

The resonant frequency is expressed as follows:

$$f_{nm} = k_{nm} \cdot c / \left(2\pi\sqrt{\varepsilon_r}\right) \tag{1.4}$$

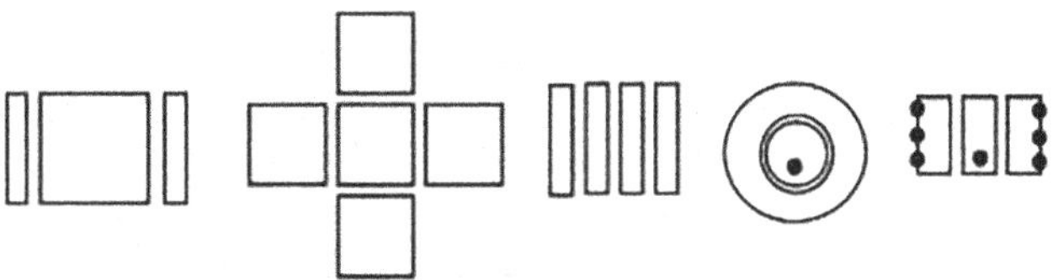

FIGURE 1.12 Different configurations of parasitic co-planar patches.

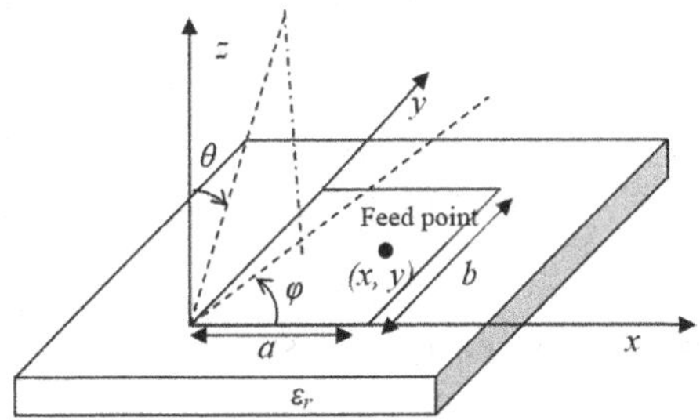

FIGURE 1.13 Rectangular patch antenna.

where k_{nm} is the mth zero of the derivatives of the Bessel function of order n and c is the speed of light. For a rectangular patch, k_{nm} is expressed by

$$k_{nm}^2 = \left(\frac{n\pi}{a^2}\right) + \left(\frac{m\pi}{b^2}\right) \tag{1.5}$$

Detailed explanation of various characteristics of a rectangular-shaped patch antenna is presented in [39–41].

1.3.2.2 The Circular Patch

As shown in Figure 1.14, the characterization of the circular patch configuration is done by a single parameter, that is radius a. With this opinion, the circular patch is the simplest geometry compared to the other shapes of patches that consist of more than one parameter to explain them. As it consists of only one parameter, its mathematical analysis is done by using Bessel functions.

The expression of electric field in a TM_{nm} resonant mode in the cavity under the circular patch is given by,

$$E_s = E_o J_n\left(k_{nm} R\right) \cos n\Psi \tag{1.6}$$

where R is the radial coordinate and ν is the azimuthal coordinate. E_o is the arbitrary constant, J_n is the Bessel function of the first kind of order n, and k_{nm} is the mth zero of the derivatives of the Bessel function of order n and, for circular patch, it is expressed by

$$k_{nm} = X_{nm}/a$$

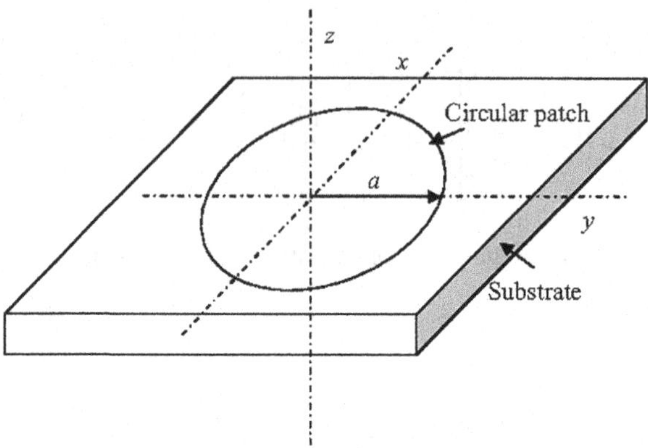

FIGURE 1.14 Geometry of a circular patch.

The resonant frequency is expressed as follows:

$$f_{nm} = X_{nm} \cdot \frac{c}{2\pi a\sqrt{\varepsilon_r}} \tag{1.7}$$

The value of X_{nm} for the TM_{11} fundamental mode is 1.841. The detailed documentation of circular patch microstrip antennas is presented in [42, 43].

1.3.2.3 The Triangular Patch

In recent times, antennas with a triangular-shaped patch have been in more demand due to the advantage of their small size. The fabrication process of a conventional triangular-shaped patch antenna is very simple. Figure 1.15 illustrates the geometry of the coaxial probe-fed triangular patch antenna.

A triangular-shaped patch antenna is designed to overcome the limitations of the earlier rectangular and circular patch antennas. The development of geometrical theory of the resonance frequencies and Q-factors for different types of antennas with triangular-shaped patches is presented in [44]. This theory is based on the electric field inside the cavity under the triangular patch. The analysis of these triangular-shaped patches is performed by the cavity model.

The various triangular shapes include 45°-45°-90°, 30°-60°-90° and 60°-60°-60° equilateral triangular patches. In contrast to the basic shapes of the patches that are studied earlier, there is only a brief analysis of these triangular-shaped patches in [45–47].

Schelkunoff [48] provided all the solutions for the fields in an equi-triangular waveguide with perfect electric walls.

1.3.2.4 Annular Ring Patch

As we know, the attraction towards the rectangular- and the circular-shaped patches is more than the other shapes of the patches. However, the annular ring patch also gained great attention and is a widely studied patch [49–53]. A number

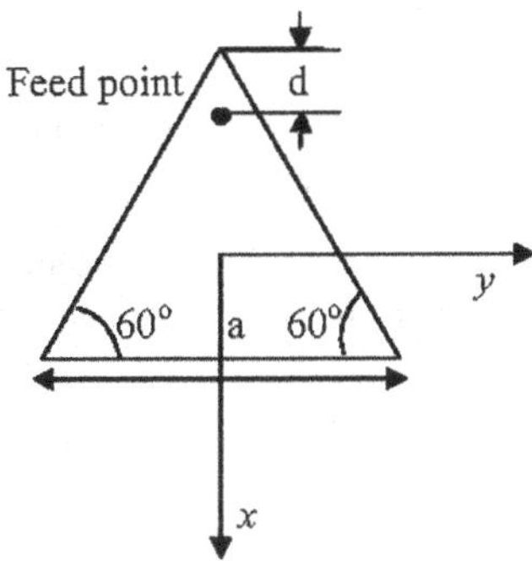

FIGURE 1.15 Triangular patch configuration.

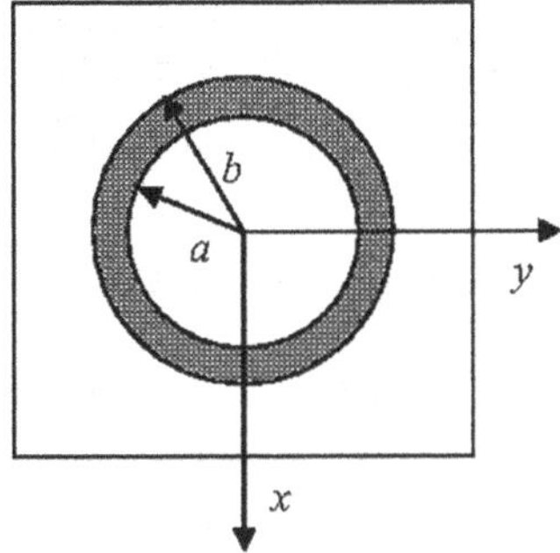

FIGURE 1.16 Annular ring patch antenna.

of attractive features are related to annular ring patches. The size is considerably small than that of the circular patch when both are operated in the lowest mode for a given operating frequency. In application to arrays, an annular ring patch reduces the problem of the grating lobe by placing its elements more densely. Another possible feature is that a dual-band compact antenna [11] can be obtained by merging an additional microstrip element, such as a circular disc, inside the aperture of the annular ring. It can be switched between different modes by adjusting the ratio of outer to inner radii of the annular ring. As compared to other patches with the same dielectric thickness, it achieves a large impedance bandwidth by operating in one of the higher-order broadside modes, such as TM_{12}. The annular ring patch antenna is analysed by using the cavity model and the spectral domain technique in the Fourier–Hankel transform domain. The geometry of an annular ring patch antenna is shown in Figure 1.16. The inner and outer radii of the annular ring patch are denoted by *a* and *b*, respectively. Various performance factors that are analysed by the cavity model for rectangular, circular, equi-triangular and annular ring patch antennas are compared and listed in Table 1.3.

TABLE 1.3
Comparison between Performance Factors of Different Shapes of Patch Antennas ($\varepsilon_r = 2.32$, $t = 1.59\,\text{mm}$, $f = 2\,\text{GHz}$)

Performance Factors	Rectangular Patch ($a = 1.5b$)	Circular Patch	Equi-triangular	Annular Ring ($b = 2a$)	
Radiation	TM_{10}	TM_{11}	TM_{10}	TM_{11}	TM_{12}
3 dB beam width: E-plane/H-plane	102°/85°	100°/80°	100°/88°	103°/81°	30°/47°
First side lobe level: E-plane/H-plane	N/A	N/A	N/A	N/A	−6 dB/N/A
Directivity (dB)	7.0	7.1	7.1	7.1	10.9
Efficiency (%)	87	94	87	86	97
Gain (dB)	6.1	6.8	6.2	6.1	10.6
Bandwidth (%), 2:1 VSWR	0.7	1.1	0.78	0.70	3.8
Physical dimensions	$a = 49.2\,\text{mm}$ $b = 32.8\,\text{mm}$	$a = 49.2\,\text{mm}$	$a = 65.7\,\text{mm}$	$b = 18.4\,\text{mm}$ $a = 9.2\,\text{mm}$	$b = 89\,\text{mm}$ $a = 44.5\,\text{mm}$
Area (mm^2)	161	243	181	106	

1.4 FIELD OF APPLICATIONS FOR PRINTED ANTENNAS

The printed antennas are mostly recognized for their robust design, high performance, ease of fabrication and wide usage in multiple applications such as medical purposes, satellite communication and military systems including rockets, aircrafts, missiles, etc. Today, printed antennas are extensively used in a number of applications in all the fields of wireless communication. Finally, after a great research on printed antennas, they have become capable of replacing the conventional antennas in a large number of applications. Some of the main applications of printed antennas are discussed below:

- **Mobile communication:**
  Mobile communication requires portable devices with lightweight, low-cost, small-sized antennas to make them compact. The selection of the best printed antenna completely fulfils these requirements. Mobile applications consist of many areas such as portable apparatus of small size, cell phones, ultra-high-frequency (UHF) pagers and vehicles such as car, ships and planes.
- **Satellite communication:**
  To meet the basic needs of radiation patterns of satellite communication, that is circular polarization, single- or double-feed points are

applied to the square or circular patch antenna. The printed antennas radiate circularly polarized waves and show flexible designing properties; hence, it is mostly applicable for satellite communication. The planar printed antenna array may be used to transmit from satellite station by replacing the usual parabolic reflector antenna in satellite communication.

- **Global positioning system:**
  Initially, Global Positioning System (GPS)-based satellite systems are applicable only for military systems, but currently, GPS is utilized for huge commercial purposes in various fields. GPS tracks the accurate path and position of vehicles such as cars, planes and ships. In L-band, dual frequencies are used to transmit and receive the signals from a GPS satellite and by several receivers on the earth, respectively. To receive a signal from a GPS satellite, a circularly polarized antenna is required. These days printed antennas with a high-permittivity substrate material for GPS are used.
- **Radio frequency identification:**
  The radio frequency identification (RFID) has many applications in multiple fields, such as identification, mobile communication, transportation, logistics systems, manufacturing process, medical purposes and many tracking applications. The RFID arrangement consists of a transponder and a transceiver with working frequency range from 30 Hz to 5.8 GHz, based on its uses.
- **Direct broadcast satellite system:**
  Several countries utilize direct broadcasting satellite system (DBS) to provide television and radio services. Because of the bulky geometry, a high-gain (~33 dB) parabolic-shaped reflector antenna that is utilized to receive the waves at ground gets affected by snow and rain. Therefore, a direct broadcast receiver requires an array of printed antennas with circular polarization characteristics. The array of printed antennas must be of small size, easy to install, less expensive and uninfluenced by atmospheric conditions.
- **Antenna for pedestrian:**
  Compact antennas for pedestrians remove a great limitation of used space. A low-profile, small-structure antenna with a low weight is normally installed in the portable pocket devices. Printed antennas are greatly applicable for pedestrians, even though it has a disadvantage that a printed antenna with a small structure leads to worst performance.
- **Radar applications:**
  Varieties of printed antennas are designed by many researchers with proper gain and beam width for the radar applications, which include naval radar, remote sensing and surveillance radar. An array of printed antennas may also be utilized to attain the required gain and beam width. A number of electrical devices for sensing the speed and direction of ocean waves and for searching the underground soil grades require synthetic aperture radar method.

1.4.1 Advantages and Disadvantages of Printed Antennas

Printed antennas have several advantages as compared to basic microwave antennas. Therefore, they are used in multiple applications over a broad range of frequency, ~100 MHz to ~50 GHz. Some great advantages of the printed antennas in contrast to basic microwave antennas are listed:

- They are of lightweight, small volume and short profile with planner configuration that is suitable for conformability.
- Antenna fabrication cost is very less, which is necessary for mass production.
- Thin designing is possible, creating no problem to the aerodynamics of host aerospace vehicles.
- Easy installation is possible on missiles, rockets and satellites without any major modifications.
- Antenna scattering region is very small.
- Linear and circular polarizations are possible with some modifications in the position of feed.
- Dual or multiple bands are easily achieved by antennas.
- They show compatibility with integrated and modern designs.
- Simultaneous fabrication of the designed antenna with feed lines and matching networks is possible.

On the other hand, the list of some disadvantages of the printed antennas in contrast to basic microwave antennas is as follows:

- Generally, the outcome of printed antennas is narrow bandwidth.
- They have lower antenna gain due to losses.
- Half-plane radiation happens in most of the printed antennas.
- The maximum gain (~20 dB) is limited due to practical reasons.
- They have degraded performances of end-fire radiation.
- Isolation between the feed and the radiating elements is inappropriate.
- Surface waves are present.
- Power handling capability is low.

However, by following some practices, we can significantly decrease the effects of these disadvantages. For example, by carefully designing and fabricating the antennas, we can suppress or eliminate the excitation of surface waves.

1.5 TECHNIQUES DEVELOPED FOR LOW-PROFILE PRINTED ANTENNAS

In the last decade, the use of parasitic patches in printed antennas gained huge attention for numerous applications in communication and navigation systems. To achieve a narrow bandwidth (>5%) and moderate gain from 15 to 25 dB, a microstrip patch antenna is the best choice. However, the necessary massive production of printed antennas for wide bandwidth and better control of radiation characteristics requires divergence in substrate parameters and manufacturing tolerances. Hence, to achieve the design specifications for an antenna, a simple and accurate analysis of radiating elements is required. The theoretical explanation of the basic-shaped patches and slots is done by the spectral domain approach, while the cavity or transmission line model is used for the analysis purpose. To design planar arrays, a detailed knowledge of various properties such as mutual coupling effect, directivity and losses of the linear sub-arrays of printed antennas is required. For designing the array of printed antennas, easy formula and analysis are explored. 2D arrays having a sub-array with non-identical elements, for example cross-fed geometry, are appropriate for cheaper antennas.

1.5.1 Features of Printed Antenna Technology

The printed antenna is the latest candidate in the field of antenna engineering. The printed antenna is fit for the latest applications because of its advanced features. Table 1.4 [12] illustrates the various performance factors that are significant and essential for designing an antenna for some specified applications. Table 1.5 illustrates all the operational and manufacturing considerations [12] which are in the equal demand, and these considerations are broadly relevant to the application. It is understandable that a new feature, i.e. production of thermal noise at the receiver-side antenna, related mainly to large lossy microstrip arrays is not appropriate for the basic antennas. Similarly, power handling, material effects, such as the use of new materials, mechanical and electrical stability of materials, and inter-modulation effects are basically related to the microstrip patch antennas.

1.5.2 Basic Issues and Design Limitations

To realize the basic issues and design limitations of printed antennas, a study of qualitative outline without examining specific designs in detail has been performed. A slotted triplate and cavity-backed printed antennas define the relation between the

TABLE 1.4
Checklist of Performance Factors for Antenna Design [12]

Performance Factor	Design Constrains
Matching impedance	Input terminals are perfectly matched with the feed
Major lobe	Gain of the antenna and beam width consideration
Side lobes	Controlled among required envelopes
Polarization	Controlled cross-polar behaviour
Circular polarization	Controlled elliptical behaviour of waves
Efficiency	Controlled power in antenna design
Aperture efficiency	Relation between illumination distribution, gain and radiation pattern
Bandwidth	Characteristics of input impedance
System demands	Profile limitation

TABLE 1.5
Manufacturing and Development Considerations [12]

S. No.	Manufacturing Considerations
1.	Effects of noise in receiver side
2.	Transmitter-side power handling capability
3.	Near-field hazardous condition for human resources
4.	Ambient temperature and humidity of environment
5.	Effects of electrostatic charge for space use
6.	Effects of wind speed, vibration, ice, snow, rain
7.	Effects of lightning strikes
8.	Sunlight exposure
9.	Controlled aerodynamic and weather shields
10.	Mechanical and electrical tolerances in manufacture
11.	Sensitivity to manufacturing tolerances for design development
12.	Effects of corrosion and creep in metals
13.	Effects of inter-modulation in materials
14.	Mechanical and electrical stability of materials

bandwidth and the height of the substrate. It defines that the bandwidth decreases with the decrease in the distance of separation between the radiating patch and the ground plane. Therefore, thick antennas have a large bandwidth. An antenna bandwidth is stated as the frequency range in which the condition of matching input is suitable, but the performance limitations of radiation pattern may also dictate the antenna bandwidth. On this basis, one may expect the bandwidth of the very thin printed antennas to be further reduced and this is definitely the case. An explanation is provided by the well-known super-gain concepts [54], which relate the antenna size to its bandwidth. The printed antenna occupies less volume than the cavity-backed antennas, and hence, a less bandwidth can be expected.

Printing of the feed structure on the substrate combined with the radiating element is the next concerning issue. A small amount of power which is coupled through one feeder to another due to surface wave action in the dielectric substrate and some additional losses are been introduced by the feeder lines. Due to this, controlling the distribution of antenna aperture and therefore side lobe level becomes difficult. However, the direct radiation from the feeders on the microstrip substrate results in the further degradation of radiation pattern. Therefore, there is a need for various screened feeder structures to be used in the requirement of modest side lobe levels until other methods are found. A substrate surrounds the microstrip radiating element by which the surface waves are supported, and a little amount of power at each radiator is injected into the substrate, which is different from that of a slot antenna. The control on the side lobes can be further exaggerated by scattering of these surface waves at the boarder of the substrate.

At last, the mechanical tolerances are the major factors in limiting the precision of the lean microstrip structures due to which the amplitude distribution and aperture phase are controlled during manufacturing. The tolerance on the electrical and mechanical parameters of the material such as substrate ageing, temperature, etc are been added to the later part. The antennas backed with a thicker cavity having air gaps require a highly intricate assembly during manufacturing, but are hardly affected by the electrical and mechanical tolerances.

The antenna designers are well known that the performance achieved from an array of waveguide travelling waves is broadly dependent on the accuracy with which the waveguide itself is designed and manufactured. A parallel feeding method is not so much critical for use in design, but at that time, some factors such as cost, size and weight are adversely affected. It is obvious that the accuracy to which the aperture distribution of all the flat-plate antennas can be designed is also dictated by the design of their respective transmission line structures. The design of printed antennas therefore largely centres on the transmission line properties of microstrip lines. However, there is a main problem with microstrip in as much that no exact design equations exist in simple closed mathematical form. For instance, the field behaviour within a rectangular metal waveguide can be rigorously expressed in terms of simple trigonometric formula; the more difficult waveguide problem of the effect of rough metal surfaces, deformed sides, the presence of holes and obstacles, etc., can be analysed with a degree of accuracy sufficient for component design [55]. To obtain the field structure in the microstrip line with a precision demanded by most design specifications, it is necessary to compute equations involving extensive mathematical series, but this in turn leaves some doubt about the resulting numerical accuracy and somewhat defeats the object from a precision design standpoint. The characterization of discontinuities in microstrip lines such as steps, tapers and bends can be assessed at a low frequency using quasi-static techniques which embody the assumption that the radiation effects can be completely neglected. Consequently, the resulting data are of very limited use in printed antenna design, giving at best some indication of how to shorten the line lengths to tune up radiating elements.

The purpose of this segment is to elaborate the existing design issues and limitations for printed antenna analysis. To our knowledge, still certain basic properties of printed antennas are concentrating on their merits and demerits, and therefore, the design challenges of the future are represented as follows.

As compared to the conventional antennas, the printed antennas consist of many differences. Some of them are having two degrees of freedom, being used in very thin topology and the ability to be designed in any shape inside the boundary of x- and y-axes. One of the most wearisome properties is the 'loss', which is encountered in the connecting elements of feeders of thin conducting strip applied in large arrays. The loss that occurs in the radiating elements is also a problem for various applications. The result of using a thin substrate is the limited bandwidth achieved by radiating elements. Actually, a thin substrate used in printed antennas acts as an intrinsic high-Q resonator. Another problem is the generation of surface waves, which is not ignored by deploying foam-type substrates. In demand of low side lobe levels and cross-polarization, the presence of surface waves distorts the radiation pattern. Some problems are generated by compromising the manufacturing of assembly of simple and single co-planar structure of printed antennas. The last one is the high cost of the substrate materials that provide mechanical and electrical stability during operation. Otherwise, the cost of the substrate is an inherent feature of a printed antenna manufacturing process. To understand the importance of future advancement scope of printed antennas, the above basic issues and limitations are highlighted.

1.6 ANALYSIS METHODS FOR SOME COMMON PATCHES

To realize the different antenna radiation characteristics such as input impedance, gain, efficiency, radiation pattern and polarization, the analysis becomes complicated due to the inherent characteristics of the printed antennas, such as heterogeneous dielectric medium, inhomogeneous boundary conditions, narrow frequency band and different types of feed technique. Thus, a trade-off is created between the complexity of the methods and the accuracy of the result. The analysis of printed antennas and consequences of its physical insight are explained by various methods. The antenna analysis may be performed by using different analysis models such as transmission line model (TLM), generalized TLM, lossy TLM, cavity model, generalized cavity model and multi-port network model (MNM). The transmission line model is the simplest technique to give a better physical insight into a designed antenna. It has been mostly useful for designing rectangular shapes of patches. It has also some fundamental drawbacks of neglecting field deviations along the radiating edges. By using the cavity model which is more complicated than the transmission line model, these fundamental drawbacks may be overcome. But it is not easy for the modelling of coupling. However, it provides an exact solution with a great accuracy.

A detailed comparison between different models is illustrated in Table 1.6 [56]. As depicted in the table, it can be easily summarized that the cavity model is the most prominent analytical method for basic configurations.

1.6.1 Analysis of Rectangular Patch Antenna by Transmission Line Model

The transmission line model is the best suited method for analysing the rectangular- or square-shaped patch antennas and will be discussed in this segment. In this method, an antenna with a basic structure including a radiating patch, a dielectric substrate and a metallic ground plane is required. This method is applied on the microstrip line-fed or probe-fed printed antennas as illustrated in Figures 1.17 and 1.18, respectively [57]. Assumed dimensions of the conducting patch for the fundamental mode are a length of 'L', a width of 'W' and a thickness of 't', as shown in figures. The patch conductor has a conductivity σ with the rms surface error e_p. This analysis method considers infinite dimensions for the dielectric substrate of length L_s, width W_s, and thickness h, which is placed in the plane of the radiating patch. The relative permittivity of the dielectric substrate and the loss tangent are denoted by ε_r and δ, respectively. The configuration of the substrate contains one homogeneous layer or multiple layers with different characteristics. Dimensions of the ground plane named with L_g and W_g may extend up to infinity for the analysis purpose. However, it has a thickness t_g with conductivity σ and rms surface error e. Figure 1.17 shows an antenna fed by a microstrip line of width W_f and length L_f.

TABLE 1.6
Evaluation of Different Analysis Models for Printed Antennas [56]

Application	Transmission Line Model	Generalized TLM	Lossy TLM	Cavity Model	Generalized Cavity Model	MNM
Patch shapes analysed	Rectangular only	Separable geometry	Arbitrary shape	Regular shapes	Separable geometry	Separable geometry
Thickness of substrate	Thin	Thin	Thin	Thick	Thin	Thin
Feed category	Microstrip edge, probe	Microstrip edge, probe	Almost all types	Microstrip edge, probe, aperture coupling	Microstrip edge, probe	Microstrip edge, probe, proximity coupling
Stacked antenna	No	No	Yes	Yes	No	No
Applicable to arrays	Yes	Yes	No	No	No	Yes
Circularly polarized antenna	No	Yes	No	Yes	Yes	Yes
Mutual coupling between edges	Explicitly included	Explicitly included	Implicitly included	Implicitly included	Implicitly included	Explicitly included

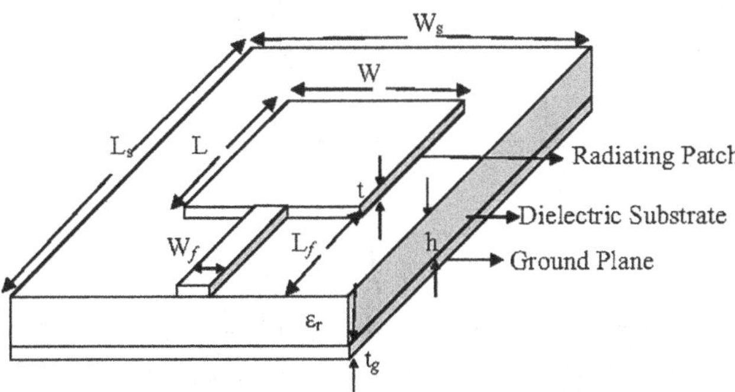

FIGURE 1.17 Detailed configuration of a rectangular patch antenna.

The aspect ratio *W*/*h* elaborates the cross-sectional geometry of the microstrip line for feed purpose. Furthermore, a high value of aspect ratio *W*/*h* considers a patch antenna as a microstrip line.

Figure 1.18 illustrates a transmission line analysis model for a printed antenna with rectangular patch. The cross sections AA′ and BB′ as shown in Figure 1.18

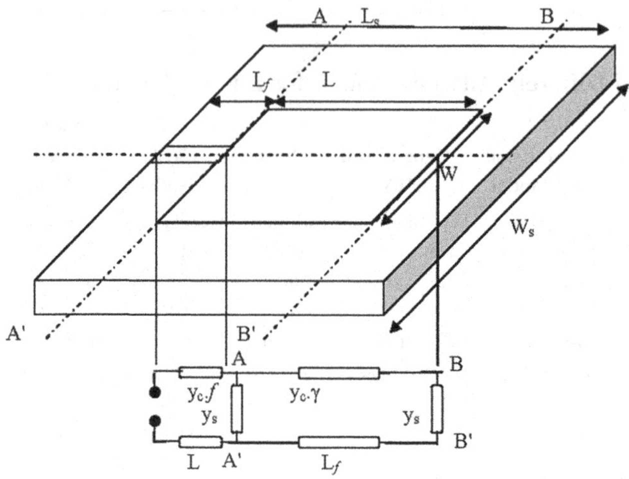

FIGURE 1.18 Rectangular patch antenna with transmission line model.

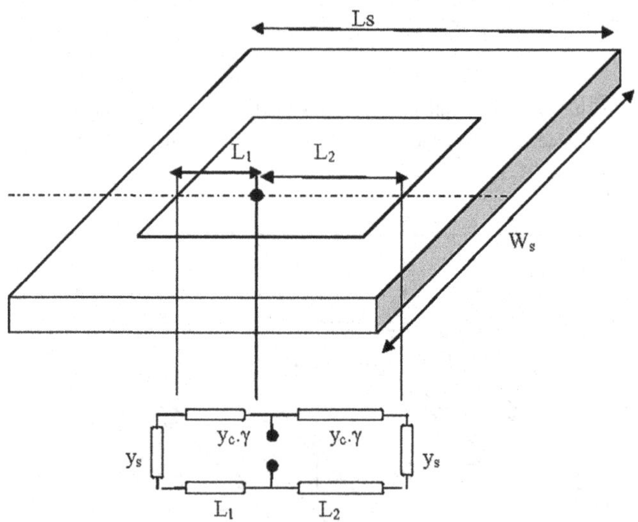

FIGURE 1.19 An equivalent transmission line model for a probe-fed rectangular patch antenna.

correspond to an open-ended termination with aspect ratio W/h of microstrip line of feed. Figure 1.19 represents a modified equivalent transmission line model in case of a coaxial probe feed.

1.6.2 Analysis of Circular Patch Antenna by Cavity Model

The basic structure of a circular printed antenna is shown in Figure 1.20. In the structure, a circular patch of radius a is supported by a substrate of height h. The

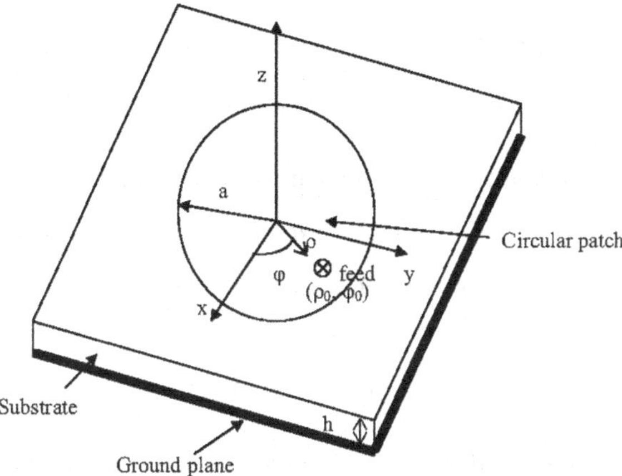

FIGURE 1.20 Geometry of a circular patch antenna.

feed is provided at (ρ_0, φ_0). ρ and φ are the radial and angular coordinates, respectively. The solution to the wave equation for a circular disc is calculated by using the polar coordinate system. Because the substrate height h is very less than λ_0, the waves do not travel towards the z-direction. As a result, the substrate consists of only the z-component of the electric field and the fundamental ρ and φ components of the magnetic field. The current vectors are normal to the edge of the circular patch and tend to zero at the edge. Therefore, the tangential component of the magnetic field vectors at the edge of the circular patch is minor. With these assumptions, the microstrip disc is modelled as a cylindrical cavity, bounded at its top and bottom by electric walls and on its edge by a cylindrical magnetic wall. Thus, the fields within the dielectric region of the microstrip cavity, corresponding to TM_{nm} modes, can be determined by solving the wave equation for a cavity.

The wave equation for a circular printed antenna without any current source is written as follows:

$$\left(\nabla^2 + k^2\right)\vec{E} = 0 \tag{1.8}$$

where $k = 2.\pi.\dfrac{\varepsilon_r}{\lambda_o}$

The electric field in the cylindrical cavity must satisfy the above wave equation and the magnetic wall boundary condition. The solution of the wave equation in cylindrical coordinates is as follows:

$$E_z = E_o J_n k_\rho \cos(n\varphi) \tag{1.9}$$

where $J_n k_\rho$ are the Bessel functions of order n. Because $\vec{E}$ has only the z-component and $\partial/\partial z = 0$, the magnetic field components become

$$H_\rho = \frac{j}{\omega\mu\rho}\frac{\partial E_z}{\partial\varphi} = -\frac{j^n}{\omega\mu\rho}E_o J_n k_\rho \sin(n\varphi) \tag{1.10}$$

$$H_\varphi = \frac{-j}{\omega\mu}\frac{\partial E_z}{\partial \rho} = \frac{-j^n}{\omega\mu} E_o J_n k_\rho \cos(n\varphi) \quad (1.11)$$

where the prime sign denotes differentiation with respect to k_ρ, the argument. The other field components E_ρ, E_φ and H_z are zero inside the cavity.

The magnetic field boundary condition at the wall is defined as

$$H_\varphi(\rho = a) = 0$$

At the edge of the circular disc, the surface current J_ρ must vanish; that is, H_φ is zero at $\rho = a$.

Hence,
$$J_n'(k_{nm}a) = 0 = J_n'(X_{nm})$$

where X_{nm} is the mth zero of the derivatives of the Bessel function of order n. Therefore, to configure every mode, a radius is defined to give a result in a resonance equivalent to the zeros of the derivatives of the Bessel function. Table 1.7 lists a few values of the lower-order modes of X_{nm} in ascending order. From the value of X_{nm} for the various modes given in the table, one infers that the mode for the value of $n=m=1$ has the smallest resonance frequency and is named as the dominant mode.

The modal expansion method is used for calculating the input impedance of the antenna as presented in [58].

The configuration of a circular-shaped patch antenna with their radiation properties among the basic printed antennas is extensively examined by designers. Assumptions for analysing circular patch antennas by this method are infinite dimensions of the substrate and the ground plane. Therefore, the obtained results are approximated and not influenced by the size of substrate and the dimension of the ground plane. The accuracy of the outcomes totally depends on the type of application. The impedance characteristic of the radiating patch depends on the parameters of the patch because of its highly resonant features. The infinite size of the ground plane has a negligible effect on the patch. Therefore, the patch itself determines the near-field radiation. On the other hand, at the back side of the antenna, manipulation in the radiation at broad angles is realized due to the fixed dimensions of the substrate and the ground plane.

1.7 SPECIAL MEASUREMENT TECHNIQUES FOR PRINTED ANTENNAS

This section describes some specific measurement techniques that are basically helpful in the development of structure and fabrication procedure of printed antennas rather than standard measurement techniques. There are two basic categories

TABLE 1.7
Roots of $J_n'(k_{nm}a) = 0$

Mode (n, m)	0,1	1,1	2,1	0,2	3,1	4,1	1,2
Root (X_{nm} or $k_{nm}a$)	0	1.84118	3.05424	3.83171	4.20119	5.317	5.331

in which the importance of these measurements exists [59, 60]. The first one is the impedance measurements, and the second one is the pattern measurement of the printed antenna. The first one covers the complex value of reflection coefficient or equivalent input impedance characteristics at the antenna terminals. Other various radiation characteristics such as gain, beam width, side lobe levels and polarization effects are describes under the second category. The noise figure and efficiency measurements are achieved by both categories. Some specific measurement techniques are recommended only to supplement the measurements of the input impedance and the radiated field. Various factors are present to stimulate the use of these techniques. One of them is the use of undefined dielectric materials or the use of multi-layer substrates with dissimilar properties. Other one is the complex analytical problems that are produced due to transition occurs into the printed network by various transmission lines or a waveguide to main radiator. The next one is applied to the network of large arrays with a complicated feed. Some specific measurement techniques such as time-domain reflectometry (TDR), probing the near field and direct efficiency measurement are used to determine the properties of such arrays.

1.7.1 Substrate Properties

The dielectric constant 'ε_r' and the loss tangent 'tan δ' are the two basic measurable physical properties based on which substrates are commercially supplied. All physical properties are generally specified at a lower value of frequency, e.g. 1 MHz, and at a specified frequency of operation such as 10 GHz. The surface resistivity 'R_s' of the metallic cladding is one more important physical property that is formulated with respect to the conductivity σ_c,

$$R_s = \left(\frac{2\pi f \mu_0}{2\sigma_c} \right)^{\frac{1}{2}} \tag{1.12}$$

where the frequency is denoted by f and the vacuum permeability is denoted by μ_0 ($= 4\,\pi \times 10^{-7}$ in SI units). There are different cases in which the measurements of the printed antenna are achieved. The first case arises at higher frequencies; at that time, the substrate is tested by the manufacturer. The second case arises in mass production that requires a high accuracy and reproducibility. Another case arises in the production of multi-layer substrates that are arranged by placing foam plates with a space between the dielectric layers.

1.7.2 Connector Characterization

The losses by generation of radiation or by surface waves occur due to the discontinuity of the currents in the transition area of the printed networks. Those transitions are formed between the printed antenna to the transmitter or receiver modules or to the coaxial or waveguide terminals which is working as a measurement tools. The main disadvantage of the transition is its reactive nature that gives unwanted reflections because this resonant frequency may be altered. However, a degradation in the performance of antenna occurs due to the passing of all radiated power through the transition. The most commonly used transition occurs in between the coaxial line and the microstrip line, and is achieved by the soldering of the extended inner conductor of the coaxial probe to the printed radiator. This type of transition is appropriate for thin substrates that contain less discontinuity, and many connectors are available commercially for this application. Some other transitions such as striplines, slotlines and transition between microstrip lines and waveguides are also of our interest.

1.7.3 Measurements of Printed Lines and Networks

Previous sections presented the properties of the antenna substrate and the effects of the feed connector briefly. This section describes the measurement of the printed lines and basic configurations of feed networks. Initially, three fundamental parameters of the printed lines as described in [61] are presented as follows:

a. The relation between the effective dielectric constant ε_{eff} and the propagation constant β is formulated by

$$\beta = K_o\left(\varepsilon_{\text{eff}}\right)^{1/2} \tag{1.13}$$

where the wave number of the free space is represented by k_o.
b. α_c and α_d are the ohmic and the dielectric parts of the attenuation factor α, respectively.
c. Z_c is the characteristic impedance of the transmission line.

After that, analyse all the characteristics related to the structures of the printed antenna applied in the feed network, such as bends, T-junctions, variation in width and cross-junctions. All these structures actually show different types of discontinuity that produces reflections and losses. These effects are generally measured by quantitative or by equivalent electrical circuit modelling. The power splitters and delay lines are the fundamental elements that are frequently considered. The main disadvantage of the measurement by network analyser is its performance achieved with or without automatic error correction in the frequency domain. The network analysis of the printed circuits consists of the main problem of coaxial or waveguide forms of test ports and standard calibration units. Therefore, printed lines are measured through transitions.

There are various solutions available for network analysis, of which one is the use of outstanding connectors (VSWR of about 1.01). To separate the resonance of connectors and the device under test, the use of a pair of tested devices is another solution. The next practical solution is the use of resonators consisting of devices under test.

1.7.4 Near-Field Probing

The near-field measurement of a printed antenna is a particular type of pattern measurement that is applied in the region of the radiating aperture of the printed antenna. In case of expensive and impossible execution of far-field measurements such as high-gain antennas and mounted antennas and secure measurement for all weather conditions, this type of measurement is applicable. In this measurement, a scanning probe identifies the near field in a given surface (planar, cylindrical or spherical). To achieve the required far-field results, analytical or numerical methods are used for the analysis of the results of near-field measurement. Some common reviews based on near-field measurements are presented in [62–65]. They consist of scanning methods, data analysis methods and error correction methods that are used to measure the errors generated by probes. To understand the distribution of the field on the aperture of an antenna, near-field scanning is essential, which means near-field measurement must be precise and probe effects are not ignored on the measured field. The near-field probing of printed antennas is one such application, which is applied at a distance of millimetres from the printed antenna. It also works as a diagnostic tool for the designing and manufacturing of the antenna. To find out local defects, asymmetric feed networks and excitation points of the radiator and to develop a report of the distribution of current, near-field probing is very helpful.

1.7.5 Efficiency Measurement

The main causes of the low efficiency of printed antennas are the dissipation losses, ohmic losses and radiation losses in the feeding network, inherent properties of the dielectric, and surface wave excitation in the substrate material. These constraints are not good for large arrays operated at high frequencies with a long and complex feed network. It is difficult to know the contribution of fraction of different losses that are responsible for the reduction in the gain of the printed antenna. A receiver with low noise is the best solution for maintaining variation between dissipation and radiation losses. At the time of dissipation losses, it works like an attenuator and decreases the overall noise figure of the system. However, radiation losses do not affect the noise figure, but noise is generated by the side lobes. Another solution to create differences between the losses is the use of a plastic cover that is placed in front of the antenna.

Here we focus on the simple technique used for describing the measurement of antenna efficiency that is achieved by power efficiency measurement. The overall power efficiency is stated as the ratio of the total radiated power to the input power at the terminals of the antenna. The formulated expression of efficiency is given in terms of gain and efficiency:

$$\eta = \frac{P_R}{P_{\text{in}}} = \frac{4\pi \int P(\theta, \varphi)\sin\theta \, d\theta d\varphi}{P_{\text{in}}} = \frac{\text{gain}}{\text{directivity}} \tag{1.14}$$

1.8 SUMMARY REMARKS

This chapter elaborated on the basics of printed antennas with a concise presentation on various feeding techniques and their essential parameters that are used in the postulation. The objective of the chapter was to discuss what new printed antennas contribute compared to others. This chapter also described the characteristics of printed antennas, features of basic patches, and advantages and disadvantages of printed antennas with their general applications. Techniques developed for low-profile printed antennas and the features of printed antenna technology were explained in detail. Fundamental issues and design limitations of printed antennas were also explored in this chapter. After explaining the design methodology of a printed antenna, this chapter was completed with a short report on the special measurement techniques for printed antennas. The techniques reported for measurement in this chapter will be very helpful in the processes of designing and manufacturing of printed antennas with different issues. These techniques are recommended for the enhancement of the basic process of measurement of far field and for improved analysis of networks.

REFERENCES

1. Greig, D. D., and Engleman, H. F. 1952. Microstrip – a new transmission technique for the kilomegacycle range. *Proc. IRE* 40:1644–1650.
2. Deschamps, G. A. 1953. Microstrip microwave antennas. *Third USAF Symp. Antenn.* 1:189–195.
3. Lewin, L. 1960. Radiatton from discontinuities in stripline. *Proc. IEE C* 107:163–170.
4. Gutton, H., and Baissinot, G. 1955. Flat aerial for ultra-high frequencies. French Patent No. 7031 13.
5. Fubini, E. G. 1955. Stripline radiators. *IRE Nat. Con. Rec.* 3:149–156.
6. Mcdonough, J. A. 1957. Recent developments in the study of printed antennas. *IRE Nat. Conv. Rec.* 5:173–176.
7. Denlinger, E. J. 1969. Radiation from microstrip resonators. *IEEE Trans. MTT* 17:235–236.
8. Howell, J. Q. 1972. Microstrip antennas. *IEEE AP-S. Int. Symp. Dig.*:177–180.
9. Munson, R. E. 1974. Conformal microstrip antennas and microstrip phased arrays. *IEEE Trans.* 22:74–78.
10. *Proceeding of Workshop* on *Printed Circuit Antenna Technology*, 17–19. Oct. 1979. New Mexico State Univ. Las Cruces. New Mexico.
11. Bahl, I. J., and Bhartia, P. 1980. *Microstrip Antennas.* Artech House, Dedham, MA.
12. James, J. R., Hall, P. S., and Wood, C. 1981. *Microstrip Antenna Theory and Design.* IEE. Peter Peregrinus, London.
13. Dubost, G. 1980. *Flat Radiating Dipoles and Applications to Arrays.* Antenna Series No 1. Research Studies Press, New York.
14. Balanis, C. A. 2005. *Antenna Theory Analysis and Design.* 3rd edition. John Wiley & Sons, Inc., Hoboken, NJ. 65–68.
15. Chen, H. D. 2002. Compact circularly polarized microstrip antenna with slotted ground plane. *Electron. Lett.* 38.13:616–617.
16. Gautam, A. K., and Kanaujia, B. K. 2013. A novel dual-band asymmetric slot with defected ground structure microstrip antenna for circular polarization operation. *Microw. Opt. Technol. Lett.* 55:1198–1201.

17. Fan, Z., and Lee, K. F. 1992. Input impedance of annular-ring microstrip antennas with a dielectric cover. *IEEE Trans. Antenn. Propagat.* 40:992–995.
18. Ramirez, R. R., Flaviis, F. D., and Alexopoulos, N. G. 2000. Single-feed circularly polarized microstrip ring antenna and arrays. *IEEE Trans. Antenn. Propagat.* 48:1040–1047.
19. Tsang, K. K., and Langley, R. J. 1994. Annular ring microstrip antennas on biased ferrite substrates. *Electron. Lett.* 30:1257–1258.
20. Waterhouse, R. B. 2001. Design and performance of large phased arrays of aperture stacked patches. *IEEE Trans. Antenn. Propagat.* 49:292–297.
21. Hsu, W. H., and Wong, K. L. 2000. A dual-capacitively-fed broadband patch antenna with reduced cross-polarization radiation. *Microw. Opt. Technol. Lett.* 26:169–171.
22. Sharma, S. K., and Vishvakarma, B. R. 1998. Agile microstrip antenna. *Int. J. Electro.* 84:53–67.
23. Lai, C. H., Han, T. Y., and Chen, T. R. 2008. Broadband aperture-coupled microstrip antennas with low cross polarization and back radiation. *PIER Lett.* 5:187–197.
24. IEEE Standards Board. 1993. IEEE standard definitions of terms for antennas. IEEE Std 145.
25. Balanis, C. A. 1998. *Antenna Theory Analysis and Design.* 2nd edition. John Wiley & Sons, Inc., New York.
26. Richards, W. F., Ou, J. D., and Long, S. A. 1984. Theoretical and experimental investigation of annular, annular sector and circular sector microstrip antennas. *IEEE Trans.* AP 12:864–866.
27. Parasnis, K., Shafai, L., and Kumar, G. 1986. Performance of star microstrip as a linearly and circularly polarised TM mode radiator. *Electron. Lett.* 22:463–464.
28. Palanisamy, V., and Garg, R. 1985. Rectangular ring and H-shaped microstrip antennas: alternatives to rectangular patch antenna. *Electron. Lett.* 21:874–876.
29. Sanford, G. G., and Munson, R. E. Conformal VHF antenna for the Apollo Soyuz test project. *IEE International Conference on Antennas for Aircraft and Spacecraft*, London: 130–135.
30. Ostwald, L. T., and Garvin, C. W. Microstrip command and telemetry antennas for communications and technology satellites. *IEE International Conference on antennas for Aircraft and Spacecraft*, London: 217–222.
31. Kerr, J. 1979. Microstrip antenna developments. *Workshop on Printed Antenna Technology*, New Mexico State University. 3.1–3.20.
32. Haneishi, M., Yoshida, S., and Goto, N. 1982. Broadband microstrip array composed of single feed type circularly polarised microstrip element. *IEEE AP-S Int. Symp. Dig.* 20:160–163.
33. Sharma, P. C., and Gupta, K. C. 1983. Analysis and optimised design of single feed circularly polarised microstrip antennas. *IEEE Trans.* 31:949–955.
34. Shen, L. C. 1981. The elliptical microstrip antenna with circular polarisation. *IEEE Trans.* 29:90–94.
35. Weinschel, H. D. 1975. Cylindrical array of circularly polarised microstrip antennas. *IEEE AP-S Int. Symp. Dig.* 13:177–180.
36. Chang, E., Long, S. A., and Richards, W. F. 1986. Experimental investigation of electrically thick rectangular microstrip antennas. *IEEE Trans.* 34:767–772.
37. Griffin, J. M., and Forrest, J. R. 1982. Broadband circular disc microstrip antenna. *Electron. Lett.* 18:26–269.
38. Fong, K. S., Pues, H. F., and Withers, M. J. 1985. Wideband multilayer coaxial fed microstrip antenna element. *Electron. Lett.* 21:497–499.
39. Hall, P. S. 1987. Probe compensation in thick microstrip patches. *Electron. Lett.* 23:606–607.
40. Lo, Y. T., Solomon, D., and Richards, W. F. 1979. Theory and experiment on microstrip antennas. *IEEE Trans.* 27:137–145.

41. Richards, W. F., Lo, Y. T., and Harrison, D. D. 1981. An improved theory for microstrip antennas and applications. *IEEE Trans.* 29:38–46.
42. Wood, C. 1981. Analysis of microstrip circular patch antennas. *IEE Proc. H* 128:69–76.
43. Shen, L. C., Long, S. A., Allerding, M. R., and Walton, M. D. 1977. Resonant frequency of a circular disc, printed-circuit antenna. *IEEE Trans.* 25:595–596.
44. Helszajn, J., and James, D. S. 1978. Planar triangular resonators with magnetic walls. *IEEE Trans.* 26:95–100.
45. Keuster, E. F., and Chang, D. C. 1983. A geometrical theory for the resonant frequencies and Q factors of some triangular microstrip patch antennas. *IEEE Trans.* 31:27–34.
46. Dahele, J. S., and Lee, K. F. 1984. Experimental study of the triangular microstrip antenna. *IEEE AP-S Int. Symp. Dig.* 22:283–286.
47. Luk, K. M., Lee, K. F., and Dahele, J. S. 1986. Theory and experiment on the equilateral triangular microstrip antenna. *Proceedings of 16th European Microwave Conference*: 661–666.
48. Schelkunoff, S. A. 1943. *Electromagnetic Waves.* Van Nostrand, New York. Chap. 10.
49. Chew, W. C. 1982. A broad-band annular-ring microstrip antenna. *IEEE Trans.* 30:918–922.
50. Ali, S. M., Chew, W. C., and Kong, J. A. 1982. Vector Hankel transform analysis of annular ring microstrip antenna. *IEEE Trans.* 30:637–644.
51. Dahele, J. S., and Lee, K. F. 1982. Characteristics of annular-ring microstrip antenna. *Electron. Lett.* 28:1051–1052.
52. Lee, K. F., and Dahele, J. S. 1985. Theory and experiment on the annular-ring microstrip antenna. *Ann. des Telecomm.* 40:508–515.
53. Mink, J. W. 1980. Circular ring microstrip antenna elements. *IEEE AP-S Int. Symp. Dig.* 18:605–608.
54. Chu, L. J. 1948. Physical limitations on omni-directional antennas. J. Appl. Phys. 19: 1163–1175.
55. Marcuvitz, N. 1951. *Waveguide Handbook.* McGraw-Hill, New York.
56. Carver, K. R., and Coffey, E. L. 1979. *Theoretical Investigation of the Microstrip Antenna.* Technical report PT-00929. Physical Science Laboratory, New Mexico State University, Las Cruces.
57. Carver, K. R. 1979. Practical analytical techniques for the microstrip antenna. *Proceedings of Workshop on Printed Circuit Antenna Technology*, New Maxico State University, Las Cruces: 7.1–7.20.
58. Chew, W. C., Kong, J. A., and Shen, L. C. 1980. Radiation characteristics of a circular microstrip antenna. *J. Appl. Phys.* 51:3907–3915.
59. Hollis, J. S., Lyon, T. J., and Clayton, L. (Eds.) 1970. *Microwave Antenna Measurements.* Scientific Atlanta, Atlanta, GE.
60. Appel-Hansen, J., Dyson, J. D., Gillespie, E. S., and Hickman, T. G. Antenna measurements.
61. Rudge, A. W., Milne, K., Olver, A. D., and Knight, P. (Eds) 1982. *The Handbook of Antenna Design.* Peter Peregrinus, Stevenage. Chap. 8.
62. Johnson, R. C., Ecker, H. A., and Hollis, J. S. 1973. Determination of far-field antenna patterns from near-field measurements. *Proc. IEEE* 61:1668–1694.
63. Paris, D. T., Leach, W. M., and Joy, E. B. 1978. Basic theory of probe-compensated near field measurements. *IEEE Trans.* 26:373–379.
64. Jory, V. V., Joy, E. B., and Leach, W. M. 1983. Current antenna near field measurement research at the Georgia Institute of Technology. *Proceedings of 13th European Microwave Conference*: 823–828.
65. Yaghjian, A. D. 1986. An overview of near field antenna measurements. *IEEE Trans.* 34:30–45.

2 Latest Trends in the Field of Printed Antennas

Dr. Anand Sharma
Motilal Nehru National Institute of Technology, Allahabad

CONTENTS

2.1 INTRODUCTION

Printed antennas are a very important part of today's wireless communication. It is because of their inherent qualities such as compactness, conformability to planar and non-planar surfaces and easy mountability to Monolithic Microwave Integrated Circuits (MMICs) [1]. Two or three decades before, researchers mainly focused on the bandwidth enhancement, generation of multiple frequency bands in a single radiator, and size reduction of printed radiators [2–4]. Now, the wireless communication technology is exponentially growing. Different wireless devices such as mobile phones, hand-held computers and multimedia devices are being used by millions of users at a time. In this era of wireless communication, we are talking about 5G or 6G technology, which requires not only a high data rate but also an efficient spectrum utilization. Now, wireless technology is moving towards high frequencies for getting more and more bandwidth as well as a high data rate. Due to the frequency increment, metallic losses are so large, which in turn reduces the antenna gain. That is why today's research is also being focused on getting a higher gain. Another important research area in the field of printed antennas is the designing of super-wideband antennas, which are also helpful to get a higher data rate in both indoor and outdoor communication systems. Designing of printed antennas with circular polarization features is a hot research topic in the field of microwave antennas. It is because of their wide requirement in the field of satellite communication for maintaining the transmitter and receiver orientation independent. In the current scenario, antenna engineers focus on the designing of an efficient MIMO antenna for getting a higher data rate and maintaining the link reliability. For that purpose, envelope correlation coefficient (ECC) parameter is targeted by researchers. Frequency-selective surface-based Fabry–Perot cavity method is highly effective in the case of ECC (using far-field parameters) reduction.

In this chapter, the latest research trends in the case of printed antennas are discussed. It is divided into six subsections: (i) discussion on the latest techniques for enhancing the gain of printed antennas; (ii) MIMO-based printed antennas with RCS reduction features; (iii) design requirements of printed antennas for 5G applications; (iv) different techniques of circular polarization development; (v) design procedure of super-wideband antennas; and (vi) ECC reduction techniques in the case of printed MIMO antennas.

2.2 LATEST RESEARCH AREAS IN THE FIELD OF PRINTED ANTENNAS

2.2.1 High-Gain Printed Antennas

Gain is the very important parameter of any antenna. One can say that it is a figure of merit for any antenna. It is a well-known fact that an antenna is a passive element. Then, the question is what the significance of the term gain of an antenna is. Actually, the antenna gain basically indicates how much input power is converted into radiated power in some specific direction. We measure the enhancement (gain) in radiated power in a particular direction with respect to hypothetical antennas such as isotropic antennas. Antenna gain is directly related to the efficiency as well as the directivity of the radiator $(G = \eta D)$ [5].

This has been a highly attractive topic of research in the field of antennas for several decades. The reason for attraction can be easily understood by the Friis equation [5]:

$$P_r = P_t G_t G_r \left(\frac{\lambda}{4\pi R} \right)^2 \tag{2.1}$$

From eq. (2.1), it can be said that the received power (P_r) can be improved by enhancing the gain of transmitting (G_t) and receiving antennas (G_r) without enhancing the transmitted power (within the given range). Nowadays, due to strict government regulations, one cannot enhance the transmitting power for getting a better range or signal power. So, gain enhancement has become a hot topic of research in the past few decades. A low value of gain (2.0–3.0 dBi) is a big problem in the case of conventional printed antennas because of high metallic losses and lesser directivity [1]. Today's communication world is moving towards higher frequencies, which makes the problem of low gain more and more dominant. Therefore, in order to get the full advantages of printed antennas (i.e. along with compactness), it is very important to develop some techniques for getting a higher gain value.

The research on gain enhancement, in the case of printed antennas, has started three to four decades ago. There are large numbers of techniques available in the literature for the enhancement of gain of printed antennas. As the relationship between directivity and gain suggests, there are two ways to get a better gain value: either by increasing the radiating power density in some specific direction or by reducing the losses associated with the antenna. In 1985, D.R. Jcakson and N.G. Alexopoulos, two US-based researchers, invented a new technique for the gain enhancement with the help of a superstrate [6]. Figure 2.1 shows the arrangement proposed by the aforementioned researchers.

In this technique, a very high permittivity substrate is used in the antenna structure, which reduces the half-power beam width of the antenna $\left(\theta_h \propto \frac{1}{\text{Gain}}\right)$ and makes the pattern more directive. It is helpful to get a better gain value. This phenomenon can be understood mathematically as follows [6]:

$$\theta_{hpbw} = \left[\frac{\pi H \varepsilon_2}{\lambda_0 \varepsilon_1 \mu_2}\right]^{-\frac{1}{2}} \tag{2.2}$$

In eq. (2.1), 'ε_2' and 'ε_1' are the permittivity of the superstrate and the antenna substrate, respectively. 'μ_2' and H are the permeability of the superstrate and the thickness of the antenna substrate, respectively. In this technique, a very high permittivity superstrate $(\varepsilon_r > 100)$ with low loss has been used. Such types of materials are impractical. Therefore, H.Y. Yang and N.G. Alexopoulos revised the technique and used a multi-superstrate arrangement in place of a single superstrate. They arranged the multiple superstrates in two ways, i.e. electric–magnetic–electric (type-1 resonance) and magnetic–electric–magnetic (type-2 resonance) [7]. Figure 2.2 shows the E-plane pattern for type-1 and type-2 resonances with two, four and six layers of dielectric. In both the cases, the gain value increases largely. Type-2 resonance provides a better gain value as compared to type-1 resonance. But, the superstrate method suffers from two drawbacks, i.e. reduction in bandwidth and increased antenna thickness. In 1988, R.Q. Lee and K.F. Lee proposed the concept of an electromagnetically coupled microstrip radiator antenna for gain improvement [8]. Figure 2.3 displays the proposed concept graphically. They took the idea from the Yagi–Uda antenna,

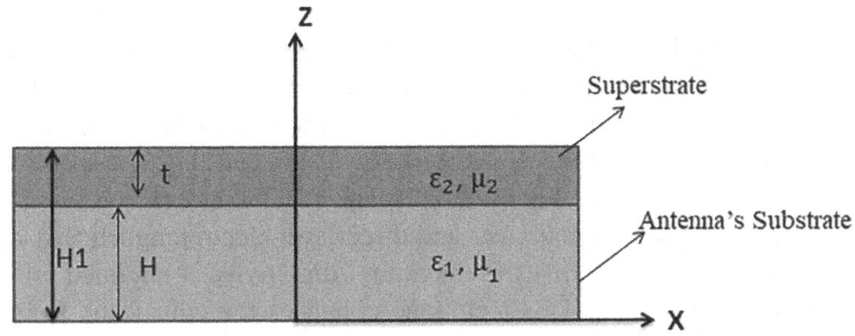

FIGURE 2.1 Gain enhancement of printed antenna with superstrate technique [6].

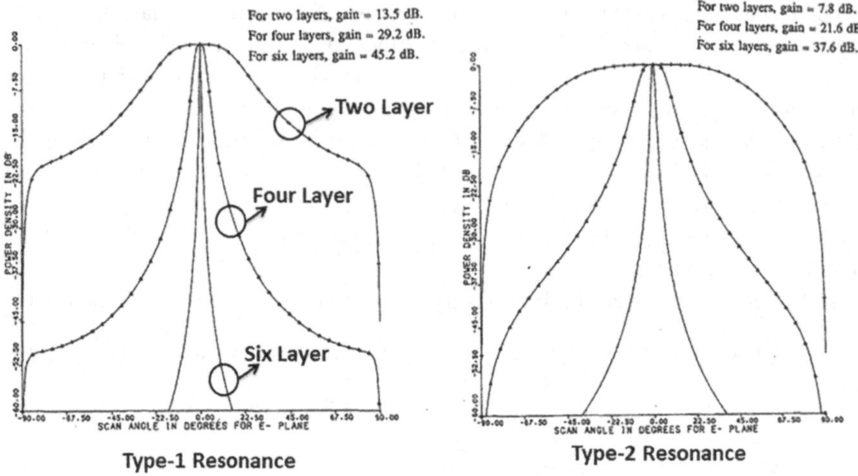

FIGURE 2.2 E-plane pattern for type-1 and type-2 resonances with two, four and six layers of dielectric [7].

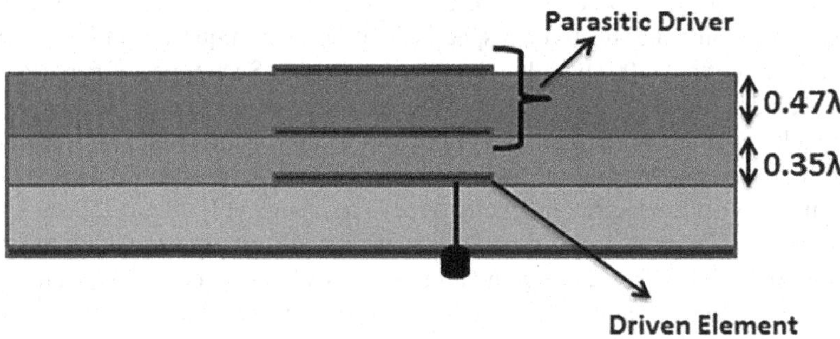

FIGURE 2.3 Multilayer electromagnetically coupled microstrip antenna [8].

where parasitic directors are used to improve the directivity and gain. An electromagnetically coupled microstrip antenna is generally divided into three regions: The first region is used to enhance the bandwidth, the second region is used for radiation purposes, and the third region is used for gain enhancement. All these classifications are based on the separation between the fed patch and the parasitic directors. The authors fed the patch with the fundamental TM_{01} mode. Two parasitic patches are placed above the driven element at a distance of 0.35λ and 0.82λ, respectively, for the enhancement of directivity. Figure 2.4 displays the 2D far-field variation in both principle planes with single-patch, two- and three-layer electromagnetically coupled antennas. Figure 2.4 clarifies that the maximum directivity is obtained with three layers in broadside direction. This technique enhances the gain value from 4.7 to 10.6 dBi with less than 1.0% reduction in bandwidth. In E-plane, the beam width is reduced from 103° to 30°, and it is reduced from 70° to 35° in H-plane.

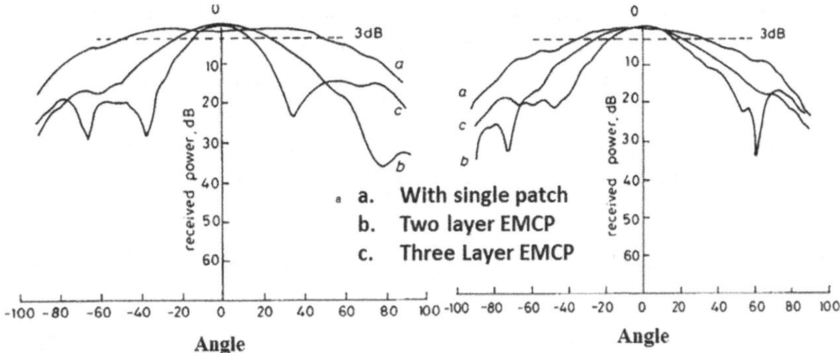

FIGURE 2.4 Far-field pattern in E- and H-planes with single-patch, two- and three-layer EMCP [8].

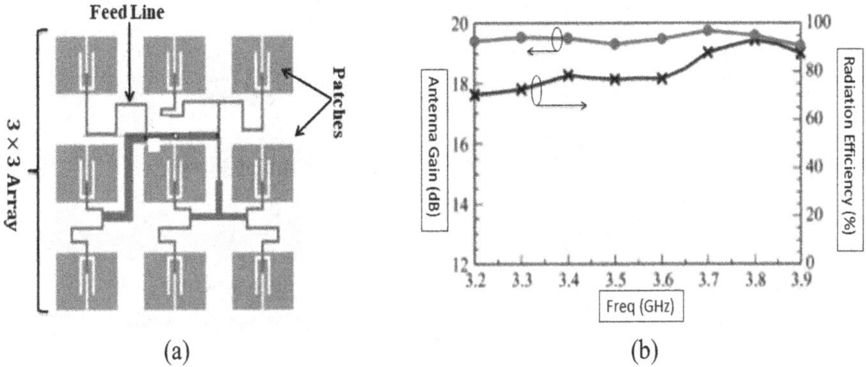

FIGURE 2.5 (a) 3×3 U-slot array antenna layout; (b) gain and efficiency variation [9].

Printed array antennas are also an important technique for gain enhancement, in which a 3-dB power divider is used to feed the different radiating elements. This feed arrangement provides the current distribution in the same phase in each radiating element. The coupling between radiating elements is also less in the array antenna design. This method is able to give a large gain value without (or sometimes minimally) affecting the impedance bandwidth.

Chen et al. proposed two different U-shaped slot-loaded patch array antenna arrangements, i.e. 3×2 array and 3×3 array. Figure 2.5 shows the geometrical layout of the 3×2 U-shaped slot array and the gain variation of the given array. As the authors move from the 3×2 array to the 3×3 array arrangement, the value of gain is increased by 1.8 dBi. Similarly, the bandwidth also increases by 14%–20%. This antenna arrangement is able to get a maximum gain value of 19.4 dBi [9]. The losses associated with the feed line are more in the array case, which in turn reduces the efficiency of the antenna at higher frequencies.

Another important way of gain enhancement is the loading of meta-material-based resonating structure with the printed radiating structure. The loading of the meta-material structure enhances the refractive index of the substrate electrically, which will bend the radiation beam in certain direction (as per Snell's law). It will enhance the directivity of the antenna, which in turn enhances the gain value. This concept was well used by Wang et al. in 2014. They used H-shaped resonating structures periodically at the top of the antenna substrate. The value of refractive index, after applying resonating structures, is calculated as follows [10]:

$$n = \frac{1}{k_0 d} \cos^{-1}\left[\frac{1}{2S_{21}}\left(1 - S_{11}^2 + S_{21}^2\right)\right] \tag{2.3}$$

The antenna design and its gain variation are shown in Figures 2.6 and 2.7, respectively. The antenna is designed vertically above the ground plane. An H-shaped

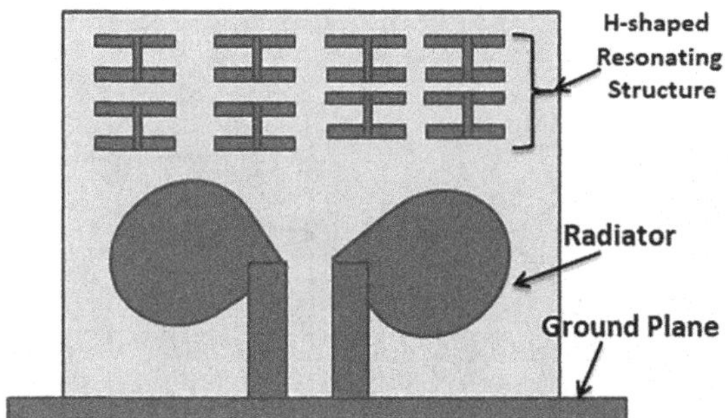

FIGURE 2.6 Vertical printed antenna with H-shaped resonating structure [10].

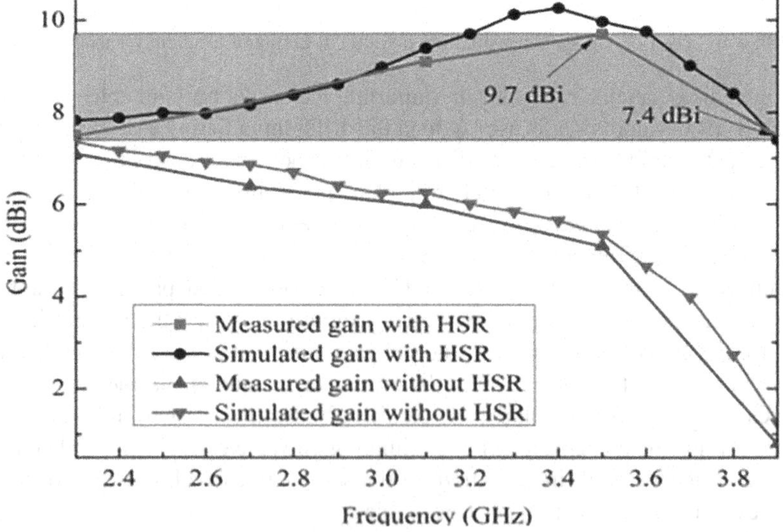

FIGURE 2.7 Gain and frequency variation with and without H-shaped resonating structure [10].

resonating structure is used to enhance the gain value. This phenomenon improves the gain value by more than 5.0 dBi. The loading of frequency-selective surfaces (FSS) over the printed antenna is also an important and latest technique in the field of gain enhancement. Frequency-selective surfaces are periodic arrangements of unit cells. This type of structure is placed over the printed antenna, which makes a cavity-like structure. EM waves strike the FSS and are reflected back to the antenna. Finally, a radiation beam leaks through the FSS. This beam has a high directivity, which in turn improves the gain value [11].

Shalini et al. proposed a FSS-loaded dual-band slot antenna with an improved gain, which is displayed in Figure 2.8. In the aforementioned antenna design, the FSS is designed in such a way that it can be used for both the operating frequency ranges. After applying the FSS structure, the value of gain is improved by 8.0 dBi in the working frequency range [12]. Figure 2.9 presents the gain variation with and without the use of frequency-selective surfaces. From Figure 2.9, it can be observed that the gain value is enhanced by 6–7 dBi in both the frequency ranges.

However, in FSS technique, the gain improvement is large as compared to other discussed methods, but it affects the main advantage of the printed radiators, i.e. compactness. Therefore, Prateek Juyal and L. Shafai proposed a printed radiator with an improved gain value with no (or very little) effect on the antenna size. The authors proposed the superposition of two modes in order to enhance the directivity and gain with less effect on physical size. Initially, they added the radiation in TM_{11} and TM_{13} modes. These modes are created by two circular discs placed in a staked configuration. The circular discs are arranged by size in such a way that both modes have the same resonant frequency. In this antenna design, the value of gain is about 13.06 dBi [13]. But, the staked configuration still creates the problem of size. In order to overcome it, the same authors proposed a new printed antenna based on the same concept, i.e. superposition of two modes (TM_{12} and TM_{14} modes), but the radiators are placed in the same plane. In this case, the value of gain is larger, i.e. 15 dBi [14]. Antenna geometry of the staked configuration and the single-layer configuration is shown in Figure 2.10.

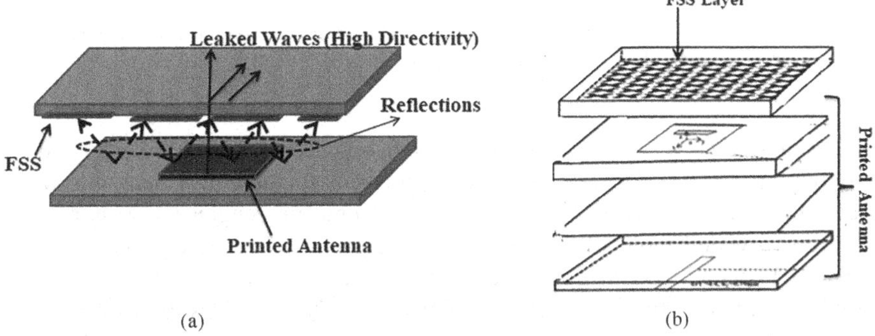

FIGURE 2.8 FSS-loaded printed antenna with improved gain value [12].

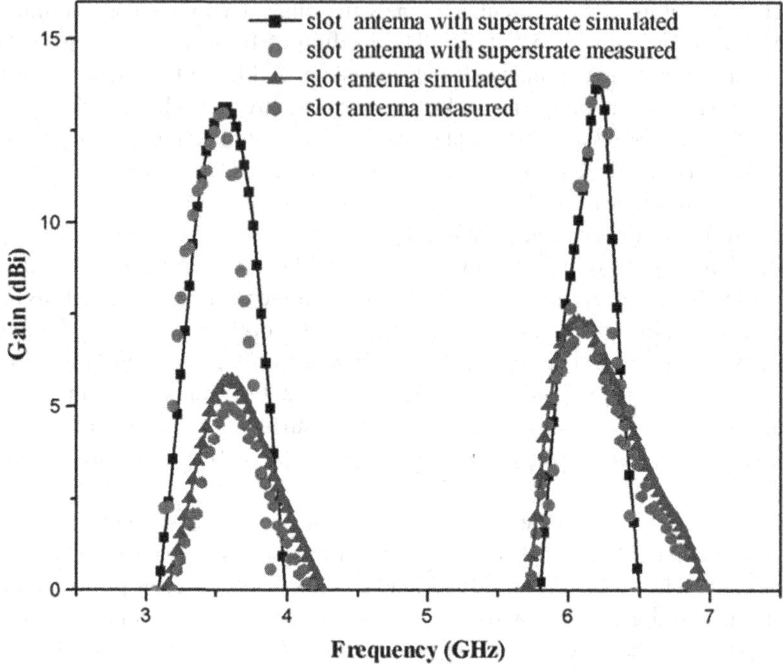

FIGURE 2.9 Gain variation with and without using frequency-selective surface over the printed radiator [12].

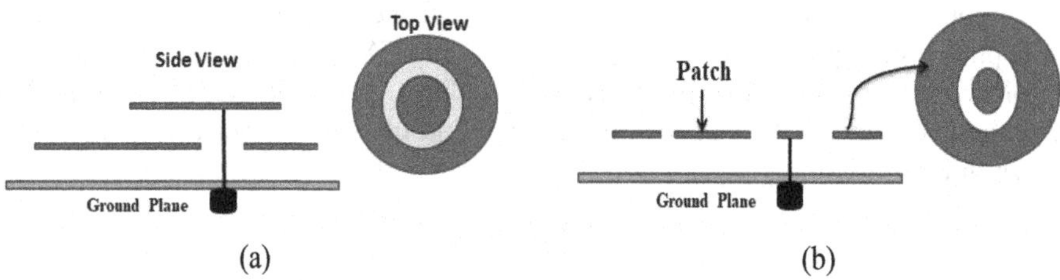

FIGURE 2.10 Dual-mode printed antenna configuration: (a) staked geometry [13]; (b) single-layer geometry [14].

The current wireless communication world has a wide requirement of a large data rate for rapidly increasing data traffic. To fulfil such type of requirement, mm-wave spectrum has been utilized by the wireless communication engineers. At mm-wave spectrum, a wide bandwidth is available as compared to lower frequencies, which is helpful to get a higher data rate. But, at mm-wave frequencies, metallic losses are very high, which in turn reduces the antenna gain effectively. As discussed above, the gain of an antenna can be enhanced by several ways, such as the use of reflectors and antenna arrays. But, at mm-wave frequencies, these techniques are not used because of certain reasons. For example, the use of reflectors along with radiators limits the impedance bandwidth and the frequency of operation is sensitive to the gap between the reflector and the antenna. Antenna arrays are not very effective at mm-wave frequency because they suffer from a large feed loss. Recently, a new technique has been developed, i.e. substrate integrated waveguide (SIW) antenna, for gain enhancement at mm-wave frequencies. SIW structure consists of two metallic plates separated by a dielectric material. The two metallic plates are connected through conducting vias. The conducting vias connect the surface current of upper and lower metallic plates in order to maintain guiding structures. SIW antennas work based on the principle of leaky wave antennas such as waveguide antenna. Leaky wave antennas generally have a high directivity. The feed network loss is minimum in the case of SIW antennas. This is the main reason for getting a high gain value. Wahab et al. proposed an aperture-coupled microstrip antenna design at 60.0 GHz. The aperture is fed by substrate integrated waveguide. The authors discussed two different antenna designs based on aperture orientation: (i) transverse slot and (ii) longitudinal slot. Figure 2.11 displays the antenna design, its reflection coefficient, and gain variation. In both the configurations, the antenna gain is about 6.0 dBi at

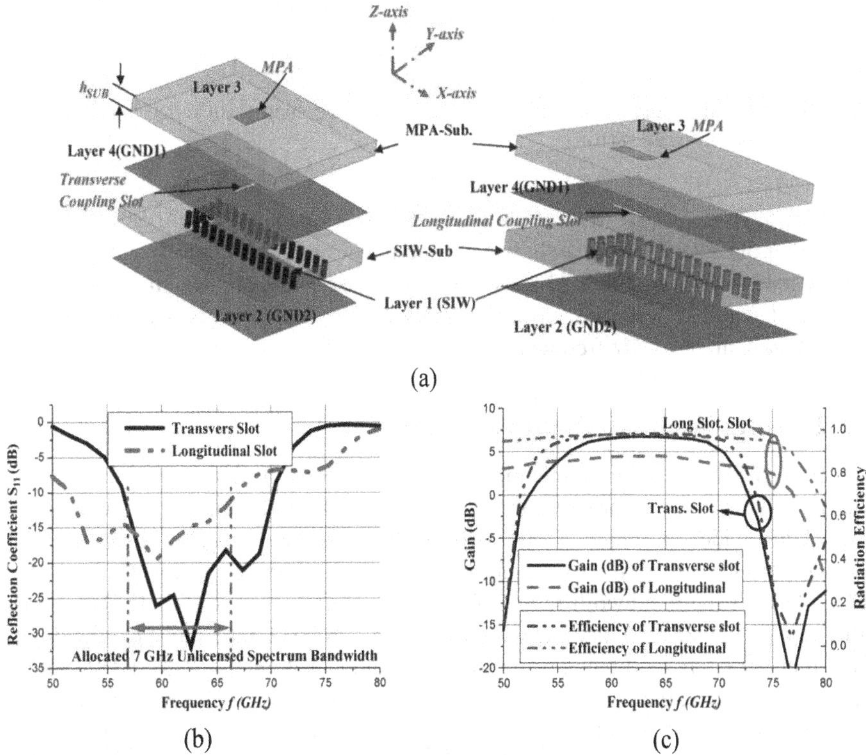

FIGURE 2.11 Millimetre-wave microstrip antenna: (a) radiator geometry; (b) reflection coefficient; (c) gain and efficiency variation [15].

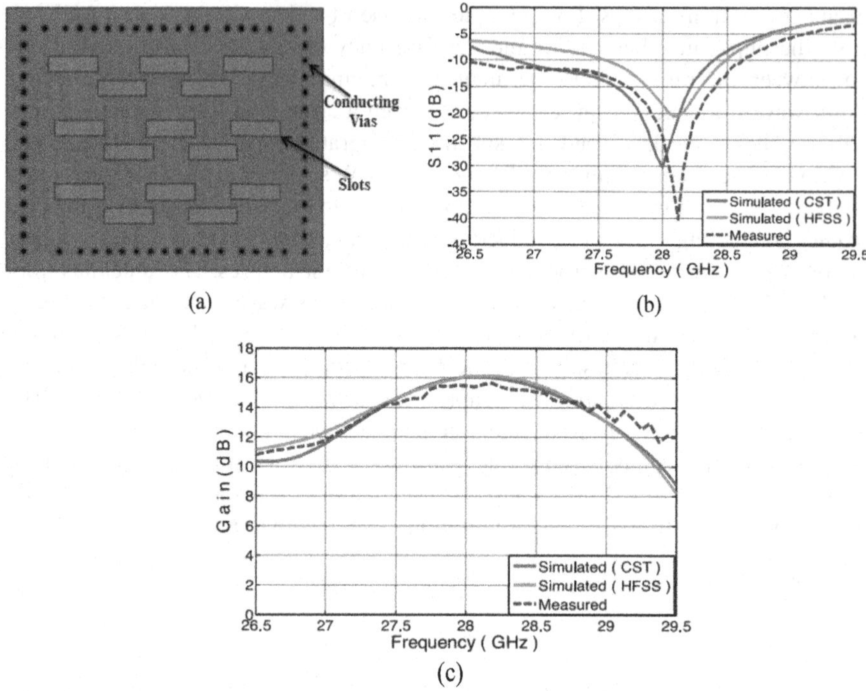

FIGURE 2.12 SIW antenna at mm waves: (a) antenna geometry; (b) reflection coefficient variation; (c) gain variation [16].

60.0 GHz [15]. Similarly, M. Asaadi and A. Sebak proposed a high-gain low-profile circularly polarized SIW antenna. In this antenna design, periodically arranged slots are excited by SIW. Figure 2.12 shows the SIW-based antenna design, its reflection coefficient features and antenna gain variation. This radiator gives approximately 16.0 dBi gain at 28.0 GHz frequency [16].

2.2.2 Super-Wideband Printed Antennas

Currently, the wireless communication mainly focuses on the development of devices with a high data rate. In order to get this feature, wideband radiators are widely utilized. Ultra-wideband antennas are also used to achieve the higher data rate with a least prerequisite of effective energy [17]. Federal Communication Commission (FCC) has set the bandwidth range of UWB antennas as 3.1–10.6 GHz [18]. Some other reports said that an antenna with more than 500 MHz impedance bandwidth or more than 20% bandwidth can be considered as an UWB antenna [19]. UWB antennas are highly used in remote sensing, imaging and wireless PAN. Though the UWB antennas are capable of supporting such a large bandwidth, the power transmission is much less for the applications at lower frequency ranges such as ISM bands. This type of antennas also suffers from low signal acquisition time [20]. That is why UWB radiators are mainly applicable for indoor or short-distance communications.

In order to overcome this drawback, the super-wideband antenna (SWB) comes into the picture. This type of radiator is applicable for both indoor and outdoor communications. In different research articles, super-wideband antennas are defined as an antenna with a decade bandwidth (VSWR < 2), or one can say that the aforementioned condition must be fulfilled for the bandwidth ratio of 10:1 [21]. Due to the support for such a large impedance bandwidth, SWB antennas are widely used as sensing radiators in cognitive radios, wireless LAN, satellite communication, personal communication service and so on. Generally, the performance of super-wideband antennas are analysed with the assistance of bandwidth dimension ratio (BDR). This parameter provides information on the percentage bandwidth provided per unit electrical area. A high value of BDR indicates that the antenna has achieved a wider impedance bandwidth within the compact area. It can be calculated by using the following formula [22]:

$$\text{BDR} = \frac{\text{B.W.}(\%)}{\lambda_L \times \lambda_w} \tag{2.4}$$

In the above equation, λ_L and λ_w are the electrical length and width of the radiator. A lower cut-off frequency is used to calculate λ_L and λ_w.

Okas et al. proposed a moderately segmented circular monopole for super-wideband applications [23]. The geometrical layout of the aforementioned printed super-wideband is displayed in Figure 2.13. In the proposed antenna design, the authors used three different concepts in order to achieve the super-wideband features:

i. The use of notch-loaded elliptical ground plane in place of simple partial ground plane reduces the effective capacitance, which is in turn helpful to give a wide impedance bandwidth.
ii. Tapering of the microstrip line gradually changes the impedance value, which is helpful to get a wider impedance bandwidth.
iii. By cutting an elliptical slot in the segmented circular monopole, the electrical length is increased, which in turn expands the BDR.

After following all these steps, the authors are capable of working within the frequency range of 0.96–10.9 GHz with a bandwidth ratio of 11.35:1. Figure 2.14

FIGURE 2.13 Antenna design of super-wideband circular monopole antenna [23].

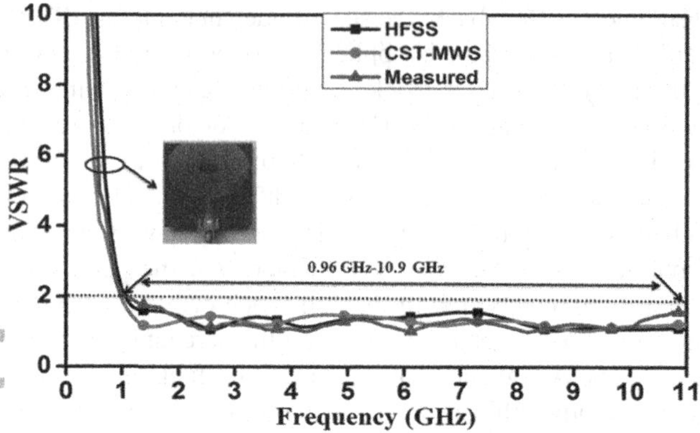

FIGURE 2.14 VSWR variation of super-wideband circular monopole antenna [23].

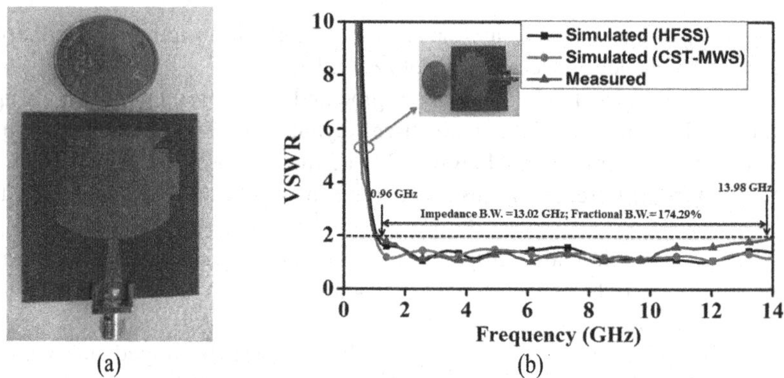

FIGURE 2.15 Modified rectangular monopole antenna: (a) antenna structure; (b) VSWR variation [24].

displays the VSWR of the segmented circular monopole. In this way, it can be considered as a super-wideband antenna.

Okas et al. designed a modified rectangular monopole for super-wideband applications. This antenna design is conceptually similar to that in Ref. [23], but in place of slot, a butterfly-shaped rectangular monopole is used to improve the electrical length of the radiator. Figure 2.15a and b shows the modified rectangular monopole for super-wideband antennas and its VSWR variation. After seeing VSWR features, it can be said that the aforementioned radiator is working in between 0.96 and 13.98 GHz [24].

The main problem of the above-mentioned radiator is that it provides an unstable far-field pattern and a poor ratio of co-polarization to cross-polarization as the frequency increases. The same research group identified this problem and solved it by proposing a perturbed square monopole stimulated by coplanar waveguide (CPW). The antenna design of the truncated square monopole is shown in Figure 2.16a.

In the aforementioned antenna design, the use of a modified square monopole and truncated ground plane is helpful to achieve SWB characteristics. Figure 2.16b displays the VSWR features of the modified square monopole antenna. This radiator works in the frequency range of 0.95–13.8 GHz. In this antenna design, the authors truncated the top corners of the square monopole. In this radiator, the current density is very less at the top corner of a super-wideband antenna. Therefore, any modification in this section of the antenna does not create much effect on the reflection coefficient features of the proposed antenna. The truncation of upper corners reduces the diffraction and scatters the radiated power, which results in a stable radiation pattern with a good ratio of co-polarization to cross-polarization even at high frequencies [25].

Singhal et al. proposed a fractal geometry-based super-wideband printed antenna, which is shown in Figure 2.17a. The authors used the Sierpinski fractal geometry on the hexagonal-shaped monopole radiator. By using fractal geometry (with increasing number of iterations), the authors achieved a better impedance matching at high frequencies. The hexagonal-shaped monopole is also helpful to obtain a gradual change in impedance and to get a better bandwidth [26]. The operating frequency range of this radiator is 3.4–37.4 GHz, which is displayed in Figure 2.17b. This article also identified and explained the very important concept

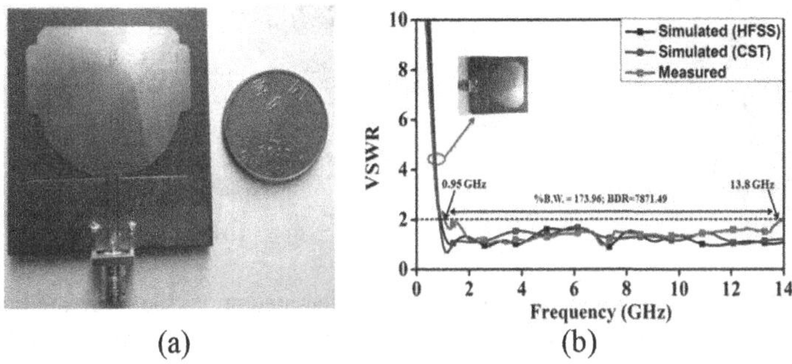

FIGURE 2.16 Truncated square monopole: (a) the fabricated antenna; (b) VSWR variation [25].

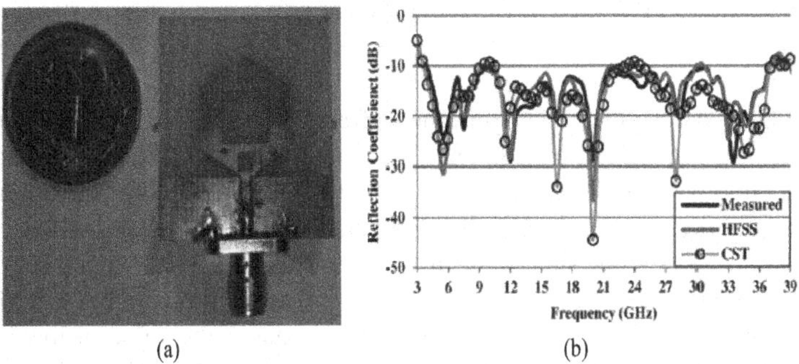

FIGURE 2.17 Sierpinski fractal geometry-based SWB antenna: (a) the fabricated prototype; (b) reflection coefficient variation [26].

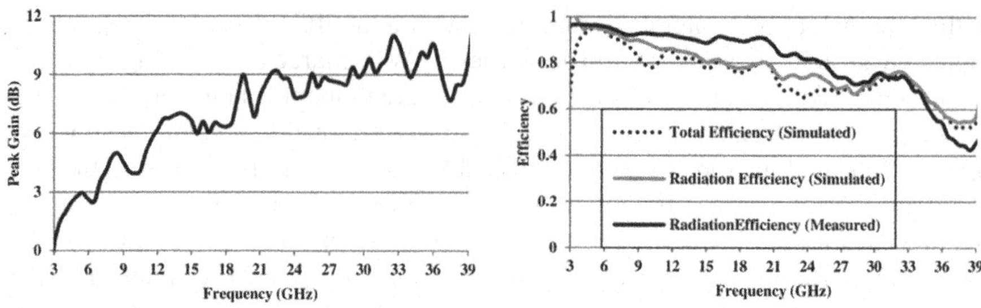

FIGURE 2.18 Gain and efficiency variation of Sierpinski fractal geometry-based SWB antenna [26].

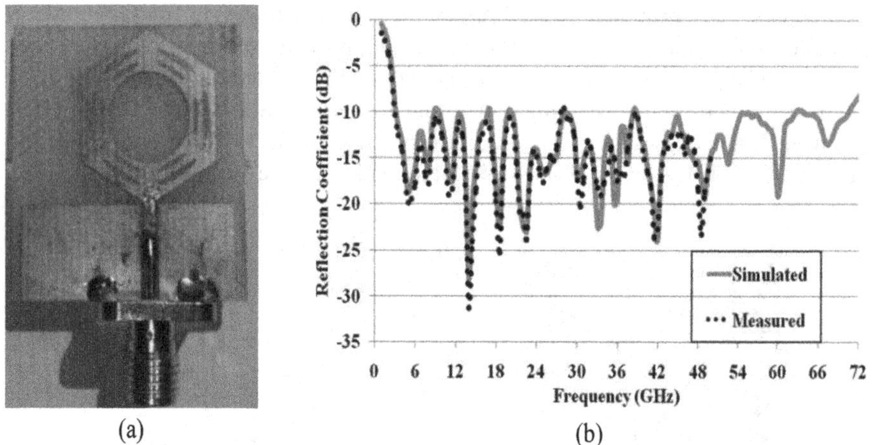

FIGURE 2.19 Hexagonal geometry-based monopole antenna: (a) the fabricated prototype; (b) reflection coefficient variation [27].

that as the frequency increases, the antenna gain increases exponentially. On the other hand, the antenna efficiency decreases at the higher frequencies. This phenomenon is shown in Figure 2.18. The efficiency decreases due to the increment in the ohmic losses (I^2R losses). The increment in gain is due to the increment in electrical aperture with frequency [1].

The same research group again designed a fractal geometry-based super-wideband antenna. It is shown in Figure 2.19a. In this antenna structure, the authors used asymmetric coplanar waveguide feeding, which allows the creation of higher-order modes in the antenna design and reduces the frequency difference between them. In this antenna design, a circular slot is cut in between the hexagonal-shaped monopoles. Due to the absence of current density at the central portion of monopole, it does not create any effect on $|S_{11}|$. This modification only reduces the value of lower cut-off frequency by enhancing the electrical length [27]. Its $|S_{11}|$ feature confirms that the given fractal antenna operates over the frequency range of 2.75–71 GHz, as shown in Figure 2.19b.

Siahcheshm et al. proposed an inverted triangular metallic monopole. The authors also used CPW feed to excite the monopole. It also includes two rectangular slots. Figure 2.20a displays the fabricated prototype of the inverted triangular metallic monopole [28]. It operates over the frequency range of 3.05–35 GHz, which is shown in Figure 2.20b. But, this article did not discuss anything about gain and time-domain analysis of the given antenna. Chen et al. proposed an egg-shaped monopole fed with an asymmetrical microstrip line in order to get a super-wide bandwidth. The authors have cut the fractal geometry on the ground plane in order to reduce the electric current on the ground plane and improve the impedance matching at lower frequencies [22]. Its geometry is displayed in Figure 2.21a. However, in this article, the antenna size was very large concerning the bandwidth obtained (1.44–18.8 GHz). Its return loss is displayed in Figure 2.21b. That is why its BDR (2735) is smaller as compared to other proposed radiators.

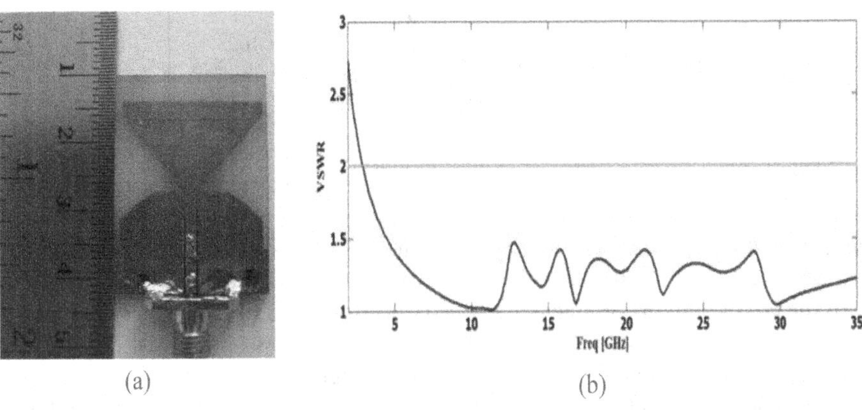

FIGURE 2.20 Inverted triangular monopole-based SWB antenna: (a) the fabricated prototype; (b) reflection coefficient variation [28].

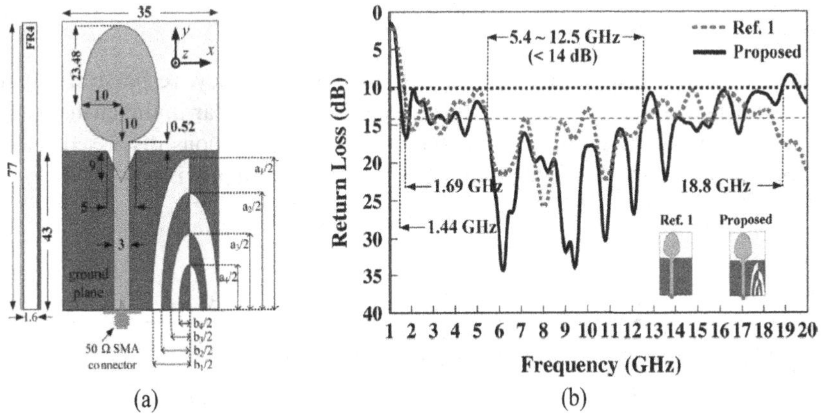

FIGURE 2.21 Egg-shaped microstrip line-fed monopole antenna: (a) antenna structure; (b) return loss variation [22].

In the case of designing of a super-wideband antenna, the time-domain (TD) analysis is also as important as the frequency-domain analysis [29]. In SWB radiators, we are dealing with pulses of picosecond order. With the help of time-domain analysis, one can easily verify that the received pulses are in synchronization with the transmitted pulses [30]. Okas et al. gave the time-domain analysis for the modified square monopole antenna [25]. For that purpose, they carried out the analysis on two different parameters: (i) fidelity factor and (ii) group delay. In order to analyze all these parameters, first, one has to design two different radiator arrangements, i.e. face-to-face and side-by-side arrangements. They are presented in Figure 2.22.

Fidelity factor indicates the similarity content between the transmitted and received pulses. It can be calculated with the assistance of the following formulation [31]:

$$\text{Fidelity Factor} = \text{MAX}\left[\frac{\int_{-\infty}^{\infty} x(t)y(t+T)dT}{\int_{-\infty}^{\infty}|x(t)|^2 dt \int_{-\infty}^{\infty}|y(t)|^2 dt}\right] \tag{2.5}$$

In eq. (2.13), x (t) and y (t) are the transmitted and received signals, respectively. The higher the value of the fidelity factor (FF) is, the more is the correlation between the transmitted and received signals. It means for a high value of FF, we get less distortion. Figure 2.23 displays the normalized transmitted and received pulses in both the configurations of antenna geometry, as discussed in Ref. [25]. After seeing Figure 2.23, it can be said that the correlation of pulses is high in the face-to-face configuration as compared to the side-by-side configuration.

Group delay is simply the time taken for the signal to reach the receiving antenna from the transmitting antenna. It gives the information about the phase distortion. It can be calculated by using the following formula [31]:

$$\tau_g(w) = -\frac{d\phi(w)}{dw} \tag{2.6}$$

'ϕ' and 'w' denote the phase response and angular frequency, respectively. If the group delay is less than 1.0 ns, then the phase variance is linear in the far-field area. Figure 2.24 shows the group delay variation for both provisions discussed in Ref. [25]. Group delay variation is approximately 1.0 ns for both provisions.

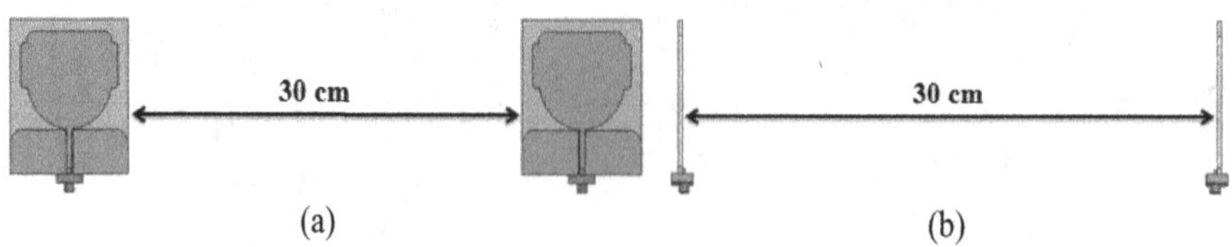

FIGURE 2.22 Arrangements of radiators for TD analysis: (a) side by side; (b) face to face [25].

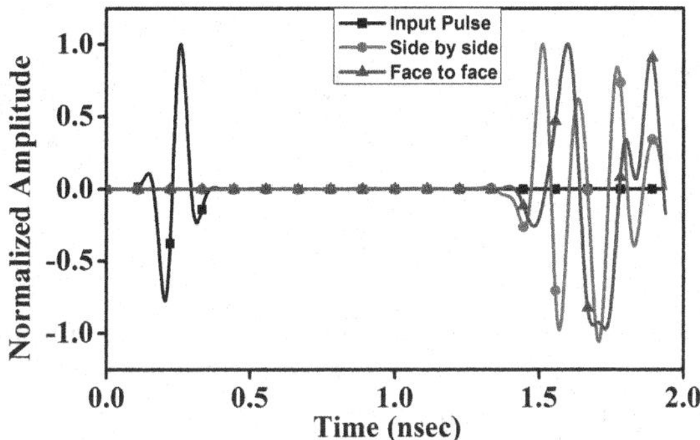

FIGURE 2.23 Transmitted and received pulses in side-by-side and face-to-face configurations [25].

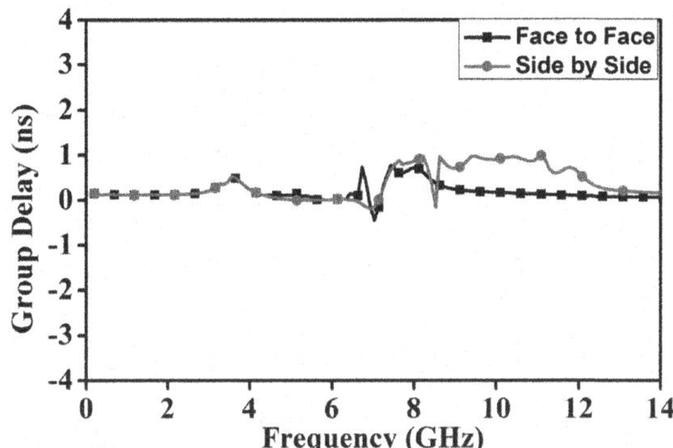

FIGURE 2.24 Group delay variation in side-by-side and face-to-face configurations [25].

In the case of super-wideband antennas, currently researchers focus on getting a higher bandwidth dimension ratio within the compact physical area. Another important current research area in the field of SWB antennas is to get a stable radiation pattern at higher frequency of operations.

2.2.3 Printed Antennas with Circular Polarization (CP) Features

Design and analysis of a wideband/multiband circularly polarized printed radiator is a hot research topic because of its capability of maintaining the orientation independency between transmitter and receiver. Before going into details of circularly polarized antennas, it is important to know about its theoretical background. Polarization is simply defined as the orientation of electric field. It can be classified into three categories: (i) linear polarization, (ii) circular polarization and (iii) elliptical polarization. Circular and linear polarizations are the

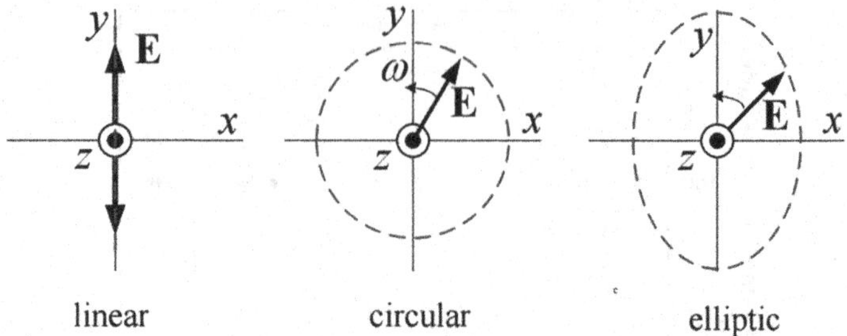

FIGURE 2.25 Different types of polarization.

special cases of elliptical polarization [1]. Figure 2.25 shows the orientation of E-field in all three cases.

For a plane wave travelling in the z-direction, the net electric field (E) will have, in general, a component in both the x- and y-directions. Let us suppose that

$$E_x = E_1 \cos(wt) \tag{2.7}$$

$$E_Y = E_2 \cos(wt + \varnothing_0) \tag{2.8}$$

After solving eqs. (2.7) and (2.8), we get the following equation:

$$\frac{E_y}{E_2} = \frac{E_x}{E_1}\cos\varphi_0 - \left(\sqrt{1 - \left(\frac{E_x}{E_1}\right)^2}\right)\sin\varphi_0 \tag{2.9}$$

Now, we take the square of eq. (2.9) and rearrange the equation:

$$\left(\frac{E_x}{E_1}\right)^2 - \left(\frac{2E_xE_y}{E_1E_2}\right)\cos\varphi_0 + \left(\frac{E_y}{E_2}\right)^2 = (\sin\varphi_0)^2 \tag{2.10}$$

From eq. (2.10), it is found that the wave is elliptically polarized in general, where the phase and magnitude change with time. If the phase is zero degree, then the polarization becomes linear polarization. On the other hand, if $\varphi_0 = \pm\ 90°$ and $E_x = E_y = E_0$, then eq. (2.10) becomes

$$(E_x)^2 + (E_y)^2 = (E_0)^2 \tag{2.11}$$

Eq. (2.11) is the equation of a circle. So, it can be said that for creating CP waves inside any radiator, it is required to generate orthogonal modes with equal amplitudes and phase quadrature. Generally, for the identification of circular polarization, one has to check the parameter known as the axial ratio. Axial ratio is defined as the ratio of the major axis to the minor axis of the locus of net electric field at a particular point when time goes and mathematically [1],

$$\text{AR} = \frac{\text{Major axis length}}{\text{Minor axis length}} = 20\log\left(\frac{E_{\max}}{E_{\min}}\right)\text{dB} \tag{2.12}$$

For circularly polarized waves, the value of axial ratio is ideally equal to zero in logarithmic scale. But, for practical cases, less than 3-dB axial ratio is preferable for CP antennas.

In the recent wireless communication world, circularly polarized antennas are more vibrantly used. It is because of their inherent qualities such as orientation independency and better performance in bad weather conditions. Currently, a wide research on multiband, wideband, dual-/triple-sense circularly polarized printed antennas is underway. While designing CP antennas, two necessary conditions must be fulfilled: (i) generating the dual orthogonal degenerated modes and (ii) 90° phase shift in between the modes [1].

Liang et al. proposed a dual-band CP printed antenna for wireless LAN and WIMAX applications. It is designed with the assistance of two eccentric rings of different sizes and fed with a metallic arc between them. Each ring is accountable for one frequency band. The resonant frequency of the eccentric ring is calculated by using the following formula [32]:

$$f_r = \frac{v_0}{3.14(r_1 + r_2)\sqrt{\varepsilon_{re}}} \tag{2.13}$$

$$\varepsilon_{re} = 1 + q(\varepsilon_r - 1) \tag{2.14}$$

In eqs. (2.13) and (2.14), v_0, r_1, r_2 and 'q' are the speed of light, the radii of the two circular rings and the correction factor, respectively. The calculation method of correction factor is radius of the ring [32]. With the help of arc, the fundamental TM_{11} mode is created inside the metallic ring. Due to radius variation of the inner and outer eccentric rings, TM_{11} mode is split into two degenerated modes in phase quadrature. The sense of polarization is also controlled by changing the radius [32]. Figure 2.26 displays the geometrical layout as well as $|S_{11}|$ and the axial ratio variation of the dual-band CP printed antenna. It is working in the 2.1–3.6 GHz frequency band.

Wang et al. proposed a tri-band CP antenna with the assistance of two non-concentric annular slots, which is fed by an L-shaped microstrip line. Figure 2.27 displays the tri-band CP slot antenna as well as its $|S_{11}|$ and the axial ratio features. It is working in three different frequency bands, i.e. 1.16–1.27, 1.55–1.59 and 2.0–2.62 GHz. This antenna design is basically used in Global Positioning System and Compass Navigation Satellite System. The inner and outer annular slots are individually accountable for the lower and middle frequency bands. On the other hand, the coupling between the inner and outer ring slots gives rise to the upper operating band. The L-shaped microstrip line with a serial step impedance transformer provides better impedance matching as well as CP waves in three different frequency bands [33]. The aforementioned antenna is a right-handed circularly polarized printed radiator. By taking the mirror image of the L-shaped feed line, RHCP is converted into LHCP. Multiband CP antennas are widely used in wireless communication because they provide better SNR as compared to a wideband antenna. In modern days, there is a wide requirement of dual-/triple-sense antennas because they provide a high transmission capacity.

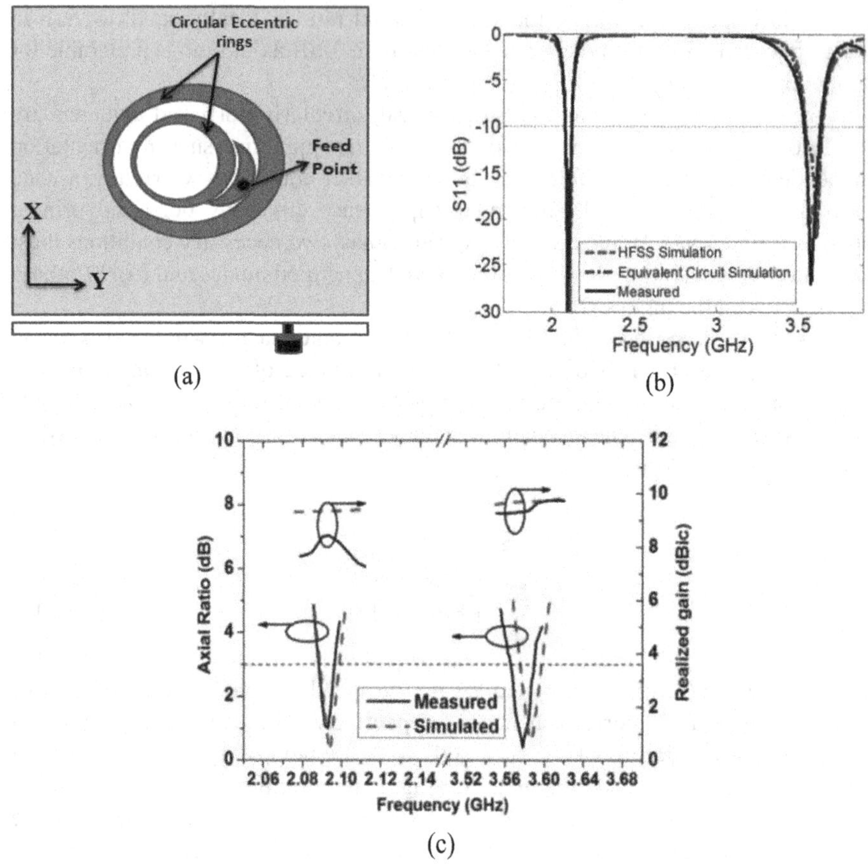

FIGURE 2.26 Dual-band CP circular eccentric rings printed antenna: (a) geometrical layout; (b) $|S_{11}|$ characteristics; (c) axial ratio variation [32].

In 2017, Xu et al. proposed a triple-sense CP slot antenna. In this antenna design, an S-shaped lot is fed by an L-shaped microstrip line. Figure 2.28 shows the geometrical layout, its $|S_{11}|$ and the axial ratio features. The aforementioned antenna structure operates over two frequency ranges, i.e. 2.34–3.66 and 4.55–9.55 GHz. CP waves are created in three different frequency ranges, i.e. 2.40–3.45, 4.65–7.27 and 8.13–8.66 GHz. Out of the three CP frequency ranges, the lower and upper frequency bands are due to the S-shaped slot, while the middle one is due to the L-shaped metallic strip. The S-shaped slot creates clockwise rotation of E-field (RHCP) in first and third bands, while the L-shaped metallic strip produces anticlockwise rotation of E-field (LHCP) in middle band. The conversion of rectangular to S-shaped slot along with the L-shaped microstrip line is helpful to create degenerated orthogonal mode with 90° phase shift. This antenna design is quite appropriate for WLAN and X-band applications [34]. Kandasamy et al. proposed a truncated square-shaped slot antenna for dual-band wireless applications. A split ring resonator was used on the bottom side of the substrate. Truncated square-shaped slot and split ring-shaped

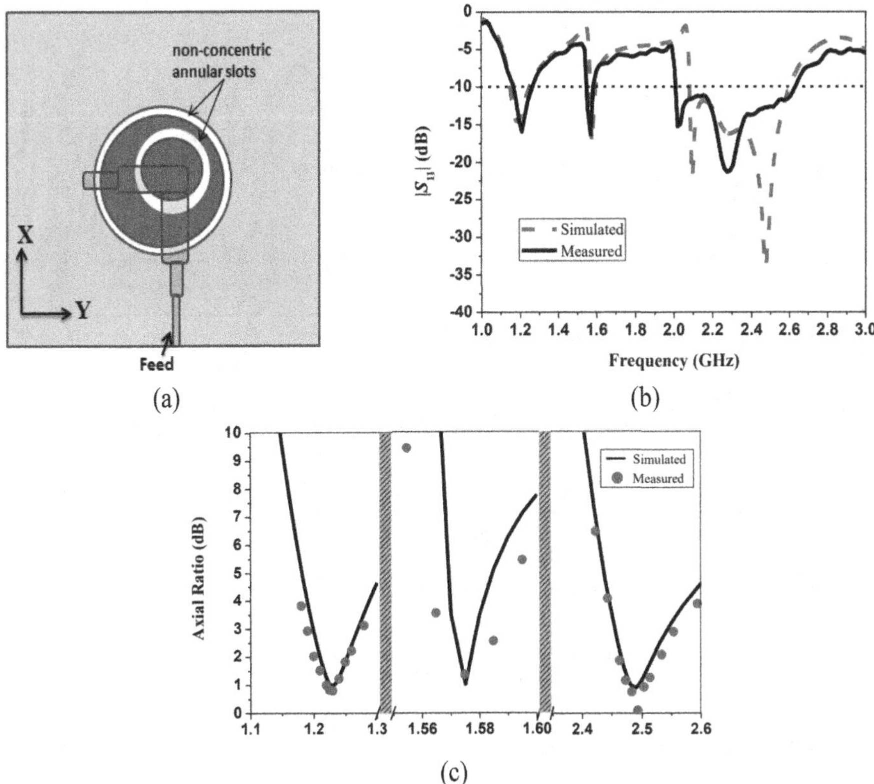

FIGURE 2.27 Tri-band CP printed antenna with two eccentric slots with L-shaped strip line: (a) geometrical layout; (b) $|S_{11}|$ characteristics; (c) axial ratio variation [33].

resonator are accountable for the lower and upper frequency bands, respectively [35]. Figure 2.29 displays the CP-based square slot antenna geometry, its $|S_{11}|$ features and gain/axial ratio variation. The resonant frequency of the slot is determined with the help of the following equation [1]:

$$f_{\text{slot}} = \frac{v_0}{2L}\sqrt{\frac{2}{1+\varepsilon_r}} \tag{2.15}$$

In eq. (2.15), 'L' is the length of the square-shaped slot. Similarly, the resonant frequency of the split ring resonator is obtained by using the following formula [36]:

$$f_{\text{SRR}} = \frac{1}{2\pi}\sqrt{\frac{L_T C_0}{4}} \tag{2.16}$$

In eq. (2.16), L_T and C_0 are the total inductance and the ring resonator. The connection between the split ring resonator and the truncated square-shaped slot is accountable for the CP waves' upper and lower frequency bands, respectively. Due to the wide slot, the back-radiation is quite high. It will reduce the gain of the antenna. To

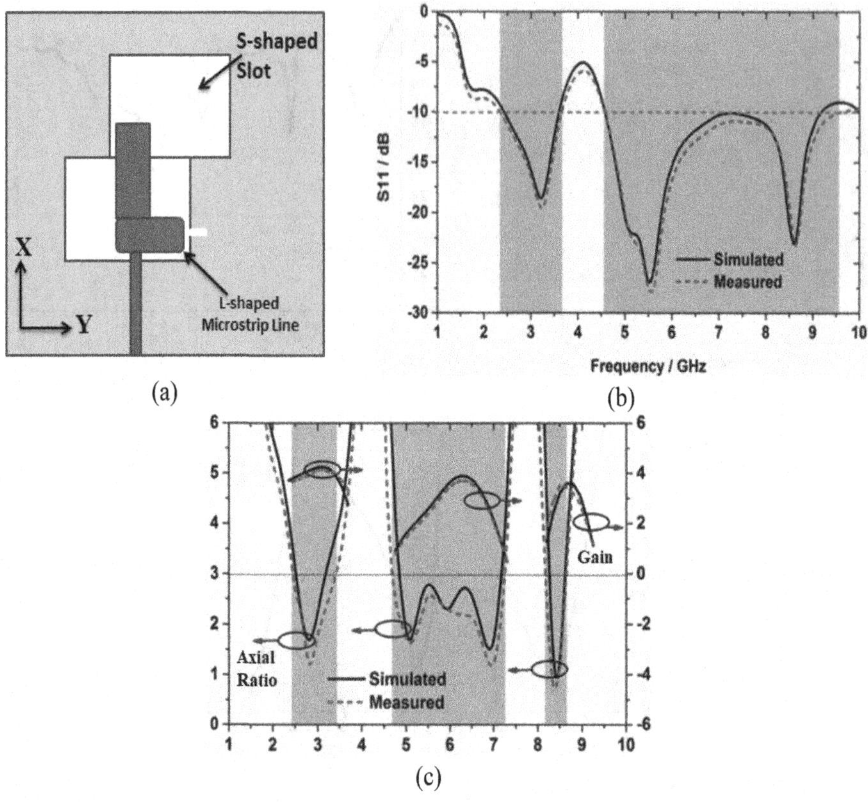

FIGURE 2.28 Triple-sense S-shaped slot antenna with CP wave: (a) geometrical layout; (b) $|S_{11}|$ characteristics; (c) axial ratio variation [34].

overcome this problem, the authors used a metallic cavity at the back. It enhanced the gain value by approximately 2.5 dBi in both the frequency bands [35].

Wideband CP antennas are also widely used because of their ability to provide a better data rate. Ellis et al. proposed a very simple wide rectangular slot antenna fed by a simple microstrip line. Figure 2.30 displays the geometrical layout and $|S_{11}|$/axial ratio features of the aforementioned antenna. In this wideband CP antenna design, a horizontal stub is used with the wide rectangular slot. Perturbing the horizontal stub from the ground plane is used to generate CP waves. This antenna structure is working over the frequency range of 3.5–9.25 GHz with a fractional bandwidth of 90.2%. It has the CP features over the frequency range 4.6–6.9 GHz [37]. Recently, Chao Sun has designed a compact UWB printed (approx. 85% fractional bandwidth) CP antenna. Figure 2.31 shows the layout, $|S_{11}|$ and the axial ratio features of the UWB CP antenna. Along with UWB features, the proposed antenna is compact in size ($0.5\times0.5\ \lambda_0$) and has unidirectional radiation property. In this antenna design, a wide impedance bandwidth is obtained by combining five different modes, i.e. TM_{10}, TM_{20}, TM_{30}, TM_{40} and TM_{50}. One to four feed networks with 0°, 90°, 180° and 270° phase differences were used to give CP waves from 2.2 to 5.5 GHz [38].

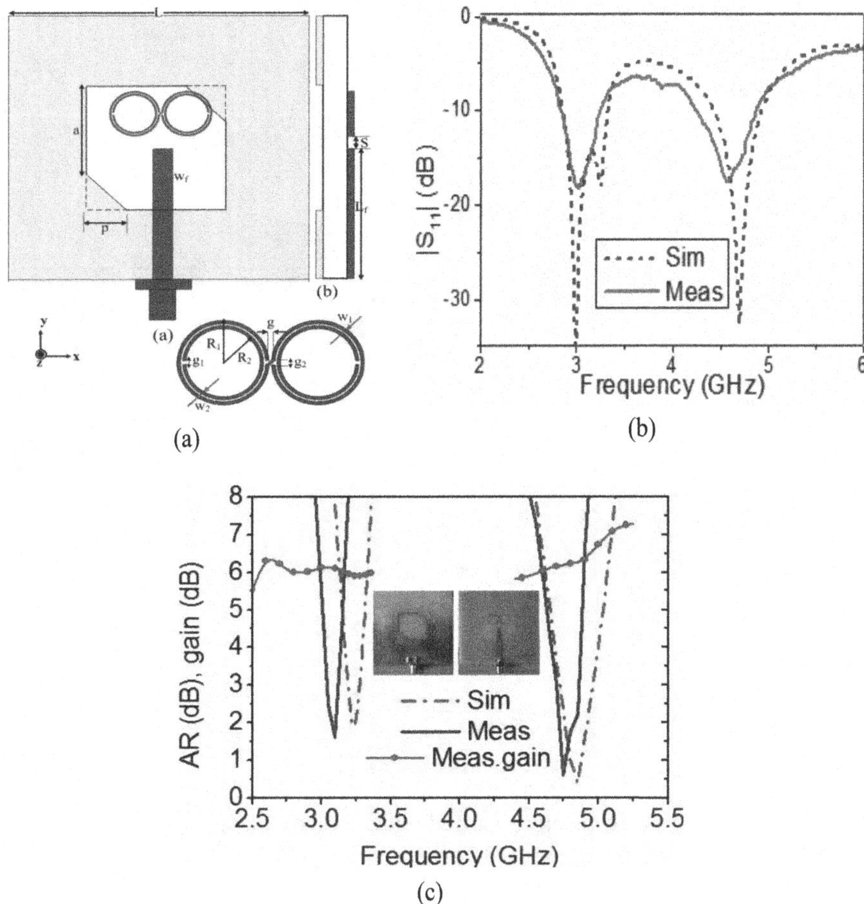

FIGURE 2.29 Square-shaped slot antenna: (a) antenna design; (b) $|S_{11}|$ variation; (c) gain and axial ratio variation [35].

2.2.4 ECC Reduction in MIMO Printed Antenna

In today's communication world, engineers widely focus on getting a higher data rate. According to the Shannon's capacity formula, it can be achieved by two different ways: (i) by enhancing the signal power level or (ii) by enhancing the bandwidth. But, both the ways are not of practical use because it is very costly to purchase a large bandwidth in the given spectrum. According to the Telecom Regulations, the transmitting signal power cannot be increased beyond a certain level. A practical way to enhance the data rate is the use of multiple antennas on transmitter and receiver sides [39]. It is nothing but a MIMO antenna system. In MIMO system, two different techniques are utilized for improving the data rate as well as link reliability: spatial diversity (the same data are transmitted and received by multiple radiators) and spatial multiplexing (different data streams are sent by different radiators), which is shown in Figure 2.32. It is quite helpful in the upcoming generations of wireless

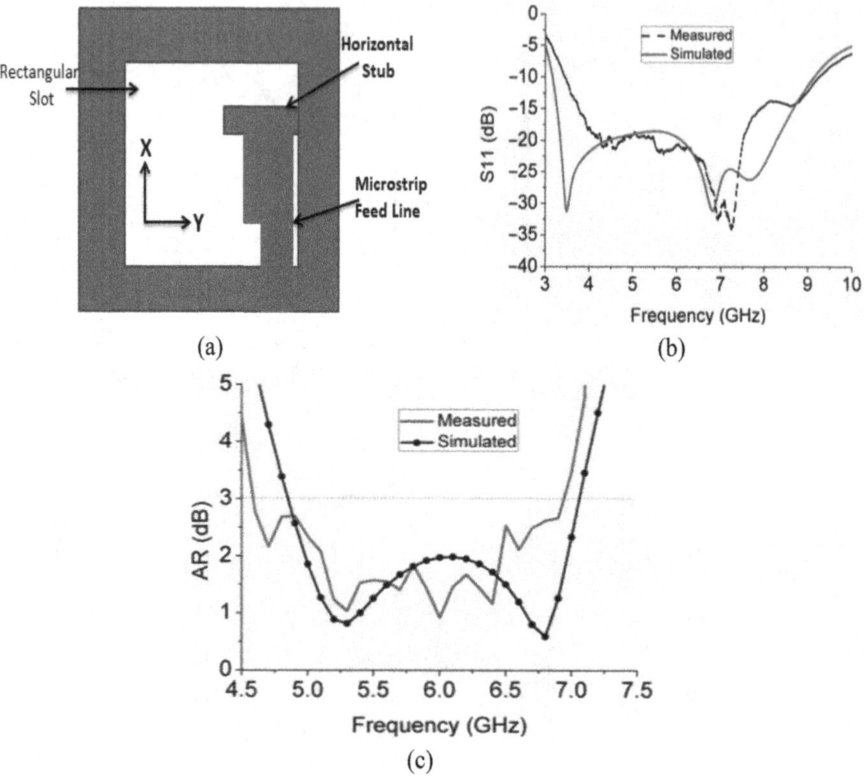

FIGURE 2.30 Wideband circularly polarized printed antenna: (a) geometrical layout; (b) $|S_{11}|$ characteristics; (c) axial ratio variation [37].

communication. In order to implement the MIMO system, it is required to design an efficient MIMO antenna.

For designing an efficient MIMO antenna, the most important parameter is the envelop correlation coefficient (ECC). It gives the information about the correlation among various antenna ports. For the case of MIMO antenna, the value of ECC (ρ) must be quite low ($\rho < 0.2$). ECC can be evaluated by two ways: (i) by using S-parameters and (ii) by using far-field parameters. The formula for calculating the ECC by using S-parameters is as follows [40]:

$$\rho_S = \frac{\left|S_{i,i}^{*}S_{i,j} + S_{j,i}^{*}S_{j,j}\right|^2}{\left(\left(1-\left(\left|S_{i,i}\right|^2+\left|S_{j,i}\right|^2\right)\right)\left(1-\left(\left|S_{j,j}\right|^2+\left|S_{i,j}\right|^2\right)\right)\right)} \tag{2.17}$$

In the above equation, 'i' and 'j' denote the antennas ports. In the literature, there are so many ways to reduce the value of ECC (by S-parameters). Wang et al. proposed a two-port printed monopole MIMO antenna for mobile terminals. In order to reduce the coupling between the ports and ECC value, the authors used a neutralized line between the antenna ports. When any antenna of the MIMO antenna

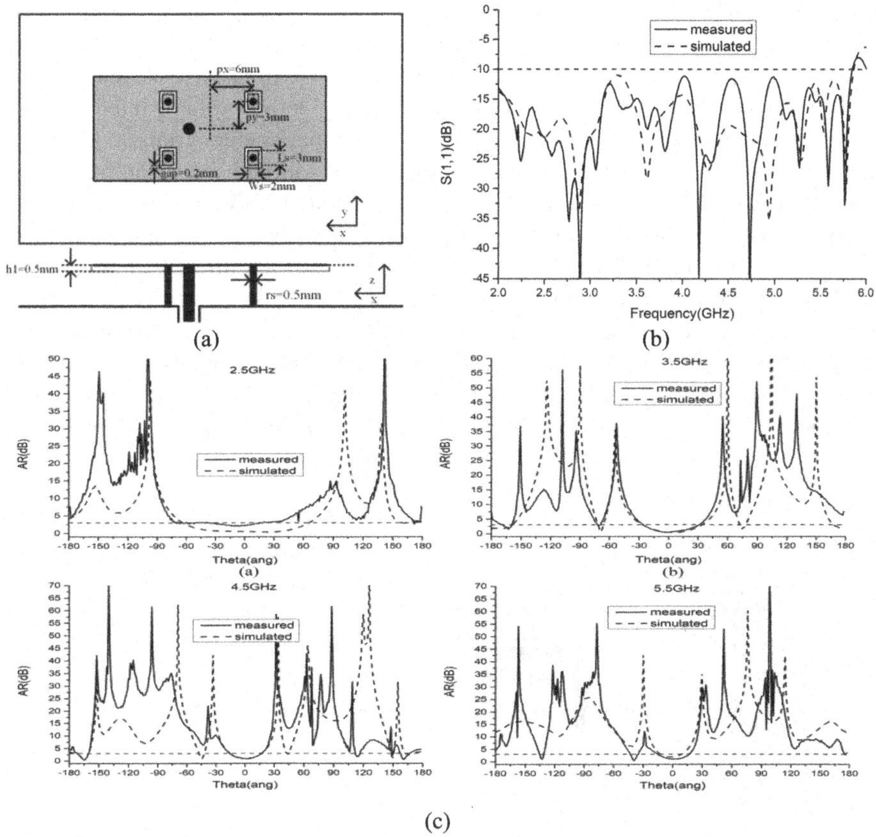

FIGURE 2.31 Circularly polarized printed antenna with ultra-wideband: (a) geometrical layout; (b) $|S_{11}|$ characteristics; (c) axial ratio variation [38].

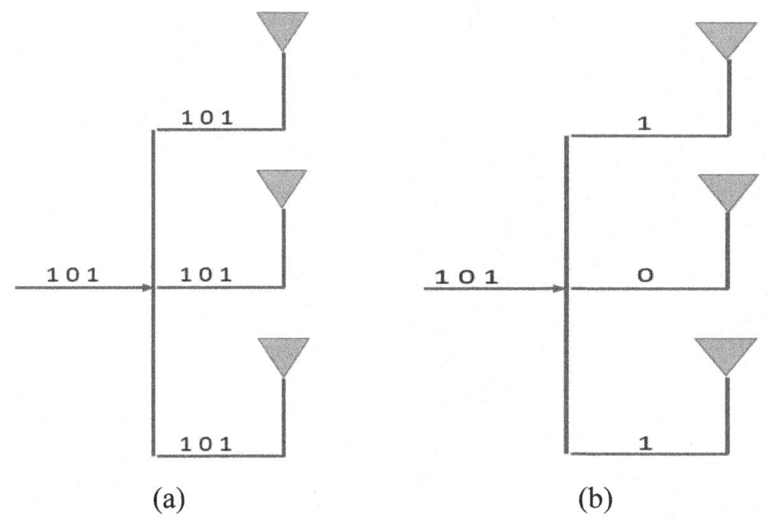

FIGURE 2.32 MIMO antenna system: (a) spatial diversity; (b) spatial multiplexing.

system is stimulated, it persuades some current in the other antennas present in the system. To reduce this current value, a metallic line is used. The length of the neutralized line is selected in such a way that the phase of electric current inverts when it moves from one port to another. Figure 2.33 shows the printed MIMO antenna geometry with the neutralized line and its S-parameter variation. By using the neutralized line, the value of ECC is less than 0.2 within the working frequency range [41]. Khan et al. also proposed a dual-port printed MIMO antenna for wireless LAN application. With the help of parasitic elements, the authors were able to get more than 15 dB isolation. Parasitic elements are not directly connected to the feed structure. These elements generate the inverted field, which will cancel each other and reduce the coupling between the antenna ports [42]. Figure 2.34 presents the two-port MIMO antenna with parasitic elements and its S-parameters. This technique provides an ECC value less than 0.2 in working frequency range. But, there are some flaws in this method of ECC evaluation. Eq. (2.17) is valid for

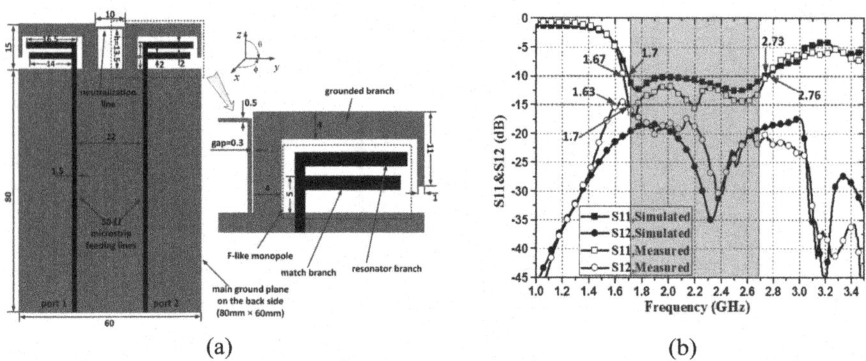

FIGURE 2.33 Two-port printed MIMO with neutralized line: (a) antenna geometry; (b) S-parameters [41].

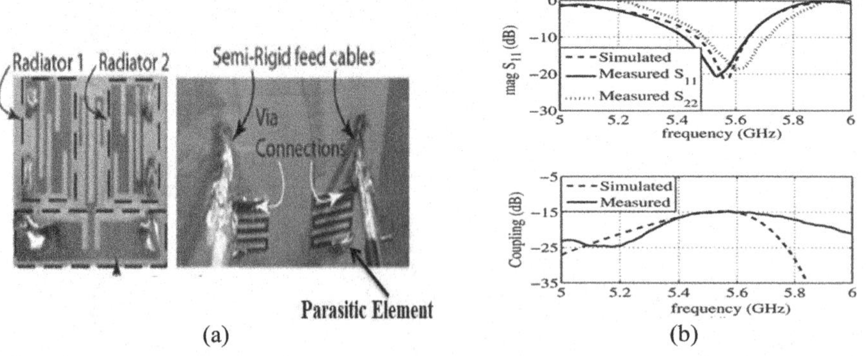

FIGURE 2.34 Two-port printed MIMO with parasitic elements: (a) antenna prototype; (b) S-parameters [42].

lossless antennas, which does not happen generally, in practice. This equation does not give the exact value of ECC if the coupling between ports is less or any other object is present in between receiver and transmitter. Recently, M.S. Sharawi has proved that the reduction in ECC (using far-field parameters) is very important in the case of MIMO antennas [43].

It can be calculated with the assistance of the following formula [40]:

$$\rho_{\text{Far-field}} = \frac{\left| \iint_{4\pi} \left[\vec{A}_i(\theta,\phi) * \vec{A}_j(\theta,\phi) \right] d\omega \right|^2}{\iint_{4\pi} \left| \vec{A}_i(\theta,\phi) \right|^2 \iint_{4\pi} \left| \vec{A}_j(\theta,\phi) \right|^2} \tag{2.18}$$

In eq. (2.18), $\vec{A_i}(\theta, \phi)$ and $\vec{A_j}(\theta,\phi)$ represent the far-field patterns for 'i' and 'j', respectively. The value of ECC (using far-field parameters) is low, if the far-field patterns obtained from different ports are complementary to each other. That means a low value of ECC by using far-field parameters indicates a large coverage area. For this purpose, Hassan et al. proposed a dual-port antenna along with a Fabry–Perot cavity [44]. This cavity is made up of partially reflecting surface (PRS) and is placed parallel to the dual-port radiating structure. PRS can be designed using FSS superstrate. If the unit cells have the same dimensions in FSS, then it will generate directive radiation pattern with no phase change. For the purpose of tilted radiation pattern, unit cells are designed with variable dimensions. In this case, the height of cavity (h) with respect to radiator is determined by the following relation [45]:

$$h = (\theta_{\text{PRS}} + \theta_G)\frac{\lambda}{4\pi} + 2N\pi \tag{2.19}$$

In the above equation, 'θ_{PRS}' and 'θ_G' are the reflection phase of PRS and the ground plane, respectively. For the purpose of tilted radiation pattern, the height of PRS with respect to radiator is about '$\lambda/2$'. Figure 2.35 displays the antenna geometry of the dual-port MIMO antenna with partially reflecting surface and its radiation pattern. Figure 2.36 displays the S-parameter and ECC variation of the aforementioned two-port MIMO printed antenna. From Figures 2.35 and 2.36, two things are very clear: (i) The use of PRS tilts the pattern by 36° in complementary direction with port-1 and port-2, respectively, and (ii) in the presence of PRS, the value of ECC is about zero in the working frequency range. Swapna et al. proposed a three-port circular-shaped printed MIMO antenna with pattern diversity. Figure 2.37 shows its layout and its S-parameter variation. FSS-based reflecting surface is designed and places in such a way that the antenna beam is tilted +25°, 0° and −25°, respectively, with different ports [46]. The isolation among the three different ports is more than 15 dB in the working frequency range. PRS structure also enhances the gain by 4.5 dBi in the working frequency range.

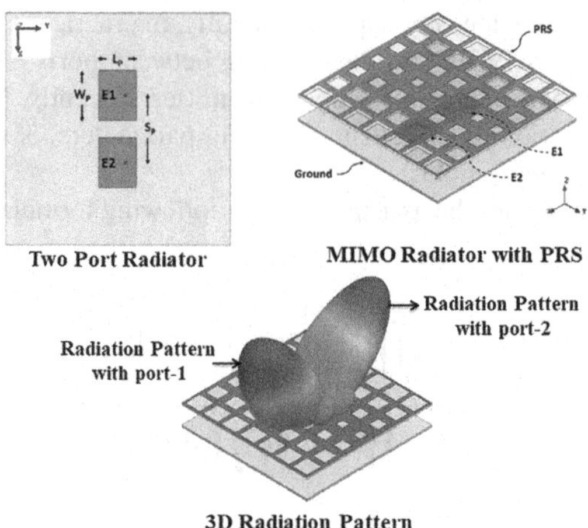

FIGURE 2.35 Design of two-port printed MIMO antenna with PRS [44].

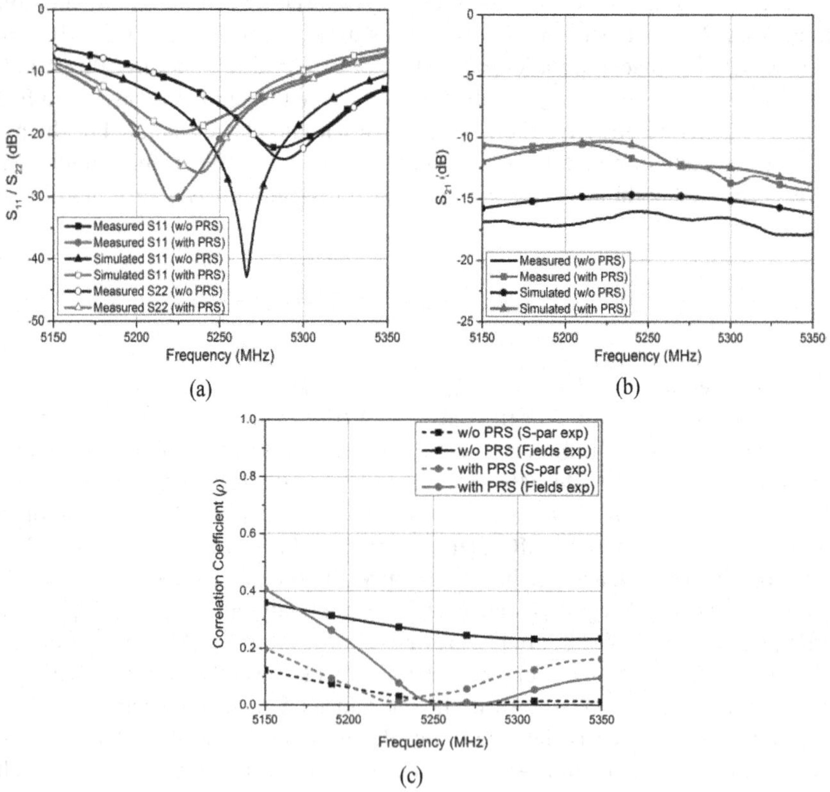

FIGURE 2.36 Different parameters of two-port MIMO antenna with PRS: (a) variation in $|S_{11}|$; (b) variation of $|S_{12}|$; (c) variation of ECC [44].

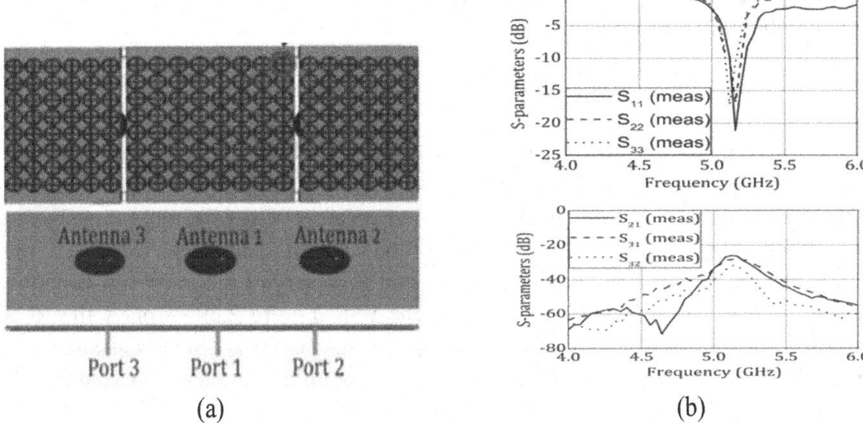

FIGURE 2.37 Three-port printed MIMO antenna with pattern diversity: (a) antenna layout; (b) *S*-parameter variation [46].

2.2.5 Printed Antenna with Low RCS Value

Radar cross section (RCS) is one of the important parameters in stealth technology which is widely used in military applications. That is why a low value of RCS is necessary in the case of antenna designing. In early days, the reduction in RCS was obtained by conventional methods such the use of ferrite substrate [47], radar-absorbing materials [48] and resistive loads [49]. In the case of RCS reduction in antennas, researchers have to face two major challenges: (i) the reduction in RCS in the complete frequency band and (ii) maintaining the radiation property and reflection coefficient properties of the radiator after making the modifications to the reference antenna. RCS highly depends on the incident angle of EM wave, so maintaining the RCS reduction approximately constant over a wide range of incident angles is another challenge in this field. The use of frequency-selective surface (FSS) is a very effective tool for reducing the RCS value along with maintaining the radiation property and reflection coefficient properties. Boundary conditions on a FSS structure are given as follows [50]:

$$\vec{E}_{\text{inc}} + \vec{E}_{\text{scat}} = Z_s \hat{I}_s \tag{2.20}$$

where $\vec{E}_{\text{scat}}$ and $\vec{E}_{\text{inc}}$ are the scattered and incident electric fields. Z_s and $\vec{J}_s$ are the surface impedance and surface current on the frequency-selective surface. In eq. (2.20), if the surface impedance tends to zero, FSS acts as a perfect conductor and reflects the entire incident wave. In a similar manner, if the surface impedance tends to infinity, the FSS becomes transparent to EM waves. The FSS also provides another advantage, i.e. less sensitivity to incident angle. In the FSS, the minimum reflection conditions under TE polarization are calculated by using the following equation [51]:

$$R_{\min} = \mu_s \varepsilon_s - \varepsilon_s^2 \sin^2\theta_i - \mu_s^2 \cos^2\theta_i \tag{2.21}$$

Similarly, the minimum reflection conditions under TM polarization are calculated by using the following equation [51]:

$$R_{\min} = \mu_s - \varepsilon_s \sin^2\theta_i - \mu_s \varepsilon_s \cos^2\theta_i = 0 \tag{2.22}$$

μ_s and ε_s are the permeability and permittivity of the FSS. θ_i and θ_t are the angles of incident and transmission. Thummaluru et al. presented a FSS-based MIMO antenna with RCS reduction feature. In this proposed antenna structure, dual-flag-shaped printed radiators are designed on one side of the substrate and on the other side, the FSS is placed. Figure 2.38 shows the fabricated geometry of the proposed structure as well as *S*-parameters with and without the FSS. Figure 2.39 displays the RCS variation of the given antenna in TE and TM polarizations at different angles of incidence. A stable RCS reduction was achieved for up to 60° incident angle for TE polarization and up to 45° incident angle for TM polarization [51].

In a similar manner, Ranjan et al. proposed a semicircular dual-port printed antenna with a low RCS value. In this article, the RCS reduction was obtained by using the frequency-selective surface at the ground plane [52]. Figures 2.40 and 2.41

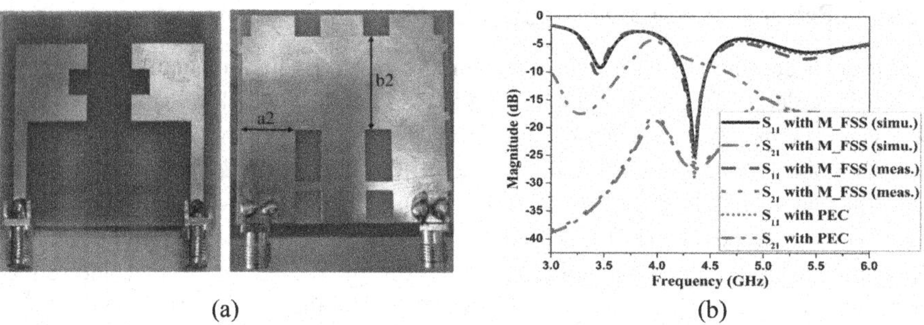

FIGURE 2.38 Dual-port printed MIMO antenna with reduced RCS: (a) geometrical layout; (b) $|S_{11}|$ characteristics [51].

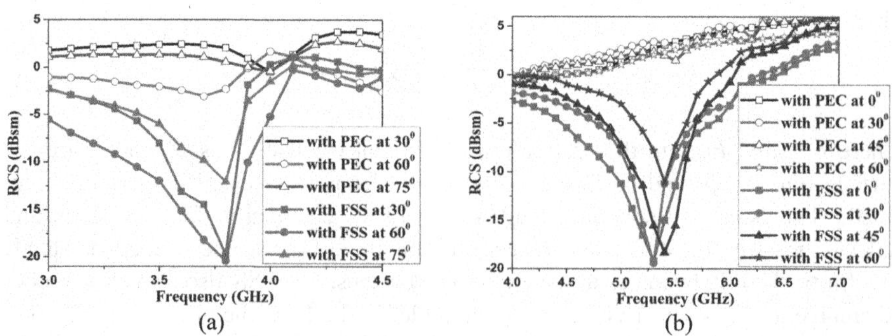

FIGURE 2.39 RCS variation of dual-port printed MIMO antenna: (a) TE polarization; (b) TM polarization [51].

FIGURE 2.40 Semicircular dual-port printed MIMO antenna's fabricated structure: (a) top view; (b) bottom view [52].

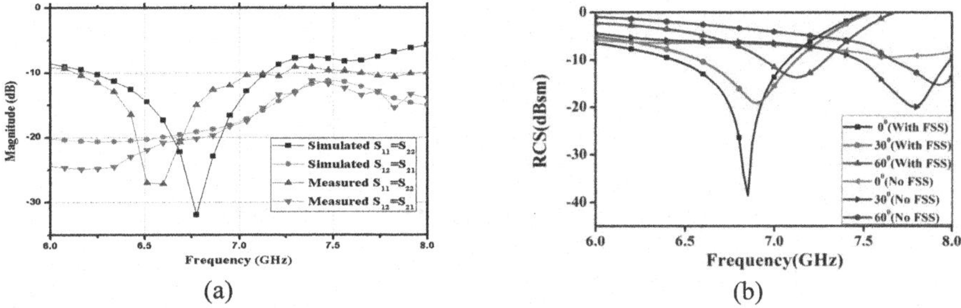

FIGURE 2.41 Analysis of semicircular dual-port printed MIMO antenna: (a) reflection coefficient; (b) RCS variation [52].

present the antenna geometry, and its *S*-parameters and RCS features, respectively. It is operating at single band of frequency, i.e. 7.5 GHz. A stable RCS reduction was achieved for up to 45° in the operating frequency range [52].

In the aforementioned MIMO antenna, the EBG structure was used in between the two radiators for getting an isolation of more than 20 dB. Gangwar et al. proposed a wideband circularly polarized flag-shaped MIMO antenna, which is shown in Figure 2.42. A FSS and a defected ground structure were used to reduce the RCS value. The given printed radiator works efficiently over the frequency range of 3.7–9.0 GHz. In the operating frequency range, the RCS was reduced by more than 25 dBsm [53]. The antenna gain is about 4.5 dBi within the operating frequency range.

2.2.6 Printed Antenna Design for 5G Applications

Fifth generation (5G) wireless communication mainly requires a high data rate, low latency, efficient spectrum utilization and high coverage area. 5G wireless communication mainly requires three different frequency bands: the lower-frequency band (<1.0 GHz), the mid-frequency band (1–6 GHz), and the higher-frequency

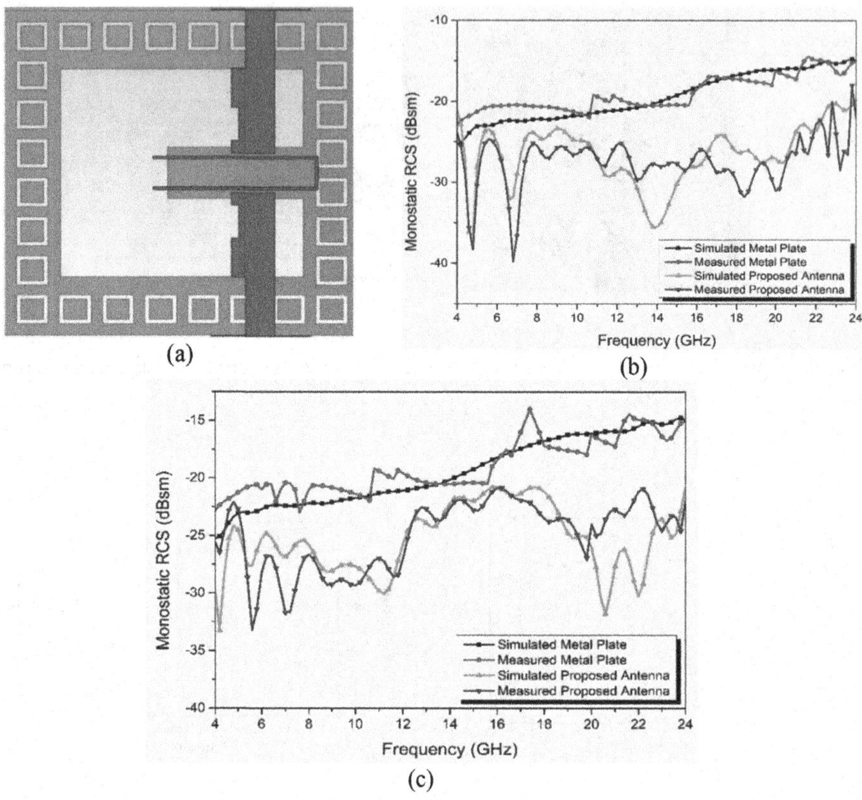

FIGURE 2.42 Dual-port printed MIMO antenna with FSS and DGS: (a) geometrical layout; (b) monostatic RCS in TE polarization; (c) monostatic RCS in TM polarization [53].

band (mm-wave frequencies). The higher data rate can be achieved by the MIMO technology by enhancing the signal-to-noise ratio without enhancing the input power. In the case of 5G technology, the data rate is not the only single concern. One has to take care about the spectral efficiency, low latency as well as maximizing the number of users [54]. In order to achieve all these constraints, different technologies are available, such as beam-forming, multi-input multi-output, smart antennas, cognitive radio (CR) and full-duplex methods [55]. The combination of CR technology along with MIMO antennas provides two important advantages: (i) high data rate and (ii) better spectrum efficiency. In the CR system, the users are divided into two parts: (i) primary users and (ii) secondary users. Primary users are those who have bought the spectrum, and they can access the spectrum at any instance of time. On the other hand, secondary users are active when primary users are not using the spectrum. The CR technology makes the system able to identify the unused part of the spectrum, so that secondary users use the idle part of the spectrum. It increases the spectrum efficiency. In the CR system, two different antennas are used: (i) sensing radiator and (ii) communicating radiator. The sensing antenna is an ultra-wideband radiator that finds the action of primary users on the spectrum. It also identifies the idle part of the spectrum that is not being used by the primary users. On the other hand, the communicating radiator is a reconfigurable narrowband radiator. It allows the secondary users to start communicating on the idle part of the spectrum. The integration of MIMO with CR technology is the effective way of achieving the aim of 5G communication systems, i.e. high data rate along with good spectral efficiency.

Hussain et al. proposed a MIMO antenna integrated with CR technology. Figures 2.43 and 2.44 show the antenna geometry for cognitive radio application and *S*-parameter variations, respectively. In this given antenna structure, the sensing antenna works over the frequency range 0.7–7.65 GHz. The communicating antenna is tuned over the frequency range of 1.77–2.51 GHz. The gain and efficiency of the proposed antenna are also good over the working frequency range, i.e. 3.2 dBi and 81%, respectively [56].

In [56], the ground plane of communicating antenna works as a sensing antenna. That is why sensing and communicating concurrently are not possible in such type of design. Thummaluru et al. proposed a four-port MIMO antenna system for 5G communication system. It also integrates cognitive radio system with MIMO. Figure 2.45 shows the geometrical layout of the proposed antenna structure [57]. In this antenna structure, sensing and communication of secondary users occur simultaneously. The authors designed a wideband sensing antenna initially, whose range is 2.3–5.5 GHz. Figure 2.46 displays the $|S_{11}|$ characteristics of the sensing and communicating antennas. After that, a narrowband communicating antenna is designed

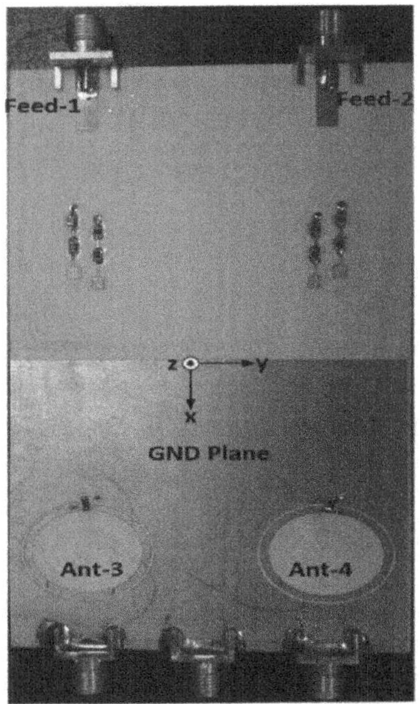

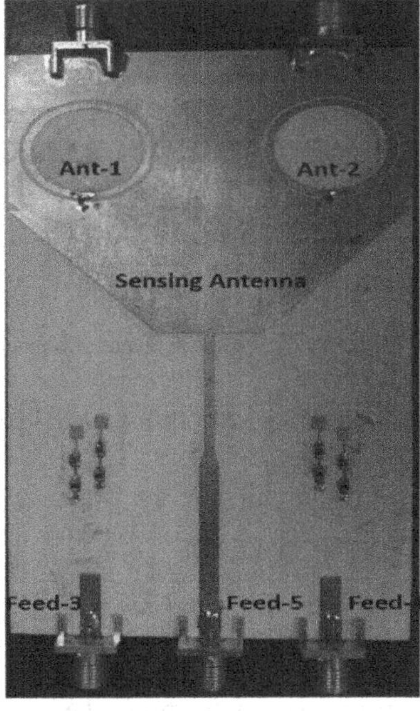

FIGURE 2.43 Fabricated MIMO antenna integrated with CR technology [56].

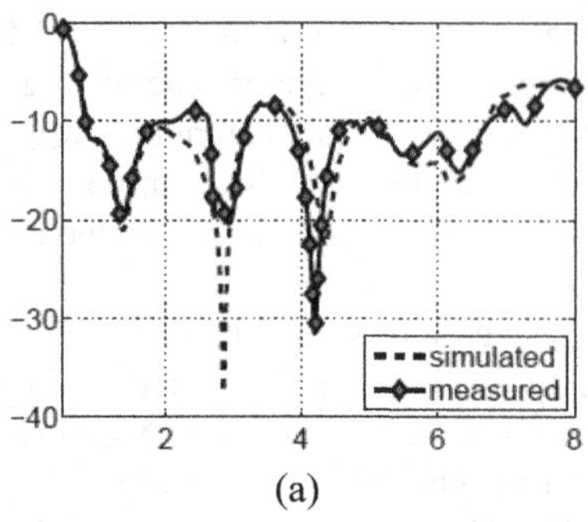

(a)

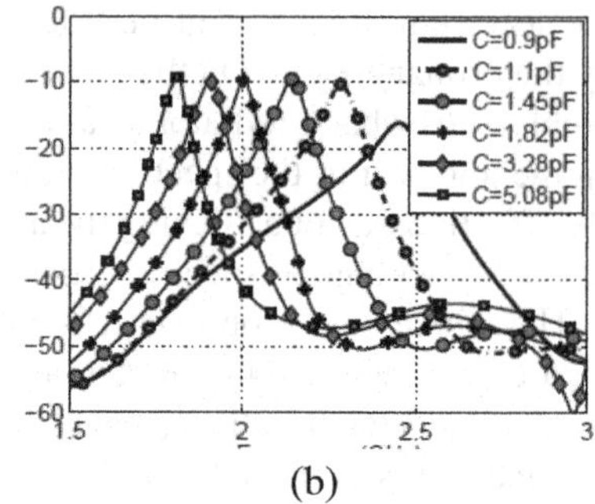

(b)

FIGURE 2.44 $|S_{11}|$ Features of MIMO antenna integrated with CR technology: (a) sensing antenna; (b) communicating antenna [56].

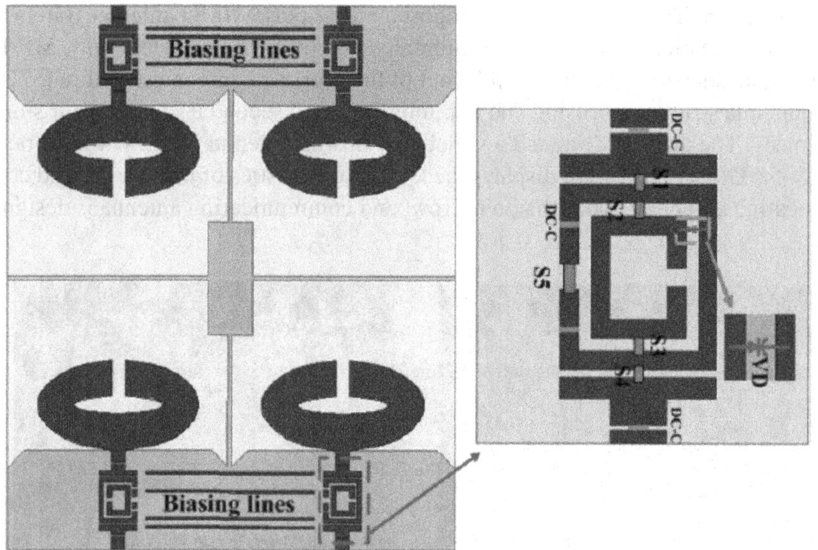

FIGURE 2.45 Geometrical layout of four-port MIMO antenna with CR system [57].

by using reconfigurable all-pass, band-pass and band-reject filters. Its range is from 2.5 to 4.2 GHz.

2.3 CONCLUSION

In this chapter, the latest research trends in the field of printed antennas were discussed. In the recent world of wireless communication, most of the applications occur at mm-wave frequencies. Due to large metallic losses, antenna gain is the main concern. So, researchers widely focus on the gain enhancement techniques. Another important area of research in the field of printed antennas is

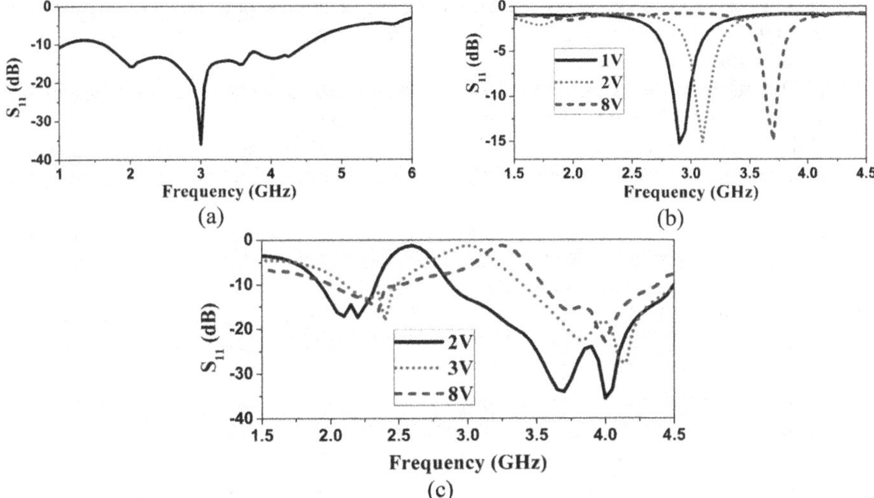

FIGURE 2.46 $|S_{11}|$ Characteristics: (a) sensing antenna; (b) interweave CR communicating antenna; and (c) underlay CR communicating antenna [57].

RCS reduction. It is greatly being used in military applications. Designing super-wideband antennas is also a hot research area because it provides a higher data rate for both indoor and outdoor communication systems. Designing of compact circularly polarized printed antennas is widely focused due to its frequent use in satellite communication applications. Modern wireless communication is moving towards 5G and 6G standards, where the communication system requires several important features such as high data rate, high spectrum efficiency and low latency. So, the integration of CR technology and beam tilting with MIMO printed antennas is an important area of research. The main design steps and important research articles related to all these latest research fields of printed antennas were discussed.

REFERENCES

1. C.A. Balanis. *Antenna Theory: Analysis and Design*. John Wiley and Sons: New York, 2005.
2. P. H. Rao, V. F. Fusco, R. Cahill. Wide-band linear and circularly polarized patch antenna using a printed stepped T-feed. *IEEE Transactions on Antennas and Propagation*, 2002; 50: 356–361.
3. Y. C. Lee, J. S. Sun. A new printed antenna for multiband wireless applications. *IEEE Antennas and Wireless Propagation Letters*, 2009; 8: 402–405.
4. G. J. K. Moernaut, G. A. E. Vandenbosch. Size reduced meander line annular ring microstrip antenna. *Electronics Letters*, 2004; 40: 1463–1464.
5. W. L. Stutzman, G. A. Thiele. *Antenna Theory and Design*. John Wiley & Sons: Hoboken, NJ, 2013.
6. D. R. Jackson, N. G. Alexopoulos. Gain enhancement methods for printed circuit antennas. *IEEE Transactions on Antennas and Propagation*, 1985; 33: 976–987.

7. H. Y. Yang, N. G. Alexopoulos. Gain enhancement methods for printed circuit antennas through multiple superstrates. *IEEE Transactions on Antennas and Propagation*, 1987; 35: 860–863.
8. R. Q. Lee, K. F. Lee. Gain enhancement of microstrip antennas with overlaying parasitic director. *Electronics Letters*, 1988; 24: 656–658.
9. H. D. Chen, C. Y. D. Sim, J. Y. Wu, T. W. Chiu. Broadband high-gain microstrip array antennas for WiMAX base station. *IEEE Transactions on Antennas and Propagation*, 2012; 60: 3977–3980.
10. H. Wang, S. F. Liu, L. Chen, W. T. Li, X. W. Shi. Gain enhancement for broadband vertical planar printed antenna with H-shaped resonator structures. *IEEE Transactions on Antennas and Propagation*, 2014; 62: 4411–4415.
11. M. Bouslama, M. Traii, A. Gharsallah, T. A. Denidni. Analysis of antenna gain enhancement using new frequency selective surface superstate. *Microwave and Optical Technology Letters*, 2016; 58: 448–452.
12. S. Sah, A. Mittal, M. R. Tripathy. High gain dual band slot antenna loaded with frequency selective surface for WLAN/fixed wireless communication. *Microwave and Optical Technology Letter*, 2019; 61: 519–525.
13. P. Juyal, L. Shafai. A high gain single feed dual mode microstrip disc radiator. *IEEE Transactions on Antennas and Propagation*, 2016; 64: 2115–2126.
14. P. Juyal, L. Shafai. Gain enhancement in circular microstrip antenna via linear superposition of higher zeros. *IEEE Antennas and Wireless Propagation Letters*, 2016; 16: 896–899.
15. W. M. A. Wahab, S. S. Naeini. Wide-bandwidth 60-GHz aperture-coupled microstrip patch antennas (MPAs) fed by substrate integrated waveguide (SIW). *IEEE Antennas and Wireless Propagation Letters*, 2011; 10: 1003–1005.
16. M. Asaadi, A. Sebak. High gain low profile circularly polarized slotted SIW cavity antenna for MMW applications. *IEEE Antennas and Wireless Propagation Letters*, 2017; 10: 752–755.
17. W. Wiesbeck, G. Adamiuk, C. Sturm. Basic properties and design principles of UWB antennas. *Proceedings of the IEEE*, 2009; 97: 372–385.
18. *Revision of Part 15 of the Commission's Rules Regarding Ultra-Wideband Transmission Systems*, Federal Communications Commission, Washington, DC, USA, 2002.
19. H. G. Schantz. A brief history of UWB antennas. *IEEE Aerospace and Electronic Systems Magazine*, 2004; 19: 22–26.
20. W. Balani, M. Sarvagya, T. Ali, M. M. Pai, J. Anguera, A. Andujar, S. Das. Design techniques of super-wideband antenna existing and future prospective. *IEEE Access*, 2019; 7: 141241–141257.
21. N. P. Agrawall, G. Kumar, K. P. Ray. Wide-band planar monopole antennas. *IEEE Transactions on Antennas and Propagation*, 1998; 46: 294–295.
22. K.R. Chen, C.Y.D. Sim, J.S. Row. A compact monopole antenna for super wideband applications. *IEEE Antennas and Wireless Propagation Letters*, 2011; 10: 488–491.
23. P. Okas, A. Sharma, G. Das, R. K. Gangwar. Elliptical slot loaded partially segmented circular monopole antenna for super wideband application. *AEU-International Journal of Electronics and Communications*, 2018; 88: 63–69.
24. P. Okas, A. Sharma, R. K. Gangwar. Circular base loaded modified rectangular monopole radiator for super wideband application. *Microwave and Optical Technology Letters*, 2017; 59: 2421–2428.
25. P. Okas, A. Sharma, R. K. Gangwar. Super-wideband CPW fed modied square monopole antenna with stabilized radiation characteristics. *Microwave and Optical Technology Letters*, 2018; 60: 568–575.
26. S. Singhal, A. K. Singh. CPW-fed hexagonal Sierpinski super wideband fractal antenna. *IET Microwaves, Antennas & Propagation*, 2016; 10: 1701–1707.

27. S. Singhal, A. K. Singh. Asymmetrically CPW-fed circle inscribed hexagonal super wideband fractal antenna. *Microwave and Optical Technology Letters*, 2016; 58: 2794–2799.
28. A. Siahcheshm, J. Nourinia, Y. Zehforoosh, B. Mohammadi. A compact modified triangular CPW-fed antenna with multioctave bandwidth. *Microwave and Optical Technology Letters*, 2015; 57: 69–72.
29. W. Sörgel, W. Wiesbeck. Influence of the antennas on the ultrawideband transmission. *EURASIP Journal on Advances in Signal Processing*, 2005; 2005: Art. no. 843268.
30. A. Shlivinski, E. Heyman, R. Kastner. Antenna characterization in the time domain. *IEEE Transactions on Antennas and Propagation*, 1997; 45: 1140–1149.
31. G. Quintero, J. F. Zurcher, A. K. Skrivervik. System fidelity factor: A new method for comparing UWB antennas. *IEEE Transactions on Antennas and Propagation*, 2011; 59: 2502–2512.
32. Z. X. Liang, D. C. Yang, X. C. Wei, E.-P. Li. Dual-band dual circularly polarized microstrip antenna with two eccentric rings and an arc-shaped conducting strip. *IEEE Antennas and Wireless Propagation Letters*, 2016; 15: 834–837.
33. L. Wang, Y. X. Guo, W. Sheng. Tri-band circularly polarized annular slot antenna for GPS and CNSS applications. *IEEE Antennas and Wireless Propagation Letters*, 2012; DOI: 10.1109/LAWP.2012.2200869.
34. R. Xu, J. Li, Y. X. Qi, Y. Guangwei, J. J. Yang. A design of triple-wideband triple-sense circularly polarized square slot antenna. *IEEE Antennas and Wireless Propagation Letters*, 2017; 16: 1763–1766.
35. K. Kandasamy, B. Majumder, J. Mukherjee, K. P. Ray. Dual-band circularly polarized split ring resonators loaded square slot antenna. *IEEE Transactions on Antennas and Propagation*, 2016; 64: 3640–3645.
36. A. Ishikawa, T. Tanaka, S. Kawata. Frequency dependence of the magnetic response of split-ring resonators. *The Journal of the Optical Society of America B*, 2007; 24: 510–515.
37. M. S. Ellis, Z. Zhao, J. Wu, X. Ding, Z. Nie, Q. H. Liu. A novel simple and compact microstrip-fed circularly polarized wide slot antenna with wide axial ratio bandwidth for C-band applications. *IEEE Transactions on Antennas and Propagation*, 2016; 64: 1552–1556.
38. C. Sun. A design of ultra wideband circularly polarized microstrip patch antenna. *IEEE Transactions on Antennas and Propagation*, 2019; 67: 6170–6175.
39. M. S. Sharawi. Printed MIMO antenna systems: Performance metrics, implementations and challenges. *Forum for Electromagnetic Research Methods and Application Technologies*, 2014; 1: 1–11.
40. M. S. Sharawi. *Printed MIMO Antenna Engineering*. Artech House: Boston, London, 2014.
41. Y. Wang, Z. Du. A wideband printed dual-antenna system with a novel neutralization line for mobile terminals. *IEEE Antennas and Wireless Propagation Letters*, 2013; 12: 1428–1431.
42. M. S. Khan, M. F. Shafique, A. Naqvi, A. D. Capobianco, B. Ijaz, B. D. Braaten. A miniaturized dual-band MIMO antenna for WLAN applications. *IEEE Antennas and Wireless Propagation Letters*, 2015; 14: 958–961.
43. M. S. Sharawi. Current misuses and future prospects for printed multiple-input, multiple-output antenna systems. *IEEE Antennas and Propagation Magazine*, 2017; 4: 162–170.
44. T. Hassan, M. U. Khan, H. Attia, M. S. Sharawi. An FSS based Correlation Reduction Technique for MIMO Antennas. *IEEE Transactions on Antennas and Propagation*, 2018; 66: 4900–4908.
45. G. V. Trentini. Partially reflecting sheet arrays. *IRE Transactions on Antennas and Propagation*, 1956; 4: 666–671.

46. S. P. Swapna, G. S. Karthikeya, S. K. Koul, A. Basu. Three-port pattern diversity antenna module for 5.2GHz ceiling-mounted WLAN access points. *Progress in Electromagnetics Research C*, 2020; 98: 57–67.
47. D. M. Pozar. RCS reduction for a microstrip antenna using a normally biased ferrite substrate. *IEEE Microwave and Guided Wave Letters*, 1992; 2: 196–198.
48. Y. Q. Li, H. Zhang, Y. Q. Fu, N. C. Yuan. RCS reduction of ridged waveguide slot antenna array using EBG radar absorbing material. *IEEE Antennas Wireless and Propagation Letters*, 2008; 7: 473–476.
49. J. L. Volakis, A. Alexanian, J. M. Lin. Broadband RCS reduction of rectangular patch by using distributed loading. *Electronics Letters*, 1992; 28: 2322–2323.
50. A. P. Raiva, F. J. Harackiewicz, J. Lindsey. Frequency selective surfaces: Design of broadband elements and new frequency stabilization techniques. Antenna Applications Symposium, Illinois, 2003.
51. S. R. Thummaluru, R. Kumar, R.K. Chaudhary. Isolation enhancement and radar cross section reduction of MIMO antenna with frequency selective surface. *IEEE Transactions on Antennas and Propagation*, 2018; 66: 1595–1600.
52. P. Ranjan, M. Patil, S. Chand, A. Ranjan, S. Singh, A Sharma. Investigation on dual-port printed MIMO antenna with reduced RCS for C-band radar application. *International Journal of RF and Computer Aided Engineering*, 2020; 30: 1–7.
53. D. Gangwar, A. Sharma, B. K. Kanaujia, S. P. Singh, A. L. Ekuakille. Characterization and performance measurement of low RCS wideband circularly polarized MIMO antenna for microwave sensing applications. *IEEE Transactions on Instrumentation and Measurement*, 2019; DOI: 10.1109/TIM.2019.2936707.
54. A. Gohil, H. Modi, S. K. Patel. 5G technology of mobile communication: A survey. *Proceedings of International Conference of Intelligent Systems and Signal Processing (ISSP)*, 2013; 288–292.
55. A. Gupta, R. K. Jha. A survey of 5G network: Architecture and emerging technologies. *IEEE Access*, 2015; 3: 1206–1232.
56. R. Hussain, M. S. Sharawi, A. Shamim. An integrated four-element slot-based MIMO and a UWB sensing antenna system for CR platforms. *IEEE Transactions on Antennas and Propagation*, 2018; 66: 978–983.
57. S. R. Thummaluru, M. Ameen, R. K. Chaudhary. Four-port MIMO cognitive radio system for mid-band 5G applications. *IEEE Transactions on Antennas and Propagation*, 2019; 67: 5634–5645.

3 Radiation Pattern Agility of Printed Antennas

Dr. Amit Bage and Surendra Kumar Gupta
National Institute of Technology, Hamirpur

CONTENTS

3.1 INTRODUCTION

In modern microwave and millimeter-wave communication systems, reconfigurable antennas play an important role in both academics and industry. The reconfigurable antennas increase the spectrum utilization by adjusting their frequency as well as radiation pattern characteristics according to the system and surrounding environment requirements. An antenna can be reconfigurable by many parameters such as radiation pattern, frequency and polarization or the combination of the above three. In the modern era, the radiation pattern agility antennas play an important role and are becoming more popular due to their quality and capacity increments in communication systems. The reconfigurable radiation pattern antenna has the capability to control the direction of beam, which adds more flexibility to directionality. In addition, its ability to avoid noisy environments, improve security, mitigate multipath effects, save power and better direct signals to the users' end enhances the overall performance of modern microwave and millimeter-wave communication systems. The reconfigurable radiation pattern antennas also have the capability to reduce noise coming from different directions by redirecting the null positions of their radiation pattern. These techniques can be used in a bigger coverage area by redirecting the main beam. These make the pattern-reconfigurable antenna of great demand and applicable in the field of wireless communication, satellite communication, radar, etc. The pattern agility has been achieved by using phased array antennas, which are costly, too complex and too large in size [1,2]. The radiation pattern agility has also been achieved by using microelectromechanical systems (RF MEMS) [3], PIN diodes [4], varactors [5], ferrites and liquid crystals [6]. These different tuning mechanism elements can be inserted into different antenna configurations to achieve a better performance.

In the above literature, various techniques have been found for achieving the radiation pattern agility. In 1979, a pattern agility antenna is implemented for the application of satellite communication [7]. The multiple shaped beam reconfigurable antennas provide flexible coverage for several areas. Changing the excitation, i.e., amplitude and phase, of the multiple feed arrays in offset reflector optics provides beam shaping configuration. The antenna can be reconfigured in six different beam angles. In 2006, J. Liang presented a radiation-pattern-reconfigurable antenna for the ultra-wideband range. The antenna is a combination of a monopole, a tapered slot structure and four diodes. The radiation patterns of the monopole and the tapered slot structure are controlled by the four diodes. The shape of the radiation pattern was maintained across all the operating frequency ranges. The combination of a disk, a sector coupling, a sector radiating element and four PIN diodes was used to rotate the main beam in steps of 90° in azimuthal planes with 30° deflection in elevation planes [8]. The Yagi antenna element array is also pattern reconfigurable from broadside to quasi-end-fire radiation by using PIN diodes [9]. The radiation pattern agility has been achieved by using agile metamaterials [10]. The metamaterial agile has been designed using agile lens which is a composition of two grids at which each grid composed of two regions. The first region is the focusing zone in which the parameters such as refraction index, size, shape and the position of the grid can be controlled by external switching systems. With modifications in the focusing zone, radiation pattern characteristics such as main lobe beam width and direction (pointing angle) in E- and H-planes can be controlled. The second region is the lens body. The frequency tuning was achieved using a combination of four reflective strips with the ground plane. An inverted-F driven element, complemented by four reflective strips with variable grounding, was used to design a pattern agility antenna. The radiation pattern characteristics were changed from omnidirectional to directional using grounding reflectors [11]. In 2017, N. N. Trong, L. Hall and C. Fumeaux presented a dual-pattern frequency-reconfigurable patch antenna, which provides monopolar and broadside radiation patterns, and its frequency band can be varied by using varactor diodes [12]. A single-feed double-element antenna array that is reconfigurable in terms of both frequency and pattern was introduced in [13]. The tuning mechanism was used to tune the frequency between 2.15 and 2.38 GHz and the beam scanning between ±23°. The combination of a monopole and patch antenna was also used to obtain frequency and pattern agility, and the concept was proposed in [14]. The printed patch on the upper layer was used for the lower resonant frequency, and the monopole was designed at the bottom ground plane for the higher resonant frequency. The antenna can be reconfigured between omnidirectional monopole radiator operating at 2.4 GHz and broad-side patch working at 5.5 GHz by controlling five PIN diodes. Another frequency and pattern agility technique was introduced in [15].

The antenna was fed using a coaxial port, and it was a combination of three PIN diodes and four slits. It is capable of reconfiguring its beam to different angles between ±15°. Two symmetrical hexagonal split rings with a monopole branch and eight PIN diodes were used for flexible frequency and radiation patterns in [16]. In the proposed technique, the antenna resonant frequency can be switched between 1.9 and 2.4GHz with a pattern-changing capability in two directions. In 2015 [17], an antenna having 6 symmetrical main radiators and 12 parasitic elements was designed to achieve frequency and radiation pattern agility. The three circular slots were etched in the circular ground patch and placed into the midline of the microstrip feed line. By changing the diodes' state, the radiation patterns of the antenna can be changed to different angles at different resonant frequencies. Since six radiation elements have been used, to achieve full coverage, the antenna pattern can be changed to different angles at 45° steps.

From the above literature, it can be seen that different methods for advanced microstrip antennas with an improved performance have been developed and reported. They are based on the use of with and without the lumped components, with simple resonators, which act as resonating elements. Multiple resonant frequencies were achieved by using a single microstrip antenna. The goal is to implement radiation-pattern-reconfigurable antenna with improved performance characteristics, compared to the above literature. The methods considered here for designing are three-dimensional electromagnetic modeling, optimization and miniaturization. In the design guidelines, easy implementation with simple design and experimental verification are important aspects. The radiation pattern agility antennas considered here can be used in various applications according to the requirements. The proposed antenna design methods allow further improvements according to their future use. We have investigated various microstrip antennas with radiation pattern agility, as well, which might be suitable for miniaturization.

This chapter is organized as follows: Section 3.2 will briefly cover different types of reconfigurable antennas, while Section 3.3 will provide the basic concepts of different radio frequency (RF) switching components. Section 3.4 will elaborate on the design method of a radiation pattern agility antenna based on PIN diodes, explanation of its modeling and its experimental validation. Sections 3.5 will introduce a radiation-pattern-reconfigurable microstrip antenna using varactor diodes. Section 3.6 will focus on a monopole antenna with pattern agility based on MEMS. Section 3.7 provides the conclusions.

3.2 TYPES OF RECONFIGURABLE ANTENNAS

The reconfigurable antennas can be divided into three categories:

1. Frequency reconfigurable
    - Continuous-tuning reconfigurable antenna
    - Coarse-tuning antenna
2. Radiation pattern reconfigurable
3. Polarization reconfigurable.

This section focuses on the first category, i.e., frequency-reconfigurable antenna, in which a single antenna can be operated at different resonant frequencies by using tuning elements such as varactors and PIN diodes. During the tuning of resonant frequencies from one band to another band, the antenna radiation pattern characteristics remain unchanged. The frequency-reconfigurable antennas are divided into two categories, i.e., continuous-tuning reconfigurable antenna (the continuous tuning is achieved by varying the bias voltage of the varactor diodes) and coarse-tuning antenna, which is designed using PIN diodes. In coarse tuning, the switching mechanism is used to operate at multiple frequency bands [18]. The second category is based on pattern reconfigurability. In this type, the resonant frequency of the antenna remains unchanged, while the radiation pattern can be changed according to the requirements. The antenna can control its radiation pattern, i.e., main beam, in different directions. This type of pattern reconfigurability has recently been reported by using different types of antenna feeding [19]. The third category is based on polarization reconfigurability, in which the polarization is switched between left-hand circular polarization (LHCP) and right-hand circular polarization (RHCP) [20].

3.3 MICROWAVE RF SWITCHES

The microwave and millimeter-wave antennas as well as filters can be reconfigured using different RF switches. The frequency reconfigurability and radiation pattern agility can be achieved using these RF switches:

1. PIN diode
2. Varactor diode
3. Microelectromechanical systems (MEMS)

3.3.1 PIN Diode Switches

A PIN diode is a P–N junction semiconductor diode in which an intrinsic layer of high resistivity is sandwiched between the P and N regions. At microwave and millimeter-wave, the diode operates as a variable resistor. It is most popular in microwave circuit applications due to its low radio frequency (RF) loss, variation of junction capacitance with reverse bias voltages, high power handling capability and fast switching time [21]. The construction and the symbolic representation of PIN diodes are shown in Figure 3.1a and b.

The equivalent circuit models for the forward and reverse biases are shown in Figure 3.2. In the forward- and reverse-biased circuits, the L_s is the lead inductance, R_i is the intrinsic resistance, C_i is the intrinsic capacitance, C_p is the parasitic capacitance, C_f is the fringing capacitance, and R_s is series resistance. PIN diodes have widely been used in many switchable antennas and filters. In microwave and millimeter-wave communication systems, PIN diodes are commonly used in microwave front end switching devices, because they have attractive properties such as low insertion loss, good isolation, ease of integration to PCB, fast switching speed and low cost [22]. The drawback of PIN diodes is the nonlinearity for large signals, which in turn limits their usage in high-power applications.

FIGURE 3.1 Construction of (a) PIN diode and (b) circuit symbol for PIN diode.

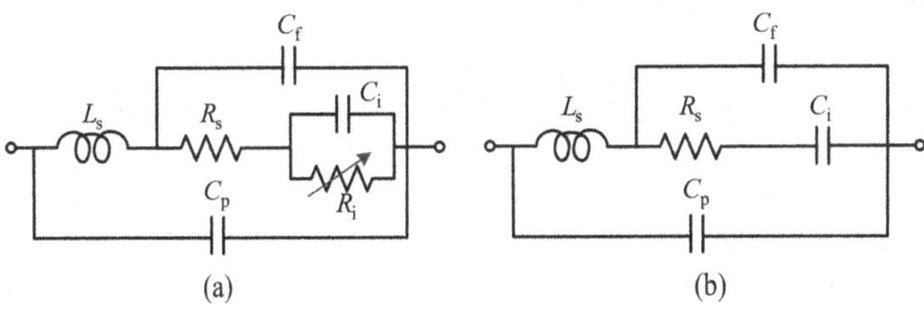

FIGURE 3.2 Equivalent circuit model for PIN diode: (a) forward bias and (b) reverse bias.

3.3.2 Varactor Diode

A varactor diode corresponds to a variable reactor in which the reactance is varied in a controlled manner varying the bias voltage. The varactor diodes have a very thin depletion layer that acts as an insulator between the two terminals of the diode, i.e., P-region and N-region. The capacitance of the diode is generally 1/∞ to the square root of the applied reverse bias voltage. The varactor diode is the most popular active component used in microstrip filters and antennas for continuous frequency tuning. In microwave and millimeter-wave communication systems, varactor diodes play an important role in radio frequency applications where tuning is needed. The most common applications of varactor diodes are parametric amplification, harmonic generation, mixing and detection. The circuit symbol and equivalent circuit model for a varactor diode are shown in Figure 3.3.

In the equivalent circuit models, R_s is series parasitic resistance, L_s is mounting parasitic inductance, C_p is packaging parasitic capacitance, and C_j is bias-dependent capacitance. The variable capacitance C_j can be decreased by increasing the reverse

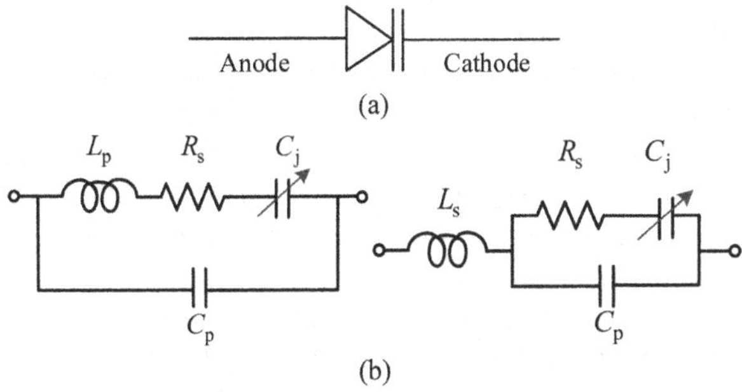

FIGURE 3.3 Varactor diode: (a) circuit symbol and (b) equivalent electrical circuit.

bias voltage across the diode, which makes the relationship nonlinear. Due to the integration of varactor diodes into microstrip antennas, their efficiency and gain are decreased, which is the major drawback of the varactor diodes. In order to improve the bandwidth, the diode is placed at the radiating edges of a microstrip patch antenna [23].

3.3.3 Microelectromechanical Systems

MEMS stands for microelectromechanical systems, which is an arrangement of electromagnetic and mechanical systems. The MEMS switches have the combined advantages of electromechanical and semiconductor technologies. The characteristics of MEMS switches are similar to the switches, i.e., low power consumption, low insertion loss and high isolation. The advantages of MEMS switches are their small size, lightweight and low cost, similar to other semiconductor switches [24]. At microwave and millimeter wave, the integration of MEMS switches provides high losses and limited power handling capability and they need expensive packaging to protect them from adverse environmental conditions. In 1990 [25], MEMS switches were proposed for use in antennas for reconfigurability. The MEMS has the ability to alter the radiation pattern. For instance, the placement of MEMS into the feed section of microstrip antennas provides beam-steering capability [26].

3.4 PIN Diode-Based Reconfigurable Patch Antenna for Pattern Agility

This section explains a single-feed square-ring patch antenna with reconfigurable pattern. The antenna's radiation pattern can be electronically reconfigured between conical and broadside radiations at a fixed frequency. In order to achieve the redirection of main beam, a square-ring patch is used and four shorting walls are placed at the edges. Two shorting walls are directly connected at the patch side, and the remaining two are connected at the patch via a PIN diode. The PIN diode's status is used to control the antenna operation modes, i.e., monopolar plate mode and normal patch [27]. The proposed antenna is shown in Figure 3.4. The proposed antenna is designed on an FR4 printed circuit board having a thickness of 0.6 mm and permittivity of 4.4. The dimensions of the square ring are 40 and 28 mm, and an inner rectangular patch is etched on the same face of the square patch. To support the patch, four shorting walls are used at a height of 7 mm above the ground plane. When there are no shorting walls, antenna radiates broadside pattern with x-directed polarization. The rectangular patch can be generated by a capacitive coupling strip and used to excite TM_{11} mode of the square-ring microstrip patch antenna [28]. When the shorting walls A and B are connected to the antenna, TM_{11} mode could not be excited. This is because the two shorting walls are located at a position where the field is null for the TM_{11} mode. These two shorting walls can be considered as a monopolar plate patch antenna fed by gap coupling, which radiates conical patterns [29]. When the antenna is operated in TM_{11} mode with dual-band characteristics, the surface current flowing along the perimeter of the square patch is maximum

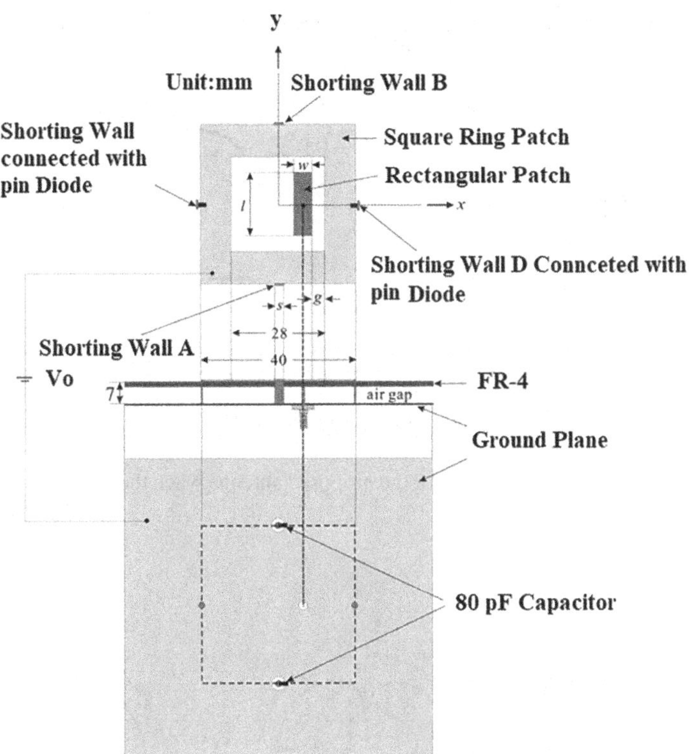

FIGURE 3.4 Geometry of the proposed pattern-reconfigurable square-ring patch antenna [27].

and minimum at the positions of the shorting walls C and D. In the monopolar plate patch with the dual-frequency characteristics, the resonant frequency can be increased when the number of shorting walls is increased [30]. This reveals that the monopolar plate patch mode has different resonant frequencies when the shorting walls C and D are switched. The addition of shorting walls C and D makes the current distribution on the square-ring patch become symmetrical, and consequently, more uniform conical patterns across the entire azimuthal plane can be obtained [31]. In order to achieve pattern reconfigurability between conical and broadside radiation patterns, a pair of plastic packaged PIN diodes is used. The PIN diodes are used to connect the square-ring patch and the shorting walls C and D. The DC bias is supplied using thin wires as the square-ring patch is connected permanently to the ground plane through the shorting walls A and B. Two coupling capacitors are required to block the DC connection. The DC bias is provided through the signal feed line when the antenna is integrated with RF circuits. When the diode is forward-biased, the diode is in ON state and each diode has an ohmic resistance of 0.9 Ω and the current starts flowing through the diode. When the diode is in ON state, the antenna operates in monopolar plate patch mode. On the other hand, when

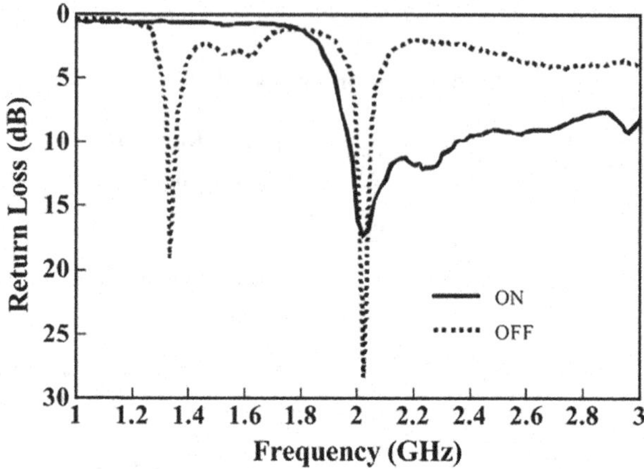

FIGURE 3.5 Measured return loss for the proposed antenna when the PIN diodes are in ON and OFF states [27].

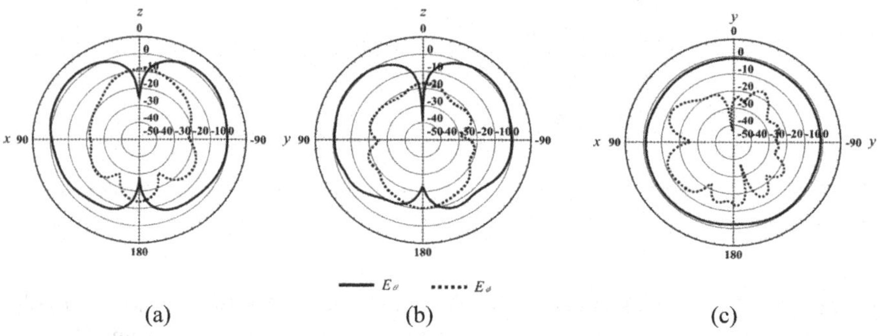

FIGURE 3.6 Measured radiation patterns at 2,020 MHz for the proposed antenna when the diodes are ON. (a) *x–z* plane, (b) *y–z* plane and (c) *x–y* plane [27].

the diode is in OFF state, the diode exhibits a capacitance of 0.2 pF that can be regarded open at 2 GHz. In this case, the antenna operates in TM_{11} mode. The measured S_{11} is shown in Figure 3.5, which reveals that the operating bandwidths of the two different modes overlap each other. The overlapped bandwidth is about 50 MHz and centered at 2.02 GHz. The measured radiation patterns of the reconfigurable patch antenna at the resonant frequency of 2.02 GHz in the *x–z* plane, (b) *y–z* plane and (c) *x–y* plane are shown in Figure 3.6. The figure shows a monopole-like conical radiation. The maximum power level is directed at an elevation angle $\theta = 50°$, and the antenna peak gain is about 2.5 dBi. Figure 3.7 shows that the main beam in the *x–z* plane is tilted by an angle of 4° off *z*-axis, which could be due to the asymmetry of the antenna structure. Apart from this, the antenna exhibits broadside radiation patterns with a peak gain of 6.8 dBi.

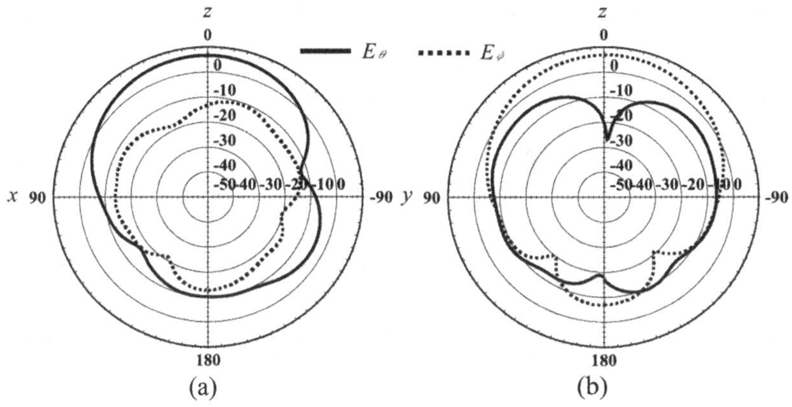

FIGURE 3.7 Measured radiation patterns at 2,020 MHz for the proposed antenna when the diodes are OFF. (a) x–z plane and (b) y–z plane [27].

The reconfigurable antennas play an important role in modern wireless communication systems. They have the ability to do multiple functions with a single component. In the modern era, frequency, pattern and polarization reconfigurabilities are becoming more popular [32]. With radiation pattern reconfigurability, a multiband antenna has the capability to operate in different radiation patterns.

This section presents a reconfigurable radiation pattern, i.e., either broadside or null-at-broadside radiation pattern in a relative continuous range of tuning frequency of more than 20% [33]. The continuous tuning of frequency is achieved by using a set of six varactor diodes. Only a single bias voltage is required to control the six varactor diodes. The antenna design starts from a square patch with a row of shorting vias in its center and is fed perpendicularly using a coaxial connector as shown in Figure 3.8. Using the proposed configuration, two resonant modes with two resonance frequencies can be excited simultaneously. In the regular TM_{100}^{z} mode resonance, the antenna radiates exactly as a conventional microstrip antenna with a broadside radiation pattern. Due to the presence of shorting vias, the antenna also resonates at quasi-radial mode [34]. The radiation pattern is approximately omnidirectional in the x–y plane due to the magnetic current loop.

The two resonant frequencies are controlled and reconfigured by varying the reactance loaded at the two opposite sides of the patch. The schematic diagram of the proposed antenna is shown in Figure 3.9. The antenna contains varactor diodes, and they are connected periodically with the open-circuited stub. The open-circuited stub is used as a transmission line and is used for additional impedance manipulation [35]. The combination of the stub and varactor diodes is used for the tuning of the antenna. The varactor diode capacitance is varied to increase or decrease the resonance frequency. With an increasing varactor diode capacitance, the resonance frequency is decreased. The first resonance frequency is for the TM_{100}^{z} mode, and the second resonance frequency is for the monopole mode.

The radius of the shorting vias is very small; the resonance frequency for conventional TM_{100}^{z} mode is almost the same as the resonance frequency for a combination

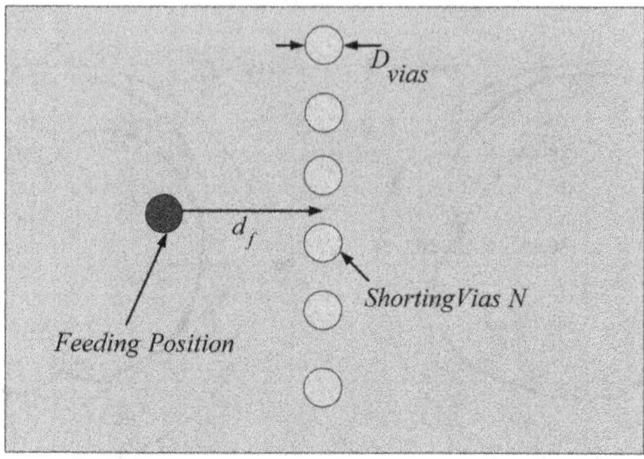

FIGURE 3.8 Microstrip patch antenna with *N* shorting vias [33].

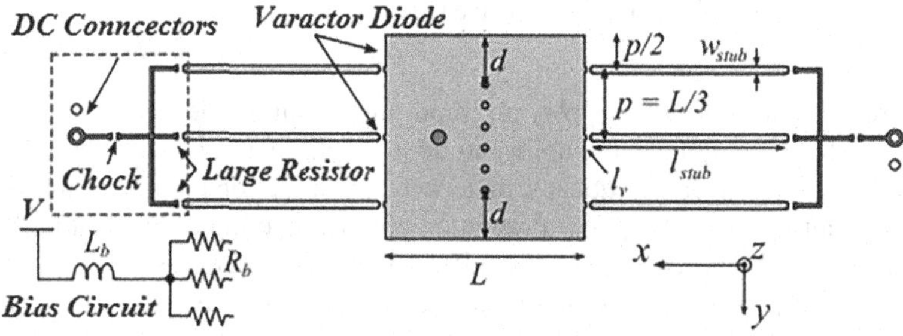

FIGURE 3.9 Proposed antenna configuration [33].

of varactors and stub-loaded single-quarter wave patch. The quarter-wave patch can be considered as a half-mode substrate integrated waveguide cavity with two open circuits in the y-direction, i.e., the travelling wave direction. Based on these analyses, the resonance frequency can be calculated using these techniques [36]. Due to the presence of shorting vias, the antenna also radiates in a quasi-radial monopole mode.

The first resonance frequency of the TM^z_{100} mode is generally not affected by the shorting vias. With an increasing or decreasing number of shorting vias, the second resonance frequency of the monopole mode decreases. The simulated reflection coefficient for different values of radii is shown in Figure 3.10. The second resonance frequency of the monopole mode decreases when the radius of the vias is increased. The radiation pattern of the antenna in monopole mode becomes more omnidirectional, which is explained by the higher isolation of the TM^z_{100} mode from the patch mode for larger frequency differences.

The normalized measured radiation patterns at three different tuning frequencies with broadside and monopole configurations are shown in Figure 3.11. The figure reveals that the patterns remain stable in the tuning range. The cross-polarization in all the different conditions are below −12 dB. In the three different resonant frequencies, the gains in the *x*–*y* plane are 5.1, 3.6 and 3 dB in omnidirectional radiation patterns.

d = 4 mm
TM$^z_{100}$ mode
d = 5 mm
d = 3 mm
$|S_{11}|$ (dB)
Frequency (GHz)
0
-10
-20
-30
3
3.5
4
4.5

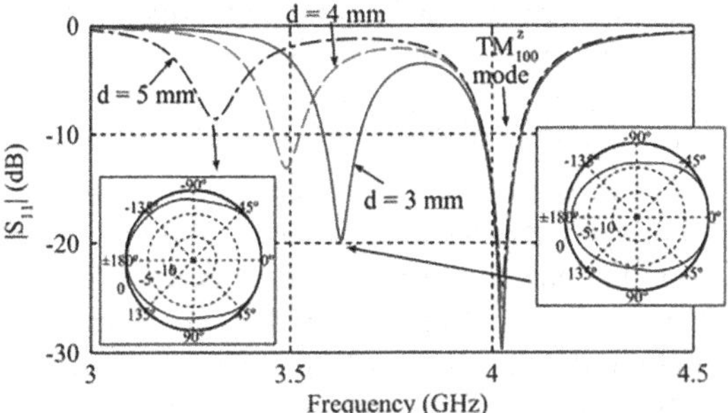

FIGURE 3.10 Simulated S_{11} of the antenna for different values of d and $N=6$ and the corresponding radiation pattern in the x–y plane for the monopole mode [33].

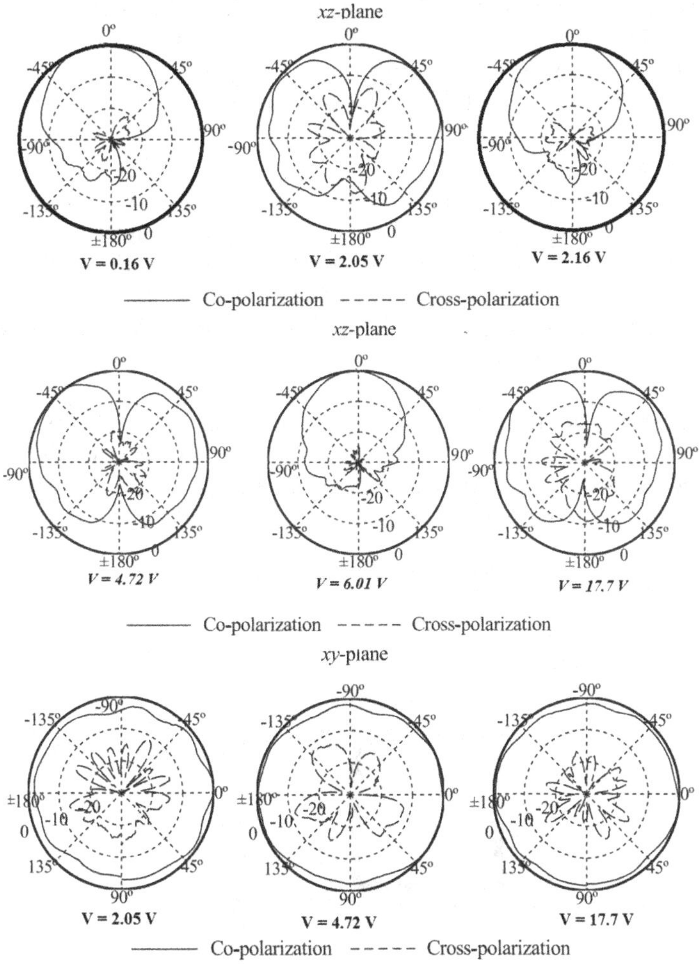

xz-plane
V = 0.16 V
V = 2.05 V
V = 2.16 V
Co-polarization ----- Cross-polarization
xz-plane
V = 4.72 V
V = 6.01 V
V = 17.7 V
Co-polarization ----- Cross-polarization
xy-plane
V = 2.05 V
V = 4.72 V
V = 17.7 V
Co-polarization ----- Cross-polarization

FIGURE 3.11 Normalized measured monopole and broadside patterns at frequencies 2.7, 3 and 3.5 GHz with different reverse voltages [33].

In the modern era, reconfigurable antennas are gaining lots of attention in microwave and millimeter-wave communication systems. In modern communication system architecture, the need for higher-data-rate, long-range wireless links is currently addressed by using ultra-wideband (UWB) and multiple-input multiple-output antenna systems, and in these systems, it is necessary to remove multipath interference and multipath fading phenomena. In wireless automotive services, an antenna with omnidirectional radiation patterns is a better option for coverage, but it is not an optimal solution. Due to this omnidirectional radiation pattern, the antenna systems are affected by the temporal fading. To overcome these problems and increase channel capacity, the MIMO systems are introduced. The pattern-reconfigurable antennas play an important role as they enable link budget optimization and provide an easy way to reduce the interference and fading in multipath environments [37,38].

This section of the chapter focuses on the design and analysis of a pattern agility antenna using microelectromechanical systems (MEMS). The MEMS switches are generally used for switching purpose at only as circuit element in the antenna reconfigurability [39]. An antenna with a single-turn square spiral is shown in Figure 3.12. The antenna consists of a single-turn square spiral. The dimensions of the antenna are 81 mm, and it is designed on Roger RT/Duroid 5880 with the resonant frequency of 6.85 GHz.

In order to execute the antenna, an SMA connector probe is incorporated at the interior end of the spiral. The outer end of the square spiral is connected to the ground using a via with a diameter of 1.23 mm. In order to alter the electric field distribution and reconfigure the radiation pattern, two MEMS SPST-RMSW100 switching elements are included in the design. To achieve end-fire configuration, switch-1 is closed and switch-2 is kept open, and to achieve broadside configuration, switch-1 is kept open and switch-2 is closed.

The operation of a MEMS switch is analogous to that of a field-effect transistor. It utilizes the gate–source–drain configuration for two different switches. A direct current (DC) of 90 V is applied as an operational voltage to the switch between the gate and source terminals. During the ON state of the switch, the insertion loss is

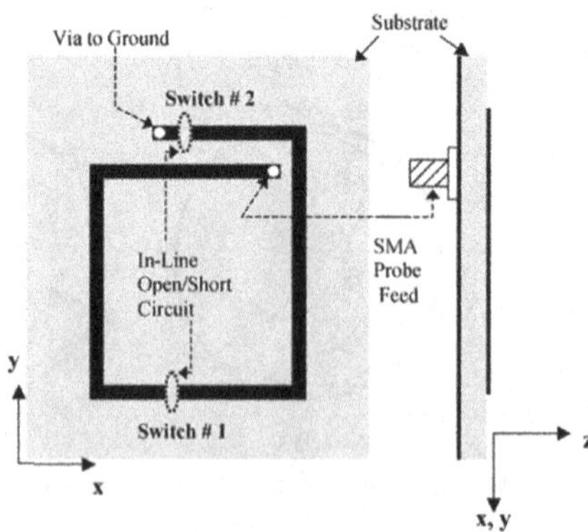

FIGURE 3.12 Antenna geometry with the locations of the switches [39].

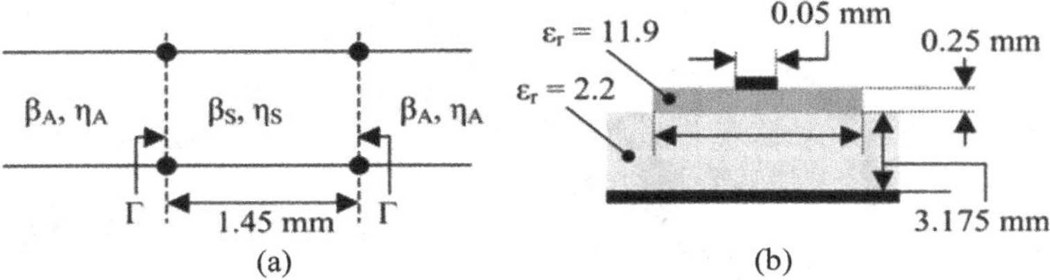

FIGURE 3.13 (a) Equivalent circuit in the locality of a switch and (b) stratified dielectric configuration of the switch [39].

about 0.225 dB and the return loss is about 24 dB. The isolation between the source and drain is about 16 dB in the OFF state. When the switch is in ON state, the device behaves as a 50 Ω microstrip transmission line, and during the OFF state of the switch, it behaves as a large capacitor.

The corresponding equivalent circuit is shown in Figure 3.13a. The cavity model is used to analyze fundamental operation, and fringing electric field is used for the radiation mechanism for the microstrip antenna [40]. The antenna structure can also be viewed from the respective of equivalent model as microstrip line can support a standing quasi-transverse electric mode and switches can be modeled as a transmission lines. Usually, the microstrip line with a characteristic impedance of $Z_0 = 145\Omega$ is necessary to provide a matching network with $Z_0 = 50\Omega$. The antenna having a microstrip line with a lower characteristic impedance correlates with the line widths, and this can limit the performance. The overall characteristic impedance of the structure can be increased by removing the ground plane of the RF MEMS switch. By removing the ground plane of the RF MEMS switch, a stratified dielectric is created beneath the switch's microstrip line with the antenna substrate, which is shown in Figure 3.13b. By making these changes, the characteristic impedance of the MEMS switch is increased, i.e., based on the mixture of dielectrics after surface mounting of the modified switch. Based on the configuration, the modified characteristic impedance of the switch becomes $Z_0 \sim 250\Omega$, which results in 50% reduction in the impedance mismatch. In order to remove the ground plane of the switch, a 400 grit sandpaper was placed on a solid work surface and the switches were held with tweezers and passed over the sandpaper, pressing gently, in a circular motion (Figure 3.14).

A proper biasing network is designed to maintain the minimum overall complexity and impedance properties. The microstrip line of the antenna connects with source and drain. These connections can be viewed as three DC-isolated microstrip lines:

1. The microstrip line connecting the probe feed line to s_1.
2. The microstrip line connecting the d_1 to s_2.
3. The microstrip lines connecting the d_2 to ground.

To isolate the actuation of the two MEMS switches, the same DC potential is applied at the source and drain of both the switches. The bias is applied independently to the

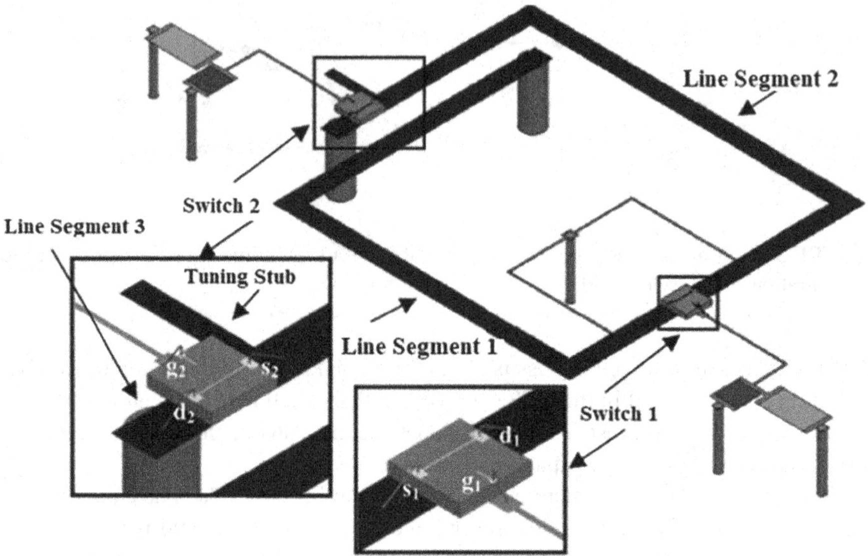

FIGURE 3.14 Antenna including vias, lumped components, tuning and bias stubs, and simplified switch model including thin wires over silicon chips to approximate the switches [39].

MEMS switches with their respective gate–source terminals. All the three microstrip lines require a DC continuity with one another; due to these, the RF ground plane of the antenna is chosen a DC power plane that has minimum effects on the antenna performance. The first and second microstrip lines achieve DC continuity with the ground plane using two high-impedance quarter-wavelength microstrip stubs, which share a shorting via. This reduces the complexity, and these two high-impedance quarter-wavelength microstrip stubs are situated around switch-1. The third microstrip line maintains DC continuity with the ground plane through the use of a grounding via. Therefore, two quarter-wavelength isolation lines create DC continuity between the RF ground plane and all the three microstrip lines. The gate of the RF MEMS device is isolated, and only minimal effort is needed to deliver the gate voltage.

A quarter-wavelength line terminated in parallel combination of a grounded 1 pF chip capacitor and 100 KΩ resistor connected to the DC source was used. This stub is designed for the effective electrical length of the capacitor. Using the basic transmission line equations, it can be found that the parallel combination of the grounded capacitor with the high-impedance resistor and the connection to the DC bias source will be dominated by the short circuit at the capacitor.

In turn, this short circuit will appear as an open circuit at the gate terminal and limits RF leakage on the gate. This eliminates the need for extra vias that affect the radiation pattern performance. Based on this bias configuration, when the switch-1 is in ON state and the switch-2 is in OFF state, the end-fire configuration can be achieved, and when the switch-1 is in OFF state and the switch-2 is in ON state, the broadside radiation pattern can be achieved.

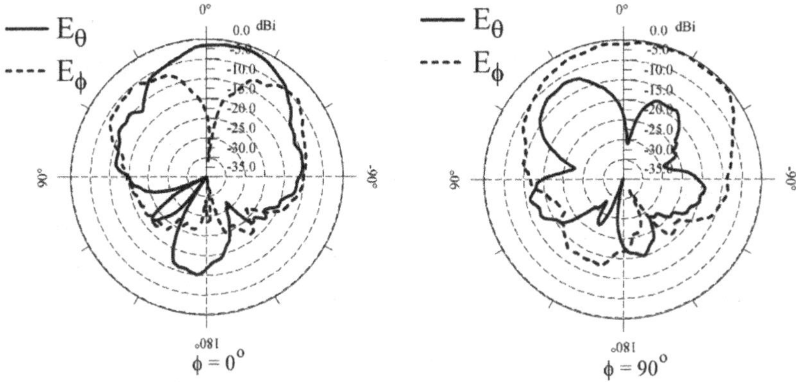

FIGURE 3.15 Measured broadside configuration radiation patterns of the two primary elevation cut-planes using the Radant RF MEMS switches [39].

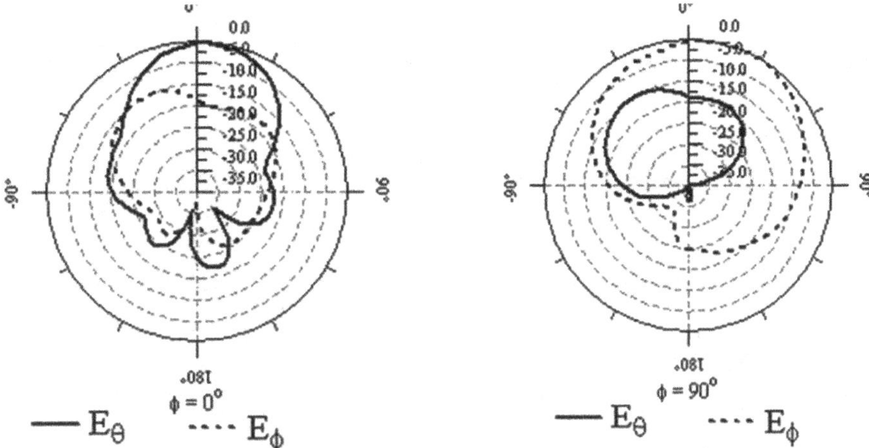

FIGURE 3.16 Simulated broadside configuration radiation patterns of the two primary elevation cut-planes using the switch model [39].

Three sets of end-fire and broadside radiation patterns are shown in Figures 3.15–3.20. Figure 3.15 reveals that the simulated and measured broadside radiation patterns are almost identical, with a simplified switch model, i.e., Figure 3.16. The figure also shows that they are very close to those of the antenna with ideal switches, i.e., Figure 3.17, with slightly higher cross-polarization levels. Figure 3.18 shows that the end-fire radiation pattern is very close to that of the simulated antenna with the simplified switch model, i.e., Figure 3.19, and is slightly distorted from the ideal end-fire while still maintaining the characteristic null at broadside, i.e., Figure 3.20.

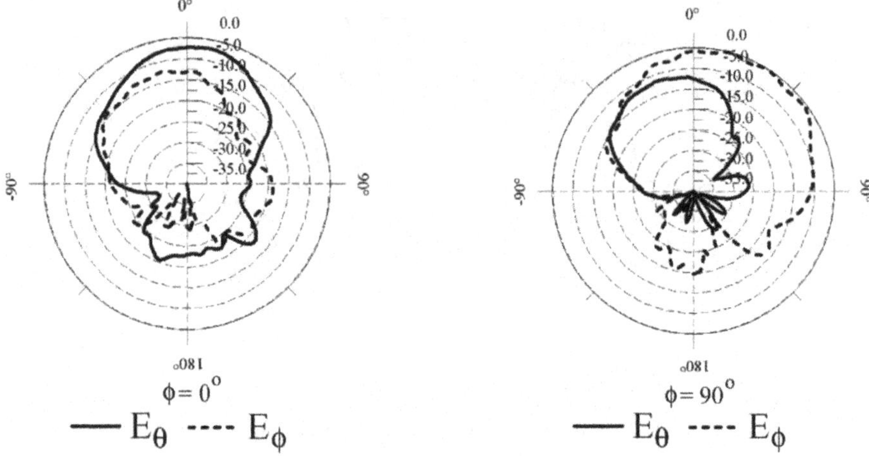

FIGURE 3.17 Measured broadside configuration radiation patterns of the two primary elevation cut-planes using ideal switches [39].

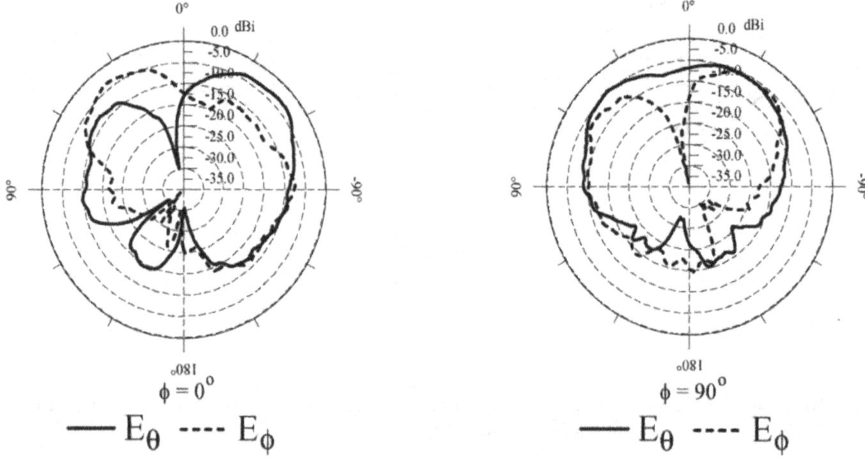

FIGURE 3.18 Measured end-fire configuration radiation patterns of the two primary elevation cut-planes using Radant RF MEMS switches [39].

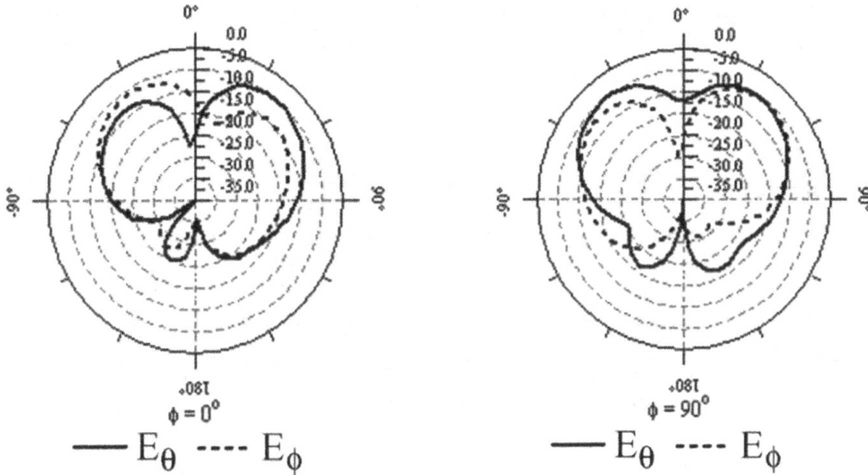

FIGURE 3.19 Simulated end-fire configuration radiation patterns of the two primary elevation cut-planes using the switch model [39].

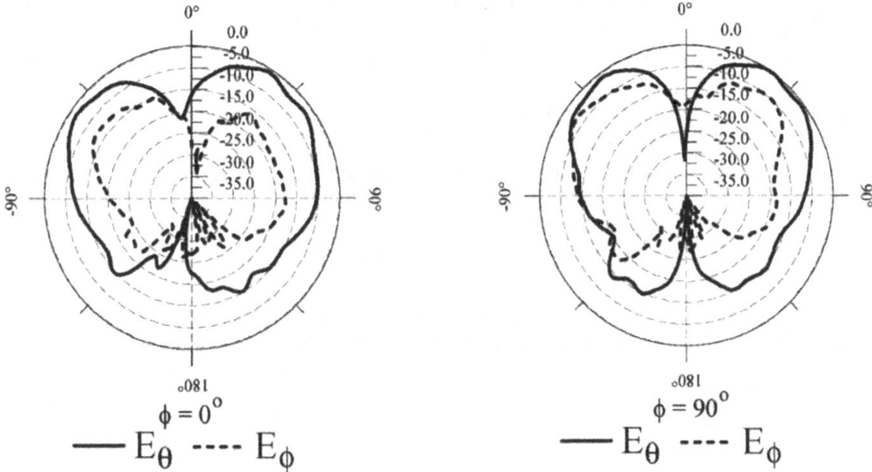

FIGURE 3.20 Measured end-fire configuration radiation patterns of the two primary elevation cut-planes using ideal switches [39].

3.5 CONCLUSION

In conclusion, an affordable chapter has been presented, which allows the reader to find an overview of the radiation pattern agility techniques, shapes and approaches appreciating the influence and the impact of the pattern agility antenna. Indeed, in this chapter, the radiation pattern agility was achieved using different reconfigurable switching components such as PIN diodes, varactor diodes and MEMS. In addition to the advantages of pattern agility described at the beginning of the chapter, a dedicated part providing the other advantages of the compact pattern agility antenna, which are desirable for emerging wireless and mobile communications, was also presented.

REFERENCES

1. K. Mori. 2004. Small beam-switched antenna with RF switches for wireless LAN. *34th Proceeding on European Microwave Conference*. 837–840.
2. S. Zhang, G. H. Huff, J. Feng and J. T. Bernhard. 2004. A pattern reconfigurable microstrip parasitic array. *IEEE Transactions on Antennas and Propagation*. 52: 2773–2776.
3. L. Petit, L. Dussopt and J. Laheurte. 2006. MEMS–switched parasitic–antenna array for radiation pattern diversity. *IEEE Transactions on Antennas and Propagation*. 54: 2624–2631.
4. W. Lin and H. Wong. 2015. Pattern reconfigurable wideband circularly-polarized quadrifilar helix with broadside and backfire radiation patterns. *9th European Conference on Antennas and Propagation (EuCAP)*. 1–4.
5. S. N. M. Zainarry, N. Nguyen-Trong and C. Fumeaux. 2018. A frequency and pattern reconfigurable two–element array antenna. *IEEE Antennas and Wireless Propagation Letters*. 17: 617–620.
6. J. Li, S. Zhu, A. Zhang and Z. Xu. 2018. Radiation pattern reconfigurable waveguide slot array antenna using liquid crystal. *Hindawi International Journal of Antennas and Propagation*. 2018: 1–9.
7. A. D. Fonzo, D. Karmel and P. Atia. 1979. Multiple shaped beam reconfigurable satellite antenna. *IEEE Antennas and Propagation Society International Symposium*. 17: 457–460.
8. S. J. Shi and W.P. Ding. 2015. Radiation pattern reconfigurable microstrip antenna for WiMAX application. *Electronics Letters*. 51: 662–664.
9. Y. Y. Bai, S. Xiao, M. C. Tang, Z. F. Ding and B. Z. Wang. 2011. Wide-angle scanning phased array with pattern reconfigurable elements. *IEEE Transactions on Antennas and Propagation*. 59: 4071–4076.
10. S. M. Chakar and M. Bouzouad. 2014. Metamaterial patch antenna radiation pattern agility. *Applied Physics A: Materials Science & Processing*. 115: 459–465.
11. M. C. L. Purisima, M. Salvador, S. G. P. Augstin and M. T. Cunanon. 2016. Frequency and pattern reconfigurable antennas for community cellular application. *Proceedings of the IEEE Conference (TENCON)*. 3767–3770.
12. N. N. Trong, L. Hall and C. Fumeaux. 2017. A dual-band dual-pattern frequency-reconfigurable antenna. *Microwave and Optical Technology Letters*. 59: 2710–2715.
13. S. N. M. Zainarry, N. N. Trong and C. A. Fumeaux. 2018. A frequency and pattern reconfigurable two–element array antenna. *IEEE Antennas Wireless Propagation Letters*. 17: 617–620.
14. P. K. Li, Z. H. Shao, Q. Wang and Y. J. Cheng. 2015. Frequency and pattern reconfigurable antenna for multi standard wireless applications. *IEEE Antennas Wireless Propagation Letters*. 14: 333–336.
15. H. A. Majid, M. K. A. Rahim, M. R. Hamid and M. F. Ismail. 2014. Frequency and pattern reconfigurable slot antenna. *IEEE Transactions on Antennas and Propagation*. 62: 5339–5343.
16. Z. Zhu, P. Wang, S. You and P. Gao. 2018. A flexible frequency and pattern reconfigurable antenna for wireless systems. *Progress in Electromagnetics Research (PIER) Letters*. 76: 63–70.

17. J. Y. Pan, Y. Ma, J. Xiong, Z. Hou and Y. Zeng. 2015. A compact reconfigurable microstrip antenna with frequency and radiation pattern selectivity. *Microwave and Optical Technology Letters*. 57: 2848–2854.
18. N. N. Trong, A. Piotrowski and C. Fumeaux. 2017. A frequency reconfigurable dual-band low-profile monopolar antenna. *IEEE Transactions on Antennas and Propagation*. 65: 3336–3343.
19. N. H. Chamok, M. H. Yılmaz, A. Arslan and M. Ali. 2016. High–gain pattern reconfigurable MIMO antenna array for wireless handheld terminals. *IEEE Transactions on Antennas and Propagation*. 64: 4306–4315.
20. R. K. Saini, S. Dwari and M. K. Mandal. 2017. CPW–fed dual–band dual–sense circularly polarized monopole antenna. *IEEE Antennas and Wireless Propagation Letters*. 16: 2497–2500.
21. K. Chang, I. Bahl and V. Nair. 2002. *RF and Microwave Circuit and Component Design for Wireless Systems*. New York: Wiley–Interscience.
22. S. L. Chen, P. Y. Qin, W. Lin and Y. J. Guo. 2018. Pattern–reconfigurable antenna with five switchable beams in elevation plane. *IEEE Antennas and Wireless Propagation Letters*. 17: 454–457.
23. P. Bhartia and I. Bahl. 1982. Frequency agile microstrip antennas. *Microwave Journal*. 25: 67–70.
24. J. R. Grau and M. J. Lee. 2010. A dual–linearly–polarized MEMS-reconfigurable antenna for NB MIMO communication systems. *IEEE Transactions on Antennas and Propagation*. 58: 4–17.
25. E. R. Brown. 1998. RF–MEMS switches for reconfigurable integrated circuits. *IEEE Trans. on Microwave Theory and Techniques*. 46: 1868–1880.
26. C. W. Jung, M. J. Lee, G. P. Li and F. D. Flaviis. 2006. Reconfigurable scan-beam single-arm spiral antenna integrated with RF–MEMS switches. *IEEE Transactions on Antennas and Propagation*. 54: 455–463.
27. S. H. Chen, J. S. Row and K. L. Wong. 2007. Reconfigurable square-ring patch antenna with pattern diversity. *IEEE Transactions on Antennas and Propagation*. 55: 472–475.
28. C. Y. D. Sim, J. S. Row and M. Y. Chen. 2005. Characteristics of superstrate-loaded circular polarization square–ring microstrip antennas on thick substrate. *Microwave and Optical Technology Letters*. 47: 567–570.
29. J. S. Row and S. W. Wu. 2006. Monopolar square patch antennas with wideband operation. *Electronics Letters*. 42: 17–18.
30. C. Delaveaud, P. Leveque and B. Jecko. 1994. New kind of microstrip antenna: the monopolar wire–patch antenna. *Electronics Letters*. 30: 1–2.
31. S. H. Yeh and K. L. Wong. 2002. A broadband low–profile cylindrical monopole antenna top loaded with a shorted cross patch. *Microwave and Optical Technology Letters*. 32: 186–188.
32. N. N. Trong, L. Hall and C. Fumeaux. 2015. A frequency and polarization reconfigurable stub–loaded microstrip patch antenna. *IEEE Transactions on Antennas and Propagation*. 63: 5235–5240.
33. N. N. Trong, L. Hall and C. Fumeaux. 2016. A frequency and pattern reconfigurable center–shorted microstrip antenna. *IEEE Antennas and Wireless Propagation Letters*. 15: 1955–1958.
34. E. W. Seeley. 1956. An experimental study of the disk–loaded folded monopole. *IRE Transactions on Antennas and Propagation*. 4: 27–28.
35. N. Nguyen-Trong, T. Kaufmann, L. Hall and C. Fumeaux. 2015. Analysis and design of a reconfigurable antenna based on half-mode substrate integrated cavity. *IEEE Transactions on Antennas and Propagation*. 63: 3345–3353.

36. L. S. Wu, X. L. Zhou, W. Y. Yin, C. T. Liu, L. Zhou and J. F. Mao. 2010. A new type of periodically loaded half–mode substrate integrated waveguide and its applications. *IEEE Transactions on Microwave Theory and Techniques*. 58: 882–893.
37. C. P. Sukumar, H. Elsami, A. Eltawil and B. A. Cetiner. 2009. Link performance improvement using reconfigurable multi–antenna systems. *IEEE Antennas and Wireless Propagation Letters*. 8: 873–876.
38. I. Dioum, A. Diallo, S. M. Farssi and C. Luxey. 2014. A novel compact dual band LTE antenna-system for MIMO operation. *IEEE Transactions on Antennas and Propagation*. 62: 2291–2296.
39. G. H. Huff and J. T. Bernhard. 2006. Integration of packaged RF MEMS switches with radiation pattern reconfigurable square spiral microstrip antennas. *IEEE Transactions on Antennas and Propagation*. 54: 464– 469.
40. G. H. Huff and J. T. Bernhard. 2004. Analysis of a radiation and frequency reconfigurable antenna. *Proceedings on Antenna Applications Symposium*, Allerton Park, Monticello, IL. 175–191.

4 Band Hopping in Printed Antennas

Dr. Surendra Kumar Gupta
Ambedkar Institute of Technology, Delhi

CONTENTS

4.1 INTRODUCTION

The printed antennas (microstrip patch antennas) are preferred to use in modern mobile communication systems such as Wireless Local Area Network (WLAN), Universal Mobile Telecommunication System (UMTS), and Worldwide Interoperability for Microwave Access (WiMAX), etc. On the other hand, microstrip antennas suffer from low efficiency and narrow bandwidth, which limit their versatility. The input impedance of an antenna tends to be sensitive to changes in frequency, and it also depends on the geometrical shape, dimensions, and the feed type of antenna. Hence, the antenna input impedance is a very important parameter that controls the radiated power and the impedance bandwidth [1]. Many efforts have been made to improve the impedance bandwidth and frequency agility/band hopping. Shorting posts are used to enhance the frequency agility. Using double posts and by adjusting them finely, a maximum tunability of 32.5% was achieved [2]. The impedance bandwidth can be increased by loading the patch with stubs, slits, and slots. Apart from loading a patch, the impedance bandwidth can be improved by increasing the substrate thickness and variable length transmission lines. Various other techniques have been suggested to improve the impedance bandwidth and frequency tunability, such as loading varactor diodes, annular ring, stacking of patches, and using an L-strip proximity coupled slot loaded patch [3–10]. Integration of devices such as varactor diodes, Gunn diodes, and impedance tuning networks into microstrip antennas could not provide sufficient band of operation. Limitations still exist on the ability of these techniques. A circular microstrip antenna with an airgap between the ground plane and the substrate was proposed in [11]. An increase in the airgap causes a decrease in the dynamic permittivity, resulting in an upward shift in the resonant frequency of the antenna. Hence, to overcome bandwidth limitations, various configurations of microstrip antennas such as the above are proposed, which provide dualband, multiband, and wideband operation. Such configurations of antennas are widely acceptable for multiservice applications and become a good substitute of wide bandwidth requirement, which motivate researchers to work on the structure of antennas for wide frequency range applications.

Printed antennas still suffer from bandwidth limitations, to overcome these limitations, wide range band hopping / frequency agile antennas are designed. In this chapter, a band hopping / tunable band / frequency agile printed antenna is discussed. In this antenna, a band can be electronically tuned in a frequency range, which is referred as frequency agility or band hopping in printed antennas.

In this chapter, active devices are loaded on the patch to achieve tunable band / band hopping / agility of antenna. A structure of a frequency agile circular microstrip antenna (CMSA) with an airgap between the ground plane and the substrate is analyzed, and to enhance the agility of antenna, a MOS device is loaded on the patch. The structures of a single MOS and a double MOS loaded patch are covered to enhance the operating frequency range of antenna. To investigate the antenna, different parameters such as resonance frequency, input impedance, frequency agility, VSWR, radiation pattern, etc. are calculated and simulated. The resonant frequency of a 10 mm radius patch antenna is upward shifted from 4.9 GHz to 6.60 GHz using a 1 mm airgap between the ground plane and the substrate, and by loading a MOS, the antenna can be tuned down to 1.127 GHz operating frequency, which leads to compactness and band hopping of the antenna. Hence, the antenna can be tuned between 1.127 GHz and 6.595 GHz frequency of operation, which makes the antenna highly suitable for a wide range of mobile communication applications, such as GPS, UMTS, WiMAX, remote sensing, and other modern communication systems. A frequency agility of 76.20% was found for a MOS loaded CMSA, and a double MOS loaded antenna possessed 82.94% frequency agility.

Also, a frequency agile BST varactor diode loaded stacked CMSA is discussed in this chapter. The antenna is analyzed using extended cavity model. One of the two bands of the antenna is tunable with the help of a BST varactor. The upper band is useful for WiMAX, and the lower band is useful for other wireless communication systems. Various antenna parameters such as return loss, resonant frequency, and frequency agility are investigated. The simulated results agreed well with the theoretical analysis. A frequency agility of 60.64% is achieved, which is better than a simple varactor diode loaded antenna. The lowest resonant frequency of 0.866 GHz is obtained, which shows a significant physical area reduction, i.e. compact size of antenna.

4.2 THEORY OF MOS LOADED CMSA WITH AN AIRGAP

The bandwidth limitation had been the major constraint in the development of microstrip antennas, in recent research, a wide frequency range band hopping/tunable antenna has been considered a good solution to this problem. Figure 4.1 presents the geometry of a wide frequency range tunable double MOS loaded circular microstrip antenna with an air gap between the ground plane and the substrate for mobile communications. The circular microstrip antenna is structured as a cavity with a magnetic wall along the edge. The structure of this antenna is a two layer cavity: the lower layer is an airgap of H_a with relative permittivity 1, and the upper layer is a dielectric substrate of thickness H with relative permittivity ε_r. The effect of the airgap below the substrate was considered for obtaining an equivalent permittivity of the medium below the patch. Using an equivalent single layer structure of total height $H_t = H + H_a$, the equivalent permittivity (ε_{re}) was calculated as in [12,18–19] and is expressed as follows:

$$\varepsilon_{re} = \frac{\varepsilon_r (H + H_a)}{(1 + \varepsilon_r H_a)} \tag{4.1}$$

where ε_r is the permittivity of the substrate and the equivalent permittivity of the CMSA without the air gap ($H_a = 0$) is $\varepsilon_{re} = \varepsilon_r$.

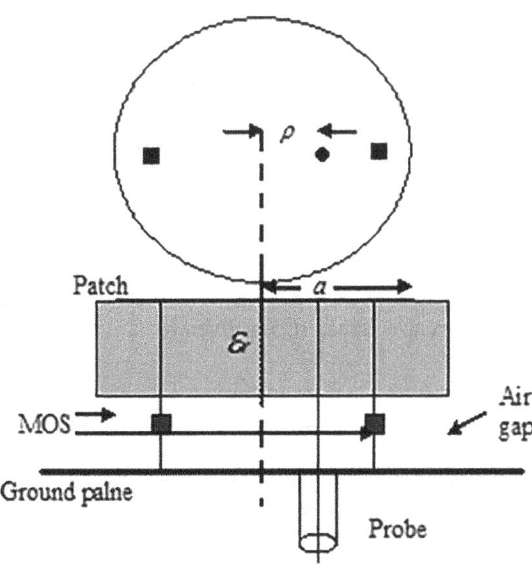

FIGURE 4.1 Geometry of a double MOS loaded circular microstrip antenna with an airgap.

Following the analytical model, an improved form of permittivity of transverse magnetic (TM) modes of the medium below the patch in the CMSA with an airgap, the effective permittivity $\varepsilon_{r,\text{eff}}$ is defined as follows:

$$\varepsilon_{r,\text{eff}} = \frac{4\varepsilon_{\text{re}}\varepsilon_{r,\text{dyn}}}{\left(\sqrt{\varepsilon_{\text{re}}} + \sqrt{\varepsilon_{r,\text{dyn}}}\right)} \tag{4.2}$$

The term $\varepsilon_{r,\text{eff}}$ is introduced to take into account the effect of the equivalent permittivity (ε_{re}) of the medium below the patch in combination with the dynamic permittivity ($\varepsilon_{r,\text{dyn}}$) to improve the model. The resonant frequency and effective radius of the CMSA with an airgap are calculated as follows:

Resonant frequency

$$f_r = \frac{\alpha_{\text{nm}} c}{2\pi a_{\text{eff}}\sqrt{\varepsilon_{r,\text{eff}}}} \tag{4.3}$$

where c is the velocity of light in free space andα_{nm} is the mth zero of the first kind Bessel function of order n.

Effective radius of patch

$$a_{\text{eff}} = a\sqrt{(1+q)} \tag{4.4}$$

where the term q arises due to the fringing fields at the edge of the patch [12].

4.2.1 Metal Oxide Semiconductor

Figure 4.2 shows a typical Metal-Insulator-Semiconductor (MIS) or Metal-Oxide-Semiconductor (MOS) and its associated capacitances. It is evident that it contains two capacitances: depletion layer capacitance (C_d), which varies with bias voltage, and insulator capacitance (C_i), which is fixed. The MOS under consideration is Au-Si_3N_4-Si. The total capacitance of the MOS (C_{mos}) was calculated as in [13]:

$$C_{\text{mos}} = \frac{C_i C_d}{(C_i + C_d)} \tag{4.5}$$

where the insulation layer capacitance (C_i) is given by

$$C_i = \frac{\varepsilon_0 \varepsilon_{\text{ro}}}{d} \tag{4.6}$$

where ε_{ro} is relative permittivity and d is the thickness of the insulation layer.

The depletion layer capacitance (C_d) is given by

$$C_d = \frac{\varepsilon_0 \varepsilon_{\text{rsi}}}{X} \tag{4.7}$$

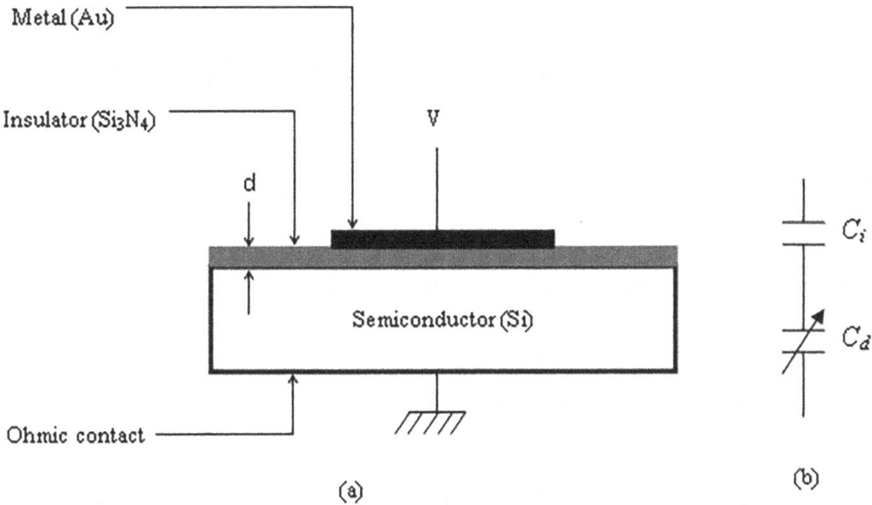

FIGURE 4.2 MOS capacitor: (a) schematic (b) equivalent circuit.

where ε_{rsi} is the relative permittivity of silicon and X is the width of the depletion layer and is given by

$$X = \frac{\varepsilon_{rsi}\varepsilon_0}{C_i}\left\{-1+\sqrt{\left[1+\left(\frac{2V_g C_i^2}{\varepsilon_{rsi}\varepsilon_0 Q N_a}\right)\right]}\right\} \tag{4.8}$$

where V_g is the bias voltage, N_a is the acceptor concentration of the doping material, and Q is the charge of an electron.

Combining eqs. (4.5)–(4.8), C_{mos} is defined as

$$C_{mos} = \frac{C_i}{\left[\sqrt{1+\left(\frac{2V_g C_i^2}{\varepsilon_{rsi}\varepsilon_0 Q N_a}\right)}\right]}A \tag{4.9}$$

where A is the cross-sectional area of the MOS device.

4.2.2 Double MOS Loaded Circular Microstrip Antenna with an Airgap

The antenna is a coaxial fed circular microstrip patch with an airgap and loaded with a MOS. The antenna is analyzed in TM_{11} mode. A circular microstrip antenna is characterized by the parallel combination of resonance resistance R_0, inductance L, and capacitance C, as presented in Figure 4.3(a). These parameters were calculated from the theory of modal expansion and cavity model as in [14].

Resonance resistance

$$R_0 = \frac{1}{G_T}\frac{J_n^2(k\rho)}{J_n^2(ka)} \tag{4.10}$$

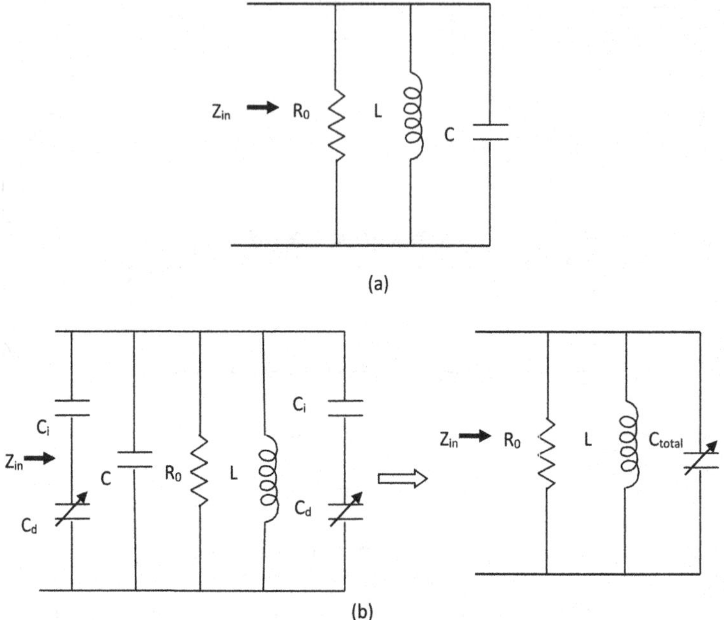

FIGURE 4.3 Equivalent circuit of (a) a circular patch microstrip antenna; (b) a double MOS loaded circular patch microstrip antenna.

where J_n is the first kind of Bessel function of order n, ρ is the probe position, and G_T is the total conductance associated with dielectric loss, radiation loss, and conduction loss [15].

Capacitance associated with antenna

$$C = \frac{Q_T}{2\pi f_r R_0} \tag{4.11}$$

and Inductance

$$L = \frac{R_0}{2\pi f_r Q_T} \tag{4.12}$$

where Q_T is the total quality factor, which includes radiation loss, dielectric loss, and conductance loss.

The equivalent circuit of a double MOS loaded circular microstrip antenna with an airgap is presented in Figure 4.3(b). The total capacitance of the patch can be calculated as in [16]:

$$C_{\text{total}} = C + 2C_{\text{mos}} \tag{4.13}$$

where C_{mos} is the capacitance of the MOS.

The total input impedance of a double MOS loaded CMSA with an airgap is calculated as

$$Z_{\text{in}} = \frac{1}{\left\{\left(\frac{1}{R_0}\right) + \left(j\omega C_{\text{total}}\right) + \left(\frac{1}{j\omega L}\right)\right\}} \tag{4.14}$$

The reflection coefficient (Γ) of the circular patch is given by

$$\Gamma = \frac{Z_{\text{in}} - Z_0}{Z_{\text{in}} + Z_0} \tag{4.15}$$

where Z_0 is the impedance (50 Ω) of the coaxial feed.

The voltage standing wave ratio (*VSWR*) of the patch is given as

$$\text{VSWR} = \frac{1+|\Gamma|}{1-|\Gamma|} \tag{4.16}$$

The return loss (*RL*) of the antenna is given by

$$\text{RL} = 20\log\left(|\Gamma|\right) \tag{4.17}$$

where Γ is the reflection coefficient of the circular patch.

4.2.3 Specifications of Double MOS Loaded CMSA with an Airgap

The design specifications of the MOS device and CMSA are given in Tables 4.1 and 4.2, respectively. The same specifications are considered for the theoretical analysis and simulation of the proposed antenna. These specifications provide a good performance of the antenna as discussed later in this chapter.

TABLE 4.1
Specifications of MOS

Parameter	Value
MOS capacitor structure	Au-Si_3N_4-Si (n + 0.0005 Ω cm)
Cross section area of device (A)	$1.6\times10^{-8}\,m^2$
Relative permittivity of oxide layer (ε_{ro})	7.5
Relative permittivity of semiconductor (ε_{rsi})	11.9
Acceptor concentration (N_a)	$1.45\times10^{22}\,m^{-3}$
Bias voltage range (V_g)	0-5 V
Thickness of oxide layer (d)	100, 200, 300, 400, 500 A°
Peak values of C_{mos}	106.2, 53.1, 35.4, 26.5, 21.2 pf

TABLE 4.2
Specifications of CMSA

Parameter	Value
Substrate material	Beeswax
Radius of circular patch (a)	10 mm
Substrate thickness (H)	1.5748 mm
Airgap (H_a)	1.00 mm
Relative dielectric constant of substrate material (ε_r)	2.35
Loss tangent (tan δ)	0.005
Probe position (ρ)	3.1 mm

4.2.4 Radiation Pattern of CMSA

The expression for radiation pattern of a circular microstrip antenna in TM_{11} mode is given as follows [17]:

$$E_\theta = -\frac{jVak_o}{2}\frac{e^{-jk_o r}}{r}\cos(\varphi)J_1'(k_o a\sin(\theta)) \tag{4.18}$$

$$E_\varphi = \frac{jVak_o}{2}\frac{e^{-jk_o r}}{r}\frac{J_1(k_o a\sin(\theta))}{k_o a\sin(\theta)}\cos\theta\sin\phi \tag{4.19}$$

where V is the edge voltage, k is the wave number, and r is the observation location that may be taken randomly large compared to the antenna size.

The radiation pattern of the antenna is calculated as

$$R = |E_\theta|^2 + |E_\varphi|^2 \tag{4.20}$$

For E-plane pattern $\phi = 0°$, and for H-plane pattern $\phi = 90°$.

4.2.5 Properties of Double MOS Loaded CMSA with an Airgap

In this section, a double MOS loaded circular patch microstrip antenna with an airgap for wide frequency range characteristics is discussed. The theoretical and simulated results for input impedance, MOS capacitance, frequency agility, VSWR, and radiation pattern in respect of the performance of the antenna are discussed in this section. The theoretical results are plotted using MATLAB, and the simulation is done on IE3D simulator [20]. Figure 4.4(a) and (b) show the variation of the real part of input impedance with the frequency of a circular microstrip patch antenna for different airgaps. The resonance frequency of the antenna increases with increasing airgap, due to the fact that the effective value of permittivity decreases. The resonant frequency of the antenna with a 1mm airgap is upward shifted to 6.8 GHz as compared

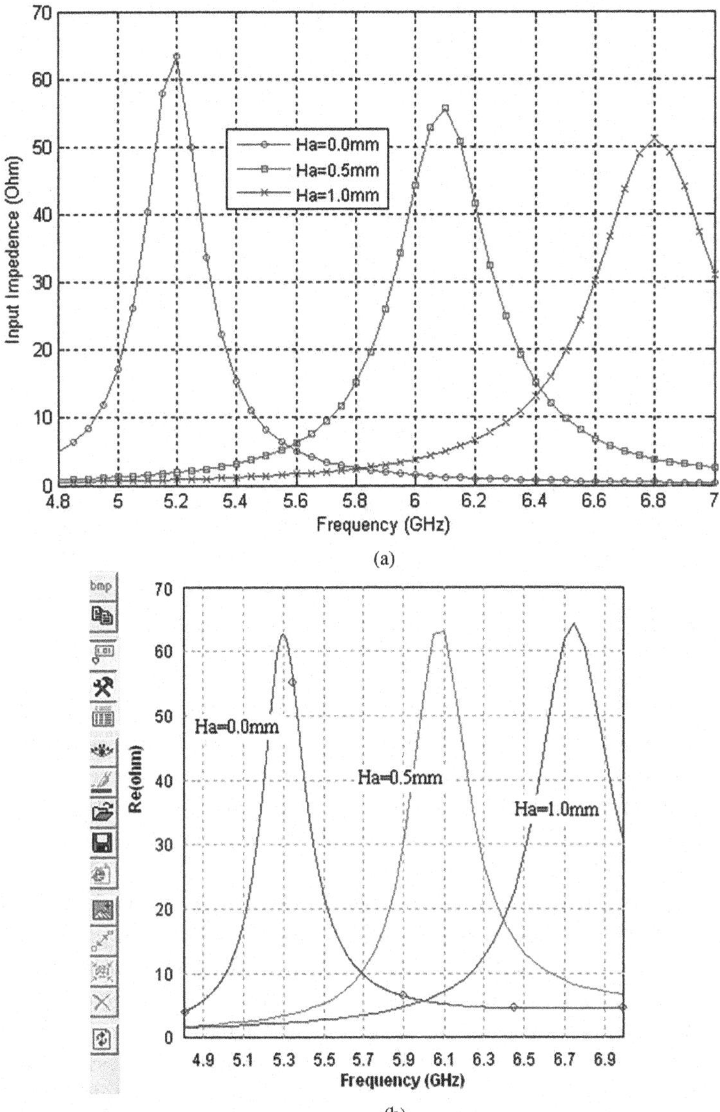

FIGURE 4.4 Variation of the real part of input impedance with the frequency of CMSA with various airgaps: (a) theoretical; (b) simulated.

to 5.2 GHz for a circular microstrip antenna without an airgap. This upward shift of frequency due to the airgap between the ground plane and the substrate has broadened the operating range of antenna. Theoretical results match with the simulated results, as seen from Figure 4.4(a) and (b); the simulated real part of input impedance for all three airgaps (H_a= 0.0, 0.5, and 1.0 mm) are in good match with the theoretical real part of input impedance. The simulated results well verify the matching of

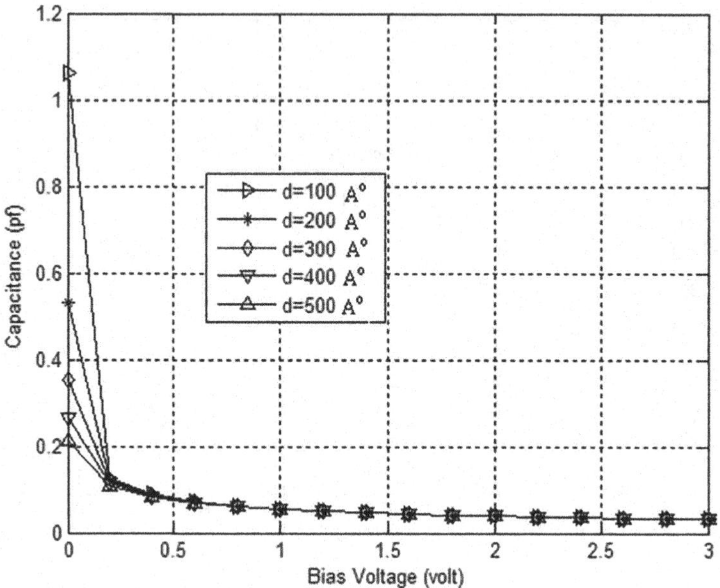

FIGURE 4.5 Variation of the capacitance of a MOS with bias voltage.

input impedance of the antenna over its wide frequency range. Figure 4.5 shows the variation of capacitance with bias voltage for a typical MOS. It is found that there is a sharp variation in capacitance near zero bias voltage. The capacitance is almost constant for all values of thickness for the bias voltage above 1volt. The peak values of capacitances are 106.2, 53.1, 35.4, 26.5, and 21.2 pf for oxide layers of 100, 200, 300, 400, and 500 A°, respectively.

The variation of resonant frequency with bias voltage for single MOS loaded and double MOS loaded circular microstrip antennas with an airgap is shown in Figure 4.6(a), and (b), respectively. The variation of resonance frequency is steep at lower bias voltage compared to higher bias voltage. The variation of resonance frequency is negligible for all oxide layers for higher bias voltage. To observe the behavior of the agility of antenna, it is simulated for various thicknesses of the oxide layer of MOS. Table 4.3 shows the percentage agility for the single MOS loaded antenna for different oxide layer thicknesses. Table 4.4 shows the percentage agility for the double MOS loaded antenna for different thicknesses of the oxide layer. It is observed from the response that the maximum agility for a double MOS loaded patch is 82.94% (for a 100 A° oxide layer). Also, it is evident from Tables 4.3 and 4.4 that the frequency agility decreases as the oxide layer thickness increases and a double MOS loaded antenna has more frequency agility than a single-MOS-loaded antenna. Antennas with different patch radii and a 1mm airgap are simulated to see the effect on agility. The theoretical and simulated variations of frequency agility with the radius of patch are in good agreement, as shown in Figure 4.7. The simulated results of agility are shown for various patch radii (a=5.0, 10.0, 15.0, 20.0, 25.0, and 30.0 mm) and are close to the theoretical results for the respective patch radius.

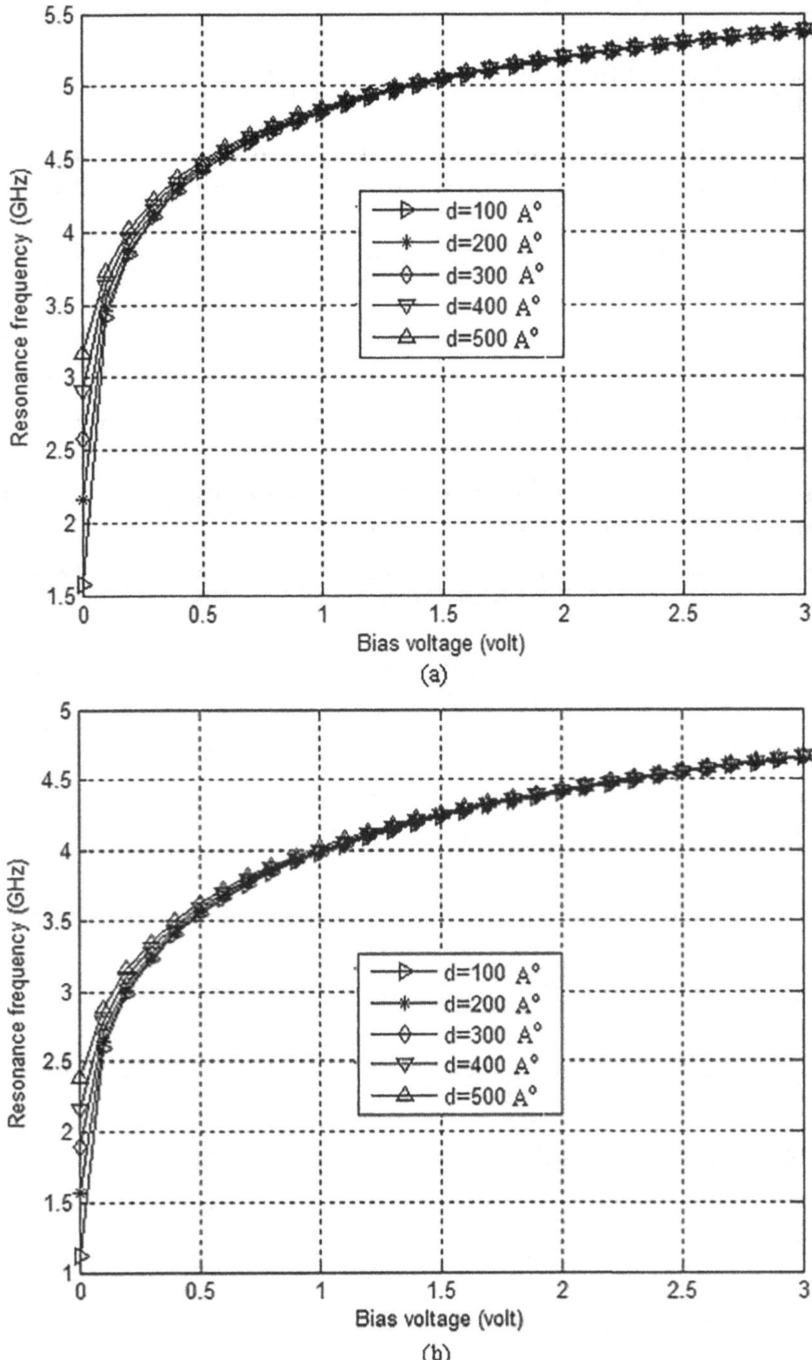

FIGURE 4.6 Variation of the resonant frequency with bias voltage of (a) a single MOS loaded CMSA with an airgap and (b) a double MOS loaded CMSA with an airgap.

TABLE 4.3
Frequency Agility of Single MOS Loaded Circular Microstrip Antenna with an Airgap for Different Thicknesses of Oxide Layer

Sr. No.	Oxide Thickness (A°)	Minimum Achievable Frequency (GHz)	Total Frequency Agility (GHz)	Percentage Agility (%)
1	100	1.571	5.032	76.20
2	200	2.161	4.442	67.27
3	300	2.579	4.024	60.94
4	400	2.905	3.698	56.00
5	500	3.172	3.431	51.96

TABLE 4.4
Frequency Agility of Double MOS Loaded Circular Microstrip Antenna with an Airgap for Different Thicknesses of Oxide Layer

Sr. No.	Oxide Thickness (A°)	Minimum Achievable Frequency (GHz)	Total Frequency Agility (GHz)	Percentage Agility (%)
1	100	1.127	5.476	82.94
2	200	1.571	5.032	76.20
3	300	1.897	4.706	71.27
4	400	2.161	4.442	67.27
5	500	2.385	4.218	63.88

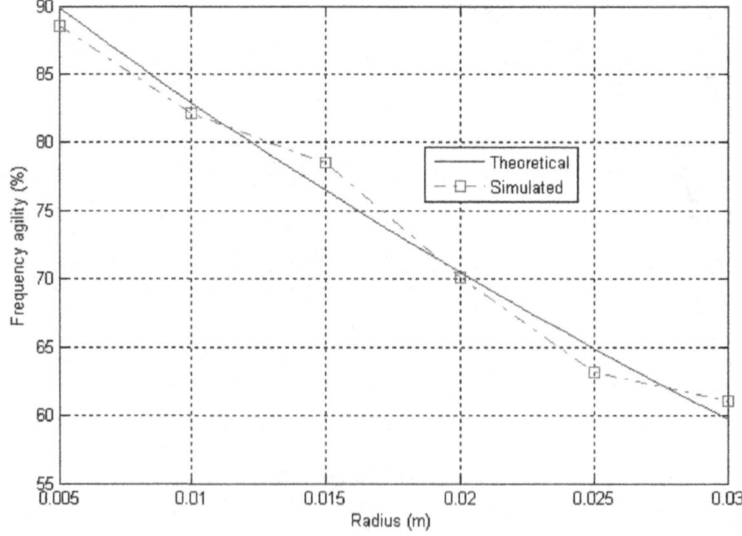

FIGURE 4.7 Theoretical and simulated variations of frequency agility with the radius of patch of a double MOS loaded CMSA with an airgap ($H_a = 1.0$ mm).

Also, it is verified by the simulated results that the agility is inversely proportional to the patch radius of the proposed antenna. The frequency agility decreases with an increase in the radius of patch; this may be understood by the fact that the contribution of MOS capacitance in the antenna decreases as the radius of patch increases. It is evident that an antenna has a better frequency agility for a low patch radius; the frequency agility of an antenna with a 10 mm radius of patch is 82.94%.

Figure 4.8 shows the variation of the real part of input impedance with frequency for different bias voltages for the described antenna. The resonant frequency of the antenna shows an interesting downward shift for different values of thickness of the oxide layer. It is found that for a 0 volt bias voltage and a 100 A° oxide layer, the resonant frequency of the antenna is downward shifted to 1.27 GHz from 6.80 GHz design frequency, which is obvious from the fact that the maximum capacitance is added into the antenna system due to the MOS. Hence, the resonant frequency of the described antenna can be tuned by varying the bias voltage. It is

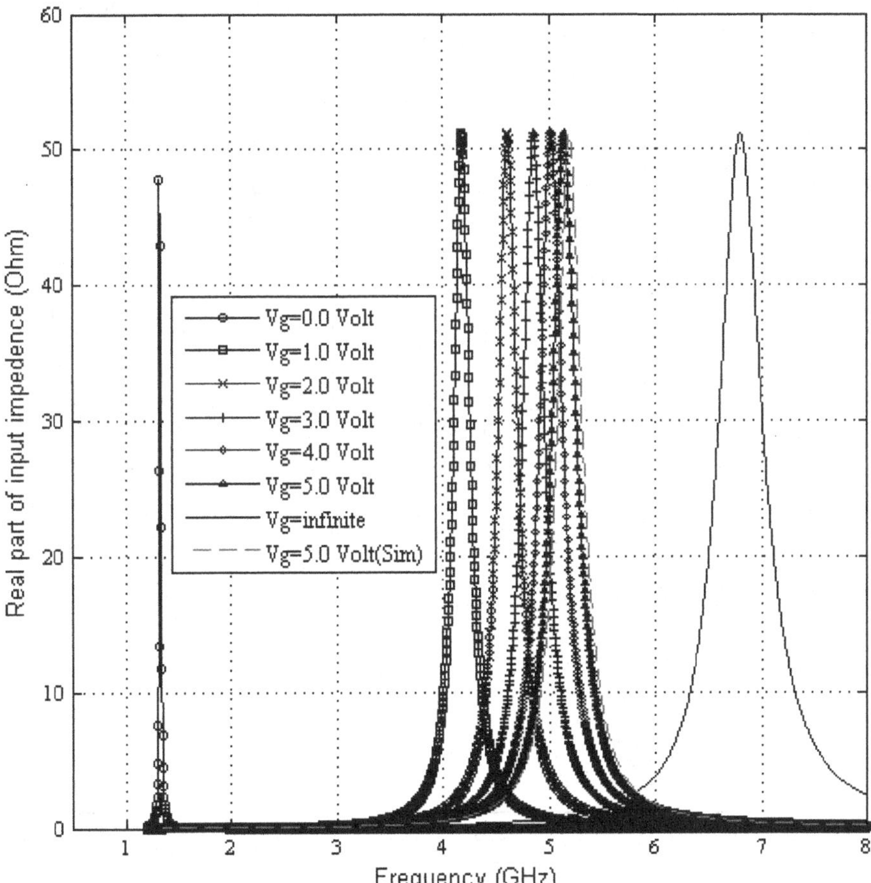

FIGURE 4.8 Theoretical and simulated variations of input impedance with frequency with various bias voltages (V_g) of a double MOS loaded CMSA with an airgap ($H_a = 1.0$ mm).

found that the variation of the resonance frequency with bias voltage provides a good tunability of the operating frequency of antenna and widens the frequency range and agility of antenna. These findings are equally verified by the simulated results. The theoretical results are in close agreement with simulated results as shown in Figure 4.8 for variation of real parts of input impedance with frequency for 5.0 volt bias voltage.

Figure 4.9 presents the theoretical and simulated results for the VSWR of antennas with and without a MOS as a function of frequency at $V_g = 1$ volt. The theoretical and simulated graphs show that the resonance impedance at lower frequency remains the same, as on 6.8 GHz, which is obvious from the fact that only the imaginary component is added into its equivalent. The theoretically calculated VSWR gives a close match with the simulated result. The E-plane and H-plane radiation patterns for a MOS loaded CMSA are shown in Figures 4.10 and 4.11, respectively. The theoretical radiation patterns are plotted for different bias voltages ($V_g = 1.0$, 2.0, 3.0, 4.0, and 5.0 volt), and the simulated radiation patterns for $V_g = 5.0$ volt for both the planes. The simulated results are close to the respective theoretical results. It is found that the beam width of an antenna decreases with an increase in bias voltage for the

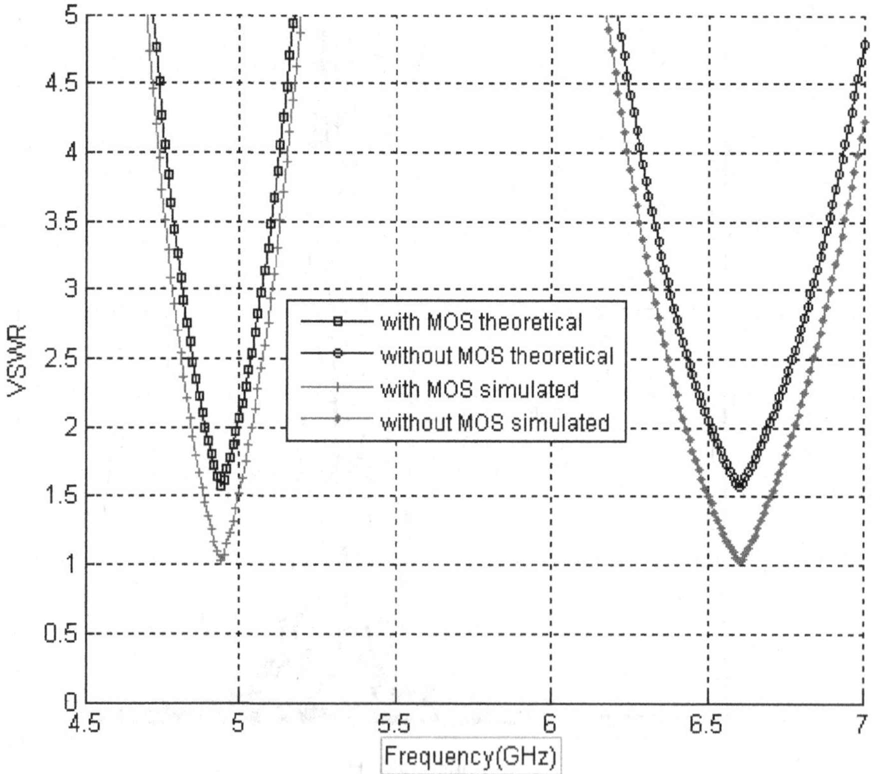

FIGURE 4.9 Theoretical and simulated variations of VSWR with frequency of a CMSA with an airgap ($H_a = 1.0$ mm) with and without a MOS.

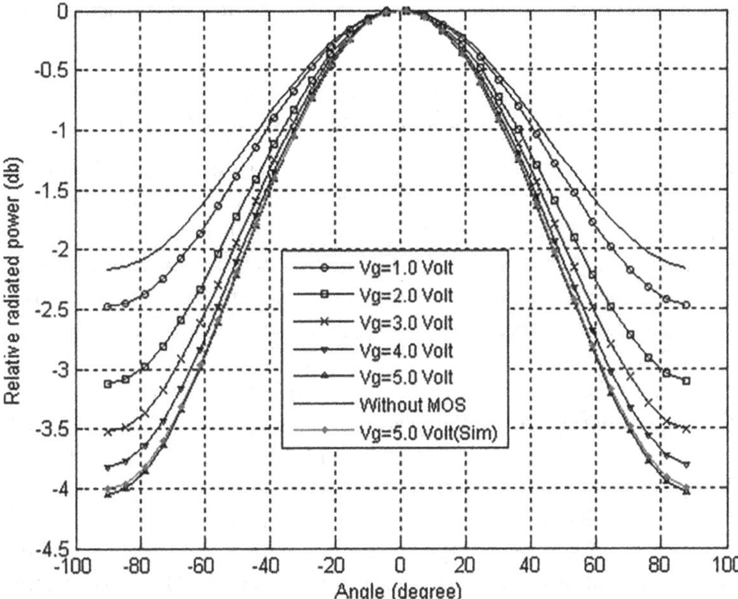

FIGURE 4.10 Theoretical and simulated E-plane radiation patterns of a double MOS loaded CMSA with an airgap.

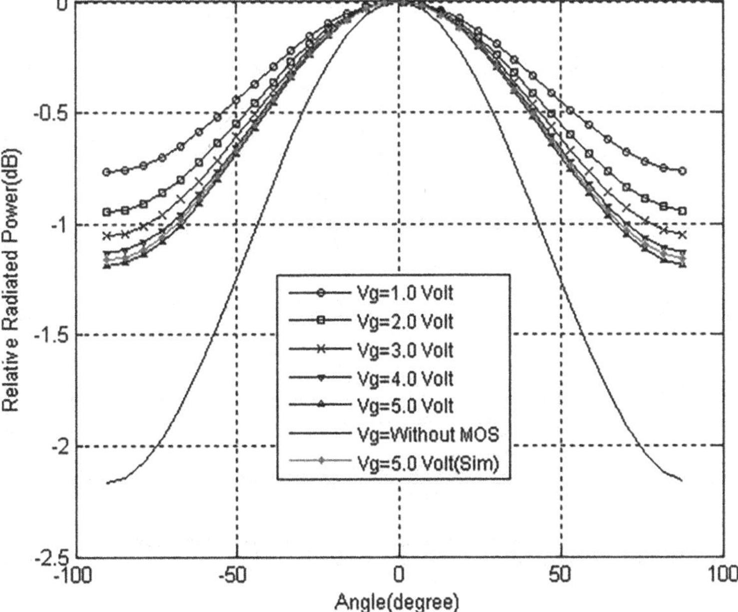

FIGURE 4.11 Theoretical and simulated H-plane radiation patterns of a double MOS loaded CMSA with an airgap.

entire range of operation, this is due to the fact that operational frequency of antenna increases with bias voltage.

4.3 BST VARACTOR DIODE LOADED STACKED CMSA

In this section, a frequency agile BST varactor loaded stacked CMSA is described. The antenna is analyzed using an extended cavity model. One of the two bands of the antenna is tunable with the help of a BST varactor. The upper band is useful for WiMAX, and the lower band is useful for other wireless communication systems. Various antenna parameters such as return loss, resonant frequency, and frequency agility are investigated. A frequency agility of 60.64% is achieved, which is better than that of a simple varactor diode loaded antenna.

The structure of the proposed antenna is shown in Figure 4.12. It consists of two layers of circular microstrip patches placed vertically with an aligned center [21]. The lower patch with a radius of a_1 is supported by a substrate of dielectric constant ε_{r1}, and the upper one with a radius of a_2 is placed on a substrate of dielectric constant ε_{r2}. The thickness of the lower substrate layer is h_1, and that of the upper substrate is h_2. In the present design, both layers are filled with PTTE substrate with a relative dielectric constant of 2.2. A BST varactor diode is embedded between the lower circular patch (LCP) and the ground plane. The center conductor of coaxial probe is electrically connected to the upper circular patch (UCP) through a hole in the lower patch. Figure 4.13 shows a 3D view from ADS. The numerical analysis of antenna is presented in four parts. The first part describes the design of the upper cavity assuming the lower patch as ground plane. The lower cavity is analyzed with a superstrate, neglecting the effect of the upper patch, in the second part of analysis. In the third

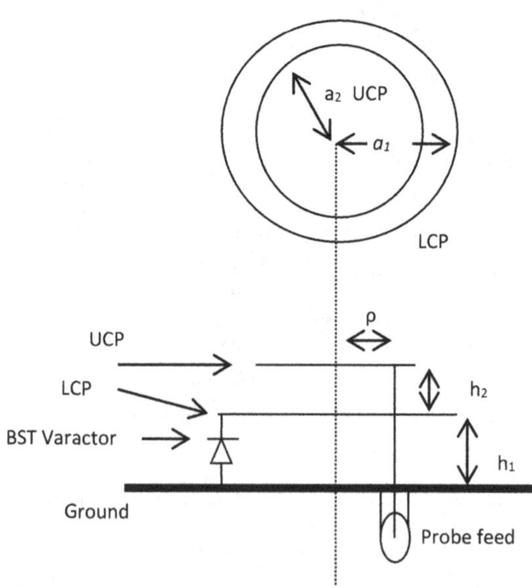

FIGURE 4.12 Structure of a BST varactor loaded CMSA.

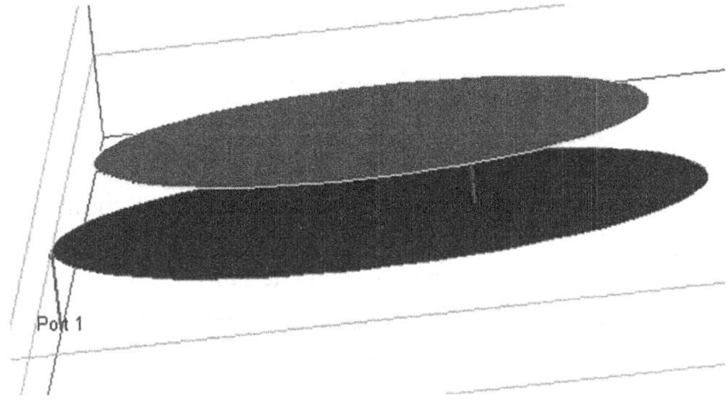

FIGURE 4.13 3D view of a stacked CMSA using ADS.

part, an analysis is given for a stacked antenna. Lastly, the theory and effect of integration of BST into a stacked circular microstrip antenna has been investigated. The theoretical and simulated results are found and are in good agreement.

4.3.1 Analysis of Upper Patch

The analysis of the upper patch is performed similar to a simple microstrip antenna with h_2 as thickness and ε_{r2} as dielectric constant [22]. Reverse current induced at lower patch, which acts as ground plane. But the current gets reflected from the edges of the lower patch due to its finite size. That is, the ground plane is not finite, which is taken in the calculation. That is why some error is expected in the results.

The input impedance of the upper cavity is given as

$$Z_{\text{in}2} = \frac{1}{\left\{\left(\frac{1}{R_2}\right) + (j\omega C_2) + \left(\frac{1}{j\omega L_2}\right)\right\}} \tag{4.21}$$

where resistance R_2, capacitance C_2, and inductance L_2 are equivalent circuit components for a circular microstrip antenna expressed as parallel combination for TM_{np} mode.

The resonance resistance (R_2) of the upper patch at feed location ρ is given by [23]

$$R_2 = \frac{1}{G_T} \frac{J_n^2(k\rho)}{J_n^2(ka_2)} \tag{4.22}$$

where J_n is the first kind of Bessel function of order n, k is the wave number at the operating frequency, with argument $k\rho$ or ka, and G_T is the total conductance of the upper cavity that includes dielectric loss, radiation loss, and conduction loss.

The capacitance associated with the upper cavity patch is given by

$$C_2 = \frac{Q_T}{2\pi f_{\text{res}2} R_2} \tag{4.23}$$

and the inductance L_2 of the upper layer is given as

$$L_2 = \frac{R_2}{2\pi f_{\text{res}2} Q_T} \tag{4.24}$$

where Q_T is the total quality factor [15] of the upper cavity, which includes radiation loss, dielectric loss, and conductance loss. The resonant frequency ($f_{\text{res}2}$) of the upper cavity of the circular microstrip antenna is calculated as [12]

$$f_{\text{res}2} = \frac{c\alpha_{\text{np}}}{2\pi a_{\text{eff}2}\sqrt{\varepsilon_{\text{reff}2}}} \tag{4.25}$$

where c is the velocity of light in free space, α_{np} is the pth zero of first kind Bessel function of order n, $a_{\text{eff}2}$ is the effective radius of the upper patch, and $\varepsilon_{r\text{eff}2}$ is the effective permittivity [12] of the upper substrate considering fringing effect of the upper patch.

4.3.2 Analysis of Lower Patch

The lower circular patch is analyzed similar to a circular microstrip antenna with a superstrate neglecting the effect of the upper patch. One or more dielectric layers above the radiating patch disturb the fringing fields, thus changing the effective radius of the lower patch. The resonant frequency of a rectangular microstrip antenna with a superstrate was calculated in [23] taking filling fraction into consideration. In this, an antenna system with one or more superstrates is represented as antenna with one substrate with the same radiation characteristics. Moreover, the formulation provided in [11] along with the analysis carried out in [5] is used to analyze the present problem more accurately.

The effective dielectric constant of the equivalent substrate is given as

$$\begin{aligned}\varepsilon_{\text{reff}} = {} & \varepsilon_{r1}p_1 + \varepsilon_{r1}(1-p_1)^2 \times \left[\varepsilon_{r2}^2 p_2 p_3 + \varepsilon_{r2}\{p_2 p_4\right. \\ & + (p_3+p_4)^2\left.\right]\left[\varepsilon_{r2}^2 p_2 p_3 p_4 + \varepsilon_{r1}(\varepsilon_{r2}p_3 + p_1)\right. \\ & \left.\times(1-p_1-p_4)^2 + \varepsilon_{r2}p_4\} p_2 p_4 + (p_3+p_4)^2\right]^{-1}\end{aligned} \tag{4.26}$$

where

$$p_1 = 1 - \frac{h_1}{2w_e}\ln\left(\frac{\pi w_e}{h_1} - 1\right) - p_4 \tag{4.27}$$

$$p_2 = 1 - p_1 - p_3 - 2p_4 \tag{4.28}$$

$$p_3 = \frac{h_1 - g}{2w_e} \ln\left[\frac{\pi w_e}{h_1} \frac{\cos\left(\frac{\pi g}{2h_1}\right)}{\pi\left(0.5 + \frac{h_2}{h_1}\right) + \frac{g\pi}{2h_1}} + \sin\left(\frac{g\pi}{2h_1}\right)\right] \tag{4.29}$$

$$p_4 = \frac{h_1}{2w_e} \ln\left(\frac{\pi}{2} - \frac{h_1}{2w_e}\right) \tag{4.30}$$

$$g = \frac{2h_1}{\pi} \arctan\left[\frac{\frac{\pi h_2}{h_1}}{\left(\frac{\pi}{2}\right)\left(\frac{w_e}{h_1}\right) - 2}\right] \tag{4.31}$$

$$w_e = \sqrt{\frac{\varepsilon_r'}{\varepsilon_{\text{reff}}}} \left[\begin{array}{l} \left\{w + 0.882h_1 + 0.164h_1 \frac{(\varepsilon_r' - 1)}{\varepsilon_r'^2}\right\} + \\ h1 \frac{(\varepsilon_r' - 1)}{\pi\varepsilon_r'} \left\{\ln\left(0.94 + \frac{w}{2h_1}\right) + 1.451\right\} \end{array}\right] \tag{4.32}$$

$$\varepsilon_r' = \frac{2\varepsilon_{\text{reff}} - 1 + \left(1 + \frac{10h_1}{w_e}\right)^{-0.5}}{1 + \left(1 + \frac{10h_1}{w_e}\right)^{-0.5}} \tag{4.33}$$

$$w = a(\pi - 2) \tag{4.34}$$

The parameters w_e and ε_r' are calculated by iteration method [23] with initial values $\varepsilon_r' = \varepsilon_{r1}$ and $\varepsilon_{\text{reff}} = \varepsilon_{r1}$. In eq. (4.34), w is the width of the RMSA equivalent to a CMSA with the same radiation characteristics [24]. These equations may be calculated by assuming equal fringing field for both the structures. When the relative dielectric constant of the superstrate is greater than that of the substrate, the surface waves are reduced to a certain extent by choosing the appropriate thickness. To accommodate this, a new dielectric constant is defined as

$$\varepsilon_{\text{re}} = \frac{\varepsilon_{r1}}{\varepsilon_{\text{reff}}} \tag{4.35}$$

Now the effective radius of LCP is calculated as

$$a_{\text{eff}1} = a_1\sqrt{(1 + q)} \tag{4.36}$$

In this, q is calculated as given in [12], and in eqs. (4.29)–(4.34), the result of eq. (4.35) is used. Using a_{eff1} calculated in eq. (4.36), the input impedance of LCP, Z_{in1}, is calculated. The antenna is assumed to be edge fed in the above calculation.

4.3.3 Stacked Circular Patch

There is no variation of electric field in the z-direction, so the total electric field is the sum of the electric fields in LCP and UCP. Moreover, LCP is represented as a parallel combination of a resistance (R_1), an inductance (L_1), and a capacitance (C_1). The equivalent circuit of a stacked microstrip antenna may be represented as a series combination of the input impedances of the two antennas, i.e., LCP and UCP, as shown in Figure 4.14.

Hence,

$$Z_{in} = Z_{in1} + Z_{in2} \tag{4.37}$$

4.3.4 BST Varactor Diode Loaded Stacked Microstrip Patch

Compared to MEMS varactors, thin-film BST capacitors exhibit a higher tuning ratio under relatively low bias voltages. High permittivity thin film dielectrics (such as barium strontium titanate (BST) and bismuth zinc niobate (BZN)) may exhibit strong field dependence in the dielectric constant that can be exploited for voltage variable capacitors in RF circuits. Varactors made from high permittivity materials show symmetrical small signal C-V characteristics. A typical BST is shown in Figure 4.15(a), while the equivalent circuit is shown in Figure 4.15(b). The device may be modeled as a series combination of bulk capacitance, $C(v)$, and interfacial capacitances, C_i, in parallel to fringing capacitance, C_f.

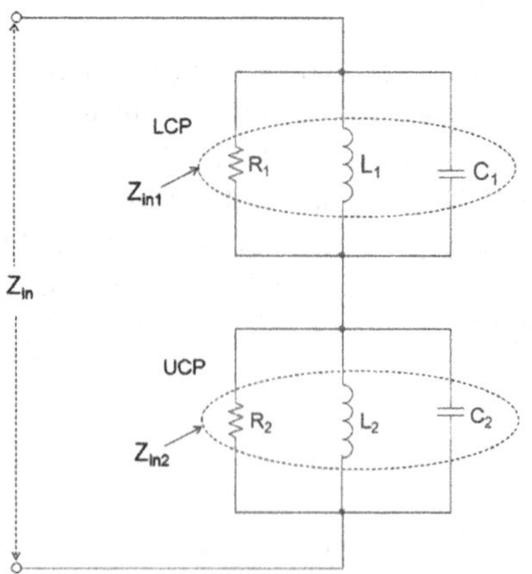

FIGURE 4.14 Equivalent circuit of a stacked CMSA.

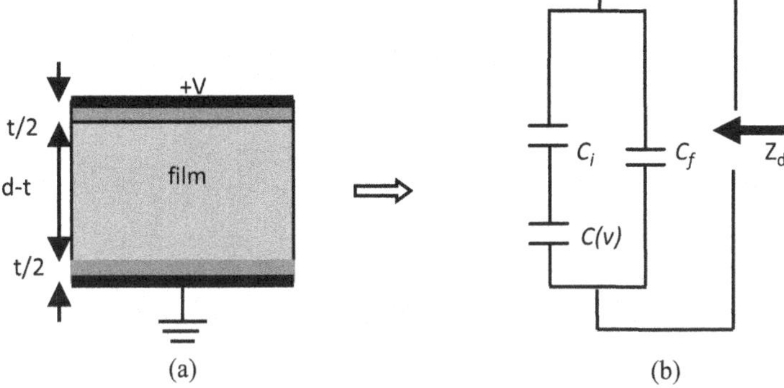

FIGURE 4.15 BST varactor device: (a) schematic; (b) equivalent circuit.

The bulk capacitance is given as [25–26]

$$C(v)=\frac{C_{\max}-C_f}{2\cosh\left[\frac{2}{3}\sinh^{-1}\left(\frac{2v}{V_2}\right)\right]-1}+C_f \tag{4.38}$$

which includes the fringing capacitance given as $C_f = k_1 P/d$. The zero bias capacitance is $C_{\max}$, and V_2 is the voltage at which the capacitance becomes half the $C_{\max}$. And k_1 is a constant.

The interfacial capacitance is given as

$$C_i=\frac{\varepsilon_i A}{t} \tag{4.39}$$

where A is the device area and t is the thickness of the "dead layer." The BST varactor is connected between the LCP and the ground plane and is positioned opposite to the coaxial feed. The diode is kept at the edge of the antenna.

The device impedance is calculated from Figure 4.15(b) as

$$Z_d=1/\left(j\omega C_{\text{eq}}\right) \tag{4.40}$$

where

$$C_{\text{eq}}=C_f+\frac{C_i C(v)}{C_i+C(v)} \tag{4.41}$$

Hence, the total input impedance of the BST varactor loaded stacked circular microstrip antenna as shown in Figure 4.16 is given as

$$Z_{\text{int}}=Z_{\text{in}2}+Z_{\text{in}1}Z_d/\left(Z_{\text{in}1}+Z_d\right) \tag{4.42}$$

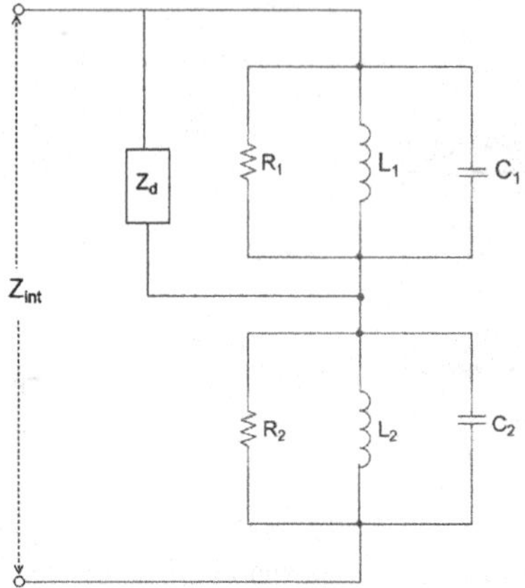

FIGURE 4.16 Equivalent circuit of a BST varactor loaded stacked CMSA.

and the return loss is given as

$$\mathrm{RL} = 20\log|\Gamma| \tag{4.43}$$

where the reflection coefficient (Γ) for a coaxial probe of 50 Ω characteristic impedance is given as

$$\Gamma = {(Z_{\mathrm{int}} - 50)}\big/{(Z_{\mathrm{int}} + 50)} \tag{4.44}$$

4.3.5 Radiation Pattern of Stacked CMSA

The radiation pattern of a single CMSA working in TM_{11} mode is given as [17, 25]

$$E_\theta = -\frac{jVak}{2}\frac{e^{-jkr}}{r}\cos(\phi)J_1'\left(ka\sin(\theta)\right) \tag{4.45}$$

$$E_\phi = \frac{jVak}{2}\frac{e^{-jkr}}{r}\sin(\phi)\frac{J_1\left(ka\sin(\theta)\right)}{ka\sin(\theta)} \tag{4.46}$$

If the induced slot voltage in LCP is k_c (coupling coefficient) times the slot voltage of the fed patch (UCP), the electric fields emanating from the lower patch may be written as

$$E_\theta' = k_c E_\theta \tag{4.47}$$

$$E_\phi' = k_c E_\phi \tag{4.48}$$

where $k_c = 1/\sqrt{Q_1Q_2}$, and Q_1 and Q_2 are the quality factors of resonant circuits associated with LCP and UCP, respectively.

Since the gap between the patches is very less compared to the minimum operating wavelength, the total electric field is taken as the sum of fields due to individual patches.

Hence,

$$E_{\theta s} = E_\theta + E'_\theta \tag{4.49}$$

$$E_{\phi s} = E_\phi + E'_\phi \tag{4.50}$$

Now the radiation pattern of the BST varactor loaded stacked circular microstrip antenna may be calculated as

$$R = E_{\theta s}^2 + E_{\phi s}^2 \tag{4.51}$$

4.3.6 Specifications of BST Varactor Diode Loaded Stacked CMSA

The stacked circular microstrip antenna is designed to operate at two resonant frequencies 2.2 and 3.5 GHz. The objective is to design an antenna that could be operated in WiMAX and one of the PCS, DCS, and UTMS bands. For these operating frequencies, the antenna design parameters are given in Table 4.5. The radius for the UCP is calculated assuming the LCP as ground plane. The device parameters corresponding to 30/70 BST with Pt electrode were taken to simulate the design. This is a large area (2,000 μm^2) device. The values of V_2, film thickness (d), and k_1 were taken to be 13 V, 210 nm, and 0.6 fF, respectively. Though for large area devices fringing capacitance and non-tunable "dead layer" capacitance could be ignored, but for more accurate results, these values were accommodated in the calculation. The fringing capacitance of 5 fF and thickness of "dead layer" of 30 nm were taken.

TABLE 4.5
Specifications of Stacked CMSA

Parameters	Value
Radius of LCP (a_1)	26 mm
Radius of UCP (a_2)	16.25 mm
Height of LCP (h_1)	1.6 mm
Height of UCP (h_2)	1.6 mm
Relative dielectric constant of substrate and superstrate ($\varepsilon_{r1} = \varepsilon_{r2}$)	2.2 (PTTE)
Probe position for UCP	5.5 mm
Loss tangent (tan δ)	0.0012

4.3.7 Properties of BST Varactor Diode Loaded Stacked CMSA

As mentioned above, a primary stacked dual band antenna was designed to operate at 3.5 and 2.2 GHz. The lower frequency was kept slightly larger than the desired wireless frequency band of operation. The antenna can operate for WiMAX and one of the above two bands simultaneously. The calculations were carried out using descriptions given in Sections 4.2 and 4.3, and the simulation was carried out using ADS simulation software [27]. Figure 4.17 shows the simulated and calculated return loss variations with frequency for a simple stacked CMSA. The calculated result from Figure 4.17 (a) shows two resonances at 2.2 and 3.5 GHz, which are in close agreement with the simulated results shown in Figure 4.17 (b). The simulated values of resonances found are 2.16 and 3.47 GHz.

A BST varactor of $Ba_{0.5}Sr_{0.5}TiO_3$ [25] was used to integrate between the LCP and the ground plane. A low pass filter was used to prevent the RF signal from reaching the DC supply used for biasing of the diode. The behavior of the diode was obtained by varying the bias voltage from −40 to +40 V. Figure 4.18 shows

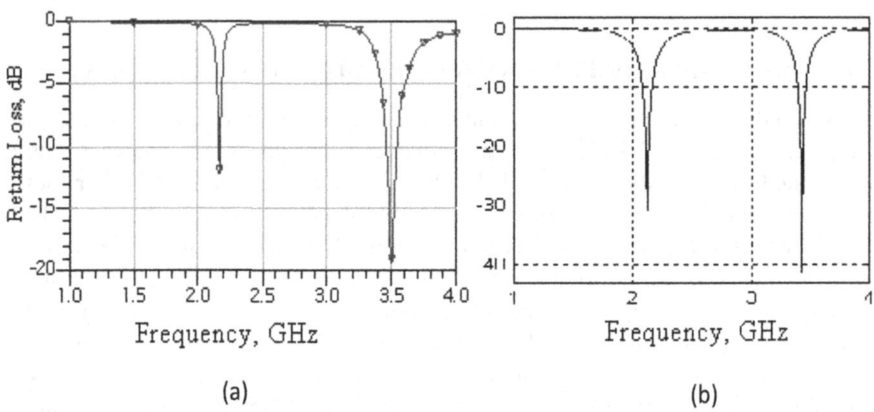

FIGURE 4.17 Return loss of a stacked CMSA: (a) calculated; (b) simulated.

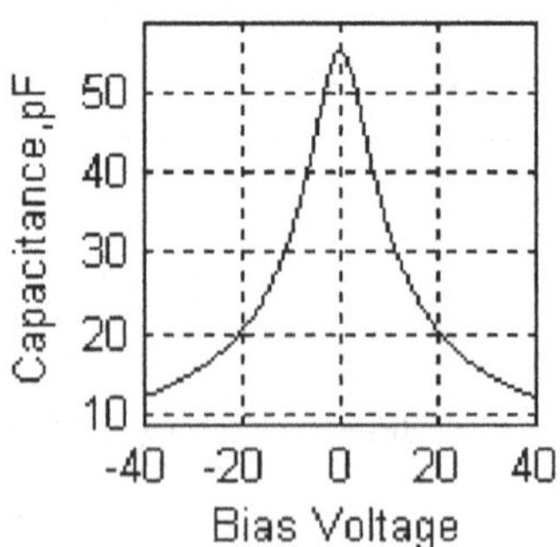

FIGURE 4.18 Variation of the capacitance of a BST varactor device with bias voltage.

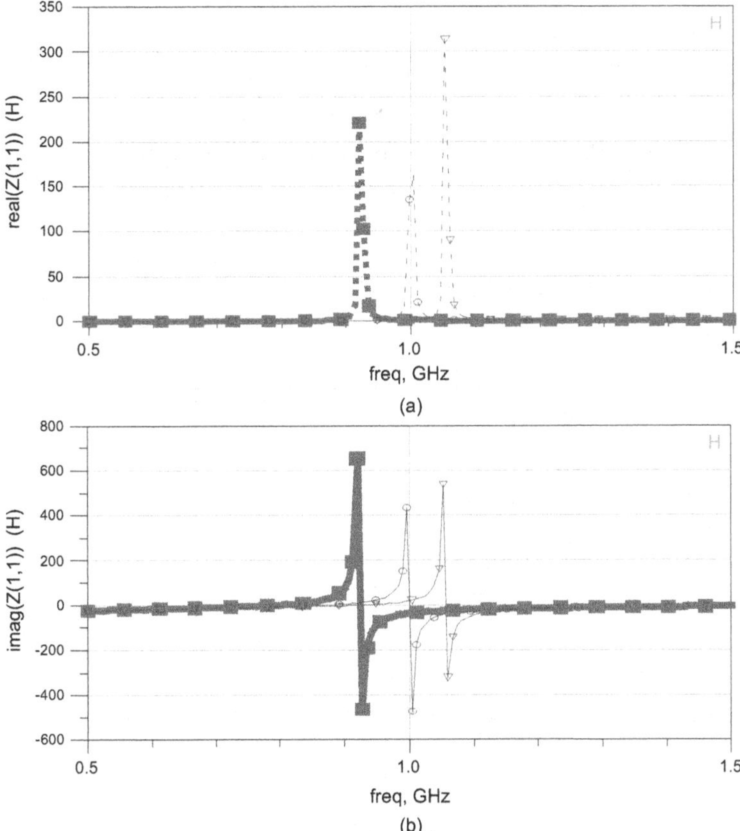

FIGURE 4.19 Variation of the real and imaginary impedances with frequency of a BST varactor loaded stacked CMSA: (a) calculated; (b) simulated.

the variation of bulk capacitance of a BST varactor. Similar monotonic variation is seen for both polarities of bias voltage. A monotonic variation is observed around zero bias.

Figure 4.19 shows the calculated and simulated return loss variations of a BST loaded stacked circular microstrip antenna. The three bands in the lower spectrum correspond to 0, 8.7, and 13 V. A close resemblance in the simulated and calculated results may be observed. The calculated and simulated resonance frequencies corresponding to these bias voltages are given in Table 4.6. The slight mismatch in the result is due to the fact that the LCP is not providing proper ground for the UCP. It is also seen that the higher resonant frequency remains unchanged, while the lower bands are electrically selectable by varying the bias voltage of the diode. The same is obvious from the equivalent circuit shown in Figure 4.16. The variation of the resonant frequency with bias voltage is shown in Figure 4.20. A minimum frequency of 0.866 GHz was achieved, giving 60.64% frequency agility.

TABLE 4.6
Comparison of Resonant Frequencies

	Resonance Frequency	
Bias Voltage (V)	**Calculated (GHz)**	**Simulated (GHz)**
0	0.866	0.92
8.7	1.038	1.00
13	1.13	1.05

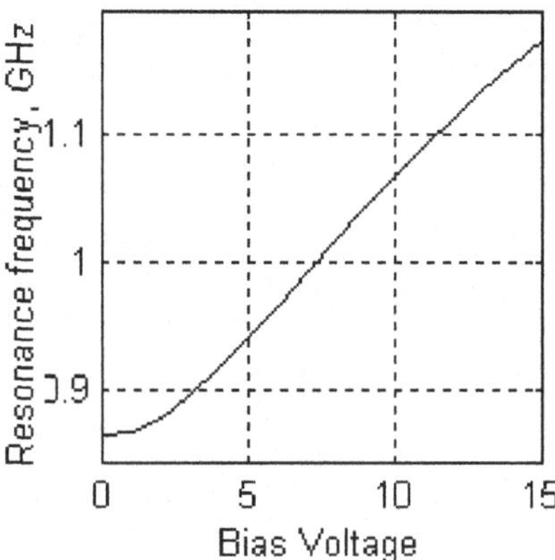

FIGURE 4.20 Variation of the resonant frequency with bias voltage of a BST varactor loaded stacked CMSA.

4.4 CONCLUSION

In this chapter, active device loaded tunable band / band hopping / frequency agile printed antennas are discussed. In the structure of antennas, active devices are loaded on the patch to achieve tunable band / band hopping / agility of antenna. Looking into the bandwidth constraints of the microstrip antennas, the discussed antennas provided a wide frequency range tunability.

Firstly, a MOS loaded CMSA with an airgap between the ground plane and the substrate is discussed. It was found that the presence of an air gap between the ground plane and substrate enhanced the operating frequency range of the MOS loaded CMSA. The minimum operating frequency achieved for a MOS loaded CMSA with 10 mm radius of patch and 1 mm airgap is 1.571 GHz, and it can be tuned to any frequency up to the design frequency of 6.60 GHz. A percentage agility of 76.20% was achieved for a MOS loaded CMSA with an airgap. The minimum operating frequency achieved for a double MOS loaded CMSA with a 10 mm radius of patch and a 1 mm airgap was 1.27 GHz, and it can be tuned to any frequency up to the design frequency of 6.80 GHz. A percentage agility of 81.30% was achieved for a double

MOS loaded CMSA with an airgap. Further, it was seen that the antenna operated at a higher frequency and agility for lower values of the radius of patch, and vice – versa, and the antenna operated up to 18 GHz frequency and 90% agility for a 3 mm radius of patch.

Secondly, a BST varactor loaded stacked CMSA is discussed using an extended cavity model. One of the two bands of the antenna is tunable by loading a BST varactor; hence, the tuning of a band was achieved using a BST varactor. A significant reduction in the physical area of the antenna and a frequency agility of 60.64% were achieved.

To investigate the antennas, different parameters such as resonance frequency, input impedance, frequency agility, VSWR, and radiation pattern were calculated. Also, various parameters such as resonant frequency and beam width can be electronically tuned. The antennas discussed are useful for modern communication systems such as WLAN, GPS, UMTS, WiMAX, and remote sensing.

REFERENCES

1. A. Kaya, & E. Yesim Yuksel (2007). Investigation of a compensated rectangular microstrip antenna with negative capacitor and negative inductor for bandwidth enhancement. *IEEE Trans on Antennas and Propagation*, 55, 1275–1282.
2. T. Chakravorty, & A. De (2003). A novel method of extending tunability of circular patch using two shorting pins. *IEEE Antennas and Propagation Society International Symposium*, 2, 736–739.
3. B. K. Kanaujia, A. K. Singh, & B. R. Vishvakarma (2008). Frequency agile annular ring microstrip antenna loaded with MOS capacitor. *Journal of Electromagnetic Wave and Application*, 22, 1361–1370.
4. A. K. Gautam, & B. R. Vishvakarma (2006). Frequency agile microstrip antenna symmetrically loaded with tunnel diodes. *Microwave and Optical Technology Letters*, 48(9), 1807–1810.
5. D. Guha, & J. Y. Siddiqui (2003). Resonant frequency of circular microstrip antenna covered with dielectric superstrate. *IEEE Transactions on Antennas and Propagation*, 51, 1649–1652.
6. H. T. Chen, H. D. Chen, & Y.-T. Cheng (1997). Full wave analysis of the anular ring loaded spherical circular microstrip antenna. *IEEE Transactions on Antennas and Propagation*, 45, 1581–1583.
7. A. K. Gautam, & B. R. Vishvakarma (2007). Analysis of varactor loaded active microstrip antenna. *Microwave and Optical Technology Letters*, 49, 416–421.
8. V. Gupta, S. Sinha, S. K. Koul, & B. Bhat (2003). Wideband dielectric resonator loaded suspended microstrip patch antenna. *Microwave and Optical Technology Letters*, 37, 300–302.
9. B. K. Kanaujia, & B. R. Vishvakarma (2004). Analysis of gunn integrated annular ring microstrip antenna. *IEEE Transaction on Antenna and Propagation*, 52, 88–97.
10. G. P. Pandey, B. K. Kanaujia, A. K. Gautam, & S. K. Gupta (2013). Ultra – wideband l-strip proximity coupled slot loaded circular microstrip antenna for modern communication systems. *Wireless Personal Communications (Springer)*, 70, 139–151.
11. K. F. Lee, K. Y. Ho, & J. S. Dahele (1984). Circular disk microstrip antenna with an airgap. *IEEE Transaction Antenna Propagation*, 32, 880–884.
12. D. Guha (2001). Resonant frequency of circular microstrip antennas with and without air gaps. *IEEE Transaction on Antenna and Propagation*, 49, 55–59.
13. S. K. Sharma, & B. R. Vishvakarma (1999). MOS capacitor loaded frequency agile microstrip antenna. *International Journal of Electronics*, 86, 979–990.

14. I. J. Bahl, & P. Bhartia. *Microstrip Antennas*. Dedham, MA: Artech House, 1980.
15. F. Abboud, J. P. Damiano, & A. Papiernik (1990). A new model for calculating the input impedance of coax-fed circular microstrip antennas with and without air gaps. *IEEE Transaction on Antenna and Propagation*, 38, 1882–1885.
16. G. P. Pandey, B. K. Kanaujia, & S. K. Gupta (2009). Double MOS loaded circular microstrip antenna for frequency agile. *IEEE Applied Electromagnetic Conference*, Kolkata; 978-1-4244-4819-7/09.
17. R. Garg, P. Bhartia, I. Bahl, & A. Ittipiboon. *Microstrip Antenna Design Hand Book*. Boston, London: Artech House, 2001.
18. S. K. Gupta, B. K. Kanaujia, G. P. Pandey, & A. K. Gautam (2012). MOS loaded circular microstrip antenna with airgap for modern communication. *UACEE International Journal of Advances in Electronics Engineering*, 2(3), 86–91.
19. S. K. Gupta, B. K. Kanaujia, & G. P. Pandey (September 2012). Double MOS loaded circular microstrip patch antenna with airgap for mobile communication. *Wireless Personal Communications (Springer)*, 71, 987–1002.
20. Zeland software co. IE3D v14.0, California, USA.
21. G. P. Pandey, B. K. Kanaujia, S. K. Gupta and A. K. Gautam (2013). BST varactor loaded frequency agile stacked circular microstrip radiator. *Wireless Personal Communications (Springer)*, 72, 1157–1172.
22. J. Gomez-Tagle, & C. G. Christodoulou (1997). Extended cavity model analysis of stacked microstrip ring antennas. *IEEE Transactions on Antennas and Propagation*, 45(11), 1626–1635.
23. I. J. Bahl, & P. Bhartia. *Microstrip Antennas*. Dedham, MA: Artech House, 1980.
24. J. T. Bernhard, & C. J. Tousgnant (1999). Resonant frequencies of rectangular microstrip antennas with flush and spaced dielectric superstrates. *IEEE Transaction on Antenna and Propagation*, 47(2), 302–308.
25. J. R. James, & P. S. Hall. *Handbook of Microstrip Antennas*. London, UK: Peter Peregrinus, 1989.
26. D. R. Chase, L. Y. Chen, & R. A. York (October 2005). Modeling the capacitive nonlinearity in thin film BST varactors. *IEEE Transaction on Microwave theory and Techniques*, 53(10), 3215–3220.
27. ADS Simulation software v2011.

5 Pattern and Polarization Diversity in Antennas

Dr. Ashwani Kumar
University of Delhi and Jawaharlal Nehru University, Delhi

Prashant Chaudhary
University of Delhi South Campus, Delhi

CONTENTS

5.1 INTRODUCTION

Diversity means a range of different things. In a harsh environment, a wireless communication system faces many problems. Several diversity schemes are available to improve the reliability and link quality of a wireless network for LTE/smartphones and the Internet of things (IoT). Diversity system means to increase the throughput without increasing the bandwidth. Typically, in antenna diversity, two or more antennas are used. In the outdoor environment, a signal takes multipath to reach the destination by bouncing off from different objects and suffers from fading, phase delay, attenuation, etc. Multipath fading and phase delay are the primary factors that affect communication in the indoor environment. To resolve these problems, several diversity schemes are used, among which three are very common. They are spatial diversity, pattern diversity, and polarization diversity. In spatial diversity, antennas occupy different locations in space; in polarization diversity, antennas are differ in polarization; and in pattern diversity, main beams are directed over a large angular area. Antenna diversity systems provide a better link quality and higher data rates as compared to a single antenna. In the real environment, most of the communication is through non-line of sight, in the diversity system, if one antenna is in deep fade area. In contrast, another antenna receives a strong signal to improve the channel capacity fading has to reduce. Antenna diversity also improves the impedance matching. The polarization diversity scheme reduces the polarization mismatch loss and enhances the mismatching and signal-to-noise ratio (SNR) up to 12 dB [1], which will improve the overall communication efficiency. So, these various diversities schemes can be used for the enhancement of the overall quality of the communication.

In the age of fifth generation (5G) and Internet of things (IoT), there is an urgent need for a high data rate to improve the signal-to noise-ratio (SNR) for the current wireless communication systems and there is also the need for pattern diversity (fingerprint) antennas to enhance the channel capacity further. The multipath fading in the indoor and urban environment degrades the channel capacity. Both the signal-to-noise ratio and channel capacity is affected by the multipath fading. To overcome the above-mentioned problem, different techniques are employed using different diversity schemes. These are the switching diversity and selection diversity are used to improve the SNR and channel capacity, and many antennas are used to accomplish these diversity schemes. In recent times, some of the new effective methods have been suggested by using many antennas at the transmitter and receiver sides, and this scheme is known as the MIMO antenna schemes. In MIMO antenna configurations, different antennas have different radiation patterns (fingerprint). They can be used to improve the channel capacity and to detect the fingerprint. The basic structure of a MIMO system is shown in Figure 5.1a. Each antenna in the MIMO system at the transmitting and receiving sides has a different radiation pattern (distinct fingerprint). To decorrelate the signals received from different antennas in the MIMO scheme, the antennas are placed separately with a minimum distance and this is known as spatial diversity. A diversity scheme that employs both pattern and polarization diversities schemes in the MIMO antenna, are preferred instead of spatial diversity because in spatial diversity the spacing between the antennas must be half of wavelength which increases the size of the system. By employing the pattern and polarization diversity scheme, the size of the MIMO system can be reduced and simultaneously, the channel capacity can be improved with a distinct fingerprint. MIMO schemes provide an exciting performance in 5G and IoT-enabled devices to get a very high quality of transmission and high throughput. In MIMO schemes, each antenna has different fingerprints (signature), which is used to merge the information-carrying bits coming from each of the numerous transmitting antennas to achieve accurate identification and authorization. Many new design methodologies are suggested in the literature to take advantage of the different diversity schemes for fast, accurate, and secure data transmission.

The assets provided by the MIMO system in the field of communication can be extended to various levels. Today, the whole world is talking about the new complex communication system between smart devices called Internet of things (IoT). IoT is a system of interconnected devices that have the ability to communicate and transfer data. IoT devices need to be connected with 5G cellular networks, which require a

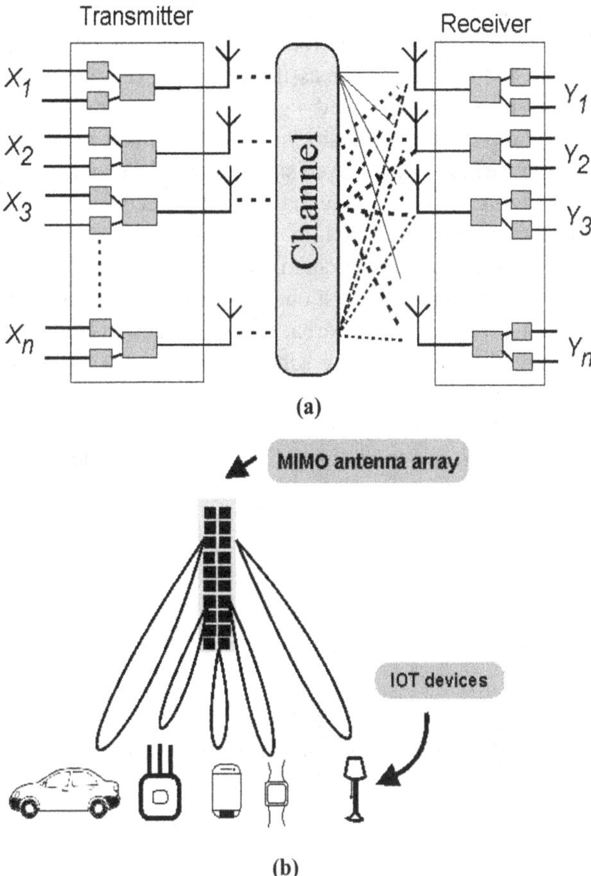

FIGURE 5.1 (a) Basic structure of a MIMO antenna; (b) a MIMO antenna with IOT devices.

high data transfer rate with a high efficiency. A MIMO scheme provides all these essential features. Generally, a MIMO system consists of two or more antennas at the receiver side and the similar number of antennas at the transmitter side. However, in IoT, a large number of devices are connected, which requires a large number of antennas at the base station or receiver, so this kind system is called Massive MIMO. Massive MIMO antennas for IoT is there search area of the current study. In Massive MIMO systems, a vast number of antenna arrays are used for both indoor and outdoor communications. A MIMO antenna array with some IoT devices is shown in Figure 5.1b. IoT-based devices are also especially useful in human health care. These devices can be implanted within the patients to monitor their health. In fact, nowadays people are using small devices to track their daily activity, heart rate, burnt calorie, sleeping time, etc., with the help of their smartphones. These are excellent example of IoT. A new smart devices such as light bulbs, air-condition, television, and other smart household devices with their virtual assistant with artificial intelligence-enabled devices are based on IoT devices.

Consider the example of the drone, as shown in Figure 5.2. It has four dipole antennas; each antenna is placed at its four edges. The motion of the drone is random, and multipath fading will occur as the drone moves. By using the diversity concept, the quality of a weak received signal can be significantly improved. This type of diversity is called spatial diversity, since the antennas are separated in space. Figure 5.3 shows the variation of received signal power level concerning the position of receiver. In this figure, only one antenna is used to receive the signal, and at that point, the received signal power is meagre, which means no reception of the signal. Now consider the spatial diversity scheme in which the antennas are spatially separated; out of the three antennas, the first one receives a higher signal power than the second one that is in the low-intensity area, as can be seen in Figure 5.4. It is clearly seen from the figure that the diversity scheme will increase the link quality and reduce the multipath fading.

The system that uses diversity schemes results in an improvement in performance compared to those that do not make use of the diversity principle. In order to use

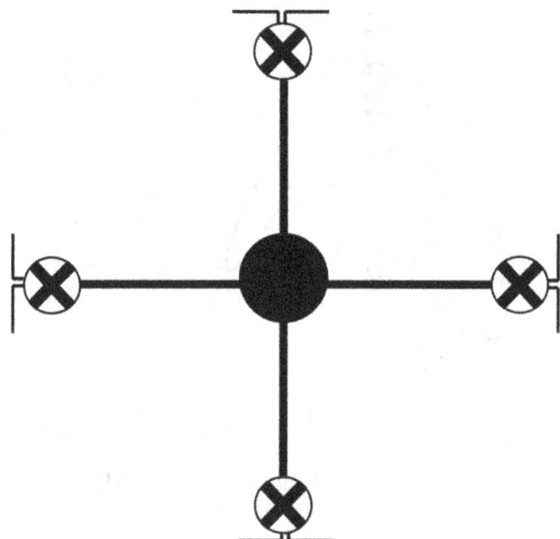

FIGURE 5.2 Four dipole antennas placed at the four edges of the drone blades.

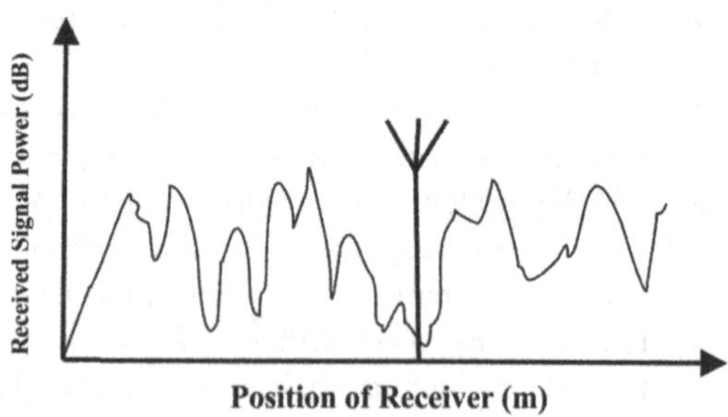

FIGURE 5.3 Power intensity of the received signal with different positions of the receiving antenna.

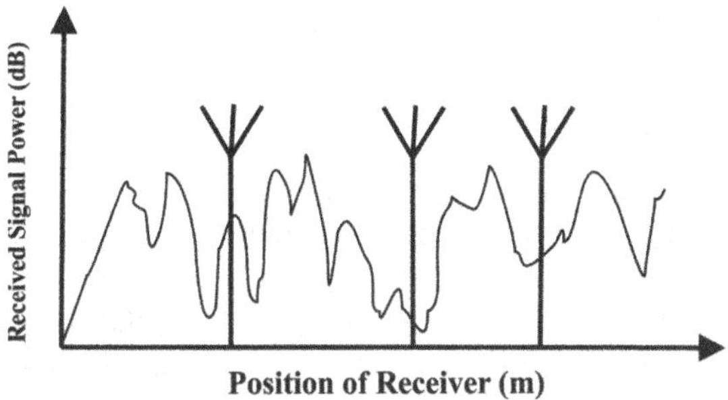

FIGURE 5.4 Power intensity of the received signal with different positions of the multiple receiving antennas.

multiple antennas simultaneously, antennas should have sound isolation between them and their radiation pattern should be uncorrelated. Cross-envelope correlation coefficient (ECC) is a parameter that tells about the effectiveness of the diversity performance of an antenna. The value of ECC(ρ_e) lies between 0 and 1. Ideally, when its value is 0, it means the system is entirely isolated, whereas when the value is 1, it means the system is highly coupled. Typically, for an effective system, the ECC value should be less than 0.5. Generally, the ECC decreases with the increase in spacing, different polarizations, various beam directions, etc. The ECC also depends upon the type of diversity used. For spatial diversity, the ECC is a function of antenna spacing, larger the antenna spacing lower the ECC; generally, the spacing is larger than half wavelength (λ). For polarization diversity, the ECC depends upon the co-polarization and cross-polarization field components, while for pattern diversity, it is a function of the direction of the main beam $f(\theta, \varphi)$. In [2], the expression for the relation between envelope correlation coefficient (ECC) and antenna separation (d) was derived and is given in eq. (5.1).

$$\rho_e \cong J_o^2\left(\frac{2\pi d}{\lambda}\right) \tag{5.1}$$

Where J_o is the Bessel function of the first kind of order zero and λ is the wavelength. This expression is valid only for a uniform distribution of angle and the same polarization with omnidirectional radiation pattern. Although all components coming through the multipath are assumed to lie in the horizontal plane, the ECC(ρ_e) between the radiation patterns in the far-field can be computed numerically by $\vec{G}(\theta, \varphi)$ of the radiation patterns of the two antennas [3].

$$\rho_e = \frac{\left|\int\int \vec{G_1}(\theta, \varphi) \bullet \vec{G_2}(\theta, \varphi) d\Omega\right|^2}{\int\int\left|\vec{G_1}(\theta, \varphi)\right|^2 \int\int\left|\vec{G_2}(\theta, \varphi)\right|^2 d\Omega} \tag{5.2}$$

$\vec{G_1}(\theta, \varphi)$ and $\vec{G_2}(\theta, \varphi)$ represent the Hermitian product of radiation patterns of antenna-1 and antenna-2 in eq. (5.2). The ECC (ρ_e) can also be obtained for two element MIMO system by using the S-parameters as given in eq. (5.3). This relation is valid only for uniform field distribution with lossless condition.

$$\rho_e = \frac{\left|S_{11}^{*} S_{12} + S_{21}^{*} S_{22}\right|^2}{\left(1-\left(\left|S_{11}\right|^2+\left|S_{21}\right|^2\right)\right)\left(1-\left(\left|S_{22}\right|^2+\left|S_{12}\right|^2\right)\right)} \tag{5.3}$$

Diversity gain (DG) is a parameter that measures the level of the signal that exceeds for some fraction of time, say 90%. According to this criterion, we can examine the system output for the same amount of the available time. For the same amount, the increase in the level of signal for the possible time is called the diversity gain (DG). It measures how reliable the system is. Diversity gain is a function of two parameters – ECC and relative amplitudes. If the ECC is very low, and the amplitude levels of signals are the same, the diversity gain will be significant. When the magnitude of one channel is much smaller than the other one, then the DG falls to zero.

Multiple antenna systems are an exciting solution to increase the throughput without increasing the bandwidth [4]. Here, in the next sections, we will discuss pattern and polarization diversity in detail.

5.1.1 Pattern Diversity

In a pattern diversity system, signals are received through directional antennas, which are separated by an angular space. Pattern diversity is also called angle diversity because radiation patterns are directed within a certain angle in a particular direction in an area. In a real environment, the signal takes multiple paths to reach the destination and the multipath arrival spreads over a narrow angle. This propagation also introduces signal distortion and reduces the SNR, resulting in poor demodulation of signal and a high bit error rate in the digital communication system. In a pattern diversity communication system, all the antennas are placed in the same place pointing to different directions in space or have a phase shift between the co-located antennas, so that their radiation patterns are distinctly directed in space, as illustrated in Figure 5.5. This figure shows three antennas placed distinctly at the same location, which are radiating in three different directions. Spacing between these antennas could be less than 0.25 λ. Since their radiation patterns are directional, there is no coupling between the antennas, which further improves the envelope correlation coefficient (ECC). The envelope correlation between the received signals is different if the directions of the patterns are different because the relative amplitudes of the received signals are different for each antenna even though these antennas are co-located. The radiation pattern can be distorted because of the mutual coupling between the antennas. The effectiveness of the pattern diversity in handsets should be high because, in a mobile system, signals can come from any direction over a narrow-angle spread, and this could be improving the performance of the mobile handsets. This method reduces the antenna spacing, still maintaining a high diversity gain.

Various methods available in the open literature to achieve pattern diversity, such as pattern combination, beam switching between smart antennas, and directionally oriented antennas. In [5,6], beam switching was done by switching the state of PIN diodes between ON and OFF state. The direction of the main beam in the far-field

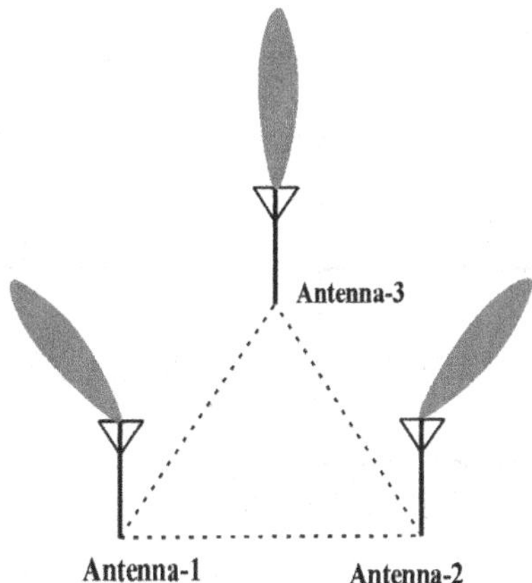

FIGURE 5.5 Antenna configuration showing pattern diversity.

can be varied with ON and OFF states. Switching load impedance produces different far-field patterns [7,8]. Suppose a parasitic antenna X_1 is connected to a two-way switch and impedance Z_1 is connected to provide field pattern $f_1(\theta_1,\varphi_1)$. In the second state, impedance Z_2 is connected to X_1 to produce another field pattern $f_2(\theta_2,\varphi_2)$, by optimizing the load impedance. The pattern direction can be optimized as shown in Figure 5.6. In [9], a Vivaldi slot antenna of travelling wave type was excited by two microstrip feeds passing through the slots. When the feed was in phase with the signal, then it produces two radiation patterns that could be beam-steered in two lateral sides. When the feed was excited 180° out of phase, then it produced a beam along the antenna direction. Pattern diversity antenna configuration is also an effective solution for RF harvesting, because the real environment is full of multipath signals [10].

There are mainly two methods that can improve the efficiency of a RF harvesting circuit. First design should have more sensitive and highly efficient RF rectifying circuit, while the second is diversity schemes to collect more RF power using multiple antennas, in RF harvesting antenna is referred to as rectenna. Generally, a rectenna has a wide band to collect power from most instrumental bands such as GSM, GPS, 4G, and WiFi/WiMax. At a specific point in space, the signals coming from different directions may be of different frequencies; thus, for RF harvesting, pattern diversity is one of the best choices. In [11], a four-element PIFA was used at 2.45 GHz in two configurations. In configuration-1, feeds had a mirror image of each other, which provided omnidirectional radiation patterns with pattern diversity. In contrast, in configuration-2, antennas were arranged in a cross-polarized manner, giving additional polarization diversity. The second configuration gives a lower ECC and higher diversity gain than the first configuration, which make the antennas more isolated. Several antennas reported in literature used common aperture with numerous feeds to generate an orthogonal pattern of radiation [12,13]. In [14], the pattern diversity was measured in an indoor scenario. The result showed that the proposed diversity system created multiple channels with minimum correlation that will improve the

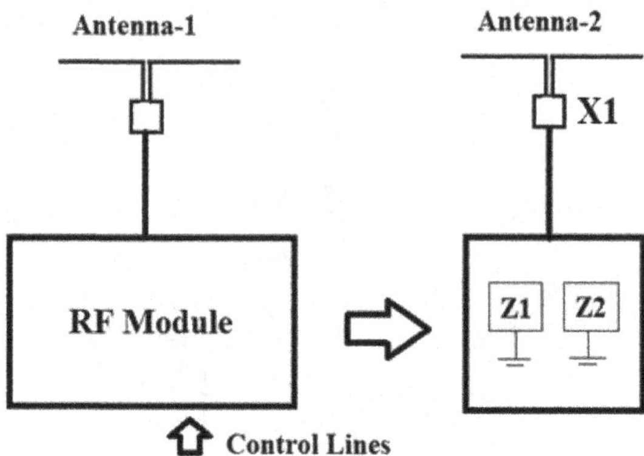

FIGURE 5.6 Antennas are located at the same place with different load impedances Z_1 and Z_2.

channel reliability in a multipath environment. The proposed analysis also effectively determined the propagation model near the ground, along with the complex gain of the transmitting and receiving antennas.

5.1.2 Effect of Pattern Diversity on Diversity Gain (DG) and Cross-Envelope Correlation Coefficient (ECC)

The diversity gain is high if the envelope cross-correlation between the antennas is low. In pattern or angular diversity, the correlation between the antennas is a function of the angle of radiation pattern $r(\theta, \varphi)$. The diversity gain can be clarified by the fact that different collocated radiation patterns receive different sets of multipath. This gain will be high if the patterns of the antennas are entirely independent. Theoretically, for a high diversity gain, the radiation patterns of the antennas should have completely zero overlapping. Envelope correlation coefficient (ECC) has been used to determine whether the radiation patterns overlap entirely or not. Various forms of pattern diversity antennas have been used in the open literature. Here, we present some of the examples of the pattern diversity antennas to make the subject clear. These antennas provide a high diversity gain and envelope correlation coefficient (ECC); they also have a high isolation. These antennas are located on the same plane.

Simple dipole antennas with PEC and EBG reflectors have been used to illustrate the phenomenon of pattern diversity. We also present different types of antennas to demonstrate the pattern diversity further; these include planar inverted-F antenna (PIFA), microstrip antenna, planar monopole, and dielectric resonator antenna (DRA). Figure 5.7 shows a single-dipole antenna resonating at 3.98 GHz; its radiation pattern is omnidirectional in the X–Y plane. The pattern diversity can easily be obtained by placing two dipoles parallel to each other and inserting a perfect electric conductor (PEC) plate of size 200×200 mm between them.

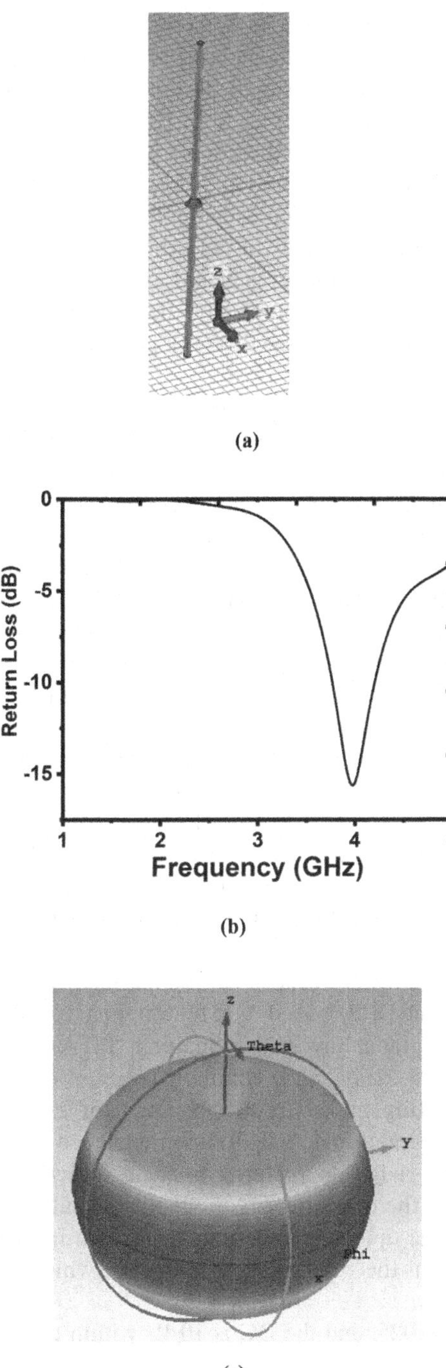

FIGURE 5.7 (a) Dipole antenna, (b) S11 response (frequency from 1 to 5 GHz), (c) radiation pattern at $f = 3.98$ GHz.

Both the dipoles are placed above $\lambda/4$ at $f = 3.98$ GHz from the PEC plate. The PEC plate behaves as a perfect reflector and shifts the radiation pattern in a particular direction. Their radiation patterns become directional as can be seen in Figure 5.8. A PEC plate having a reflection coefficient of $\Gamma = -1$ reflects the radiation pattern normal to the surface in the opposite direction with 180° phase shift. The radiations of the two dipoles are reflected back and in different directions.

The PEC plate also provides a high isolation between the two dipoles as there is no interference between the far-fields of both the dipoles. Dipole antennas with a PEC show pattern diversity; the radiation beams are in the opposite directions. Since the radiation patterns are isolated completely, there is no mutual coupling between the antennas. The ECC is zero, and the diversity gain is 10 dB at 3.98 GHz, which is very high.

Next, we shall see the effect of an electromagnetic band gap (EBG) surface between the two dipoles. These dipole antennas are linearly polarized. Figure 5.9 presents a two dipole antenna with an EBG, and the EBG surface is placed between the two diagonally orthogonal dipole antennas. The EBG acts as a reflector having a reflection coefficient of $\Gamma = +1$, and it behaves as a perfect magnetic conductor (PMC) [15]. The EBG surface reflects the radiation pattern normal to the surface; the omnidirectional radiation pattern becomes directional. The dipole antennas with the EBG shows the pattern diversity phenomena. Here, both the dipoles are placed obliquely (at ±45°) parallel to the EBG surface and it also gives circular polarization [15], which we shall discuss in detail in Section 5.2 in the name of polarization diversity. The perfect magnetic conductor changes the linearly polarized antenna to a circularly polarized one. The ECC and DG are given in Figure 5.9f and g. The EBG surface not only provides isolation, but also changes the polarization of the antennas. By using the EBG surface, a compact antenna with a high isolation and pattern diversity can be designed, and it is useful for the modern communication systems. These simple examples are presented here to become familiar with the term of pattern diversity and in what sense we can achieve it.

To further show the effectiveness of the pattern diversity, the hand-held mobile antenna and the DRA based MIMO antenna is presented. For a small hand-held device, the planar inverted-F antenna (PIFA) is generally preferred. Figure 5.10a shows two PIFA son a common large ground plane of dimension 65 mm × 55 mm operating in the frequency range of 2.27–2.70 GHz. To make the radiation pattern directional, the PIFAs are placed opposite to each other in space; hence, their radiation patterns become directional. Since these two antennas are placed in the opposite directions, the radiation patterns of both the antennas are also in the opposite directions, as shown in Figure 5.10c. There is no overlapping between the two radiation patterns, which also provides a high isolation between them.

The ECC is below 0.025, and the DG is 10 dB within the operating bandwidth, as shown in Figure 5.10d and e. The pattern diversity in hand-held mobile handsets is used to increase the signal coverage and channel capacity. Sometimes, a mobile with a single antenna has a chance of signal fading, which will break the

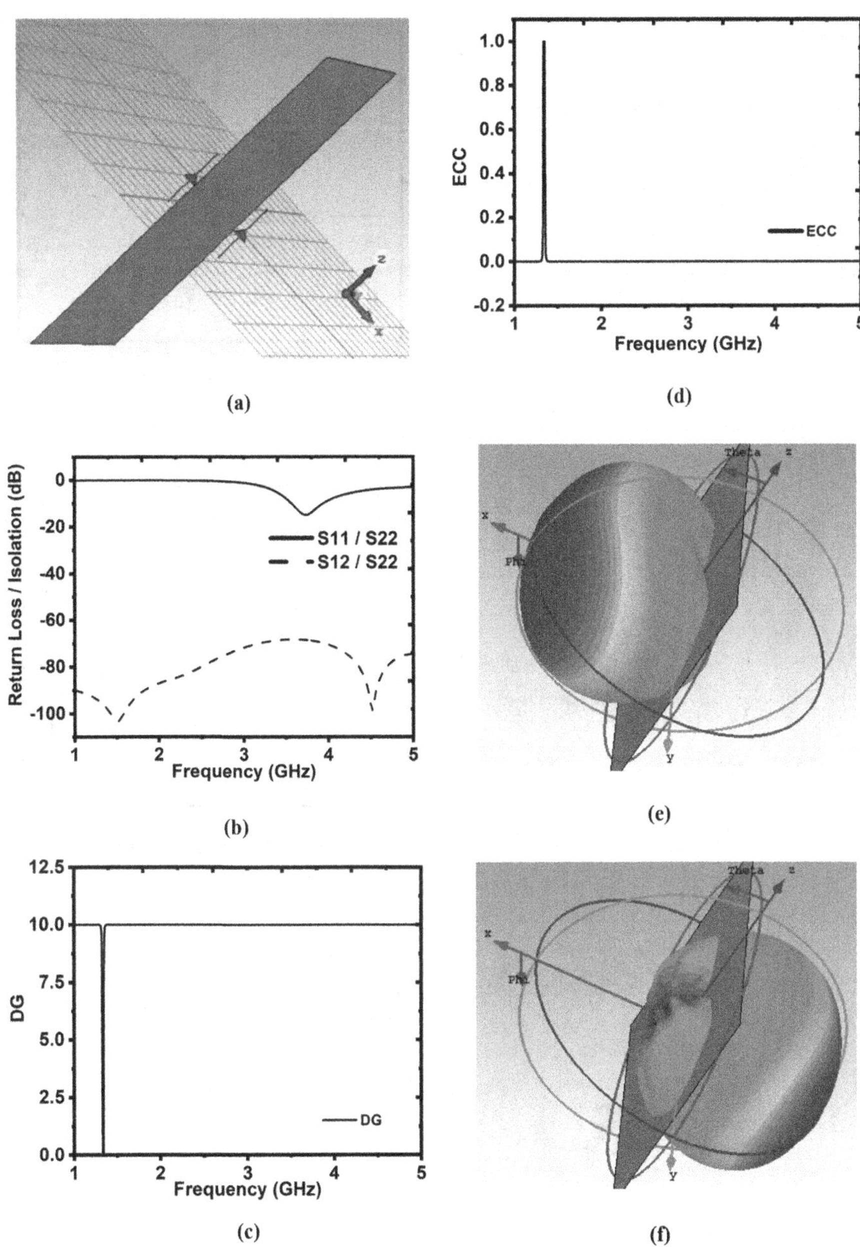

FIGURE 5.8 Dipole antennas with a PEC plate at $\lambda/4$ at $f = 3.98$ GHz: (a) two parallel dipoles with a PEC plate, (b)S_{11}, S_{22}, and S_{21} response, (c) DG, (d) ECC, (e)radiation pattern of dipole-1, (f)radiation pattern of dipole-1.

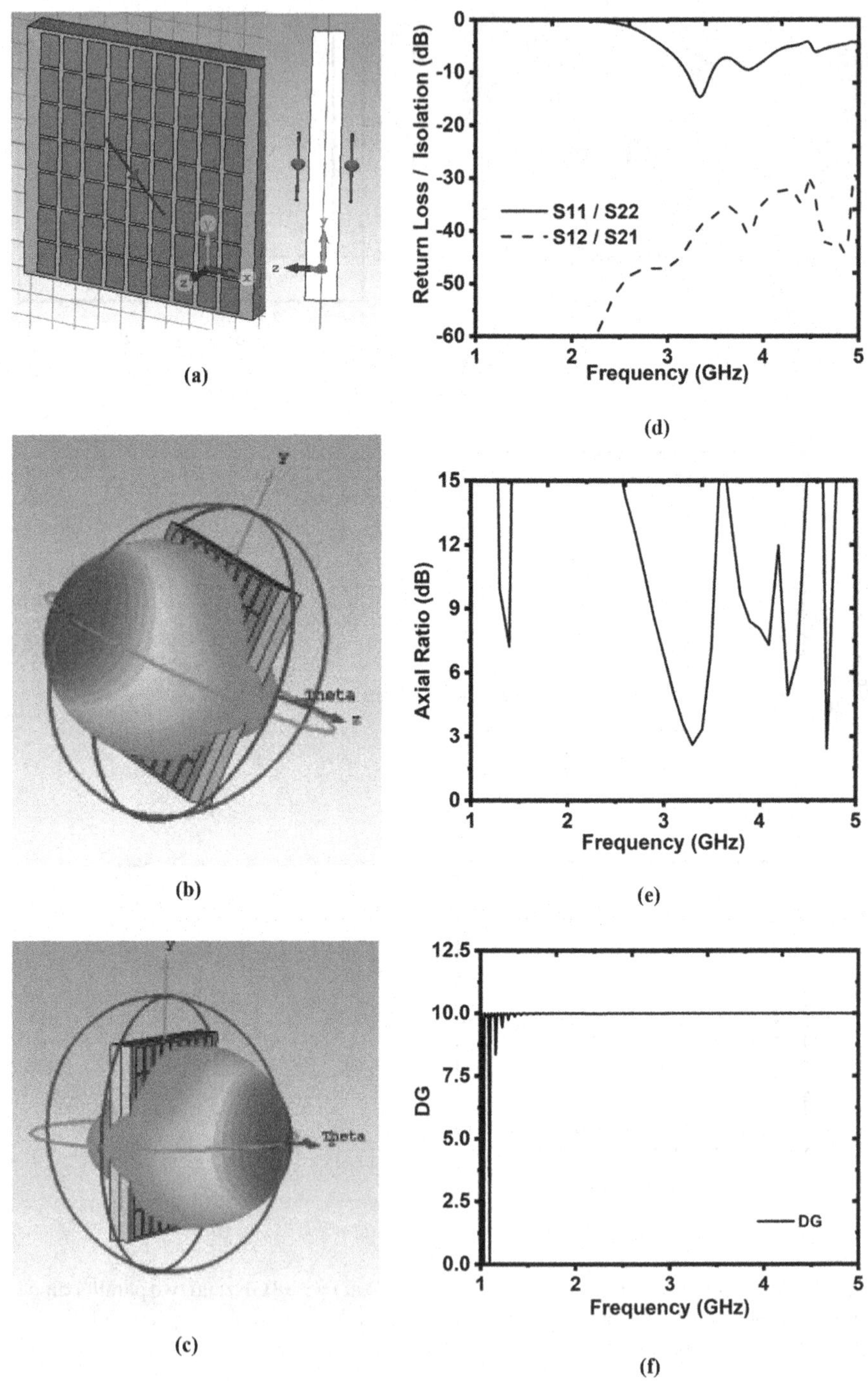

FIGURE 5.9 Dipole antennas with an EBG surface: (a) two dipoles with an EBG, (b)radiation of dipole-1, (c) radiation of dipole-2, (d) S_{11}, S_{22}, and S_{21}, (e)axial ratio, (f) DG, (g) ECC.

(Continued)

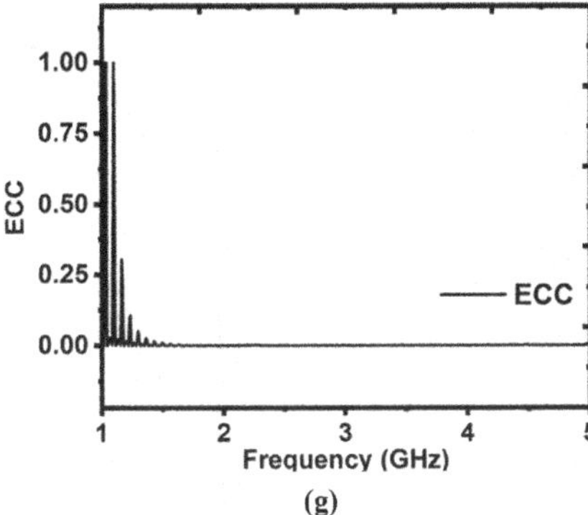

FIGURE 5.9 (CONTINUED) Dipole antennas with an EBG surface: (a) two dipoles with an EBG, (b)radiation of dipole-1, (c) radiation of dipole-2, (d) S_{11}, S_{22}, and S_{21}, (e)axial ratio, (f) DG, (g) ECC.

communication link. To reduce the fading and to increase the channel capacity, two or more antennas are used in mobile handsets to keep the communication link established in adverse weather conditions. In adverse weather conditions, if one antenna does not receive any signal, the communication link will not break since the other antenna can receive the signal, improving the communication performance of the device. These types of multiple-input multiple-output (MIMO) antennas in handheld devices are extremely useful for defence personnel working in adverse conditions. The polarization diversity can also be achieved in such antennas to improve the device performance further.

Nowadays, DRA and substrate integrated waveguide (SIW) technologies are widely adopted to meet the requirement of high performance. The DRA and SIW technologies provide a high gain and easily compatible with the planar technology. In terrible weather conditions, the communication systems demand high-performance pattern diversity antennas with a high gain. Dielectric resonator antenna (DRA) is one of the useful candidates that have been very attractive for millimetre-wave communication because it offers some attractive features such as no intrinsic conductor loss, which leads to a high efficiency.

The DRAs are much easily coupled to all the transmission lines at microwave and millimetre-wave frequencies, which make them very suitable for integration in planar technology, etc. Figure 5.11a shows an alumina ceramic-based DRA having the dimensions of 30×30×8.1 mm^3 fed by a microstrip line at opposite ends. This MIMO antenna is operating in the range of frequency of 5.1 GHz onwards with an isolation more significant than −11 dB, as shown in Figure 5.11b. The ECC is less than 0.025, while its DG is 10 dB all over the operating band, as illustrated in Figure 5.11c. Figure 5.11d and e shows the pattern diversity; the direction of the radiating beam is shifted concerning port excitation. DRA based MIMO pattern diversity antennas are simple, and they can be designed by merely exciting the dielectric resonator at different locations. The excitation at different locations provides the pattern diversity.

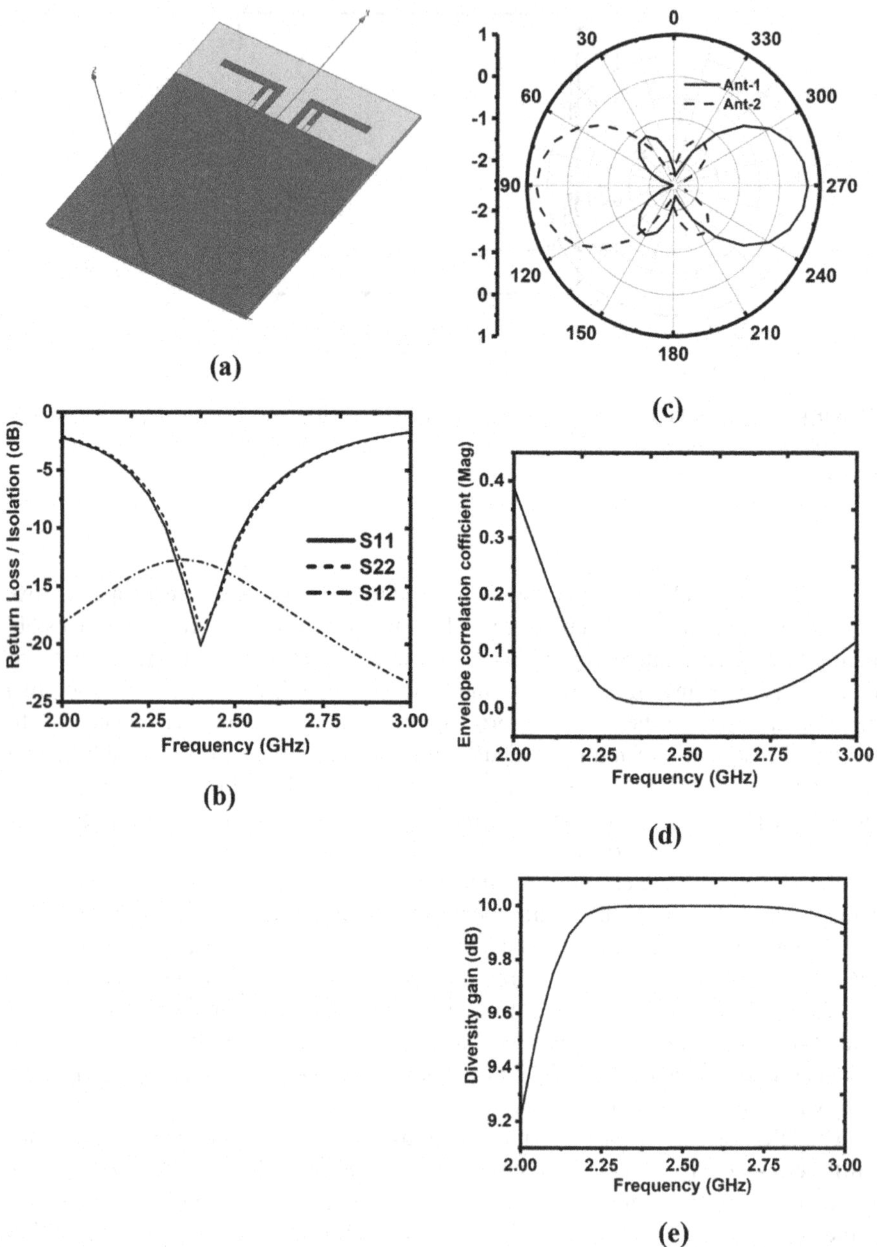

FIGURE 5.10 PIFA with pattern diversity: (a) two PIFAs on a common large ground plane, (b) S_{11}, S_{22}, and S_{21}, (c) radiation pattern of the two PIFAs, (d) ECC, (e) DG.

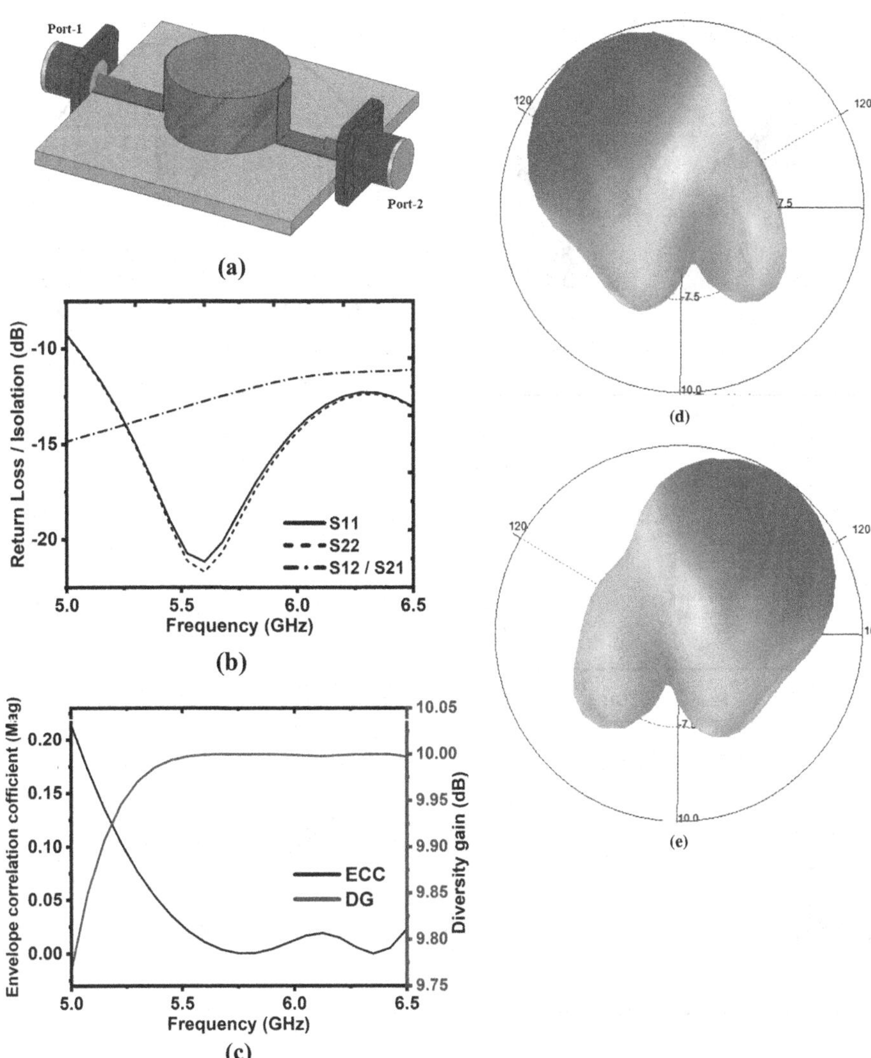

FIGURE 5.11 DRA showing pattern diversity: (a) a two-port DRA MIMO antenna, (b) return loss and isolation, (c) ECC and DG, (d) port-1 excited, (e) port-2 excited.

As the radiation patterns of the DRA MIMO antennas are uncorrelated, they have a high isolation. Since these antennas are designed using a high permittivity substrate, they provide a high diversity gain, and such kinds of antennas are useful in terrible environmental conditions.

Another exciting pattern and polarization diversity MIMO antenna with wideband circular polarization is presented in Figure 5.12(a). The capacity of any wireless communication system can significantly be improved by employing new technology multiple-input multiple-output (MIMO) antennas. The need for

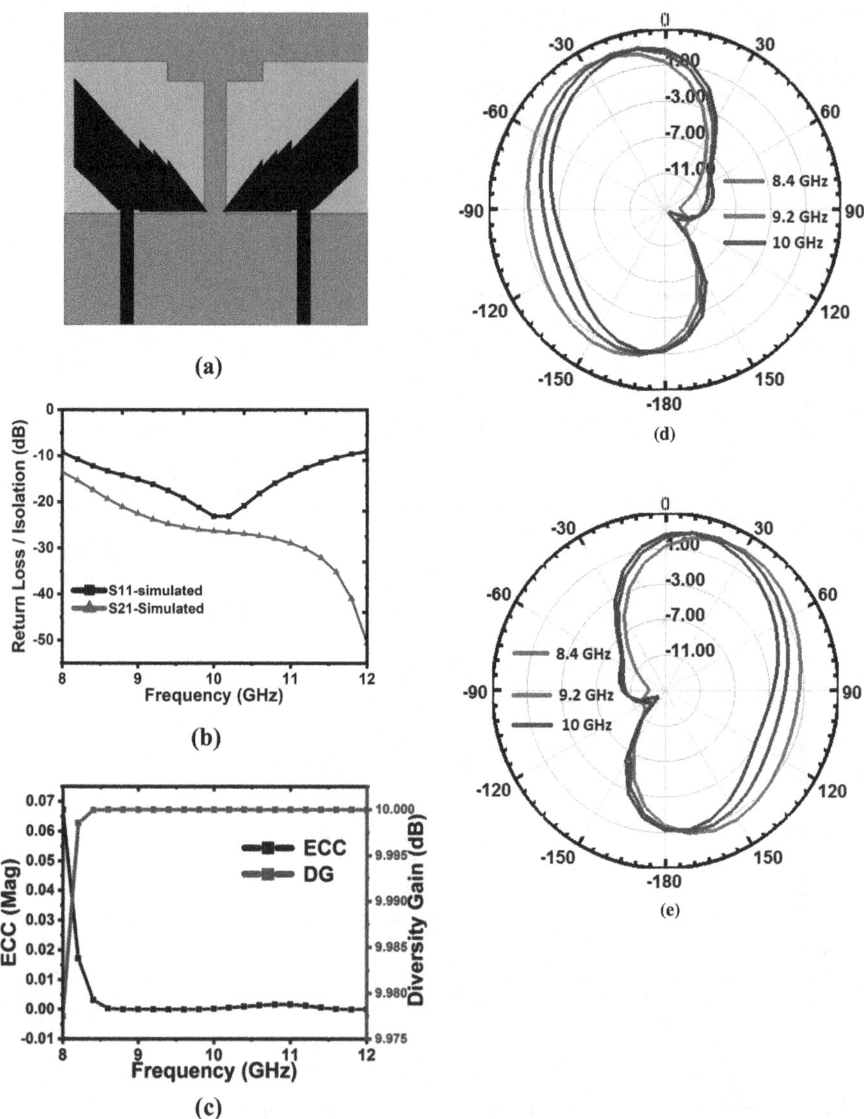

FIGURE 5.12 Quadrilateral-shaped wideband antenna with pattern diversity: (a) quadrilateral-shaped antennas, (b) S_{11} and S_{21}, (c) DG and ECC, (d) radiation pattern of antenna-1, (e) radiation pattern of antenna-2.

high data rates with this emerging technology can easily be achieved without employing an additional or external power source. Consequently, MIMO antennas have become a promising choice for the new generation wireless communication systems. Generally, in MIMO systems, transmitters and receivers employ more than one antenna to receive and send numerous signals over the same channel. By applying this methodology, we can significantly boost the spectral efficiency, channel capacity, and radio link for communication specifically in a deep multipath situation. In a compact MIMO system, the isolation between the antennas is a big issue and achieving it is a challenge; it will have undesirable effects on the system response. Extensive research in the current scenario is going on to achieve the increased demand for high capacity, to realize compact size, and also to alleviate the problems relevant to the antenna correlation. The performances of MIMO systems are continuously improved by implementing various methods.

In this section, we shall discuss only pattern diversity, while the polarization diversity is discussed in detail in the next section. This antenna is working in the X-band (8.05–11.62 GHz). The isolation between the antennas is more significant than 20 dB, as shown in Figure 5.12b.

The spacing between antenna-1 and antenna-2 is 0.16 λ. The ECC of the MIMO antenna is almost zero, while the DG is 10 dB. Antenna-1 radiates in the left-hand direction, while Antenna-2 radiates in the right-hand direction, exhibiting pattern diversity as shown in Figure 5.12d and e for different frequencies. The pattern diversity in the whole operating X-band. This type of antennas could be useful for defence applications. Truncation in monopole and stub in the ground plane directs the radiation pattern in the opposite direction and stops the overlapping of the radiation pattern, which improves the isolation and directive gain. To receive signals from any direction and reduce the multipath fading or to improve the channel capacity, we can place many antennas in different directions. This type of arrangements provides pattern diversity in each direction without overlapping of the radiation patterns; hence, a high directive gain and high isolation can be achieved. Suppose if the antennas are radiating normally or directionally, they can be placed in the form of a cube, hexagon or any other shape, so that their radiation patterns point to different directions. Figure 5.13 shows a pair of antennas placed on the sides of a cube. In total, eight antennas are placed on the cube and one PEC cuboid is placed in the centre of the cube to reduce the mutual coupling. The operating frequency range of the antenna is from 8 to 11.5 GHz with isolation, as shown in Figure 5.14a and b. Each of the antenna radiation patterns are directed in space with nearly 45° from the normal of surface covering the whole 360°, so the signal coming from any direction can be detected by the receiver, which improves the channel capacity and link stability. The pattern diversity is presented in Figure 5.14c. The next section presents the polarization diversity, which is also used to improve the channel capacity of the system. Figure 5.15 shows the radiation pattern in 3D to further illustrate the radiation pattern diversity for different port excitations.

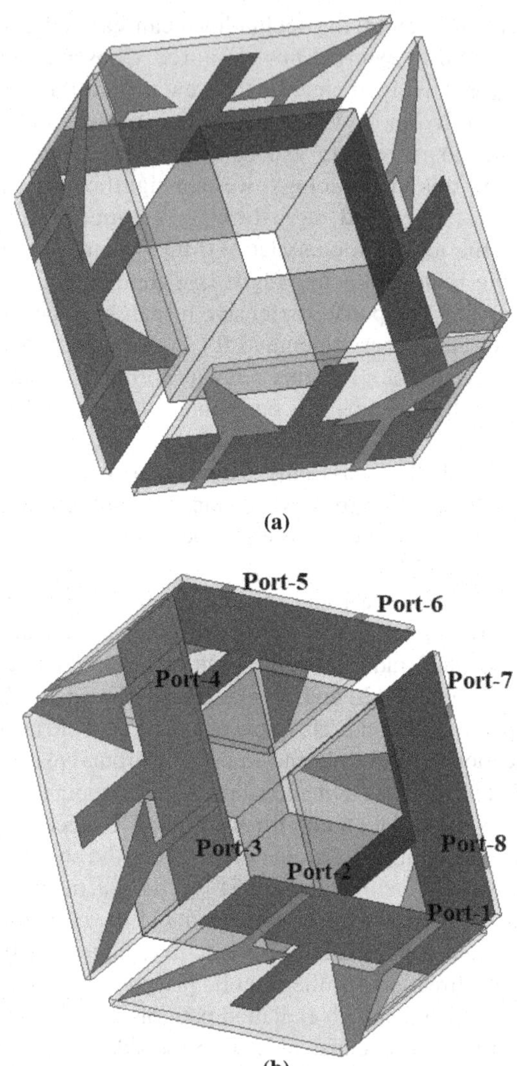

FIGURE 5.13 Eight antennas placed in form cube with PEC cube in the middle: (a) isotropic view; (b) bottom view.

5.2 POLARIZATION DIVERSITY

Usually, polarization diversity in MIMO antennas is achieved with linearly polarized vertical and horizontal antennas The linear polarized antennas have high losses in their path, and rarely suitable for the current era. The LP antennas have many losses such as absorption losses, and the change in polarization of incoming signal with the Faraday rotation effects, and these antennas are more sensitive to misalignment and require alignment at transmitter and receiver sides [37–40].

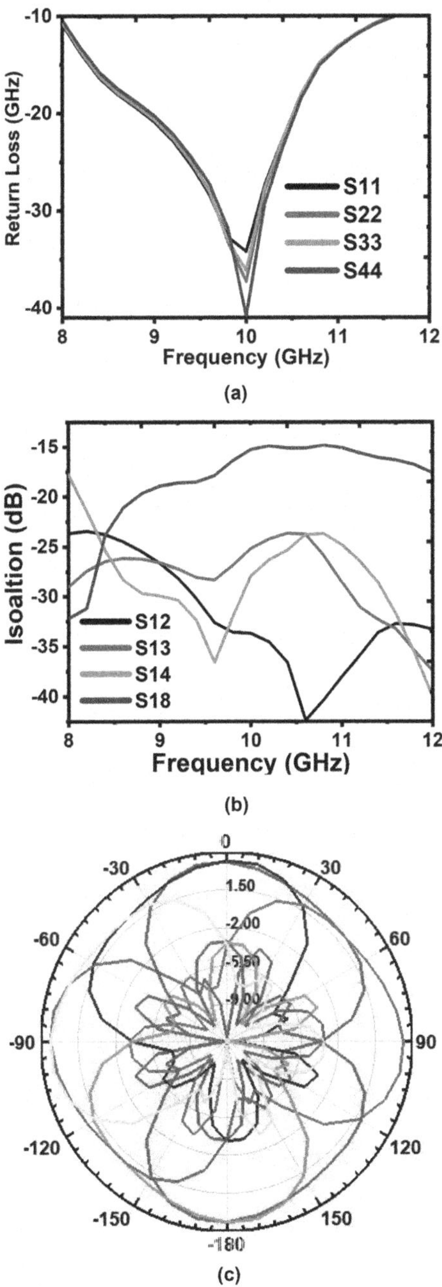

FIGURE 5.14 Performance of cube antenna: (a) return loss, (b) isolation, (c) 2D radiation pattern in X–Z plane at $f = 9.2\,\text{GHz}$.

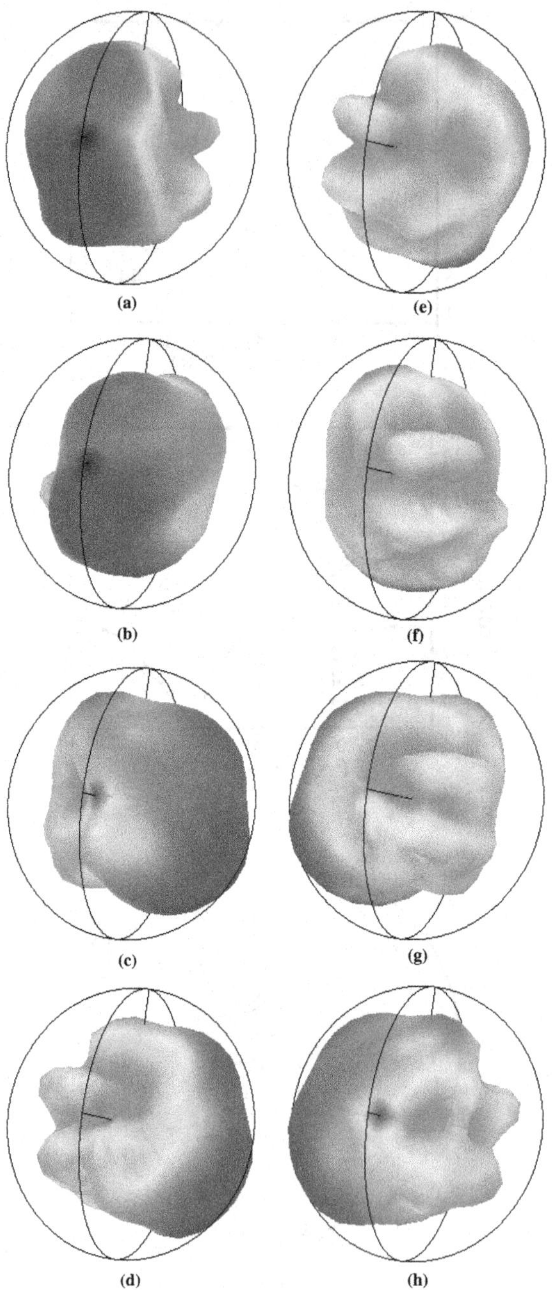

FIGURE 5.15 3D radiation patterns showing pattern diversity at frequency $f = 9.2\,\text{GHz}$, (a) port-1, (b) port-2, (c) port-3, (d) port-4, (e) port-5, (f) port-6, (g) port-7, (h) port-8.

However, the circular polarization (CP) shows several benefits in terms of sending a signal from one place to the other place or by propagation. One of the several prominent aspects of circular polarization includes decreased multipath losses, and it is less sensitive to the spatial arrangement of different antennas at the transmitter and receiver sides of the communication setup. In [41], dual CP was used in MIMO antennas to improve the link capacity dramatically in dense multipath area by utilizing low far-field correlations. Numerous MIMO antennas were proposed to implement the circular polarization (CP) diversity schemes [42–44]. The dual polarization in MIMO systems can be achieved by adding circuit components externally, such as hybrid coupler and diode. These circuit elements make system complex, and they require an extra bias (DC-biasing), which will increase the cost and real size. These make the realization of these antennas difficult in consumer electronics devices. Furthermore, these antennas have a limited bandwidth and do not fulfil the bandwidth requirements of the communication systems of the current era and also the complexity in the excitation further increases the polarization mismatch. Reconfigurability in the polarization of the MIMO schemes is another exciting technique of correlation reduction for nearby antennas. Nevertheless, this methodology is not simple to apply and requires complicated feeding techniques and high power circuit elements, which hence makes it less appealing in the perspective of size reduction, reduction in cost, and improvement in power requirements.

In the polarization diversity system, the signal is received at the same place, but with orthogonal polarization or linear components of different phases. The exciting thing about the polarization diversity is that the antennas can be physically placed at the same location and easily embedded in a small space. This technique lowers the cost and provides compactness. It is the strategy to utilize the condition that the two antennas are responding independently for the incoming signal with two different orthogonal components. The immediate polarization of the received signal at the receiver is determined from the polarization of the received signal from the base station and from the scattering parameters of the environment. The signal travelling through the environment gets reflected by various obstacles and takes multiple paths to reach the destination. Its polarization may get changed throughout the path. Suppose the transmitted signal is vertically polarized and through the transmission, it becomes horizontally polarized. But, there is no single antenna that simultaneously receives both horizontally and vertically polarized signals. Thus, the polarization diversity scheme required to receive these two orthogonal horizontally and vertically polarized signals. Figure 5.16 shows a simple example of a polarization diversity antenna; here, the loop antenna provides horizontally polarized radiation pattern, whereas the dipole has an omnidirectional pattern in azimuth plane with vertical polarization. But, in the real environment, the polarization of the signal is elliptical with a random orientation of the electric field.

When a linearly polarized (LP) signal travels through obstacles, it gets changed to orthogonal polarization, resulting in elliptical polarization which is the summation of transmitted and reflected signals. It is impossible to receive all the transmitted power using a single antenna. To receive the elliptically polarized signals, circularly

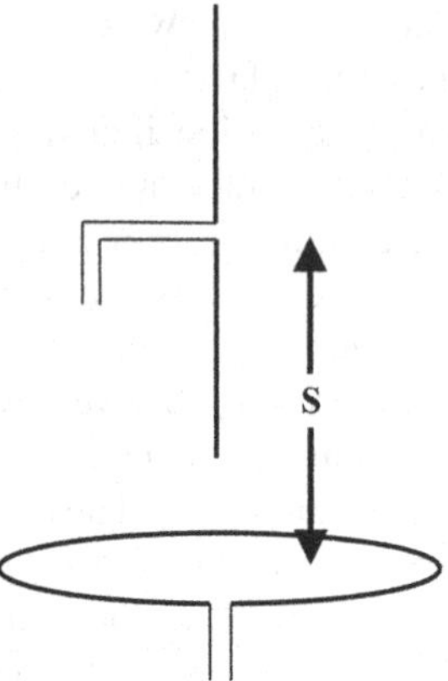

FIGURE 5.16 Antenna having polarization diversity.

polarized (CP) antennas are often used. Antennas with right-hand circular polarization (RHCP) and left-hand circular polarization (LHCP) diversity are being used for random orientation of the signals. The use of polarization diversity can reduce the polarization mismatch and improve the signal-to-noise ratio (SNR) up to 12 dB even in the line-of-sight signal (LOS) channels [1]. The polarization mismatch due to the random orientation of the mobile handset can be reduced by implementing polarization diversity schemes. In [16], ultra-wideband (UWB) antennas employing orthogonal feeding were used to get polarization diversity. These antennas have two vertical monopoles loaded with an open slot, CSRR, ground with a rectangular stub, and an asymmetrical ground plane to achieve a high isolation. The differential feed method is also suggested for polarization diversity [17]. For polarization diversity, a differential feed has been used in UWB antennas. These antennas consist of an octagonal-shaped slot with four monopoles placed symmetrically. A novel feeding network consisting of two quadrature hybrid couplers, two Wilkinson power dividers, and one 180° phase shifter was used to generate circular polarization with polarization diversity [18]. Polarization diversity reduces the bit error rate in RFID, which makes the RFID tags highly sensitive. In [19], a PIFA composed of a 1×2 sub-array was used, which is placed orthogonally at quarter wavelength, and two different feeding networks were used to control the horizontal and vertical currents in the array.

Dielectric resonator antennas (DRAs) have certain advantages compared to conventional conductor-based antennas, such as high radiation efficiency, small size, and versatility. A DRA with broadside CP pattern and omnidirectional pattern with vertical polarization was presented in [20]. They show polarization diversity. The DRA antenna has two independent orthogonal modes. In the DRA antenna, the quasi-TM_{111} mode provides linearly polarized omnidirectional radiation, while circularly polarized radiation can be obtained by exciting in TE_{111} mode. Active devices such as PIN diodes are also used for switching the polarization state. The polarization state of an antenna can be switched between different states by controlling the bias voltage of a PIN diode [21]. The polarization state can easily be switched between CP and LP by simply changing the shape of a slot by switching the PIN diode ON and OFF without changing the geometrical shape [22], which provides four different polarization states for WLAN applications. In [23], a pair of L-shaped slits were etched on the opposite side of a square patch, two PIN diodes are placed at the end of the slit, and the polarization is switched between linear and circular by switching the diodes. A circularly polarized antenna is realized by a shorting pin between the circular patch and the ground plane, and the polarization diversity between RHCP and LHCP can be obtained by two shorting pins, which are placed between the patch and the ground plane [24]. In [25], an antenna structure consisting of a balun-fed cross-dipole and a switchable feeding network was described; this cross dipole provides quad polarization diversity. Different polarization states were generated by controlling the states of PIN diodes integrated with the feeding network, which consisted of a power divider, two SPDT, and a 90° delay line to excite these four cross dipoles, to generate two LP at ±45°, and two orthogonal polarization RHCP and LHCP. The concept of polarization switching using PIN diodes [26] was also employed in an aperture coupled patch antenna. The given antenna provided quadri-polarization states of two orthogonal linear polarizations and two orthogonal circular polarizations by just switching eight PIN diodes embedded in a feeding network. Biasing of a PIN diode requires a RF choke inductor to provide high RF impedance and a capacitor to DC current.

5.2.1 Diversity Gain of Polarization Diversity System

The effectiveness of a polarization diversity system can also be determined in a similar way as we have explained in Section 5.1 for a pattern diversity system by using diversity gain. Now, we shall use some examples of polarization diversity systems to explain the effect on diversity gain and polarization mismatch.

5.2.2 If Two Orthogonal Components are Transmitted

Suppose that two signals are transmitted from the base station simultaneously and they are orthogonal to each other. Since the transmitted signals are orthogonal to each other, no coupling will take place between them. Hence, this is an effective way of using bandwidth. These transmitted signals are scattered by buildings and other obstacles, and the polarization may get changed. At the same time, the ground will have minimal effect on the signal polarization, which means the signal remains the same after reflection from the ground. The signals take multiple paths to reach the receiver, which will be the cause for the fading phenomenon; however, there is no significant coupling between the two polarized components. If the receiver has a pair of orthogonally polarized antennas, each antenna will receive different orthogonal components. Since these orthogonal components have a phase difference of 90° in between them, there is no coupling; thus, the correlation between them is very low. If the envelope correlation coefficient is very low, then the diversity gain will be more. The system can provide a diversity gain of up to 10dB.

5.2.3 If a Single Linearly Polarized Component Is Transmitted

If a single linearly polarized (LP) signal is transmitted through the channel, throughout the path, it may get depolarized. Suppose the transmitted signal is a vertically polarized signal. It becomes horizontally polarized in an environment, which is generally referred to as Rayleigh fading.

If in this case, we use a pair of horizontally and vertically polarized antennas at the receiver, then the horizontally polarized antenna will receive most of the signal power of the channel and the vertically polarized antenna will receive much lower power level. Since orthogonal polarization has no coupling between the signals, the samples are uncorrelated, which provides an improved diversity gain.

Figure 5.17 shows two printed dipole antennas with linear polarization diversity. These antennas radiate two orthogonally polarized signals. Antenna-1 directs the radiation along the Y-direction as shown in Figure 5.17c, and antenna-2 directs the radiation along the X-direction as shown in Figure 5.17d. Both the antennas radiate Y-polarized (vertically polarized) and X-polarized (horizontally polarized) signals. Figure 5.18a shows the response of this dual-band antenna, which is operating at 4.25 and 6.5 GHz with an isolation higher than 15 dB. The ECC is almost zero at the operating frequencies, and the DG is 10 dB, as shown in Figure 5.18b and c.

5.2.4 If a Circularly Polarized Antenna Transmits a Right-Hand Circular Polarization (RHCP) Component

As in the above illustrations, a similar case is observed for CP antenna. If the transmitted signal is RHCP, while travelling through the environment, its polarization may get changed to LHCP. If the receiver has antennas that support both RHCP and LHCP polarizations, whatever the polarization of the incoming signal is, the receiver can detect it. The multiple-input multiple-output (MIMO) antennas support both LHCP and RHCP. Two monopole antennas truncated diagonally and fed by a quarter-wave feed is shown in Figure 5.19a. The radiation patterns of both the antennas are in broadside direction. Thus, they have an identical performance, but with different polarizations. Figure 5.21 shows the 3D radiation pattern, which is in the broadside. Figure 5.19b shows that the antennas have proper matching and an excellent isolation between them. Figure 5.19c shows that the axial ratio is below 3 dB within the matching bandwidth. Figure 5.19d and e shows the ECC and diversity gain. The ECC is 0.005, and the DG is 10 in the operating range of the antenna. Figure 5.20 shows the current distribution on both the antennas. The direction of current in both the antennas rotated clockwise and anticlockwise, which shows LHCP and RHCP with respect to phase ωt. These kinds of antennas are useful for the modern communication systems to reduce the multipath fading and polarization mismatch.

The next examples are based on the dipole antennas with an EBG surface discussed in Sections 5.2 and 5.3. The EBG surface behaves as a perfect magnetic conductor and has a positive reflection coefficient. These dipoles are placed at angles of ±45° with the EBG surface, and both are parallel to each other. The EBG surface is in between the two dipole antennas. The role of the EBG here is to change the linear polarization to circular polarization. These examples have exciting performances. They have both polarization and pattern diversities. The EBG surface between them behaves as a polarization converter, and the EBG surface as a perfect magnetic conductor plays the role of a reflector.

The EBG acts here as a perfect magnetic conductor instead of a perfect electric conductor. The perfect magnetic conductor first changes the polarization and then

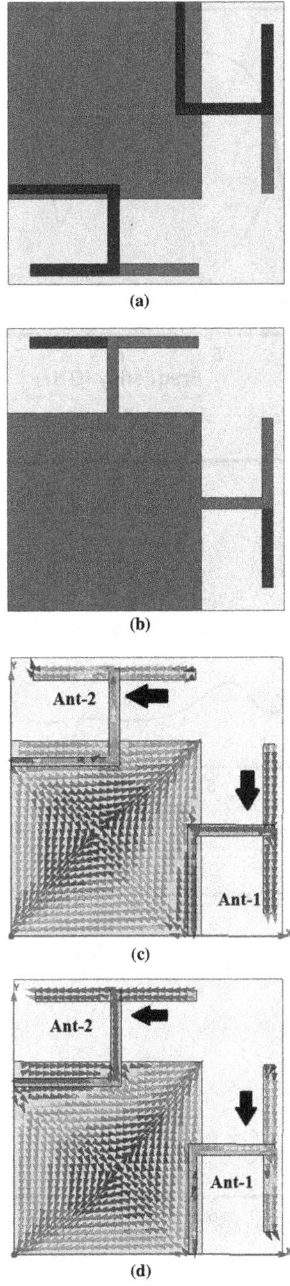

FIGURE 5.17 Two printed dipole antennas with linear polarization diversity: (a) top view, (b) back view, (c) antenna-1 is excited, (d) antenna-2 is excited.

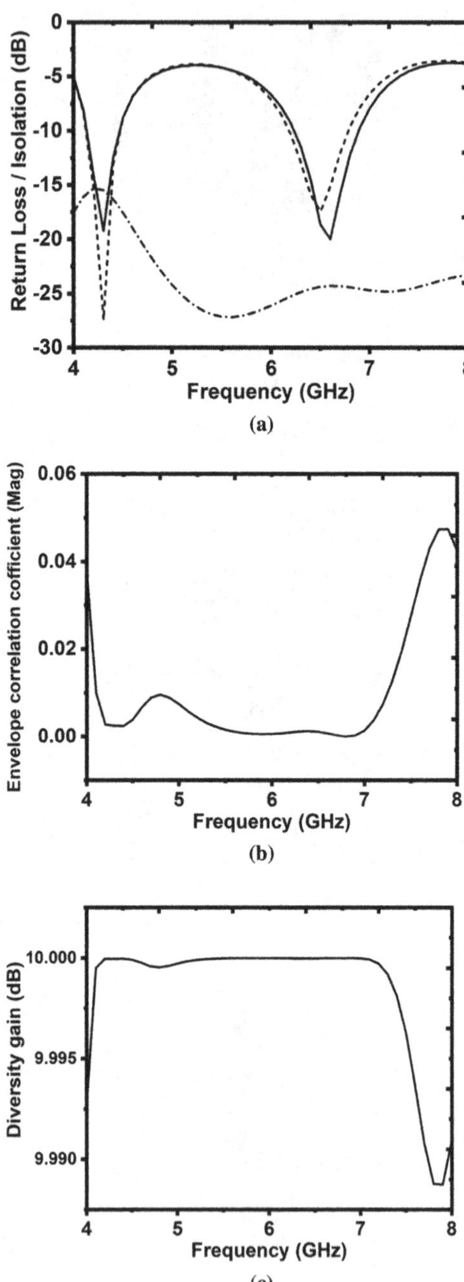

FIGURE 5.18 Dual-band antennas with polarization diversity (vertical and horizontal polarization): (a) return loss, (b) ECC, (c) DG.

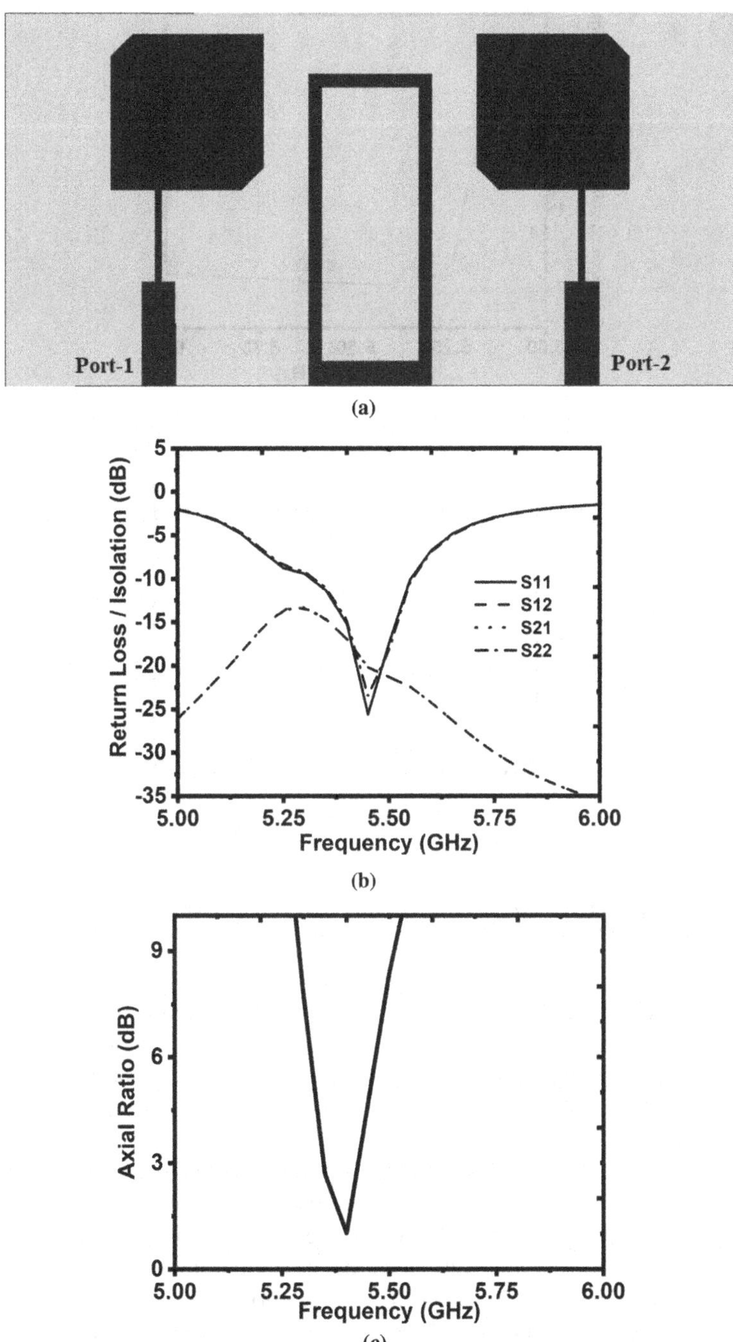

FIGURE 5.19 Antenna with LHCP and RHCP polarization diversity: (a) two monopole CP MIMO antennas, (b) S_{11}, S_{22}, and S_{21}, (c) axial ratio, (d) ECC, (e) DG.

(Continued)

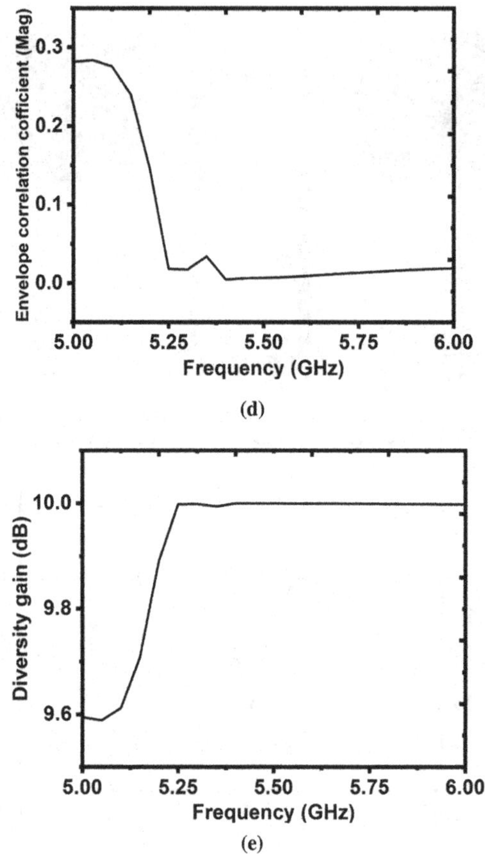

FIGURE 5.19 (CONTINUED) Antenna with LHCP and RHCP polarization diversity: (a) two monopole CP MIMO antennas, (b) S_{11}, S_{22}, and S_{21}, (c) axial ratio, (d) ECC, (e) DG.

changes the omnidirectional signal to directional signal; both the dipoles have radiations in the opposite directions, which provides the pattern diversity as can be seen in Figure 5.9 in Section 5.1.2. One dipole generate LHCP and other dipole generates RHCP with pattern diversity because of orthogonal antenna placing with axial ratio shown in Figure 5.9e.

Another example is based on quadrilateral shaped antennas, which are used for the generation of LHCP and RHCP, with pattern diversity; these antennas produce wideband circular polarization (CP). Figure 5.12a shows the layout of this antenna. This antenna has both pattern and polarization diversities and has an improved channel capacity and improved isolation. Figure 5.22 shows the surface current distribution on the antenna surface and the ground plane. By observing the surface current distribution in both the planar monopoles, the direction of current in one antenna is rotating anticlockwise, illustrating that the antenna is radiating LHCP, while the direction of current in the other antenna is rotating clockwise, illustrating that the antenna is radiating RHCP. They are wideband circularly polarized antennas, and their surface current distributions confirm

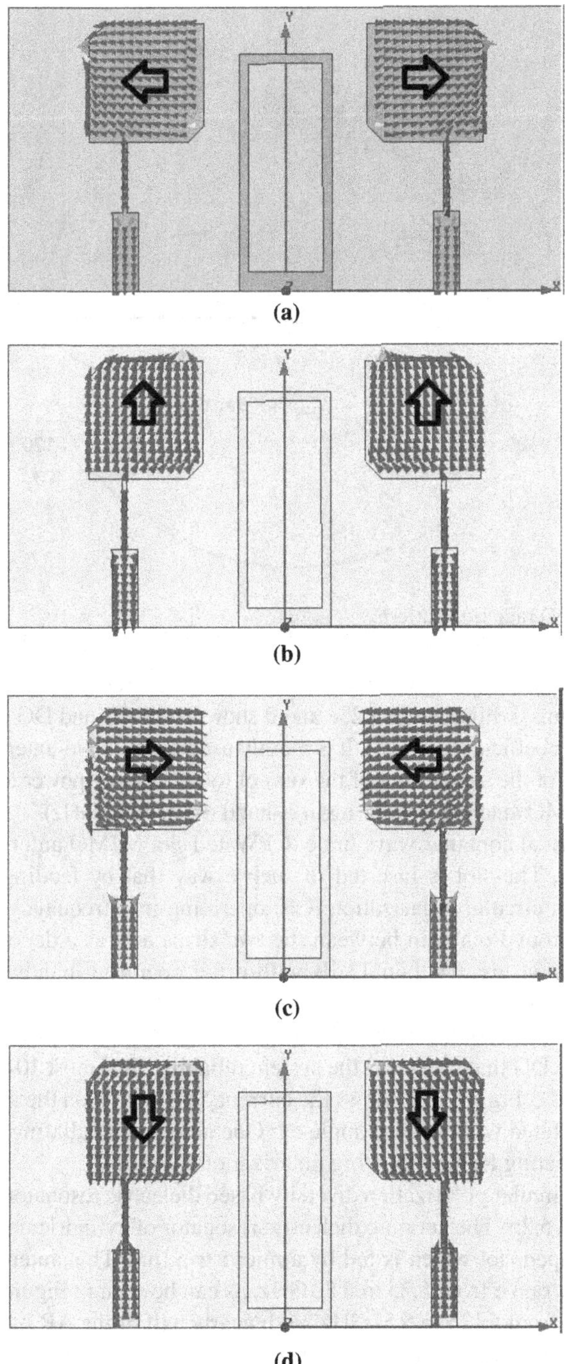

FIGURE 5.20 Surface current distribution on both the antennas with different phases: (a) $\omega t = 0°$, (b) $\omega t = 90°$, (c) $\omega t = 180°$, (d) $\omega t = 270°$.

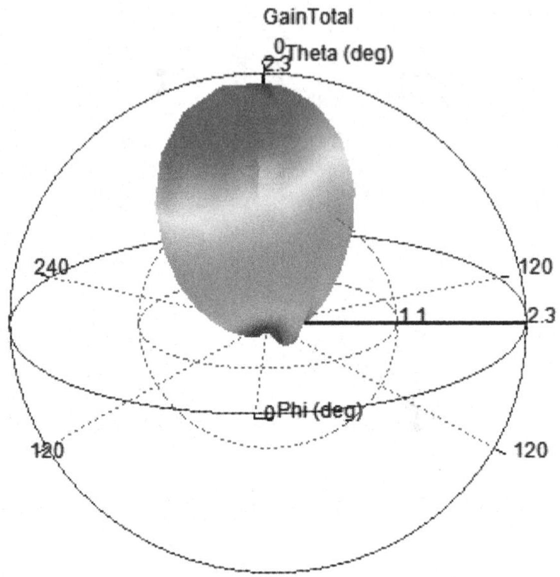

FIGURE 5.21 3D radiation pattern.

the polarization diversity. Figure 5.23a and b shows the return loss, axial ratio, and gain of the antenna, while Figure 5.23c and d shows the ECC and DG along with total active reflection coefficient (TARC). It is mainly used for multiple-antenna systems and defines the ratio of the square root of the sum of total outgoing power to the total input power. This MIMO antenna is working in X-band (8.05–11.62 GHz).

Another unusual coplanar waveguide (CPW)fed slot MIMO antenna is presented in Figure 5.24a. The slot is inserted in such a way that by feeding with a centre strip, it excites a circular polarization (CP) operating in a frequency band from 4.8 to 7.6 GHz. A ground plane in between the two strips acts as a decoupler or a filter having an isolation greater than 13 dB within the operating bandwidth, as shown in Figure 5.24b. The axial ratio bandwidth of the slot antenna is 300 MHz. The ECC is 0.015, which shows the effectiveness of diversity performance of the MIMO antenna, and the DG that measures the system reliability is almost 10 dB, as shown in Figure 5.24d and e. Figure 5.25 shows the current distribution on the antenna surface; the current is rotated with a phase angle ωt. One antenna is radiating LHCP, and the other one is radiating RHCP, showing polarization diversity.

Similarly, a circular polarization diversity based dielectric resonator antenna is presented in Figure 5.26. The ceramic dielectric resonator of cylindrical shape is excited by a tilted I-shaped slot which is fed by a microstrip line. This antenna is operating in the frequency range from 5.25 to 5.80 GHz, as can be seen in Figure 5.27a, with an axial ratio band from 5.125 to 5.52 GHz, with nearly half of the AR band lying within the impedance matching, as can be seen in Figure 5.27b. The isolation between the antennas is still satisfactory without any decoupling structure between them.

The ECC varies from 0 to 0.212, which is much smaller than 0.5, and the DG varies from 9.76 to 10 dB. The E-field distribution is shown in Figure 5.28. With respect

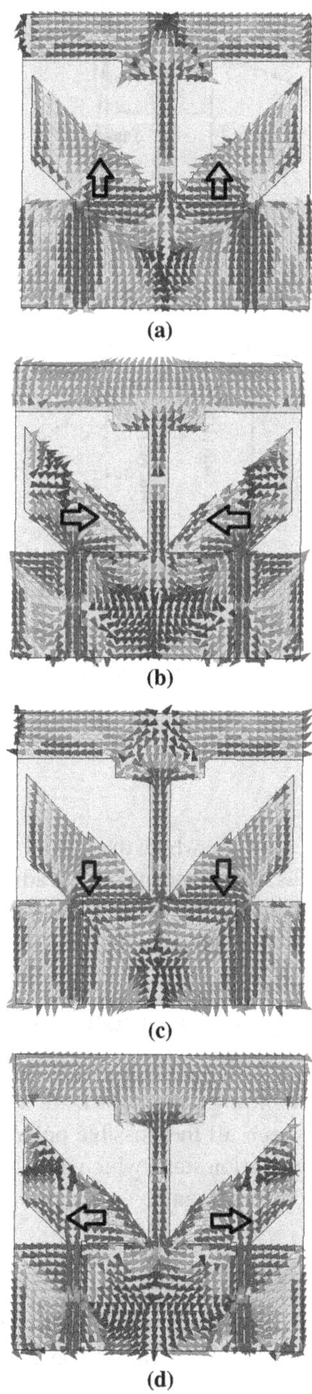

FIGURE 5.22 Surface current distribution showing polarization diversity at $f = 10\,\text{GHz}$: (a) $\omega t = 0°$, (b) $\omega t = 90°$, (c) $\omega t = 180°$, (d) $\omega t = 270°$.

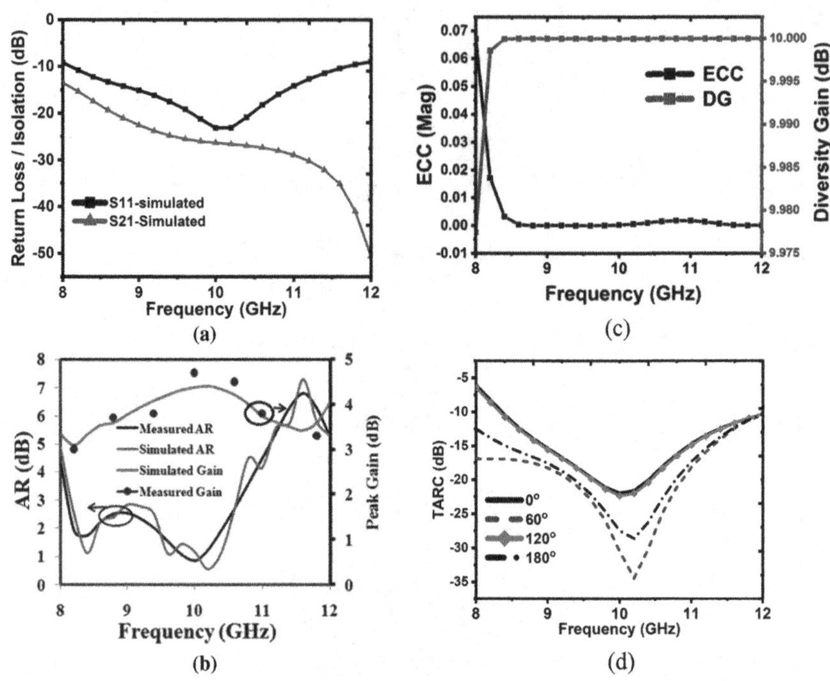

FIGURE 5.23 *S*-parameters, gain, axial ratio, DG, ECC, and TARC: (a) return loss, (b) gain and axial ratio, (c) DG and ECC, (d) TARC.

to the phase angle ωt, the directions in which the E-field vectors rotate are opposite to each other in both the antennas, which means one is radiating LHCP, while the other is emitting RHCP, as illustrated by the surface current distribution. This antenna represents the polarization diversity phenomenon in DRAs.

5.2.5 If the Radiated Signal Is Obliquely Polarized

If the incoming signal from the base station is polarized obliquely, after each reflection from the obstacles and buildings, the polarization of the signal gets changed. It will result in coupling between all the possible polarization states. The received signal has a time-varying polarization state, which increases the envelope correlation coefficient and decreases the diversity gain.

5.3 MASSIVE MIMO ANTENNAS

As the demand for wireless communication is exploding, it is very much necessary that the operators have to increase the efficiency of their network. Massive MIMO systems are gripping the most technology for future wireless communication. "Massive" simply refers to an array of a large number of antennas at the base

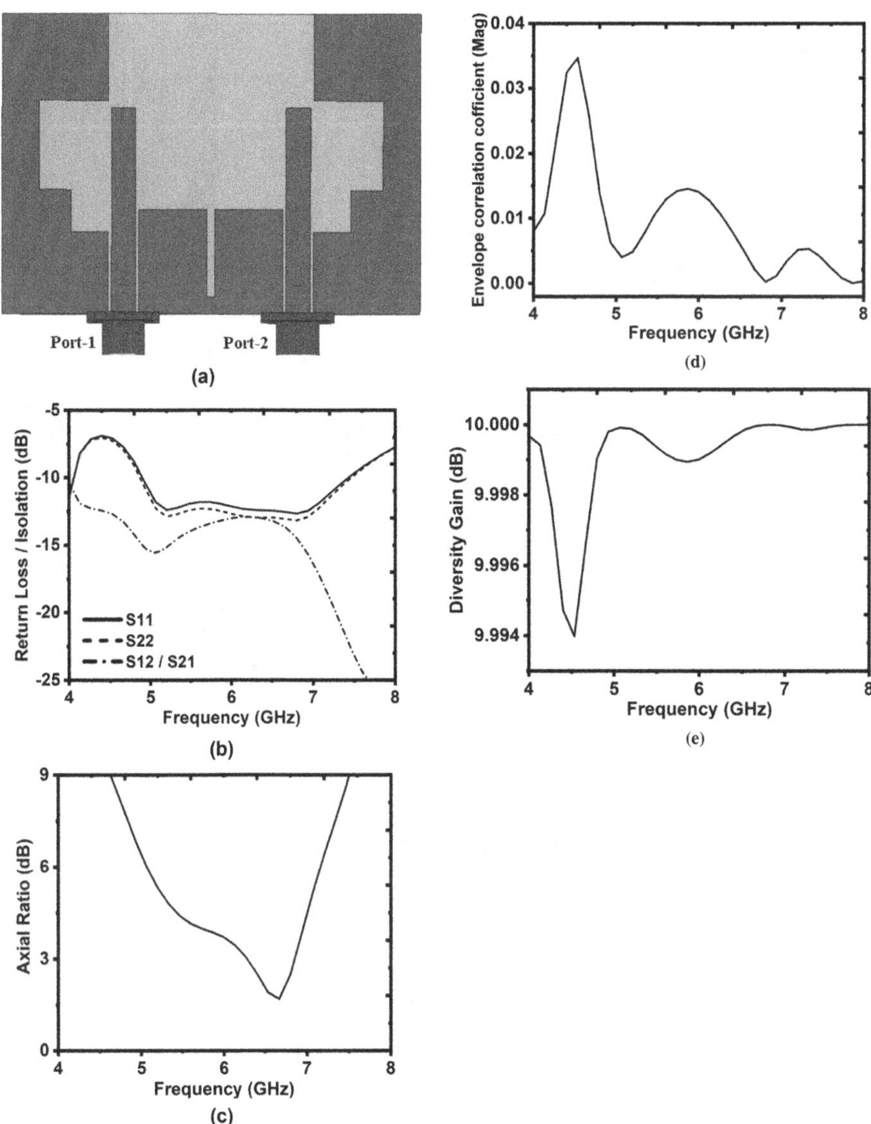

FIGURE 5.24 CPW-fed slot MIMO antenna showing polarization diversity: (a) geometry of the CPW-fed MIMO antenna, (b) return loss and isolation, (c) axial ratio, (d) ECC, (e) DG.

station, as shown in Figure 5.29. This concept of the large antenna array is to serve multiple users, providing much better throughput and spectral efficiency. Basically, the Massive MIMO system is an extension of MIMO system.

The idea behind Massive MIMO system is having transmitter/receiver equipped with a large number of antennas provides better throughput, stable data link and

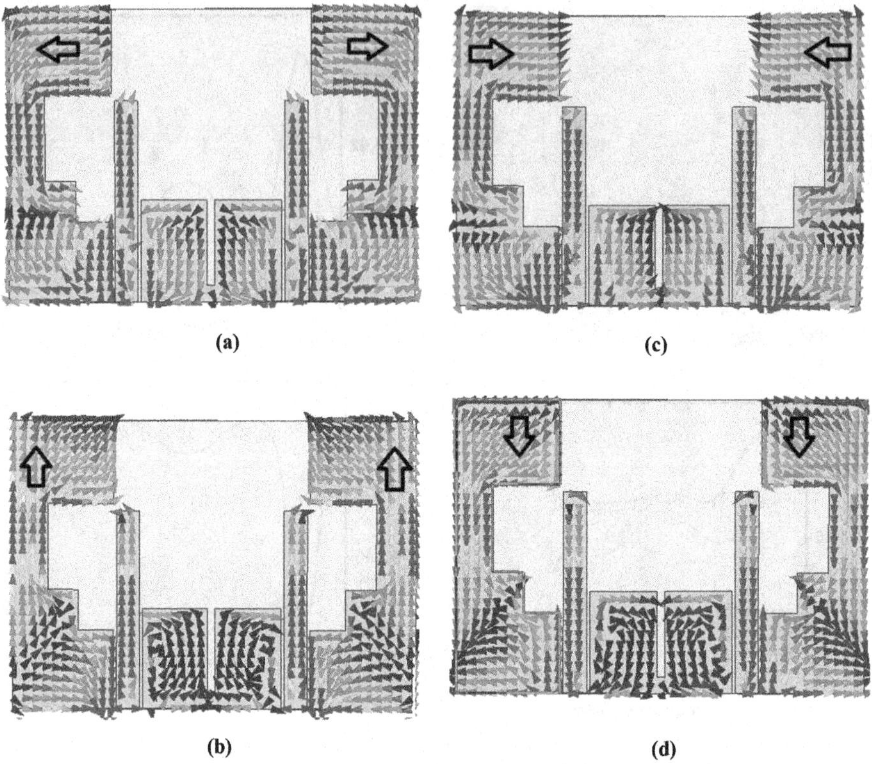

(a) (c)

(b) (d)

FIGURE 5.25 Surface current distribution at 6.5 GHz showing polarization diversity at $f = 6.5$ GHz: (a) $\omega t = 0°$, (b) $\omega t = 90°$, (c) $\omega t = 180°$, (d) $\omega t = 270°$.

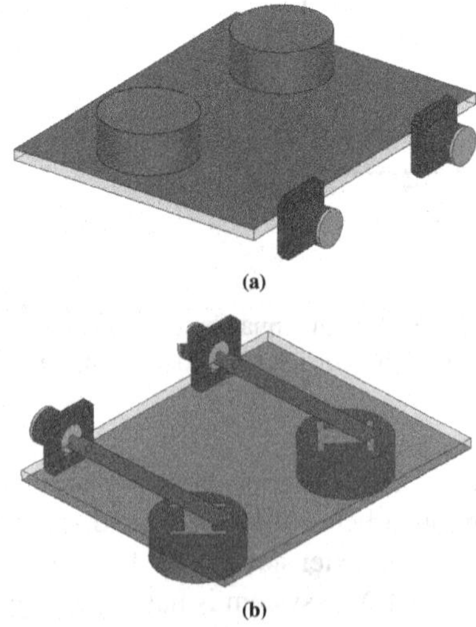

FIGURE 5.26 MIMO antenna with a DRA: (a) top view; (b) bottom view.

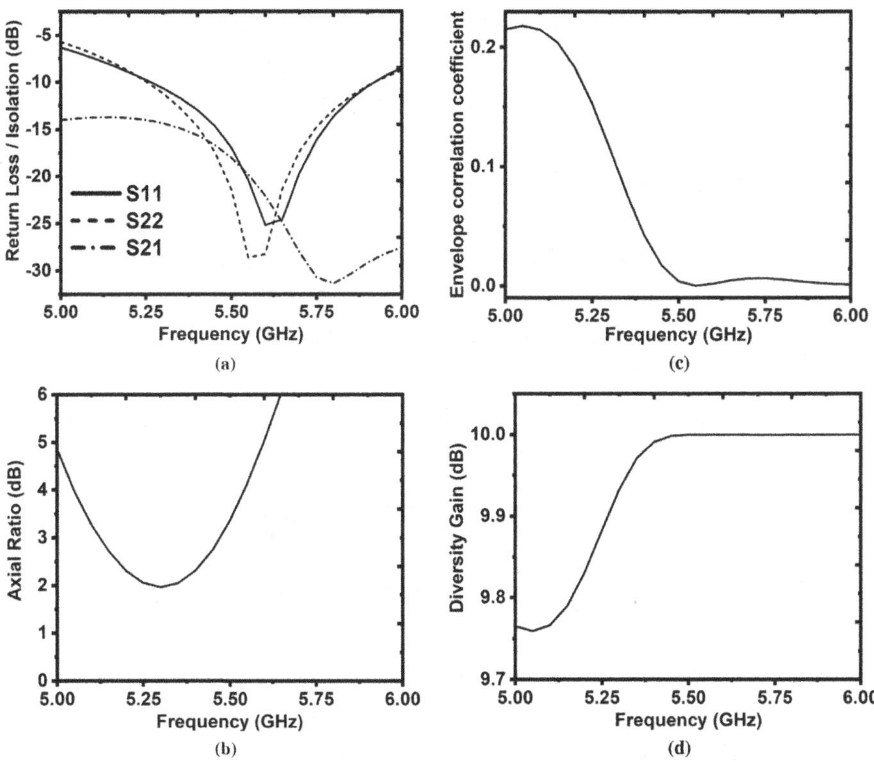

FIGURE 5.27 Response of a DRA having polarization diversity: (a) return loss, (b) axial ratio, (c) ECC, (d) DG.

high data rate. The Massive MIMO concept was first introduced in [27] with the theoretical assumption of an infinite number of antennas at the base station in a multicellular environment where the base station is selectively transmitting data to terminals. Likewise, the base link combines reverse link signal for transmission to its terminals. In 2018, the Federal Communication Commission (FCC) [28] had approved the first Massive MIMO product, which is Ericsson AIR 6468 [29]. This designed product consists of 64 antennas connected with 64 transmitter/receiver channels. This baseband signal is entirely digitally processed and designed for 4G LTE. A different version of AIR 6486 is also designed for LTE band, i.e. 2,496–2,690 and 3,400–3,600 MHz. Size and weight of unit $988 \times 520 \times 187\,mm^3$ and 60.4 kg. Generally, there is no difference between a low-gain Massive MIMO antenna array and high gain antenna in terms of link availability, but in terms of other performance factors such as throughput, link stability, and data rate, the Massive MIMO antenna array is much better. The MIMO or Massive MIMO system antennas are arranged in 2-D configuration, which makes the design compact and easy to fabricate. The Massive MIMO system also requires up-to-date chipset software for high spectral efficiency because such an extensive antenna array system has a complex

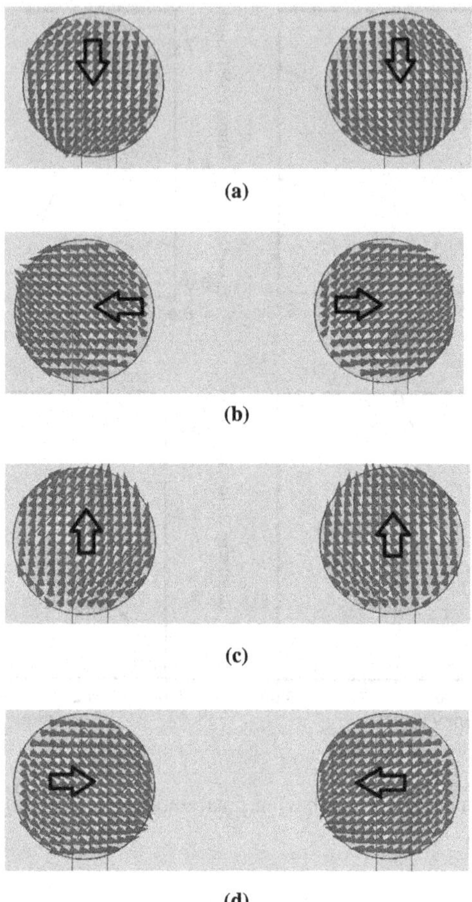

FIGURE 5.28 Electric field distribution at $f = 5.4\,\text{GHz}$: (a) $\omega t = 0°$, (b) $\omega t = 90°$, (c) $\omega t = 180°$, (d) $\omega t = 270°$.

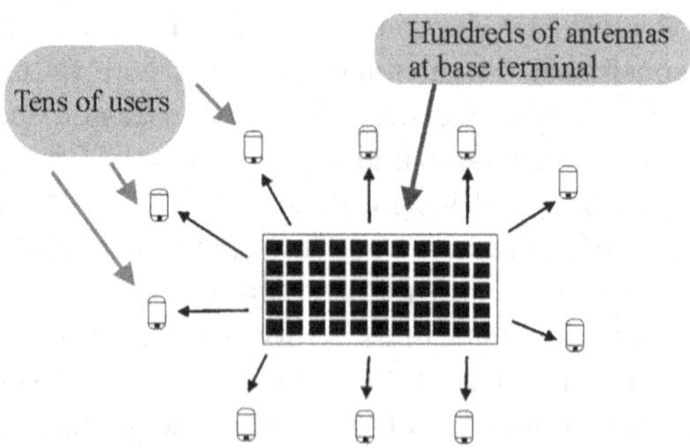

FIGURE 5.29 Massive MIMO system.

multiplexing system. In [30], an entirely digitally controlled Massive MIMO active antenna system (AAS) was developed at 28 GHz for 5G communication with a 24-antenna array. Several antennas have been reported in the literature for Massive MIMO systems. In [31], an array of 64 antenna units operating at 5G is presented. A 4 by 4 MIMO antenna array for LTE design was presented in [32]. A dual polarized planar antenna at 60 GHz was reported in [33], consisting of a circular disc fed by a proximity feed. In [34], a compact Massive MIMO antenna was arranged in a cubical form, showing high multiuser capacity as compared to planar arrays. This given design uses orthogonally polarized patches arranged at the faces of the cube. An array of dual polarized cavity-backed printed dipole antennas was presented in [35]. Dual polarize multi beam antenna [36] with arranged in 4×4 configuration for the base station can be used for a Massive MIMO application. In [36], a wideband dual polarized Vivaldi antenna in 4×4 configuration with multibeam for 5G was described. Antennas arranged in the form of a cube, as shown in Figure 5.30, can also be used as a Massive MIMO receiver antenna, and the 2D radiation pattern is shown in Figure 5.31, which shows the different radiation patterns in different directions. The main radiation beam depends on each antenna; by exciting a particular antenna, we can control the directional radiation pattern. Massive multiple-input multiple-output (MIMO) systems are compelling in the signal transmission that uses a large number of transmitter (TX) and receiver (RX)antennas. High spectral and energy efficiency (EE) in Massive MIMO systems can be obtained by increasing the excessive gain of the channel [45–48]. The Massive MIMO systems in industrial wireless communication networks have gained considerable attention and are widely adopted because of its high level of reliability [49–51]. In the current era, Internet in industrial or Internet of things for industrial use (IIoT) has got acknowledged. So, we require a new type of wireless communication network technology that can uphold the hyper wireless connectivity between the data centre at a specific location and distributed user entities (UEs) and/or IIoT-enabled devices. In this respect, a Massive MIMO technology can be the main answer for the fulfilment of the Internet in industry.

Pattern and polarization diversity techniques can also be used to improve the channel capacity and link reliability in case of Massive MIMO systems [52–54]. To elaborate on the Massive MIMO concept, here we present the cubical Massive MIMO antenna with eight antenna elements. Here each antenna is radiating in a specific direction within the same operating frequency bands. This cubical Massive MIMO antenna has wideband impedance matching and wideband axial ratio. The pattern diversity is helpful in the beamforming. The beam can be made by shifting the radiation pattern in a particular direction. Different polarized antennas can be used to get the polarization diversity. In Massive MIMO polarization, diversity can be useful to improve the overall performance. Since this antenna is working in wideband with both pattern and polarization diversities, one face has two antennas placed side by side with a small distance, generating both polarization RHCP and LHCP.

The expansion in the Massive MIMO system is currently in the product departments of many leading communication industries. Many optimization algorithms and signal processing techniques have been developed over the years and are still

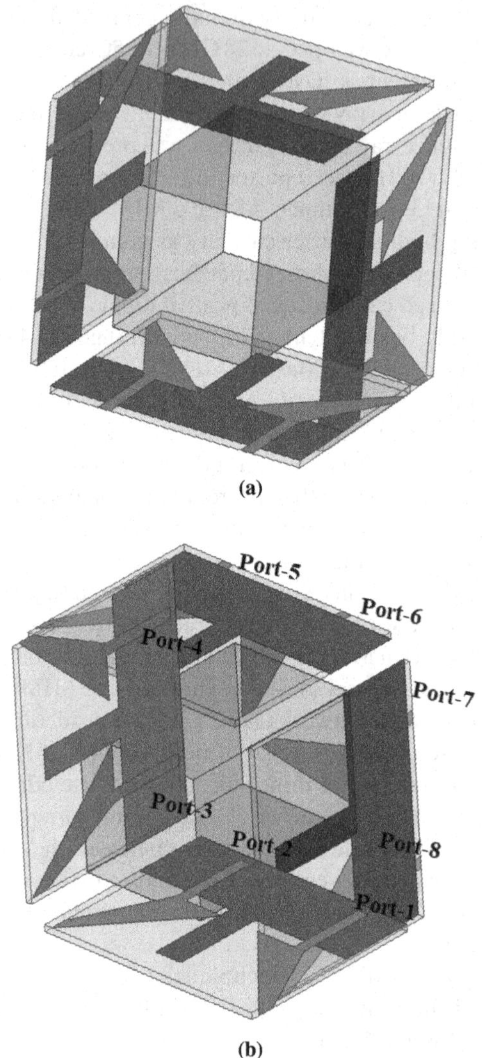

FIGURE 5.30 Eight antennas of a Massive MIMO system with a PEC cube: (a) isotropic view; (b) bottom view.

being developed to meet the current demands. Commercial success in Massive MIMO systems still requires a lot of effort.

5.4 CONCLUSION

This chapter used different examples to explain the concept of pattern and polarization diversity in antennas. Linearly polarized and circularly polarized antennas were used for both diversity. Simple dipole antennas with PEC and EBG reflectors

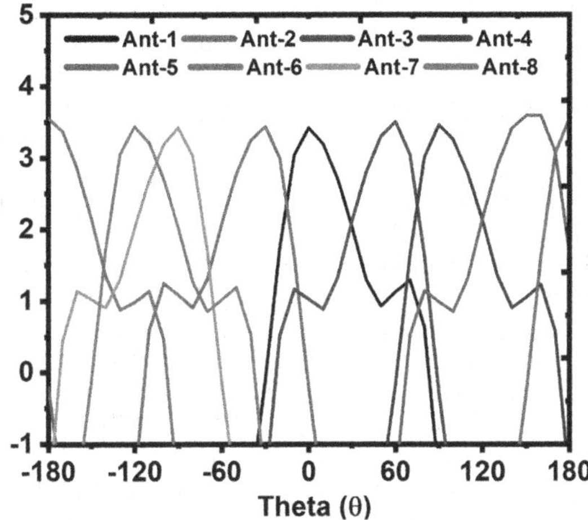

FIGURE 5.31 2D radiation pattern of a Massive MIMO antenna at 9.2 GHz.

were used to illustrate the phenomenon of pattern diversity. Planar inverted F antenna (PIFA), microstrip antenna, planar monopole, and dielectric resonator antenna (DRA) were used to present the pattern diversity. Two dipoles with an electromagnetic bandgap (EBG) surface were used to generate the left-hand circular polarization (LHCP) and right-hand circular polarization (RHCP) with pattern and polarization diversity. A brief discussion about Massive MIMO antenna systems and industrial Massive MIMO systems with pattern and polarization diversity was also presented with some published literary works. This chapter could be useful for the industry as well as academics to understand the phenomena of pattern and polarization diversities and to design new MIMO antennas for the modern communication systems.

REFERENCES

1. C. B. Dietrich Jr., K. Dietze, J. R. Nealy, and W. L. Stutzman, "Spatial, polarization, and pattern diversity for wireless handheld terminals," *IEEE Transactions on Antennas and Propagation*, vol. 49, no. 9, pp. 1271–1281, September 2001.
2. R. H. Clarke, "A statistical theory of mobile-radio reception," *Bell System Technical Journal*, vol. 74, pp. 957–1000, July–August 1968.
3. F. T. Dagefu, J. Oh, J. Choi, and K. Sarabandi, "Measurements and physics-based analysis of co-located antenna pattern diversity system," *IEEE Transactions on Antennas and Propagation*, vol. 61, pp. 5725–5734, November 2013.
4. L. Mouffok and F. Ghanem, "Wideband collocated antennas for radiation pattern diversity applications," *IEEE International Symposium on Antennas and Propagation (APSURSI)*, 2016. doi: 10.1109/APS.2016.7695977.
5. S. Nikolaou, R. Bairavasubramanian, C. Lugo, I. Carrasquillo, D. C. Thompson, G. E. Ponchak, J. Papapolymerou, and M. M. Tentzeris, "Pattern and frequency reconfigurable annular slot antenna using PIN diodes," *IEEE Transactions on Antennas and Propagation*, vol. 54, No. 2, pp. 439–448, February 2006.

6. G. H. Huff, J. Feng, S. Zhang, and J. T. Bernhard, "A novel radiation pattern and frequency reconfigurable single turn square spiral microstrip antenna," *IEEE Microwave and Wireless Components Letters*, vol. 13, no. 2, pp. 57–59, February 2003.
7. S.-H. Chen, J.-S. Row, and K.-L. Wong, "Reconfigurable square-ring patch antenna with pattern diversity," *IEEE Transactions on Antennas and Propagation*, vol. 55, no. 2, February 2007.
8. L. Low and R.J. Langley, "Single feed antenna with radiation pattern diversity," *Electronic Letters*, vol. 40, no. 16, pp. 975–976, August 2004.
9. Y. Dong, J. Choi, and T. Itoh, "Vivaldi antenna with pattern diversity for 0.7 to 2.7 GHz cellular band applications," *IEEE Antennas And Wireless Propagation Letters*, vol. 17, no. 2, pp. 247–250, February 2018.
10. N. Rezazadeh and L. Shafai, "A pattern diversity antenna for ambient RF energy harvesting in multipath environments," *18th International Symposium on Antenna Technology and Applied Electromagnetics (ANTEM)*, 2018.
11. S. Ghosh, T.-N. Tran, and T. Le-Ngoc, "Miniaturized four-element diversity PIFA," *IEEE Antennas and Wireless Propagation Letters*, vol. 12, pp. 396–400, 2013.
12. A. Elsherbini and K. Sarabandi, "UWB high-isolation directive coupled sectorial-loops antenna pair," *IEEE Antennas and Wireless Propagation Letters*, vol. 10, 2011.
13. D. C. Cox, "Antenna diversity performance in mitigating the effects of portable radiotelephone orientation and multipath propagation," *IEEE Transactions on Communications*, vol. 31, pp. 620–628, May 1983.
14. J. Oh and K. Sarabandi, "Compact, low-profile, common aperture polarization and pattern diversity antennas," *IEEE Transactions on Antennas and Propagation*, vol. 62, pp. 569–576, 2013.
15. F. Yang and Y. Rahmat-Samii, *Electromagnetic Band Gap Structures in Antenna Engineering*, RF and Microwave Engineering Series, Cambridge, 2008.
16. H. Huang, Y. Liu, S. Zhang, and S. Gong, "Uniplanar ultrawideband polarization diversity antenna with dual band-notched characteristics," *IEEE Antennas and Wireless Propagation Letters*, vol. 13, pp. 1745–1748, 2014.
17. H. Huang, Y. Liu, S. Zhang, and S. Gong, "Uniplanar differentially driven ultrawideband polarization diversity antenna with band-notched characteristics," *IEEE Antennas and Wireless Propagation Letters*, vol. 14, pp. 563–566, 2014.
18. J.-H. Han and N.-H. Myung, "Novel feed network for circular polarization antenna diversity," *IEEE Antennas and Wireless Propagation Letters*, vol. 13, pp. 979–982, 2014.
19. J.-S. Kim, K.-H. Shin, S.-M. Park, W.-K. Choi, and N.-S. Seong, "Polarization and space diversity antenna using inverted-F antennas for RFID reader applications," *IEEE Antennas and Wireless Propagation Letters*, vol. 5, pp. 265–268, 2006.
20. L. Zou and C. Fumeaux, "A cross-shaped dielectric resonator antenna for multifunction and polarization diversity applications," *IEEE Antennas and Wireless Propagation Letters*, vol. 10, pp. 742–745, 2011.
21. M. S. Nishamol, V. P. Sarin, D. Tony, C. K. Aanandan, P. Mohanan, and K. Vasudevan, "An electronically reconfigurable microstrip antenna with switchable slots for polarization diversity", *IEEE Transactions on Antennas and Propagation*, vol. 59, no. 9, pp. 3424–3427, September 2011.
22. S.-Y. Wang, D.-Y. Lai, and F.-C. Chen, "A low-profile switchable quadri-polarization diversity aperture-coupled patch antenna," *IEEE Antennas and Wireless Propagation Letters*, vol. 8, pp. 522–524, 2009.
23. Y. J. Sung, "Reconfigurable patch antenna for polarization diversity," *IEEE Transactions on Antennas and Propagation*, vol. 56, no. 9, pp. 3053–3054, September 2008.
24. A. Khaleghi and M. Kamyab, "Reconfigurable single port antenna with circular polarization diversity," *IEEE Transactions on Antennas and Propagation*, vol. 57, no. 2, pp. 555–559, September 2009.

25. J.-S. Row and Y.-H. Wei, "Wideband reconfigurable crossed-dipole antenna with quad-polarization diversity," *IEEE Transactions on Antennas and Propagation*, vol. 66, pp. 2090–2094, 2009.
26. Y.-F. Wu, C.-H. Wu, D.-Y. Lai, and F.-C. Chen, "A reconfigurable quadri-polarization diversity aperture-coupled patch antenna," *IEEE Transactions on Antennas and Propagation*, vol. 55, no. 3, pp. 1009–1012, September 2007.
27. https://www.fcc.gov.
28. T.L. Marzetta, "Noncooperative cellular wireless with unlimited numbers of base station antennas," *IEEE Transactions on Wireless Communications*, vol. 9, no. 11, pp. 3690–3600, 2010.
29. E. Björnson. A look at an LTE-TDD Massive MIMO product, http://ma-mimo.ellintech.se/2018/08/27/a-look-at-an-lte-tdd-massive-mimo-product/
30. N. Tawa, T. Kuwabara, Y. Maruta, M. Tanio, and T. Kaneko, 28 GHz downlink multi-user MIMO experimental verification using 360 elements digital AAS for 5G Massive MIMO, *European Microwave Conference*, EuMC, 2018.
31. H. Yuan, C. Wang, Y. Li, N. Liu, and G. Cui, "The design of array antennas used for Massive MIMO system in the fifth generation mobile communication," *11th International Symposium on Antennas, Propagation and EM Theory (ISAPE)*, 2016, Guilin, China.
32. L. Zhao, A. Chen, J. Zhang, S. Zheng, and Y. Yin, "A Single Radiator with Four Decoupled Ports for Four by Four MIMO Antennas and Systems," *IEEE International Symposium on Antennas and Propagation & USNC/URSI National Radio Science Meeting*, San Diego, CA, USA, 2017.
33. R. P. Astuti and B. S. Nugroho, "Dual polarized antenna decoupling for 60 GHz planar Massive MIMO," *International Conference on Signals and Systems (ICSigSys)*, Sanur, Indonesia, 2017.
34. K. Oikawa, K. Yuri, N. Honma, and K. Nishimori, "Compact Massive MIMO antenna using cubic arrangement suitable for indoor base station," *International Symposium on Antennas and Propagation (ISAP)*, Busan, Korea (South), Korea, 2018.
35. M. Vamshikomandla, G. Mishra, and S. K. Sharma, "Dual slant polarized cavity backed Massive MIMO panel array antenna with digital beamforming," *IEEE International Symposium on Antennas and Propagation & USNC/URSI National Radio Science Meeting*, San Diego, CA, USA, 2017.
36. L.-P. Shen, H. Wang, W. Lotz, and H. Jamali, "Compact Dual Polarization 4×4 MIMO Multi-Beam Base Station Antennas," *IEEE International Symposium on Antennas and Propagation & USNC/URSI National Radio Science Meeting*, Boston, MA, USA, 2018.
37. H. H. Tran and T. T. Le, "Ultrawideband, high-gain, high-efficiency, circularly polarized Archimedean spiral antenna," *AEU-International Journal of Electronics and Communications*, vol. 109, pp. 1–7, 2019.
38. P. Mohammadi, M. Rezvani, and T. Siahy, "A circularly polarized wideband magneto-electric dipole antenna with simple structure for BTS applications," *AEU-International Journal of Electronics and Communications*, vol. 150, pp. 92–97, 2019.
39. U. Ullah, I. B. Mabrouk, and S. Koziel, "A compact circularly polarized antenna with directional pattern for wearable off-body communications," *IEEE Antennas and Wireless Propagation Letters*, vol. 18, no. 12, pp. 2523–2527, December 2019.
40. N. Hussain, H. H. Tran, and T. T. Le, "Single-layer wideband high-gain circularly polarized patch antenna with parasitic elements," *AEU-International Journal of Electronics and Communications*, vol. 113, pp. 1–8, 2020.
41. P. Qin, Y. J. Guo, and C. Liang, "Effect of antenna polarization diversity on MIMO system capacity," *IEEE Antennas and Wireless Propagation Letters*, vol. 9, pp. 1092–1095, 2010.

42. Z.-X. Liang, D.-C. Yang, X.-C. Wei, and E.-P. Li, "Dual-band dual circularly polarized microstrip antenna with two eccentric rings and an arc-shaped conducting strip," *IEEE Antennas and Wireless Propagation Letters*, vol. 15, pp. 834–837, 2016.
43. D.S. Chandu and S. S. Karthikeyan, "A novel broadband dual circularly polarized microstrip-fed monopole antenna," *IEEE Transactions on Antennas and Propagation*, vol. 65, no. 3, pp. 1410–1415, March 2017.
44. H.H. Tran, N. Hussain, and T.T. Le, "Low-profile wideband circularly polarized MIMO antenna with polarization diversity for WLAN applications," *AEU-International Journal of Electronics and Communications*, vol. 108, pp. 172–180, 2019.
45. B. M. Lee, "Calibration for channel reciprocity in industrial Massive MIMO antenna systems," *IEEE Transactions on Industrial Informatics*, vol. 14, no. 1, pp. 221–230, January 2018.
46. T. L. Marzetta, "Noncooperative cellular wireless with unlimited numbers of base station antennas," *IEEE Transactions on Wireless Communications*, vol. 9, no. 11, pp. 3590–3600, November 2010.
47. E. Bjornson, E. G. Larsson, and T. L. Marzetta, "Massive MIMO: ten myths and one critical question," *IEEE Communications Magazine*, vol. 54, no. 2, pp. 114–123, February 2016.
48. T. L. Marzetta, "Massive MIMO: an introduction," *Bell Labs Technical Journal*, vol. 20, pp. 11–22, 2015.
49. F. Tramarin, S. Vitturi, M. Luvisotto, and A. Zanella, "On the use of IEEE 802.11n for industrial communications," *IEEE Transactions on Industrial Informatics*, vol. 12, no. 5, pp. 1877–1886, October 2016.
50. L. Seno, G. Cena, S. Scanzio, A. Valenzano, and C. Zunino, "Enhancing communication determinism in Wi-Fi networks for soft real-time industrial applications," *IEEE Transactions on Industrial Informatics*, vol. 13, no. 2, pp. 866–876, April 2017.
51. P. Park, P. D. Marco, and K. H. Johansson, "Cross-layer optimization for industrial control applications using wireless sensor and actuator Mesh networks," *IEEE Transactions on Industrial Electronics*, vol. 64, no. 4, pp. 3250–3259, April 2017.
52. G. Das, N. K. Sahu, A. Sharma, R. K. Gangwar, M. S. Sharawi, "Dielectric resonator based 4-element 8-port MIMO antenna with multi-directional pattern diversity," vol. 13, no. 1, 2019. doi:10.1049/iet-map.2018.5081.
53. B. Feng, J. Lai, Q. Zeng, and K. L. Chung, "A dual-wideband and high gain magneto-electric dipole antenna and its 3D MIMO system with metasurface for 5G/WiMAX/WLAN/X-band applications," *IEEE Access*, vol. 6, pp. 33387–33398, 2018.
54. P. Chaudhary, A. Kumar, and B.K. Kanaujia, "A low-profile wideband circularly polarized MIMO antenna with pattern and polarization diversity," *International Journal of Microwave and Wireless Technologies*, vol. 12, pp. 1–7, 2019.

6 Compact Printed Antenna Designs: Need for UWB Communications

Dr. Rakesh Nath Tiwari
Uttarakhand Technical University, Dehradun

Dr. Prabhakar Singh
Galgotias University, Greater Noida

Prof. Binod Kumar Kanaujia
Jawaharlal Nehru University, New Delhi

CONTENTS

6.1 INTRODUCTION

The Federal Communication Commission (FCC) approved and allocated the frequency range from 3.1–10.6GHz for ultra wideband (UWB) communication systems [1]. In the recent years, these UWB systems have emerged as the most prominent technologies for the wireless industries. UWB designs have the advantages such as low profile, low cost for consumer electronics, compact size, high data transmission rate, easy integration with microwave circuits, and easy design. Also, UWB antennas are very useful in biomedical applications such as microwave imaging and lung cancer detection [2–4]. There are few conventional techniques to achieve the UWB characteristics categorized as the slot antenna in which different shapes of wide slots are used in the ground plane. The second method is the direct modification in the radiating elements, such as ring-shaped patch. The third approach is the creation of defects in the ground plane with CPW/strip line feeding techniques. The fourth method focuses on the modification in the feeding line to improve the antenna bandwidth and consistency in the radiation pattern. Some novel efforts are made by many researchers to design UWB printed antennas. The use of proper feeding methods such as microstrip line and W-shaped feed stub and modifying the ground plane [5–8] enable the designs to exhibit UWB characteristics. Further, a printed E-shaped slot antenna fed by microstrip line and CPW feeding for bandwidth enhancement was also reported, but the overall size of antenna was quite large [9]. A parasitically loaded CPW-fed antenna with a circular hat patch and a slot loaded in the ground plane was reported for broad bandwidth and monopole-like radiation pattern [10]. Some designs of UWB antennas such as asymmetrical slot antenna using microstrip line and modified CPW-fed antenna with a very thin substrate were investigated [11,12]. Different CPW-fed fractal-shaped antennas were also designed to improve the bandwidth [13–15]. The modifications in conventional monopole antennas provide high bandwidths, constant group delay, and stable radiation patterns [16,17]. Recently, CPW-fed symmetric ring patches embedded with slits and fork-shaped patches, a step-shaped microstrip antenna, a hexagonal-shaped antenna with additional fractal elements, and symmetric circular slots have been reported to enhance the antenna bandwidth and gain [18–21]. However, in the process of improving the bandwidth, most of the designs become large in size or complex in structure.

In this chapter, some novel designs of UWB printed antennas are described, which not only cover the modern wireless spectrum, but simultaneously provide the compatibility and miniaturized structures. The various aspects of designing monopole UWB antennas such as (i) modification of the radiating patch, (ii) modification in feed line, and (iii) modification in ground plane are covered [22–25]. All the designs are first modeled and optimized using CST Microwave Studio and then fabricated to validate the results for practical implications.

6.2 AN ASYMMETRIC U-SHAPED PRINTED MONOPOLE ANTENNA EMBEDDED WITH A T-SHAPED STRIP

In this section, a compact and planar design of an asymmetric U-shaped printed monopole antenna embedded with a T-shaped strip with a defected ground plane is described. The overall dimensions of this antenna including the ground plane are 34 (L_s)×20 (W_s)×1.6 (H) mm^3, and the antenna is printed on a low-cost FR4 material and fed by a 50 Ω microstrip line.

6.2.1 Antenna Configuration and Its Specifications

Figure 6.1a–d shows the evolution of the monopole planar antenna. A rectangular portion of dimensions 15×16 mm^2 is removed from the patch, resulting in a symmetrical U-shaped monopole antenna (Figure 6.1b). Further, the left arm of the symmetrical U-shaped monopole antenna is reduced and a center strip is used (Figure 6.1c). Finally, a horizontal strip of dimensions $L_4 \times W_4$ is placed at the top of the center strip, which makes the asymmetric U-shaped antenna embedded with a T-shaped center strip (Figure 6.1d) for wideband applications. Figure 6.2 shows the top, bottom, front, and side view geometries of the planar antenna.

Removing the ground plane conductor (dimensions $L_g \times W_s$) of constant dimensions ($L \times W$) from the two ends modifies the ground plane into a stair shape. Further, a rectangular slot is used in the ground plane to improve the impedance of the antenna. The relative position of the radiator and the ground plane is taken as $g = 4.5$ mm. The details of the antenna design are presented in Table 6.1. The return loss values for different antenna structures are depicted in Figure 6.3. Antenna-1 operates in dual-band mode with lower and higher impedance bandwidths of 2.37–3.54 and 3.99–6.73 GHz, respectively. In case of antenna-2, the higher resonance frequency shifted toward higher side, while the lower resonance frequency remains the same as in the case of antenna-1. However, the matching condition is deteriorated because of the removal of the conductor from the patch, which certainly changes the impedance of the antenna. The operating bands for antenna-2 are 2.26–3.05 GHz (lower band) and 4.97–6.37 GHz (upper band). Moreover, when the center strip is used in

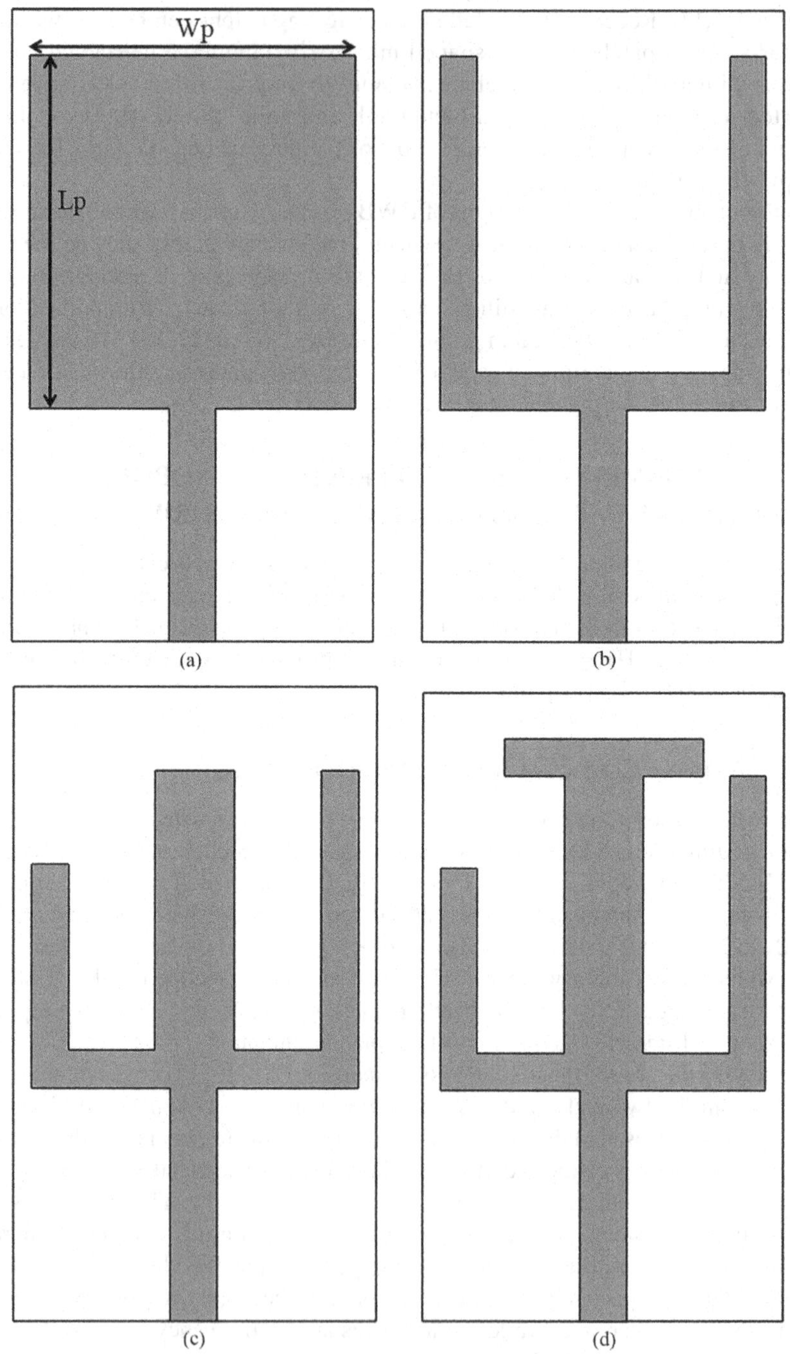

FIGURE 6.1 Evolution of the antenna:(a) antenna-1, (b) antenna-2, (c) antenna-3, and (d) antenna-4 (final).

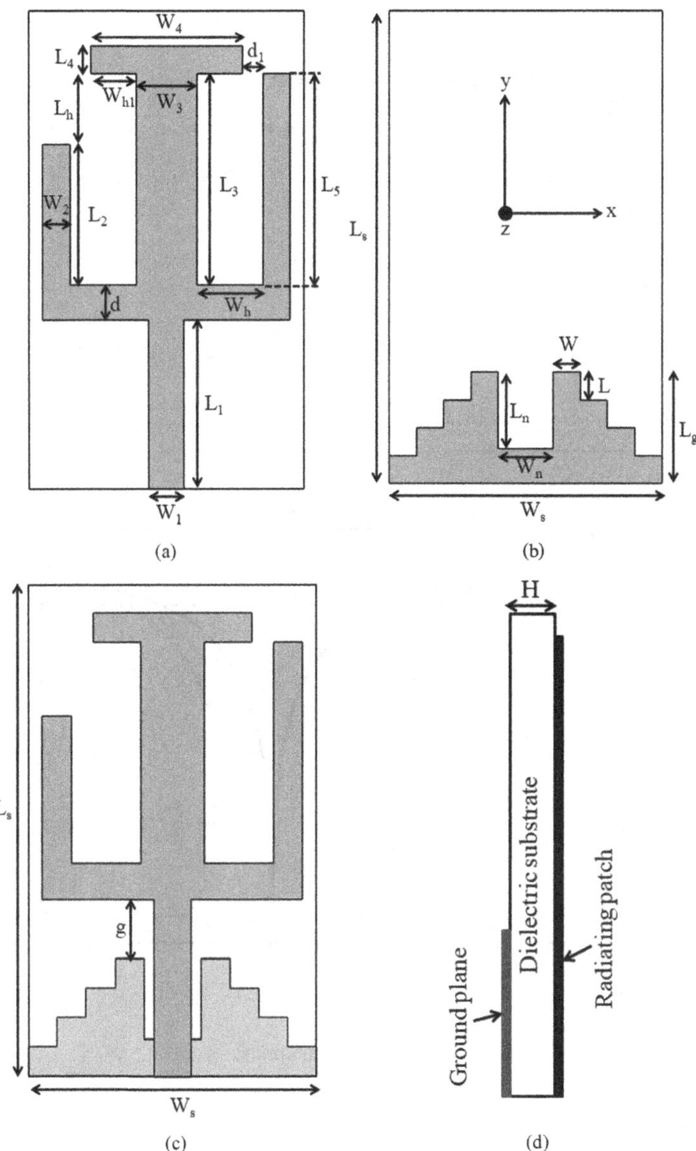

FIGURE 6.2 Geometry of the antenna: (a) top view, (b) bottom view, (c) front view, and (d) side view.

the antenna (antenna-3) again a dual response is observed, but with improved operating bands, i.e., 2.43–3.63 GHz (lower band) and 4.37–7.13 GHz (upper band), respectively. Further, to achieve wideband characteristics, a horizontal strip of dimensions ($L_4 \times W_4$) is placed at the top of the central strip. Consequently, in the final antenna-4, a wide bandwidth of 4.89 GHz (2.26–7.15 GHz) is obtained.

TABLE 6.1
Design Parameters of the Antenna

Parameter	Unit (mm)	Parameter	Unit (mm)
$L_s \times W_s$	34×20	$L_p \times W_p$	19×18
$L_1 \times W_1$	12.5×2.2	$L_2 \times W_2$	10×2
$L_3 \times W_3$	15×4.4	$L_4 \times W_4$	2×11
L_5	15	D	2
H	1.6	$L_n \times W_n$	6×4
$L_g \times W_s$	8×20	G	4.5
$L \times W$	2×2	W_h	4.8
W_{h1}	3.3	L_h	5
d_1	1.5		

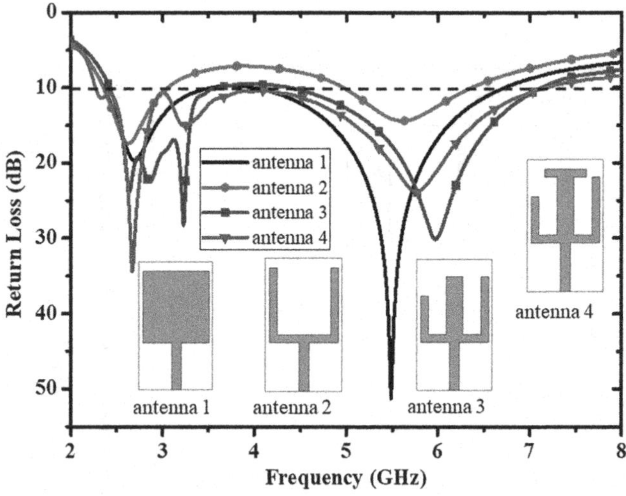

FIGURE 6.3 Return loss variation of different antennas.

6.2.2 Parametric Study of the Antenna

Various design parameters are studied to investigate their effects on the impedance bandwidth of the antenna. The variations of asymmetric U-arm strips and the T-strip of the radiator affect the antenna resonance and its bandwidth. The variation of the number of step slots (N) created in the ground plane is also studied. Finally, the variation of slit in the ground plane is investigated to optimize the impedance bandwidth.

6.2.2.1 Effect of the Radiating Patch

The simulated surface current distributions of the antenna at different resonance frequencies of 2.40, 2.68, 3.24, and 5.77 GHz are shown in Figure 6.4. From Figure 6.4a, it

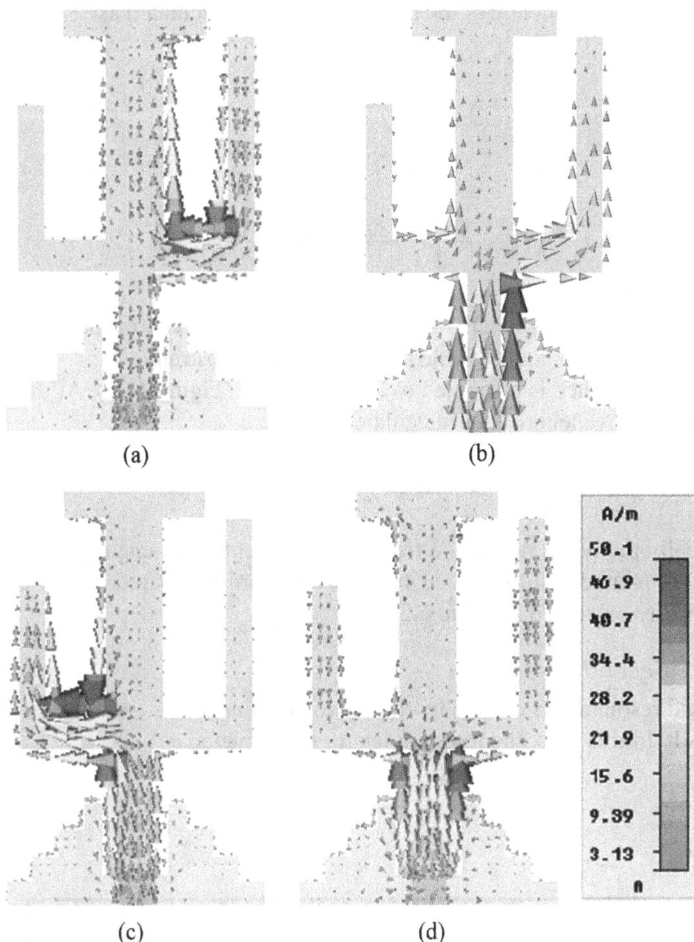

FIGURE 6.4 Surface current distribution at various sample frequencies: (a) $f=2.40$ GHz, (b) $f=2.68$ GHz, (c) $f=3.24$ GHz, and (d) $f=5.77$ GHz.

is observed that the first resonance at 2.40 GHz is contributed by both the T-strip (L_3) and right asymmetric U-arm strip (L_5) of the radiator and the second resonance (2.68 GHz) is created due to L_5 and microstrip line (Figure 6.4b). The current density concentrated at the left arm (L_2) and microstrip line reveals that the third resonance (3.24 GHz) is totally contributed by L_2 and microstrip line (Figure 6.4c). From Figure 6.4d, it is clear that the fourth resonance mode (5.77 GHz) is produced solely by the microstrip line because the current density is maximum on the strip line at this frequency.

6.2.2.1.1 First Resonance

A theoretical analysis is presented to calculate the various resonance modes created due to the effect of the arms of the asymmetric U-shaped radiator, T-strip, and microstrip line.

The resonance frequency of the radiating patch is calculated as

$$f_r = \frac{c}{2L_r\sqrt{\varepsilon_e}} \tag{6.1}$$

in which,

$$\varepsilon_e = \frac{\varepsilon_r + 1}{2} + \frac{\varepsilon_r - 1}{2}\left(1 + \frac{12H}{Wp}\right)^{-1/2}$$

where, L_r is the resonant length, ε_r is the relative permittivity of the substrate, W_p is the width of the patch, and H is the thickness of the substrate.

The first resonance is attributed to T-strip and L_5 (Figure 6.4a). At the first resonance, the effective length L_{r1} is calculated as

$$L_{r1} = L_3 + (L_5 - d_1) + (W_h - W_2) + W_{h1} + (L_4 - d) \tag{6.2}$$

Theoretically predicted resonance frequency is calculated using eqs. (6.1) and (6.2). The CST-simulated results plotted in Figure 6.5 are compared with the predicted results and are presented in Table 6.2.

6.2.2.1.2 Second Resonance

The second resonance is excited by the microstrip line and L_5, and it is evident from Figure 6.4b. In this case, the resonant length is calculated as

$$L_{r2} = L_5 + L_1 + d \tag{6.3}$$

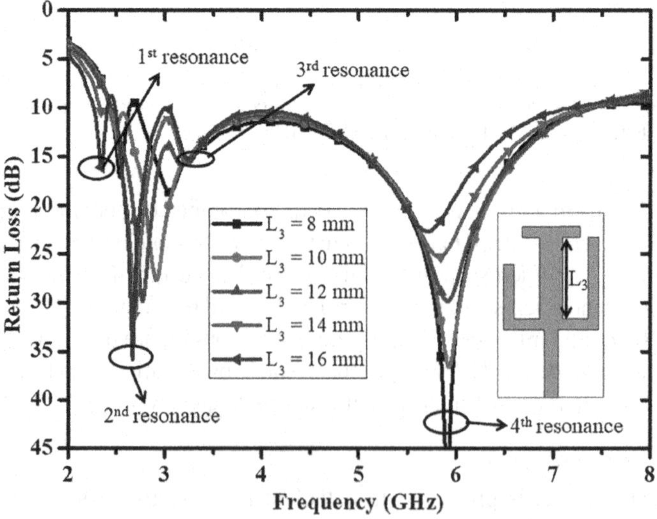

FIGURE 6.5 Return loss variation of the antenna for different values of L_3.

TABLE 6.2
Comparison for the First Resonance Frequency

		Resonance Frequency (GHz)		
L_3(mm)	L_{r1} (mm)	f_{r1} (Theory)	f_{r1} (Simulated)	% Difference
8	27.6	2.76	2.55	8.24
10	29.6	2.57	2.49	3.21
12	31.6	2.41	2.45	1.63
14	33.6	2.27	2.37	4.23
16	35.6	2.14	2.34	8.55

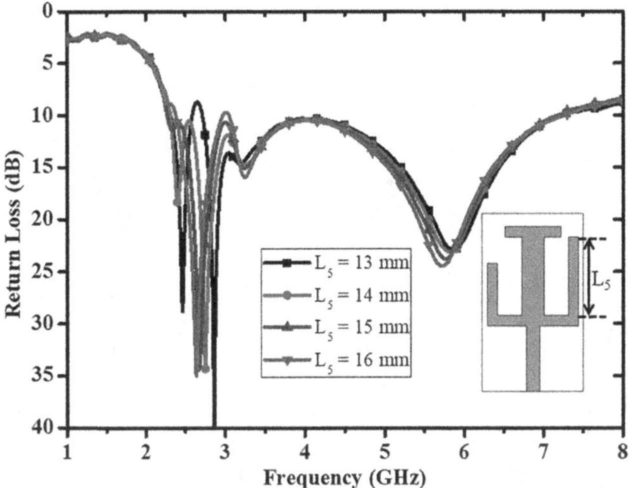

FIGURE 6.6 Return loss variation of the antenna for different values of L_5.

The resonance frequencies are calculated for different values of L_5 as shown in Figure 6.6, and a comparison with the simulated results is given in Table 6.3.

6.2.2.1.3 Third Resonance

The third resonance is observed due to the current concentration on the feed line and L_2, which is verified by Figure 6.4c. The corresponding resonant length for this frequency can be calculated as

$$L_{r3} = L_2 + L_1 + d + W_h - L_h \tag{6.4}$$

The resonance frequency calculated for different values of L_2 is validated with the simulated results, as shown in Figure 6.7 and presented in Table 6.4. A good agreement between theory and simulation results validates the proposed designed formula for the resonance frequency.

TABLE 6.3
Comparison for the Second Resonance Frequency

		Resonance Frequency (GHz)		
L_5(mm)	L_{r2} (mm)	f_{r2} (Theory)	f_{r2} (Simulated)	% Difference
13	27.5	2.76	2.86	3.50
14	28.5	2.67	2.75	2.91
15	29.5	2.58	2.67	3.37
16	30.5	2.50	2.64	5.30

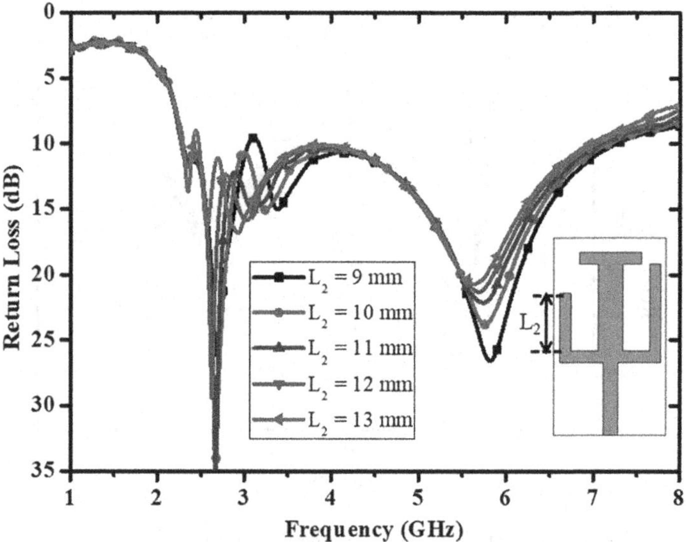

FIGURE 6.7 Return loss variation of the antenna for different values of L_2.

TABLE 6.4
Comparison for the Third Resonance Frequency

		Resonance Frequency (GHz)		
L_2(mm)	L_{r3} (mm)	f_{r3} (Theory)	f_{r3} (Simulated)	% Difference
9	23.3	3.27	3.38	3.25
10	24.3	3.13	3.23	3.10
11	25.3	3.01	3.10	2.90
12	26.3	2.90	3.03	4.29
13	27.3	2.79	2.93	4.78

6.2.2.1.4 Fourth Resonance

From Figure 6.4d, it is clear that the fourth resonance is excited due to the strong current density on the feed line. Hence, only the feed line length will give the fourth resonance and it can be calculated using the resonant length $L_{r4} = L_1$ in eq. (6.1). The predicted resonance frequency is 6.09 GHz, while the simulated one is 5.77 GHz with 5.54% difference, which are in good agreement.

It is noted that, while varying an antenna parameter to study the impedance bandwidth and resonance frequencies, the rest of antenna dimensions are fixed and given in Table 6.1. The variation of return loss for different values of L_3 is shown in Figure 6.5. From the figure, it can be observed that the first and second resonances (2.4 and 2.68 GHz) are controlled by L_3. The maximum bandwidth 4.97 GHz (2.33–7.30 GHz) is obtained for $L_3 = 14$ mm. Therefore, L_3 can be optimized for Bluetooth/WLAN applications.

The variation of return loss for different values of L_5 is plotted in Figure 6.6. From the figure, it is clear that the first resonance frequency (2.4 GHz) is controlled by varying the value of L_5. In this case, the maximum bandwidth achieved is 4.91 GHz (2.27–7.18 GHz).

Figure 6.7 shows that the third resonance (3.24 GHz) is controlled by L_2, and it is shifted toward the lower side as the value of L_2 increases. The maximum bandwidth is found to be 4.91 GHz (2.27–7.18 GHz) for $L_2 = 10$ mm.

In Figure 6.8, both the values of L_2 and L_5 are varied for the optimization and to achieve maximum impedance bandwidth. The optimized values of L_2 and L_5 are 12 and 17 mm, respectively, at which the highest bandwidth of 4.90 GHz (2.34–7.24 GHz) is obtained.

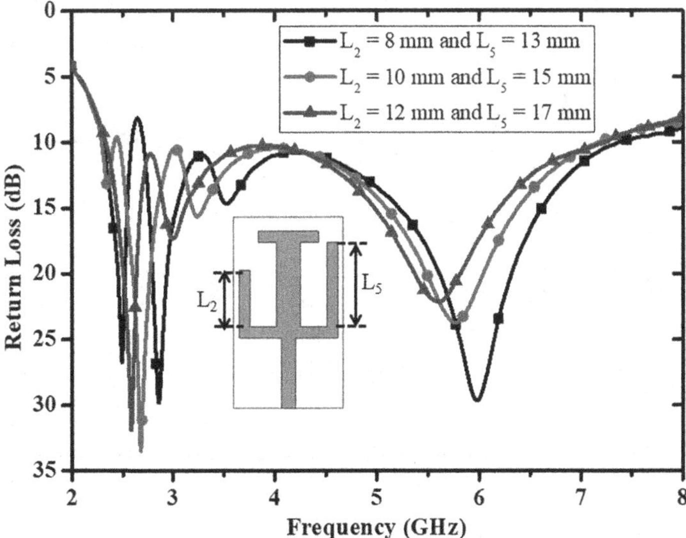

FIGURE 6.8 Return loss variation of the antenna for different values of L_2 and L_5.

6.2.2.2 Effect of the Gap between the Ground Plane and Radiating Patch

The relative gap (g) between the ground plane and the radiating patch affects the impedance bandwidth, and it acts as an impedance matching network. The return loss of the antenna at different gaps 'g' is shown in Figure 6.9. At the optimized value of $g=4.0$ mm, the maximum bandwidth of 4.90 GHz (2.28–7.18 GHz) is achieved. It is also noted that above and below $g=4.0$ mm, the antenna exhibits dual band characteristics.

6.2.2.3 Effect of the Ground Plane Structure

The dimensions of the ground plane are a very important parameter to optimize the impedance of the antenna, which ultimately improves the antenna bandwidth. In Figure 6.10, the return loss is plotted for different values of step slots (N) incorporated in the ground plane. It shows that the bandwidth significantly depends on the values of N. It is interesting to note that the lower end of the operating band is constant at 2.27 GHz, but the upper end of the band decreases abruptly as N goes to zero. The maximum bandwidth of 4.92 GHz (2.27–7.19 GHz) is observed at $N=3$. The variation of return loss for different values of L_n (Figure 6.11) shows that the bandwidth of the antenna improves upto 5.16 GHz (2.29–7.45 GHz) as L_n increases. Figure 6.12 shows the variation of return loss with different values of W_n. From the figure, it is clear that the fourth resonance is shifted toward lower side with increasing values of W_n. At the optimum value of $W_n=4.5$ mm, the maximum bandwidth of 5.26 GHz (2.27–7.53 GHz) is achieved.

6.2.3 Comparative Results of the Antenna

The return loss of the antenna is measured using Agilent N5230A vector network analyzer and compared with the simulated results (Figure 6.13). The measured and

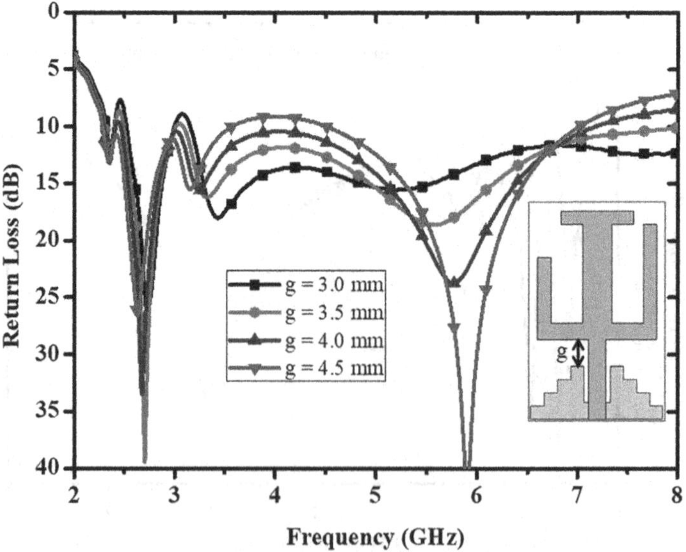

FIGURE 6.9 Return loss variation of the antenna for different values of g.

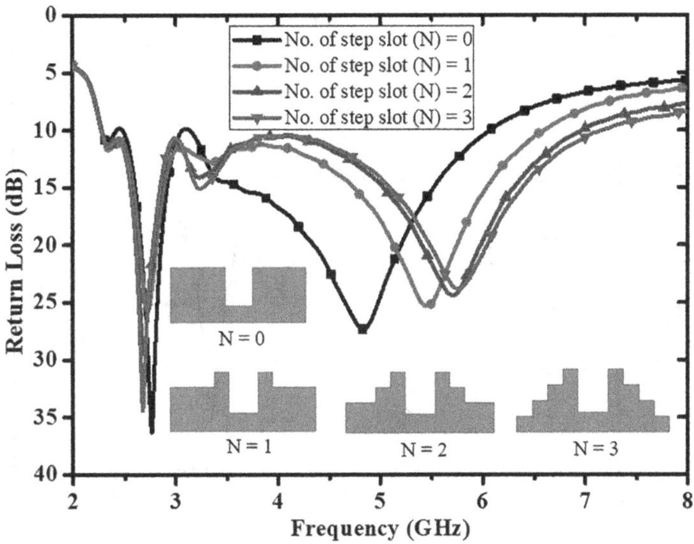

FIGURE 6.10 Return loss variation of the antenna for different numbers of step slots (N).

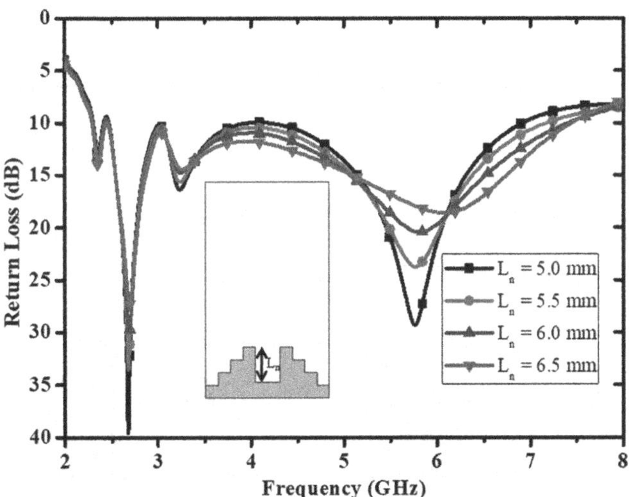

FIGURE 6.11 Return loss variation of the antenna for different values of L_n.

simulated antenna bandwidths are 4.47 GHz (2.28–6.75 GHz) and 4.92 GHz (2.27–7.19 GHz), respectively, which are in good agreement. The radiation efficiency and gain of the antenna are plotted in Figure 6.14. From the figure, it is clear that the simulated radiation efficiency is always above 70.0% throughout the entire operating band. The realized antenna gains are in good agreement and vary between 2.14–4.83 dBi (simulated) and 2.20–4.91 dBi (measured).

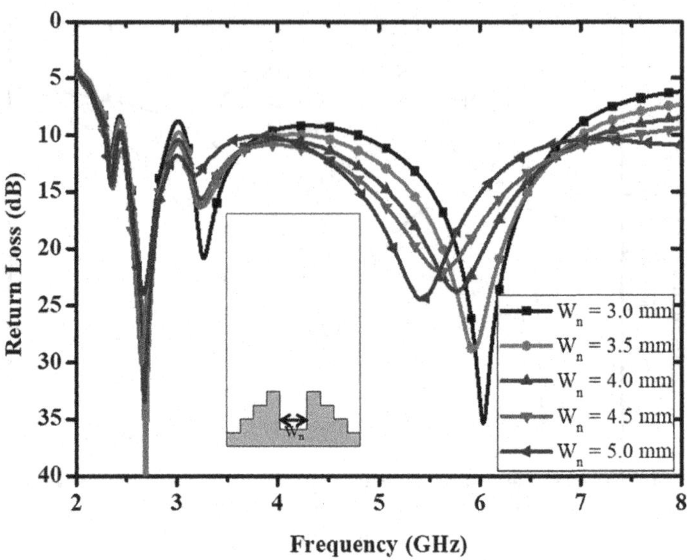

FIGURE 6.12 Return loss variation of the antenna for different values of W_n.

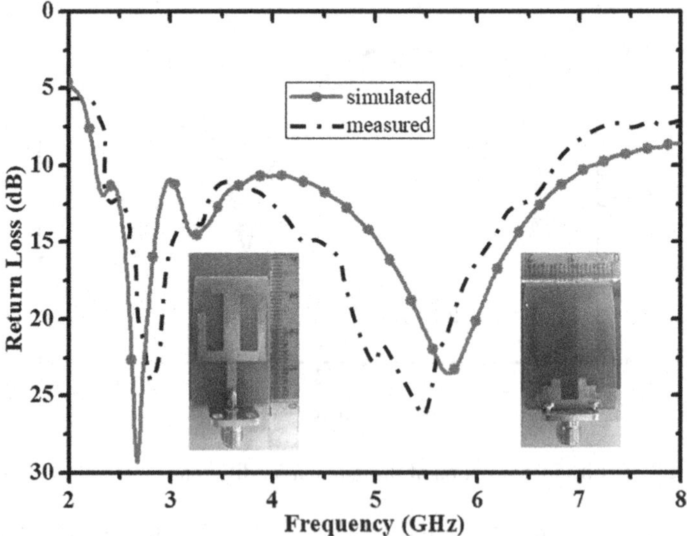

FIGURE 6.13 Simulated and measured return loss of the antenna.

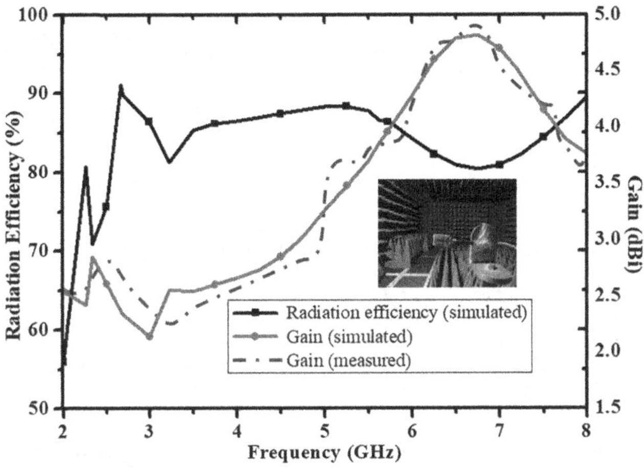

FIGURE 6.14 Measured and simulated gain and radiation efficiency.

The far-field radiation patterns are measured in a microwave-shielded anechoic chamber of dimensions $7.0 \times 5.0 \times 3.0\,m^3$ (length × width × height). The test antenna is placed at a distance of 3.2 m from the horn antenna. The simulated and measured co- and cross-polarization patterns are plotted in Figure 6.15 for frequencies 2.4, 4.0, and 5.5 GHz, respectively. Co-polarization plot at $\varphi = 0°$ (H-plane) is almost omnidirectional for the entire operating band with low cross-polarization level. At $\varphi = 90°$ (E-plane), the co-polarization shows bidirectional pattern at two sampling frequencies 2.4 and 4.0 GHz, while at 5.5 GHz, E-plane pattern shows a quasi-omnidirectional pattern due to high mode generation.

6.3 SMALL SIZE SCARECROW-SHAPED CPW- AND MICROSTRIP LINE-FED UWB ANTENNAS

This section describes a UWB printed antenna which consists of a small and compact scarecrow-shaped patch fed with CPW and microstrip line. Two half-circle and square-shaped slots are etched in the ground plane. This structure achieves a high impedance bandwidth, good gain, constant group delay, and stable radiation patterns over the entire operating frequency band. This section is organized into two parts: The first part describes the CPW-fed design, and the second part describes the microstrip line-fed UWB antenna.

6.3.1 CPW-Fed Scarecrow-Shaped Patch Antenna

6.3.1.1 Antenna Design

The geometry of the CPW-fed scarecrow-shaped patch antenna (SSPA) is shown in Figure 6.16.

The antenna structure consists of a scarecrow-shaped patch, with two half-circle slots and two square slots etched in the ground plane. The antenna is printed on a

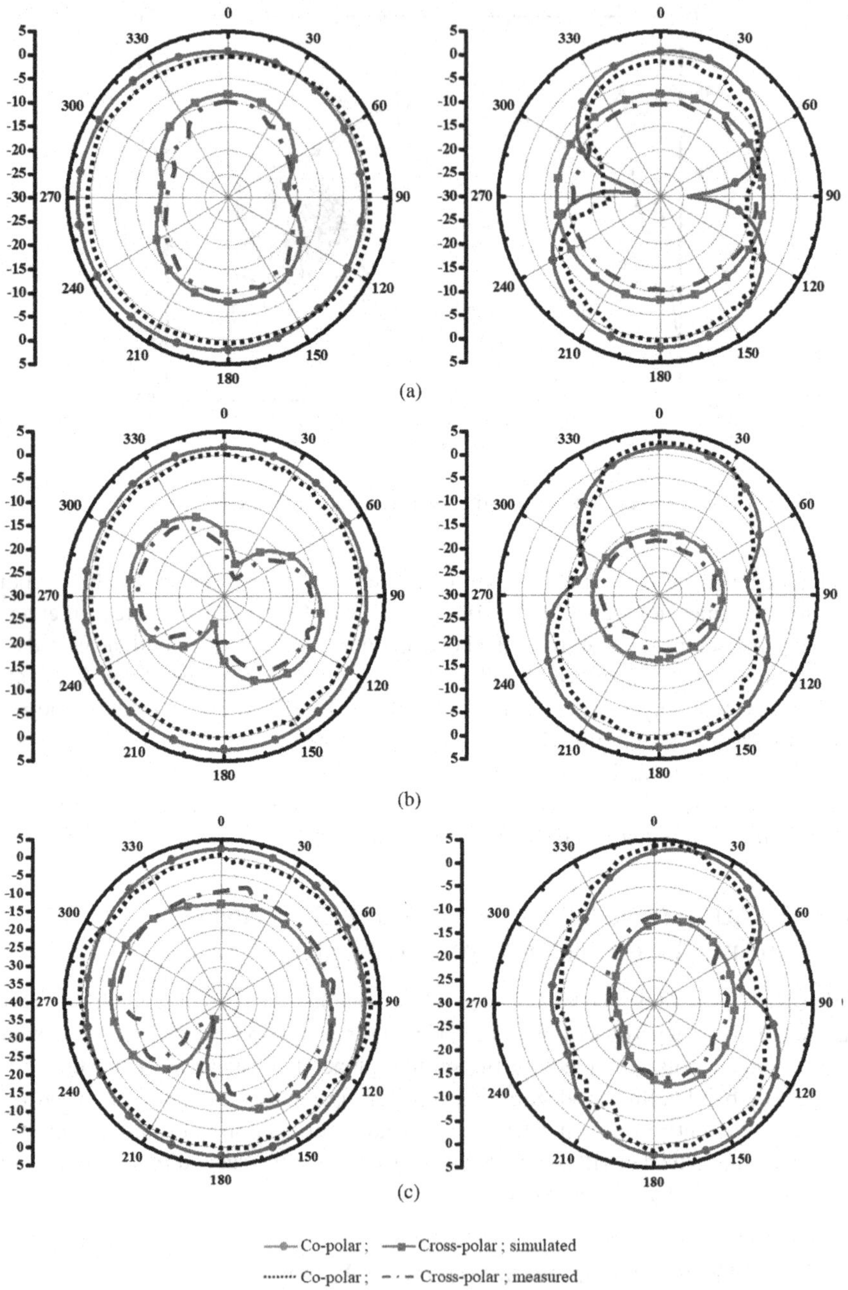

FIGURE 6.15 Simulated and measured radiation patterns of H-plane (left side) and E-plane (right side): (a) $f = 2.4\,\text{GHz}$, (b) $f = 4.0\,\text{GHz}$, and (c) $f = 5.5\,\text{GHz}$.

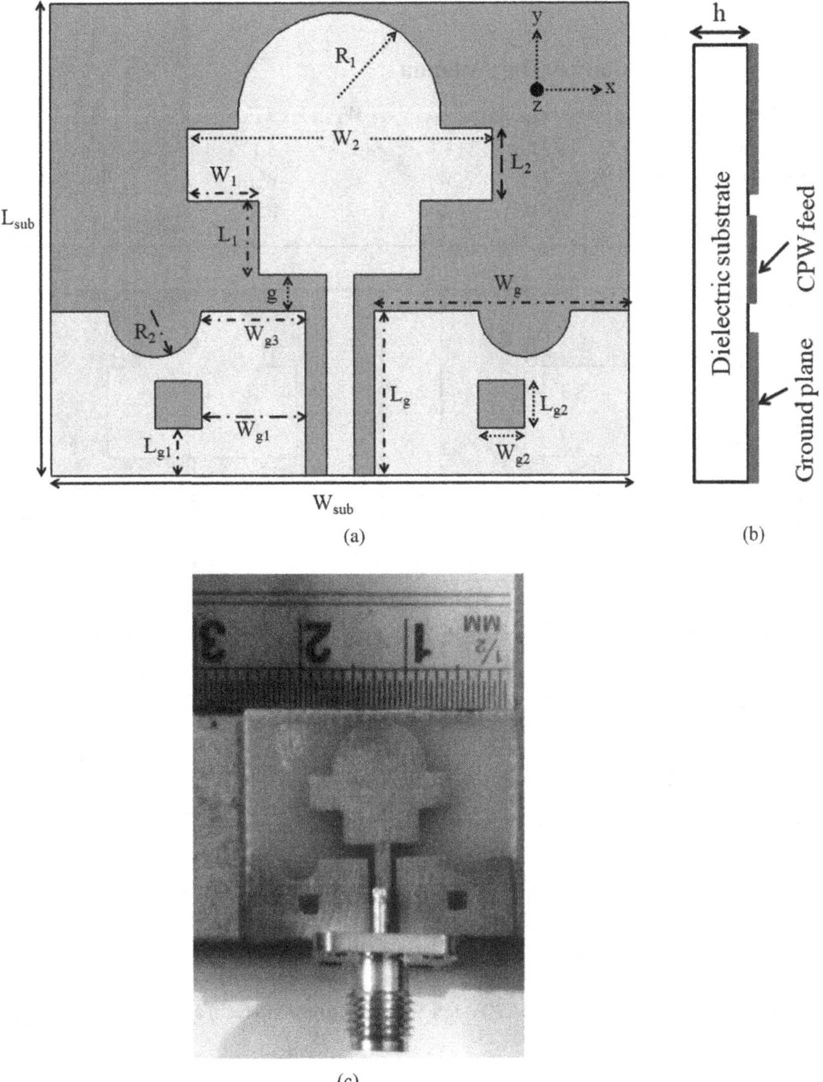

FIGURE 6.16 Geometry of the CPW-fed SSPA: (a) top view, (b) side view, and (c) the fabricated antenna.

FR4 substrate of thickness (h) 1.6 mm, dielectric constant $\varepsilon_r = 4.4$, and loss tangent 0.02. The overall size of the antenna is $25 \times 20 \times 1.6\,\text{mm}^3$. The width of the feed line is taken as 1.2 mm to achieve 50 Ω characteristic impedance. The separation between the center strip line and the ground plane is 0.88 mm. The optimized values of antenna dimensions are listed in Table 6.5.

Figure 6.17 presents the evolution of the antenna design. Initially, a conventional CPW-fed rectangular patch is designed (antenna-1). In the next step, a half-disk patch of radius R_1 is placed on the top of the rectangular patch (antenna-2). After that, a

TABLE 6.5
Physical Dimensions of the Antenna

Parameter	L_{sub}	W_{sub}	L_1	W_1	L_2	W_2	R_1	g
Unit (mm)	20	25	3.1	3.1	3.1	13.2	4.4	1.55
Parameter	R_2	L_g	W_g	L_{g1}	W_{g1}	L_{g2}	W_{g2}	W_{g3}
Unit (mm)	2	6.95	11.02	2	4.52	2	2	4.52

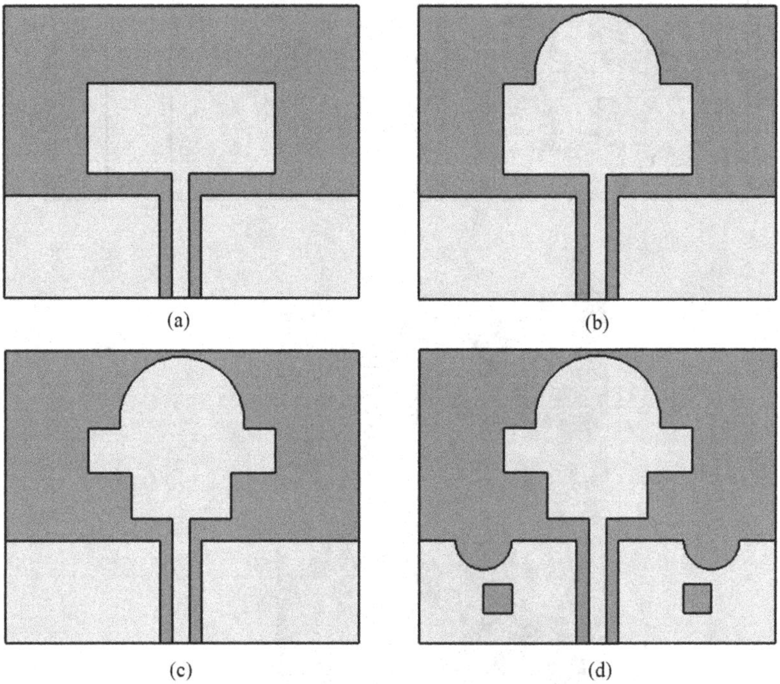

(a)

(b)

(c)

(d)

FIGURE 6.17 Steps to realize the CPW-fed SSPA: (a) antenna-1, (b) antenna-2, (c) antenna-3, and (d) antenna-4 (final).

portion of patch (of dimensions $L_1 \times W_1$) is removed from the two lower corners of the rectangular patch, which realizes a CPW-fed SSPA (antenna-3). Finally, two half-circle-shaped slots, each of radius R_2, are etched at the periphery of the ground plane and two square-shaped slots of dimensions $2 \times 2\,\text{mm}^2$ are etched within the ground plane (antenna-4). Both the slots are etched symmetrically with respect to the center feed line. The return loss graph for all the antennas are illustrated in Figure 6.18. From this figure, it is clear that the antenna-1 and antenna-3 show a dual behavior, while antenna-2 and antenna-4 exhibit UWB characteristics.

It is noted that the etching of half-circular slots in the ground plane helps in achieving UWB characteristics, while square slots incorporated in the ground plane improves the matching condition. Bandwidth comparison of these antennas is calculated and is presented in Table 6.6.

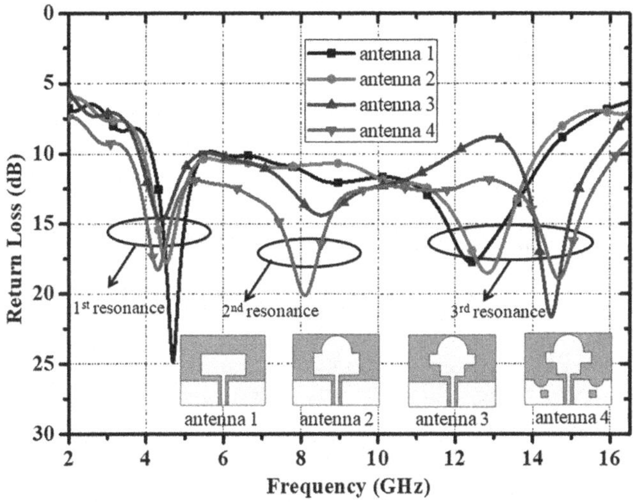

FIGURE 6.18 Variation of return loss versus frequency for different types of antenna.

TABLE 6.6
Bandwidth Comparison of Different Antennas

Antenna Design	Lower Frequency (f_{L1}), GHz	Upper Frequency (f_{H1}), GHz	Lower Frequency (f_{L2}), GHz	Upper Frequency (f_{H2}), GHz	Bandwidth (GHz)
Antenna-1	4.14	5.11	5.90	14.38	0.97 (20.97 %) and 8.48 (83.63 %)
Antenna-2	3.94	14.12	–	–	10.18 (112.74 %)
Antenna-3	3.83	11.95	13.52	15.70	8.12 (102.92 %) and 2.18 (14.92 %)
Antenna-4 (final)	3.35	16.32	–	–	12.97 (131.88 %)

Figure 6.19 depicts the simulated surface current distribution at three resonant frequencies 4.31, 8.10, and 14.67 GHz, respectively. From Figure 6.19a, it is clear that the first resonance is generated mainly due to the strong surface current flowing on the strip line. The surface current density around the half-circles and strip line contributes to the second resonance as shown in Figure 6.19b. Figure 6.19c, clearly shows that the third resonance is observed due to the current flowing on the radiating patch.

6.3.1.2 Parametric Study

A parametric study has been made to find the optimized design parameters of the patch and the ground plane. The return loss curve of the antenna for different values of L_1 is shown in Figure 6.20. It is observed that L_1 plays a significant role in

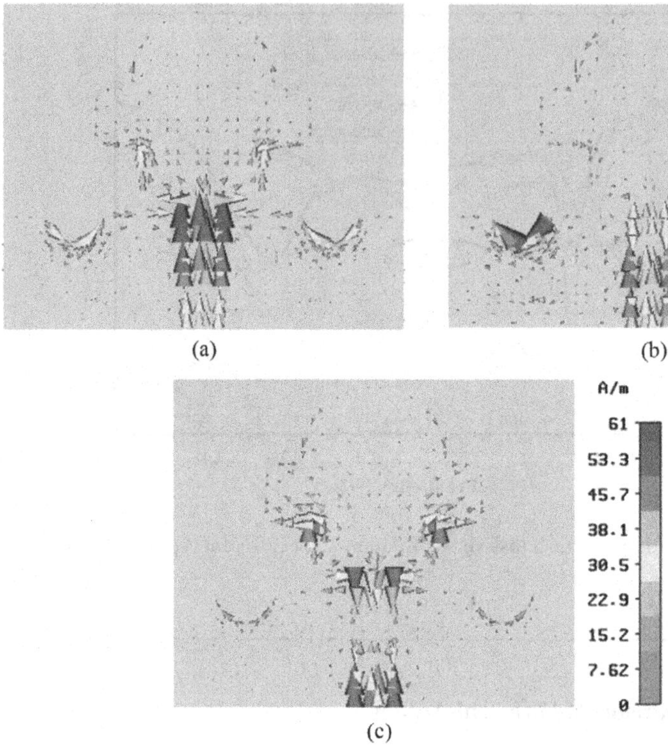

FIGURE 6.19 Surface current distribution at (a) 4.31 GHz, (b) 8.10 GHz, and (c) 14.67 GHz.

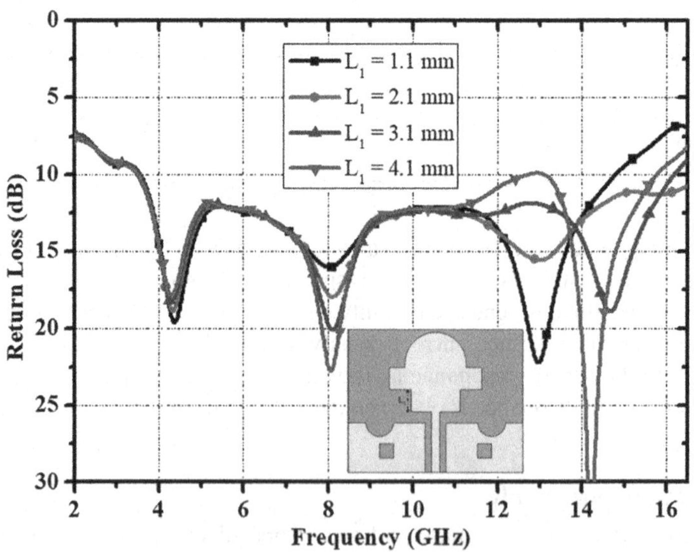

FIGURE 6.20 Variation of return loss against frequency for different values of L_1.

optimizing the antenna bandwidth. From the figure, it is found that for $L_1 = 2.1$ mm, the maximum band of 13.21 GHz (3.56–16.77 GHz) is achieved.

The effect of W_1 on the antenna bandwidth is shown in Figure 6.21. At the optimized value of $W_1 = 3.1$ mm, we get the maximum bandwidth covering the frequency range from 3.35 to 16.31 GHz. It should be noted that the variation of W_1 shifts the higher edge frequency band and keeps the lower edge frequency band unaltered. It is also observed that both L_1 and W_1 control the third resonance, which is also justified by Figure 6.19c. Figure 6.22 illustrates that the increasing value of R_1 improves the

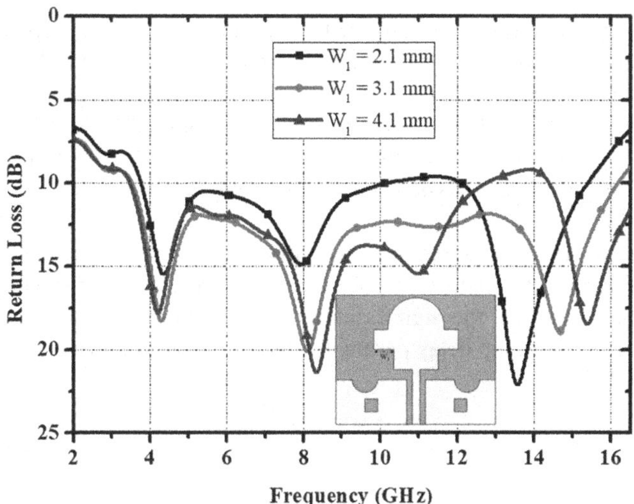

FIGURE 6.21 Return loss against frequency for different values of W_1.

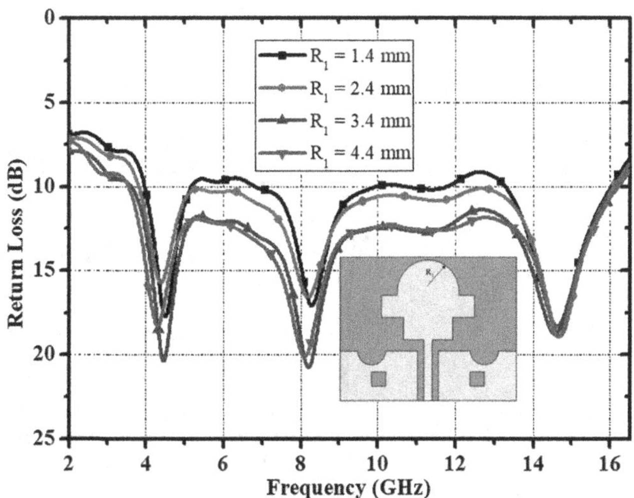

FIGURE 6.22 Return loss against frequency for different values of R_1.

impedance matching condition with a negligible effect on the bandwidth. Hence, the optimized value of radius R_1 taken in the design is 3.4 mm. The radius of the half-circular slot in the ground plane is varied to observe its effect on the antenna behavior (Figure 6.23). From the figure, it is clear that the second resonance is changed with the variation of R_2. This is also confirmed by the current distribution plot shown in Figure 6.19b.

The experimental result is calculated using Agilent N5230A vector network analyzer and compared with the simulated result (Figure 6.24). The measured and simulated bandwidths are 147.13% (2.51–16.48 GHz) and 131.88% (3.35–16.32 GHz), respectively.

The gain and the group delay are the key parameters in the design process of an UWB antenna. The simulated gain and the group delay of the CPW-fed SSPA are shown in Figure 6.25. From the figure, it is observed that the gain of the antenna varies from 1.75 to 5.27 dBi for the entire band of operation. The group delay reveals upto what extent the transmitted pulse is distorted in the UWB communication. For the acceptable pulse transmission, the group delay should necessarily be almost invariant in the UWB antenna.

The group delay of the antenna in two different orientations is calculated using two identical antennas separated by 30 cm from each other. The group delay is the parameter that describes the signal delay while propagating from receiver to the transmitter end. The group delay parameter can be calculated as

$$\tau = -\frac{d\theta(\omega)}{d\omega} \tag{6.5}$$

in which, θ= signal phase (radian), and ω= angular frequency. The calculated group delay of the antenna is almost constant (with a fluctuation of ±0.50 ns) for the whole band of operation.

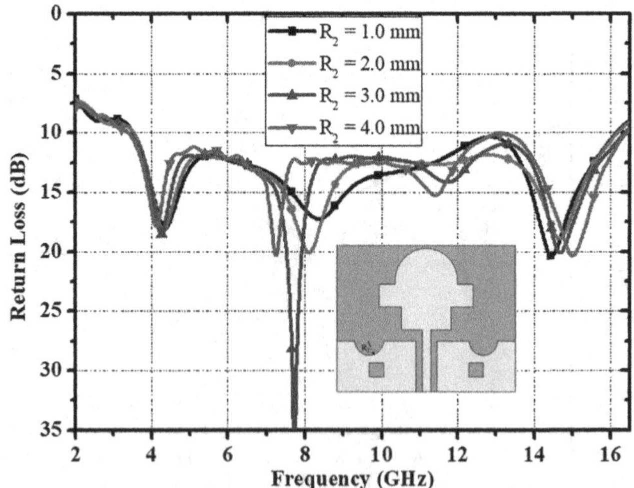

FIGURE 6.23 Return loss against frequency for different values of R_2.

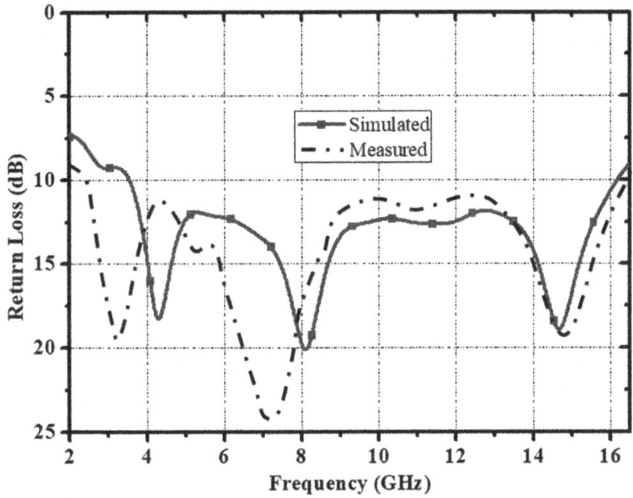

FIGURE 6.24 Measured and simulated results of the CPW-fed scarecrow-shaped antenna.

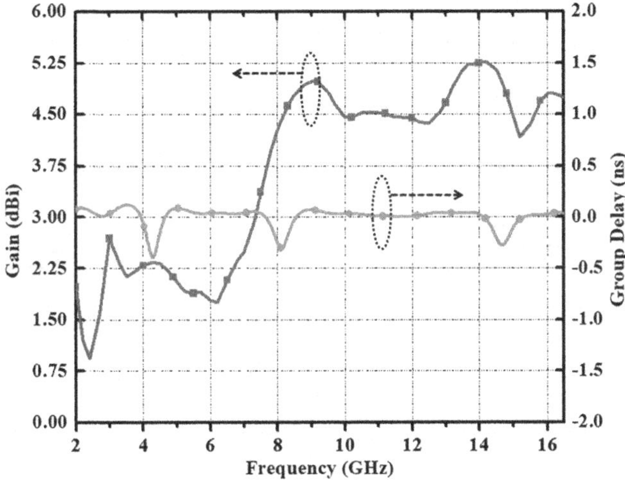

FIGURE 6.25 Variation of gain and group delay with frequency for the antenna.

6.3.1.3 Radiation Pattern of the CPW-Fed Antenna

The simulated and measured co- and cross-polarization patterns of the designed CPW-fed antenna are shown in Figure 6.26. E-plane and H-plane patterns are plotted for three sampling frequencies of 4.13, 6.5, and 10.5 GHz. The simulated results are in good agreement with the measured results. It is seen that the E- and H-planes exhibit nearly omnidirectional patterns at all given frequencies. For E-plane, the cross-polarization level is far below as compared to the co-polarization level. In case

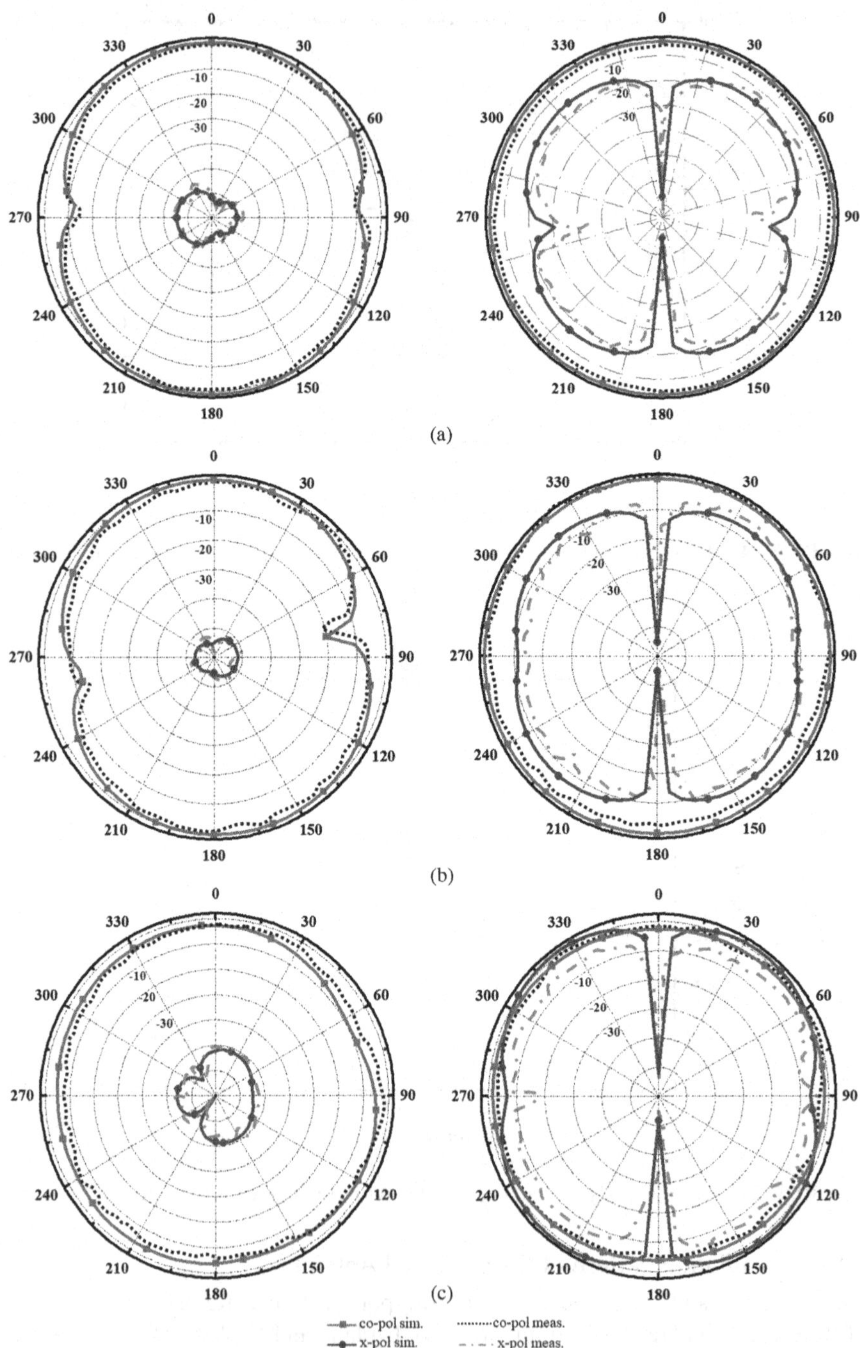

FIGURE 6.26 Measured radiation patterns for the CPW-fed antenna: E-plane (left side) and H-plane (right side) at (a) 4.31 GHz, (b) 6.5 GHz, and (c) 10.5 GHz.

of H-plane, the cross-polarization level increases at higher frequencies because of the hybrid current distribution generated on the antenna.

6.3.2 Microstrip Line-Fed UWB Antenna

The effect of microstrip line feeding is also studied for the above-mentioned design. The detailed description of this design is given in the following section.

6.3.2.1 Antenna Design

In this section, a microstrip line-fed antenna is designed and the corresponding geometry of the antenna is shown in Figure 6.27. The dimensions of this antenna are the same as in the case of CPW-fed antenna. The microstrip line and the defected ground plane are printed on either sides of a dielectric substrate. The modification in the patch and ground plane affects the bandwidth and resonant frequencies of the antenna. This can be observed by the surface current distribution for three resonant frequencies at 4.31, 7.85, and 14.53 GHz, which is shown in Figure 6.28.

6.3.2.2 Antenna Results

The comparison graph between the measured and simulated return losses is shown in Figure 6.29 for the microstrip line-fed UWB antenna. This design offers the measured bandwidth of 139.88% (2.86–16.17 GHz) and the simulated bandwidth of 125.68% (3.64–15.95 GHz), which is less than the bandwidth obtained in the case of CPW-fed antenna. Figure 6.30 shows the gain and group delay variation with frequency for this antenna. The gain varies from 1.92 to 4.90 dBi for the entire band of operation. The change in group delay for the entire band of operation is less than 0.5 ns, which is quite acceptable for UWB technology.

6.3.2.3 Radiation Pattern of the Microstrip Line-Fed Antenna

Radiation pattern of this antenna is shown in Figure 6.31 at three sampling frequencies 4.31, 6.5, and 10.5 GHz. The co-polarization plots of E- and H-planes are almost similar as in the case of CPW-fed antenna. Cross-polarization patterns for H-plane are distorted at higher frequencies due to higher mode generation. However, cross-polarization patterns for E-plane are much lower than co-polarization patterns. Thus, E- and H-plane radiation patterns show a stable and almost omnidirectional pattern.

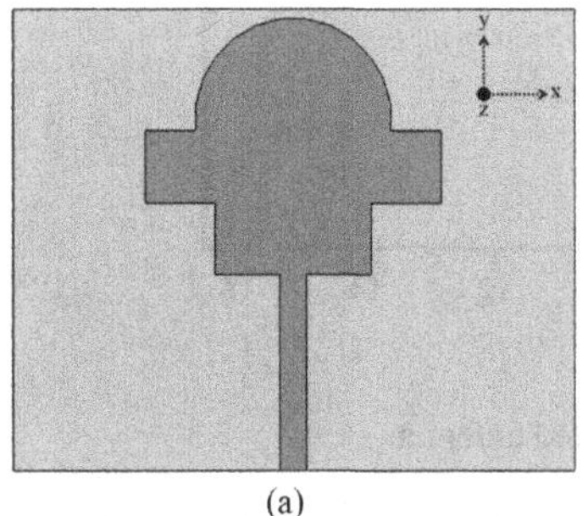

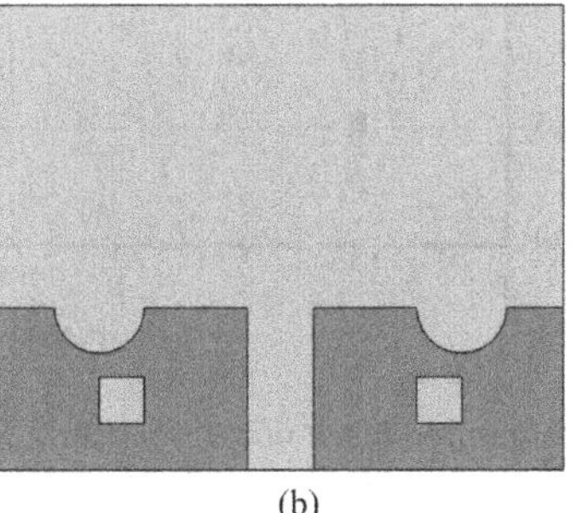

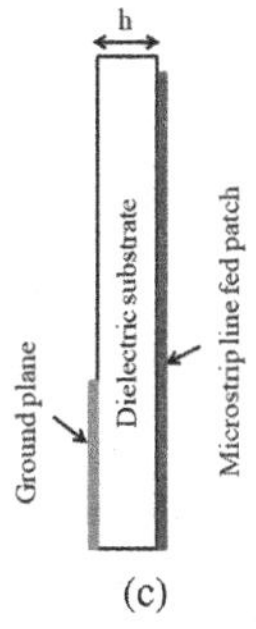

FIGURE 6.27 Geometry of the antenna: (a) top view, (b) bottom view, and (c) side view.

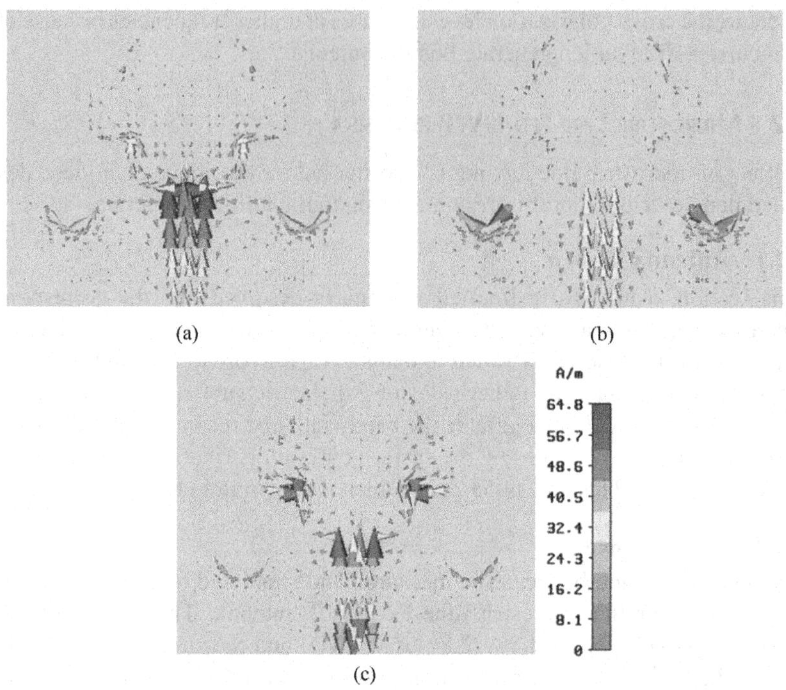

FIGURE 6.28 Surface current distribution at (a) 4.31 GHz, (b) 7.85 GHz, and (c) 14.53 GHz.

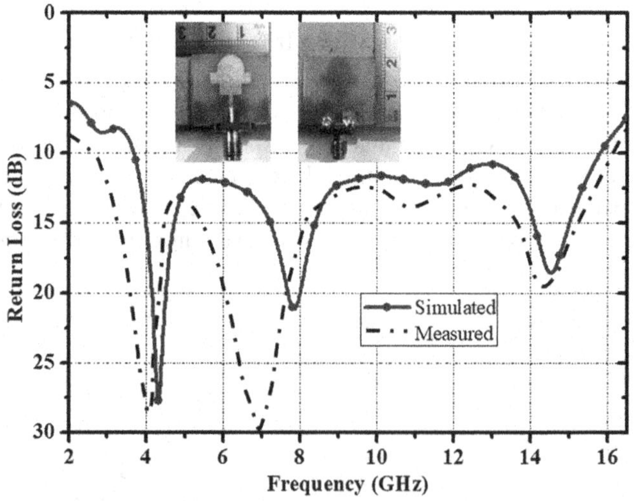

FIGURE 6.29 Measured result of the microstrip line-fed antenna.

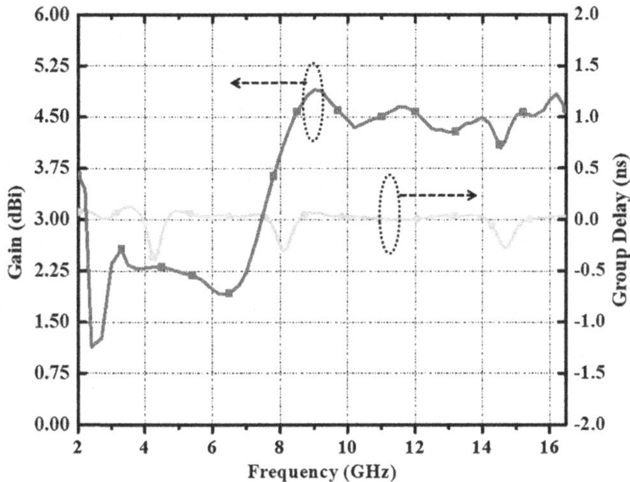

FIGURE 6.30 Variation of gain and group delay with frequency for the microstrip line-fed antenna.

6.4 A HALF-CUT DESIGN OF A LOW-PROFILE UWB PLANAR ANTENNA

The applications of small UWB antennas are in demand because of their omnidirectional radiation patterns in the near-field region, low cost, and simple geometry. These antenna designs are commonly used to cover the mobile and wireless spectrum. In this view, a small and compact planar UWB antenna is discussed using the perfect magnetic wall (PMW) technique.

6.4.1 Development of Antenna and Its Optimization

Initially, in the first stage, the unminiaturized antenna is designed by taking two kite-shaped patches in coalesced form. The radiating patch and the defected ground plane are optimized and are shown in Figure 6.32a. The overall dimensions of this full design are taken as $25\times 18 \times 1.6\,mm^3$. A microstrip line of 50 Ω is used to feed the antenna structure. In the second stage, the full antenna structure is vertically cut into half and the corresponding design is depicted in Figure 6.32b and c. Thus, the symmetrical full design, when cut into half, significantly reduces the size (≈50%) of the antenna. It is also noted that in this design, the overall impedance bandwidth is improved, fulfilling the UWB region to be used for USB dongle and other integrated devices. The ground plane structure plays a very important role in optimizing the antenna impedance and hence its bandwidth. The variation of return loss curve at different values of L_{g1} is plotted in Figure 6.33, and the figure shows that L_{g1} affects the impedance of the antenna significantly. From this figure, the maximum bandwidth of 16.27 GHz (0.82–17.09 GHz) is found at L_{g1}=3.0 mm. It is also observed that beyond L_{g1}=3.0mm, the antenna exhibits dual-/multiband operations. The effect of

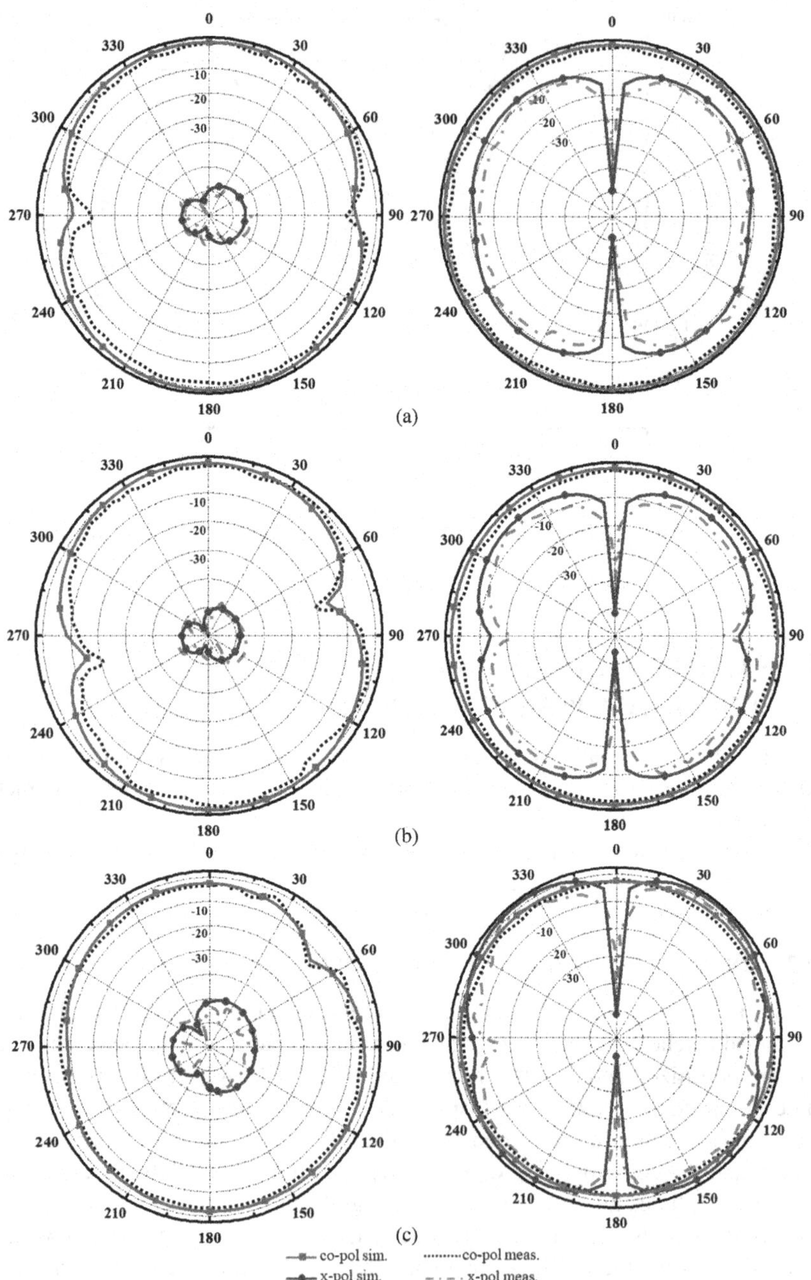

FIGURE 6.31 Measured radiation patterns: E-plane (left side) and H-plane (right side) at (a) 4.31 GHz, (b) 6.5 GHz, and (c) 10.5 GHz.

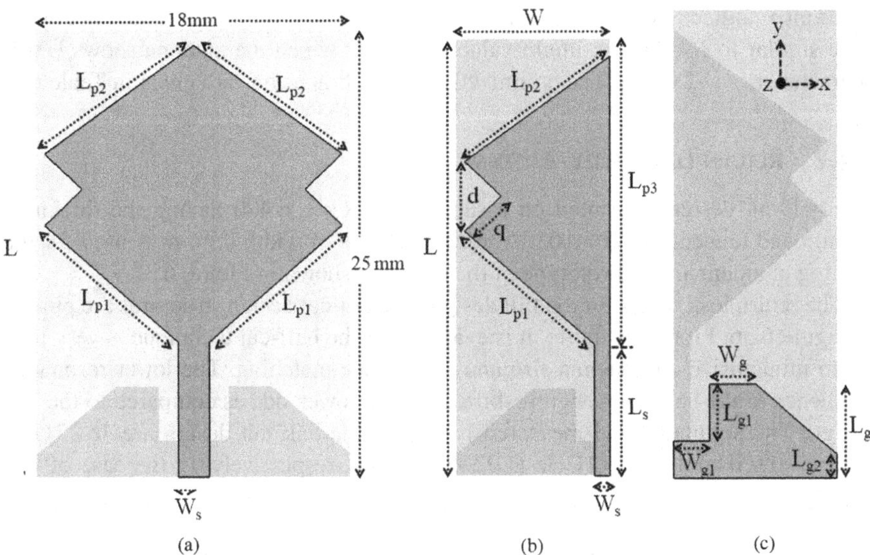

FIGURE 6.32 Development of the antenna: (a) full design, (b) half-cut design, and (c) ground plane.

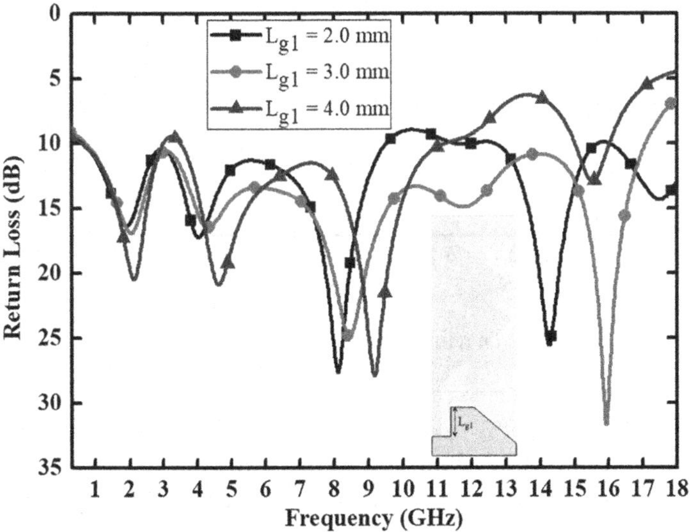

FIGURE 6.33 Effect of L_{g1} on return loss.

W_{g1} is also studied in Figure 6.34, and it is noted that W_{g1} controls the antenna impedance similar to L_{g1}. The optimum value of W_{g1} for which the antenna shows UWB characteristics is 2.0 mm. The optimized antenna dimensions are given in Table 6.7.

6.4.2 Return Loss of the Antenna

The half-cut design is printed on a substrate (FR4, $\varepsilon_r = 4.4$) having the thickness 1.6 mm and tangent loss $\delta = 0.02$. A microstrip line of width 0.9 mm is used for the feeding of antenna. The prototype of the design is shown in Figure 6.35.

The return loss values for the full design and half-cut design structures are plotted in Figure 6.36. From this figure, it is evident that the half-cut technique is very useful to miniaturize the antenna size and impedance matching. The lower resonance frequency of the half-cut design is shifted toward lower side as compared to the full design. The simulated and measured bands for the half-cut design are 16.27 GHz (0.82–17.09 GHz) and 16.12 GHz (1.02–17.14 GHz), respectively. In the case of full

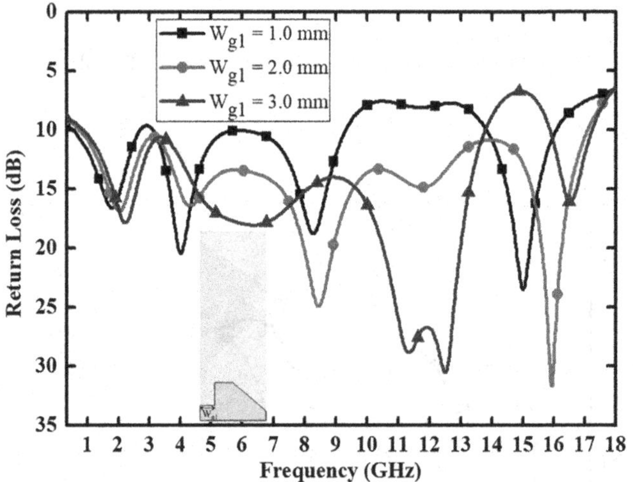

FIGURE 6.34 Effect of W_{g1} on return loss.

TABLE 6.7
Optimized Dimensions of the Half-Cut Design

Parameter	Dimensions (mm)	Parameter	Dimensions (mm)
L_{p1}	10.95	L_{p2}	10.4
L_{p3}	16.42	L_s	7.58
W_s	0.9	q	2.92
d	4	L_g	5
W_g	2.5	L_{g1}	3
W_{g1}	2	L_{g2}	1.25

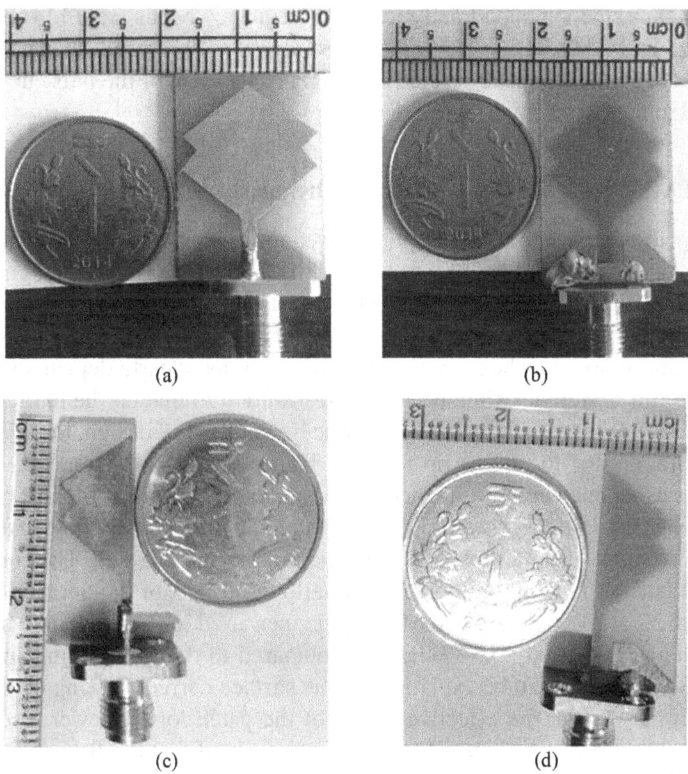

(a) (b) (c) (d)

FIGURE 6.35 Fabricated antenna– full design: (a) top view and (b) bottom view, and half-cut design: (c) top view and (d) bottom view.

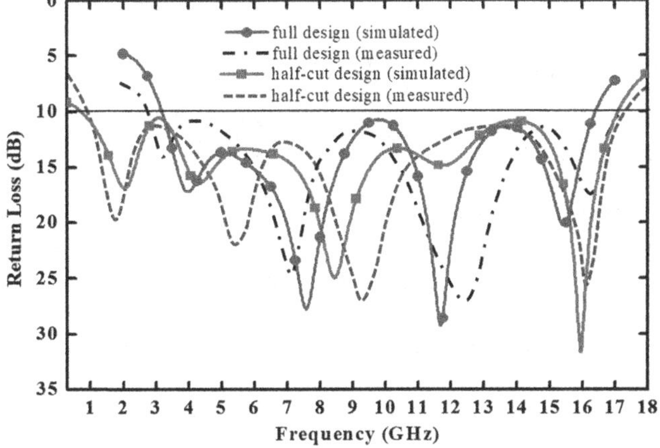

FIGURE 6.36 Return loss comparison of the full and half-cut antenna designs.

design antenna, the simulated and measured frequency bands are 13.14 GHz (3.17–16.31 GHz) and 14.2 GHz (2.8–17 GHz), respectively. The result shows the better improvement in the frequency band (1.92 GHz) in the case of the half-cut design as compared to the corresponding full design.

6.4.3 PMW Technique and Current Distribution Analysis

According to cavity model, the region between the patch and the ground plane is considered as a cavity and it is surrounded by the field which is called magnetic wall. In this design, the magnetic field distribution exists only in the vertical direction. The perfect magnetic wall (PMW) has no tangential magnetic field, but the magnetic field normal to PMW exists. The surface current distribution clearly depicts that the left part of the full design has similar current distribution to that on the half-cut design (Figure 6.37a). With respect to the symmetric line of the full design, there is an even mode excitation due to microstrip feeding. Thus, the vertical symmetric line (y–z plane) acts as an open circuit (magnetic wall) and cutting of the right part of the full design along this vertical symmetric line will not affect the current flow in the resulting design. This leads to almost 50% reduction in the antenna size. The comparison of surface current flowing on the full and half-cut designs of the antenna is shown in Figure 6.37 at 3.1, 6.86, and 10.6 GHz. The measured lower resonance frequency of the antenna is shifted at (1.73 GHz) as compared to the full design (lower resonance at 3.28 GHz). It can be explained by the surface current flowing on the patch. In the half-cut design, the effective length of the patch for the lower resonance is increased and its value becomes L_s+L_{p3}. It is also noted that the flow of current (in Figure 6.37a) is along the vertical direction (y-direction) at low frequency, which makes the antenna a good radiator.

At higher frequencies, in the case of full design, the x-direction current is modified, while for the half-cut design, the x-direction current ends at the periphery of the patch. Due to this current, the cross-polarization level of the antenna is enhanced at higher frequencies.

6.4.4 Antenna Gain and Group Delay

The variation of the simulated and measured gains of the antenna is plotted in Figure 6.38. It is observed that the gain varies from 1.28 to 3.68 dBi (simulated) and

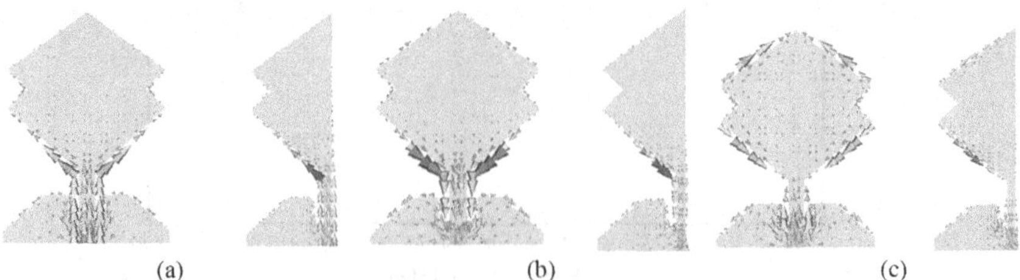

FIGURE 6.37 Current distribution comparison between full and half-cut design antennas at (a) 3.1 GHz, (b) 6.86 GHz, and (c) 10.6 GHz.

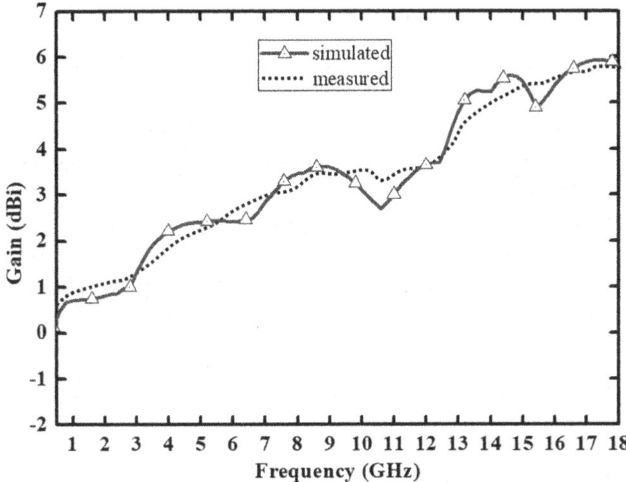

FIGURE 6.38 Variation of gain with frequency.

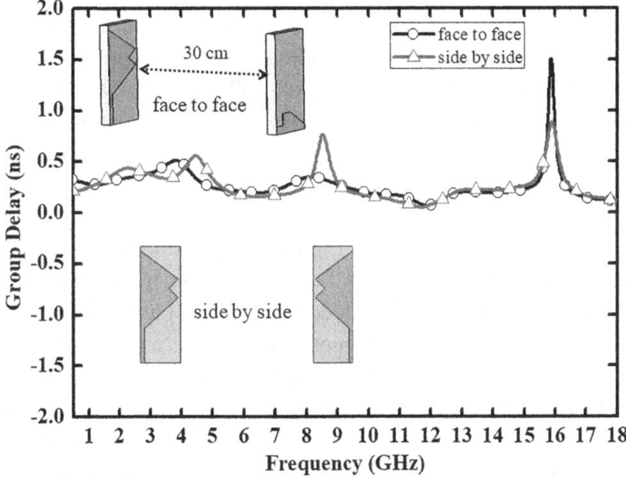

FIGURE 6.39 Variation of group delay with frequency.

1.29 to 3.62 dBi (measured) over the entire operating band. The average simulated and measured gains of the antenna are also calculated and found to be 2.92 and 3.12 dBi, respectively. The discrepancies between the simulated and measured gain results are mostly due to fabrication tolerances as the size of the antenna is small.

The group delay variation for face-to-face and side-by-side configurations is shown in Figure 6.39. From this figure, it is clear that the group delay fluctuation is almost constant throughout the band, which is quite acceptable for UWB wireless communications.

6.4.5 Radiation Characteristics

The co- and cross-polarization patterns of the antenna are plotted at frequencies 1.18, 3.1, 6.85, and 10.6 GHz (Figure 6.40).

The cross-polarization level for the *y–z* (E)-plane is very low as compared to the co-polarization at 1.18 and 3.1 GHz frequencies, but at higher frequencies (6.85 and 10.6 GHz), the cross-polarization level increases. The *x–z* (H)-plane shows the omnidirectional radiation pattern for all the given frequencies. The cross-polarization level of the H-plane also increases at higher frequencies. It is because of (i) higher mode generation at higher frequencies, (ii) the effect of asymmetric structure of the ground plane, and (iii) the *x*-component of the current on the patch that deteriorates the radiation patterns of the antenna.

6.5 A Modified Microstrip Line-Fed Compact UWB Printed Antenna

In this section, a printed antenna with a tapered microstrip line which is modified into a spanner shape, for the feeding purpose, is described for UWB applications. The novel design consists of a rectangular radiating patch with a corner slot, a defected ground plane with a mirror image 'P'-shaped slot, and a spanner-shaped feed line. The overall dimensions of the antenna are $25 \times 17 \times 1.6\,\text{mm}^3$. In order to achieve an appropriate UWB antenna for the modern wireless technology, many antenna parameters are calculated, such as return loss, VSWR, bandwidth dimension ratio (BDR), far-field radiation patterns, group delay, and fidelity factor.

6.5.1 Antenna Design

The antenna structure is illustrated in Figure 6.41. The antenna consists of a rectangular patch of dimensions $L_p \times W_p$. A tapered microstrip line with annular sector at the top is used, which modifies the feed line into a spanner shape. The modified spanner-shaped microstrip line terminates with the rectangular patch whose lower left edge is etched with a rectangular slot of dimensions $L_{p1} \times W_{p1}$. The distance of the feed line toward the right side of the substrate is W_3=4.0 mm (Figure 6.41a). In the bottom of the design, the ground plane is embedded with a mirror-imaged '*P*'-shaped slot. The dimension of the '*P*'-shaped slot consists of two different portions. The first portion has a wide rectangular slot of dimensions $a \times b$, and the second one has a narrow rectangular slot of dimensions $c \times d$ (Figure 6.41b). The optimized values of the antenna dimensions are listed in Table 6.8. The lower cut-off frequency of this antenna can be formulated with some modifications in the following equation:

$$f_L = \frac{7.2}{\left(L_p + R + L_m\right)} (\text{in GHz}) \tag{6.6}$$

where, L_m is the effective length of the microstrip line and can be calculated as

$$L_m = 1.9\,\text{mm} + A \tag{6.7}$$

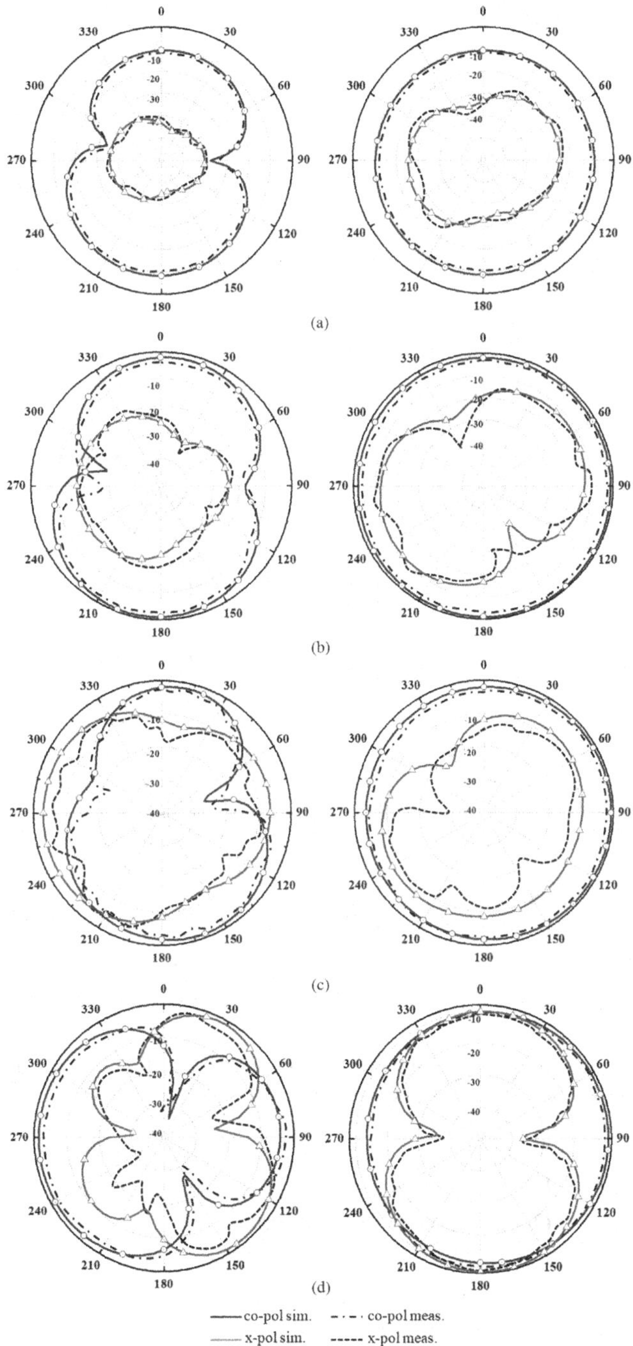

FIGURE 6.40 Measured radiation patterns of the antenna: E-plane (y–z plane, left side) and H-plane (x–z plane, right side) at (a) 1.18 GHz, (b) 3.1 GHz, (c) 6.85 GHz, and (d) 10.6 GHz.

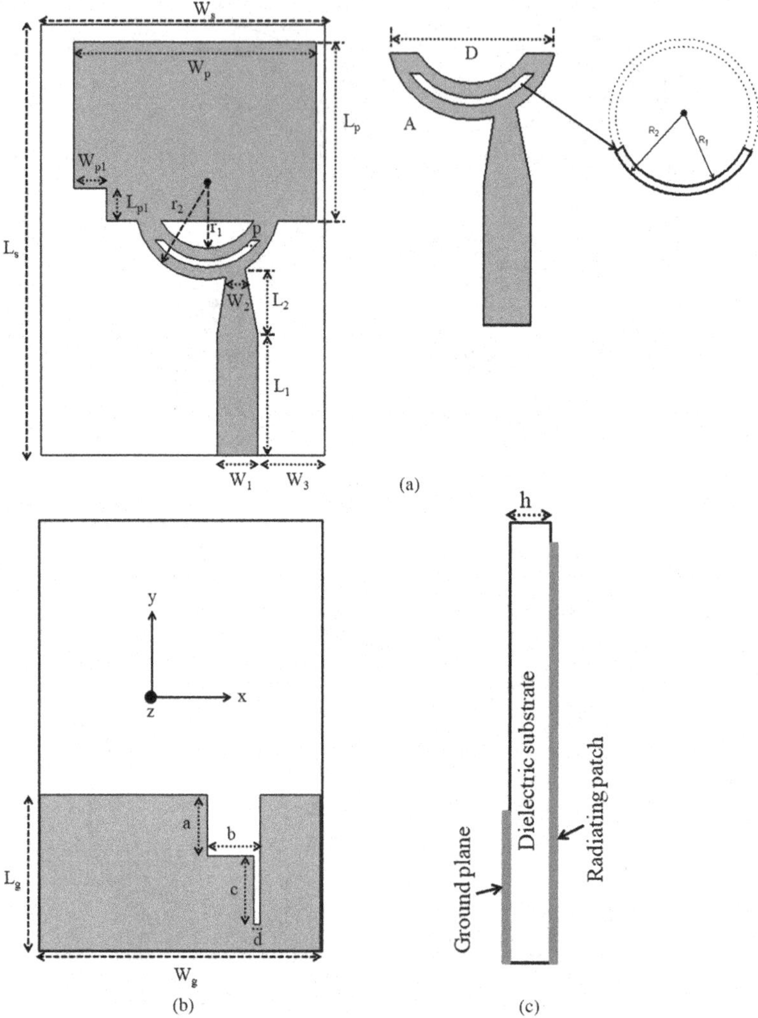

FIGURE 6.41 Geometry of the antenna: (a) top view of the design of the antenna with a modified microstrip feed line, (b) ground plane structure, and (c) side view.

TABLE 6.8
Physical Dimensions of the Antenna

Parameter	L_p	W_p	L_{p1}	W_{p1}	L_1	W_1	D
Unit (mm)	10.4	14.5	2	2	7	2.4	8.41
Parameter	L_2	W_2	R_1	R_2	r_1	r_2	p
Unit (mm)	3.9	1	3.5	3.9	3.2	4.3	0.51
Parameter	A	L_g	W_g	a	b	c	d
Unit (mm)	6.54	9.1	17	3.6	3.2	4	0.4

where, A is the arc perimeter of the annular sector used at the top of the tapered microstrip line and given by

$$A = 2\pi r_2 \left(\frac{\theta}{360^\circ} \right) \tag{6.8}$$

in which, θ is calculated using the simple geometry.

$$\theta = \cos^{-1}\left[1 - \frac{1}{2}\left(\frac{6.23}{r_2} \right)^2 \right] \tag{6.9}$$

where, the chord length of the arc A is 6.54 mm. Thus, using the value of A in eq. (6.7), we get $L_m = 8.44$ mm.

Now, the effective radius of the equivalent cylindrical monopole antenna is calculated as

$$R = \frac{W_p}{2\pi} \tag{6.10}$$

In case of planar antenna printed on dielectric material, the effective dimensions of the antenna are increased. Thus, the lower cut-off frequency is further reduced. Therefore, the modified equation for f_L is given as

$$f_L = \frac{7.2}{1.15 \times \left(L_p + R + L_m \right)} (\text{in GHz}) \tag{6.11}$$

where, L_p, R, and L_m are in cm. The calculated value of f_L obtained from the equation is ≈2.96 GHz.

The various stages of the antenna structure are depicted in Figure 6.42 along with its corresponding return loss curves. In the first step, a conventional rectangular radiating patch is taken as radiator and fed with spanner-shaped microstrip line. This antenna-1 gives a triple-band notched UWB response (3.28–24 GHz). In the second step, an arch-shaped slot is etched in the feed line and a stepped slot is cut in the bottom left corner of the patch. This design (antenna-2) gives four operating bands in the UWB range of 3.32–24.32 GHz. By cutting a rectangular slot of dimensions ($a \times b$) from the ground plane, we get antenna-3, which is again revealing triple-band notched UWB characteristics within the range of 3.07–21.14 GHz. Finally, when a narrow slot of dimensions ($c \times d$) is removed from the ground plane, antenna-4 is achieved. This spectrum gives an ultra-wide bandwidth varying from 3.04 to 22.1 GHz (151.63%).

6.5.2 Parametric Study of the Designed Antenna

Since a novel design of feeding is utilized, a parametric analysis becomes important to provide information to antenna designers about the different parameters and dimensions of the antenna and how they affect the antenna results. Here, the effects

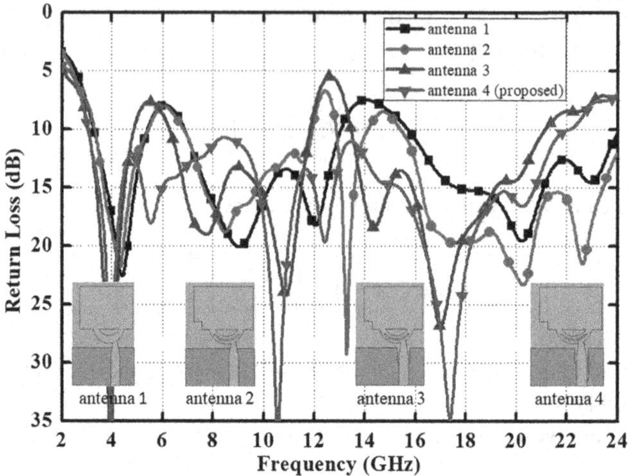

FIGURE 6.42 Antenna evolution and the corresponding return loss variation.

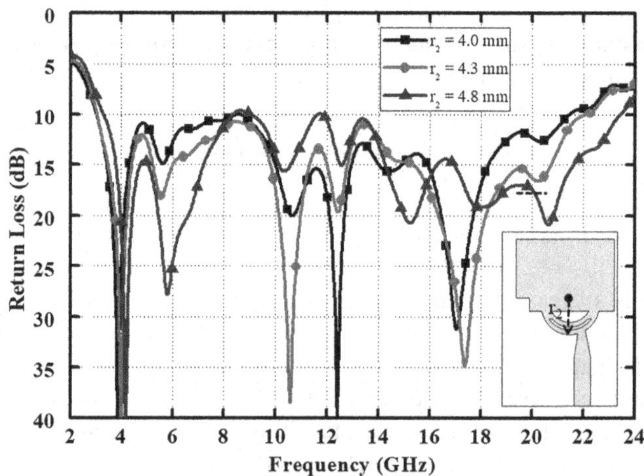

FIGURE 6.43 Variation of return loss against frequency for different values of r_2.

of r_1, r_2, a, b, and c are analyzed using the CST simulation. The variation of return loss curve for different values of r_2 is shown in Figure 6.43. It is noted that r_2 plays a significant role in improving the antenna bandwidth. From this figure, it is noted that the maximum bandwidth of 19.07 GHz (3.04–22.11GHz) is found for r_2=4.3mm. However, the UWB range is discontinued at any value of r_2 except at r_2=4.3 mm. The effect of r_1 is also studied, but it merely improves the matching conditions, hence not mentioned here. The effect of slot in the ground plane plays a very important role in achieving the UWB characteristics. The variation of the return loss of the antenna at different values of slot length (a) is calculated, and it is shown in Figure 6.44. It is observed that the value of 'a' significantly affects the antenna matching and it is

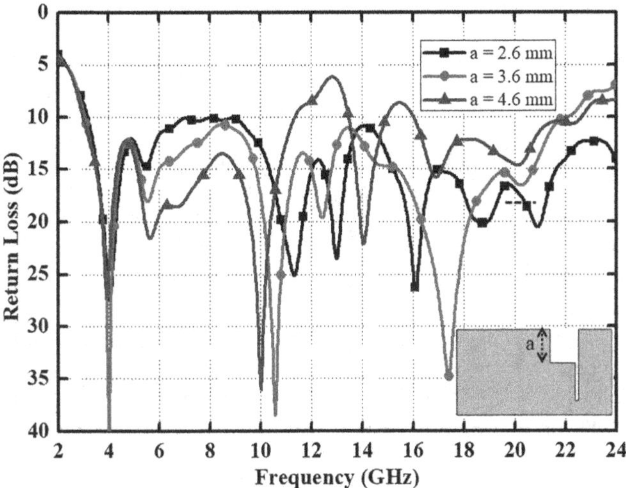

FIGURE 6.44 Variation of return loss against frequency for different values of '*a*'.

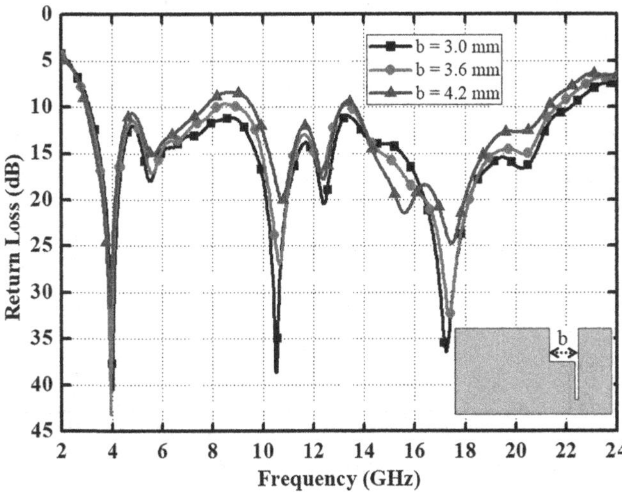

FIGURE 6.45 Variation of return loss against frequency for different values of '*b*'.

highest at $a = 3.6$ mm. Also, the lower end of the UWB band is unaffected, but the higher end of the band significantly changes with '*a*'. Again, the impedance matching can be controlled by varying the value of '*b*' (Figure 6.45) with a slight change in the higher-end frequency band.

The effect of a narrow slot length '*c*' is also studied, and it is found that '*c*' plays a very significant role in controlling the impedance of the antenna and the higher-end frequency band (Figure 6.46). The variation of '*d*' has the least contribution to the return loss value. Therefore, it is not mentioned here. Thus, the best antenna result for

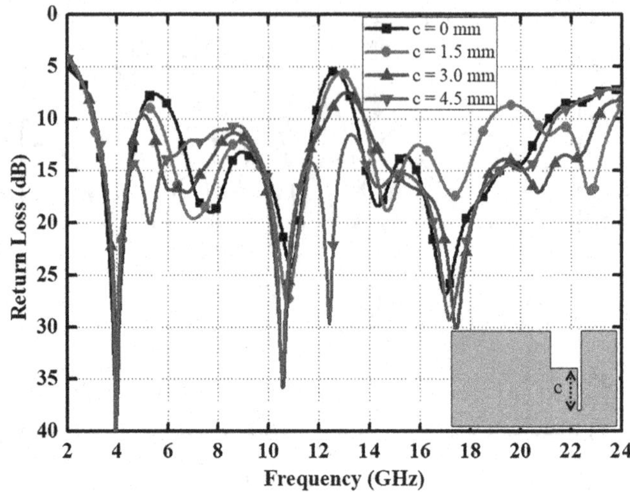

FIGURE 6.46 Variation of return loss against frequency for different values of '*c*'.

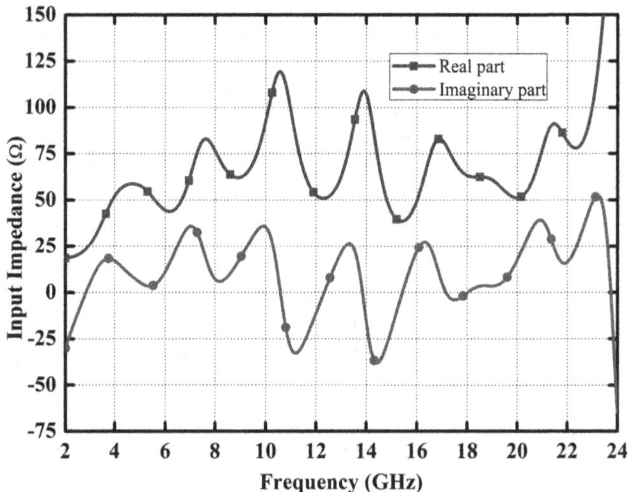

FIGURE 6.47 Input impedance plot of the antenna.

UWB response has been obtained by optimizing the dimensions of the ground plane, feed line, and radiating patch. The corresponding values are presented in Table 6.8.

The real and imaginary curves of the antenna are simulated and shown in Figure 6.47. From this graph, we can observe that the impedance of real part is approximately oscillating in between 25 and 125 Ω and the imaginary part is closely oscillating from 35 to −35 Ω. Therefore, it can be seen that the input impedance of the antenna shows good matching with the coaxial cable feed within the entire operating frequency band.

6.5.3 Antenna Fabrication and Results

The design is printed on a dielectric material (FR4) with a dielectric constant $\varepsilon_r = 4.4$, loss tangent 0.02, and thickness 1.6mm. The fabricated antenna is measured using Agilent N5230A vector network analyzer.

6.5.3.1 VSWR Measurement

The comparison between the measured and simulated VSWR is shown in Figure 6.48, which are in good agreement. The measured and simulated bandwidths are 153.22% (2.94–22.20 GHz) and 153.28% (2.97–22.46 GHz), respectively, for VSWR ≤ 2.

The impedance bandwidth (fractional bandwidth) tells only the difference in the bandwidth of various antennas, but it does not give any input regarding the size of those antennas. Therefore, BDR of the UWB antenna is presented because it gives more information as compared to the conventional impedance bandwidth. The BDR for the antenna is calculated as

$$\text{BDR} = \frac{\text{Antenna impedance bandwidth (\%)}}{\left(L_\lambda \times W_\lambda\right)} \tag{6.12}$$

where, L_λ and W_λ are electrical length and width, respectively. Thus, BDR is the bandwidth of the antenna per unit electrical area of the antenna, provided that antenna is planar. BDR calculation is useful because it not only gives the comparison of bandwidth with other designs, but also tells that which antenna designs bear a larger size.

6.5.3.2 Realized Gain and Radiation Efficiency

The simulated realized gain (which includes the impedance mismatch), directivity, and radiation efficiency of the designed antenna are plotted and are shown in Figure 6.49. From the figure, it is observed that the gain varies from −1.38 to 5.18 dB,

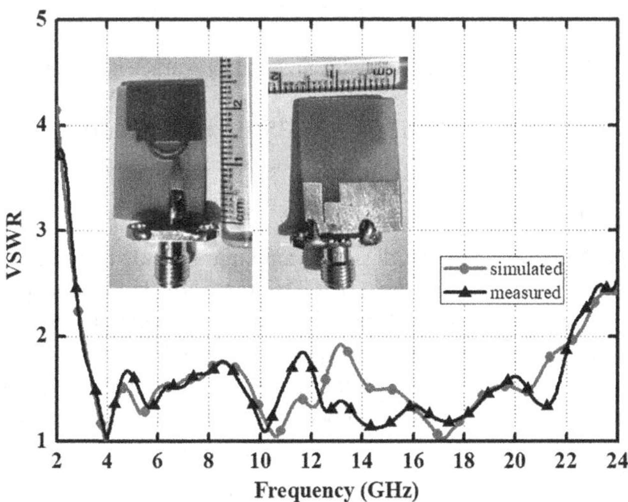

FIGURE 6.48 Measured and simulated VSWR of the antenna.

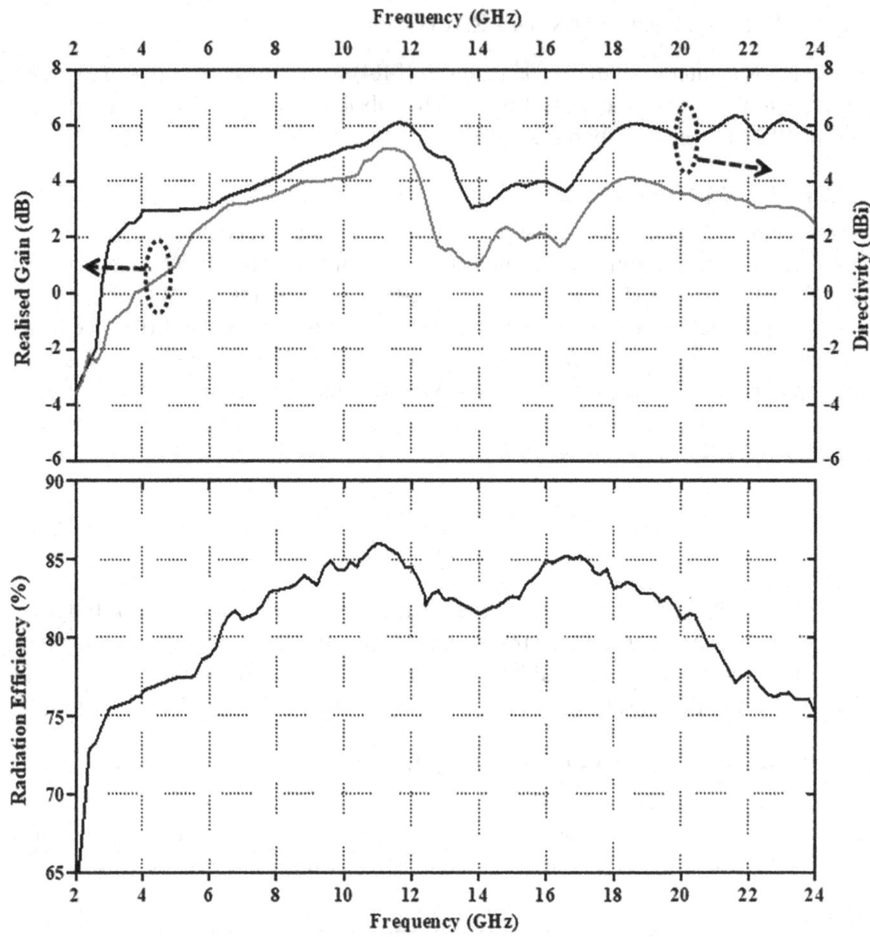

FIGURE 6.49 The realized gain, directivity, and radiation efficiency of the antenna.

while the directivity changes from 1.52 to 6.35 dBi for the entire band of operation. The radiation efficiency is found to be more than 70% throughout the operating band.

6.5.3.3 Radiation Characteristics

Far-field radiation patterns of the designed antenna are calculated at four different frequencies, i.e., 3.1, 6.85, 10.6, and 18 GHz, respectively (Figure 6.50a–d). The co-polarization patterns for *x*–*z* (H-plane) and *y*–*z* (E-plane) planes are omnidirectional and of dumbbell shape, respectively, at 3.1 and 6.85 GHz. However, the cross-polarization component increases at higher frequency, particularly beyond 10.6 GHz. It is due to the fact that the additional modes are generated at higher frequencies.

6.5.3.4 Time-Domain Analysis

The time-domain analysis is performed by placing two identical antennas in two different orientations, i.e., face to face and side by side. The first antenna is transmitting

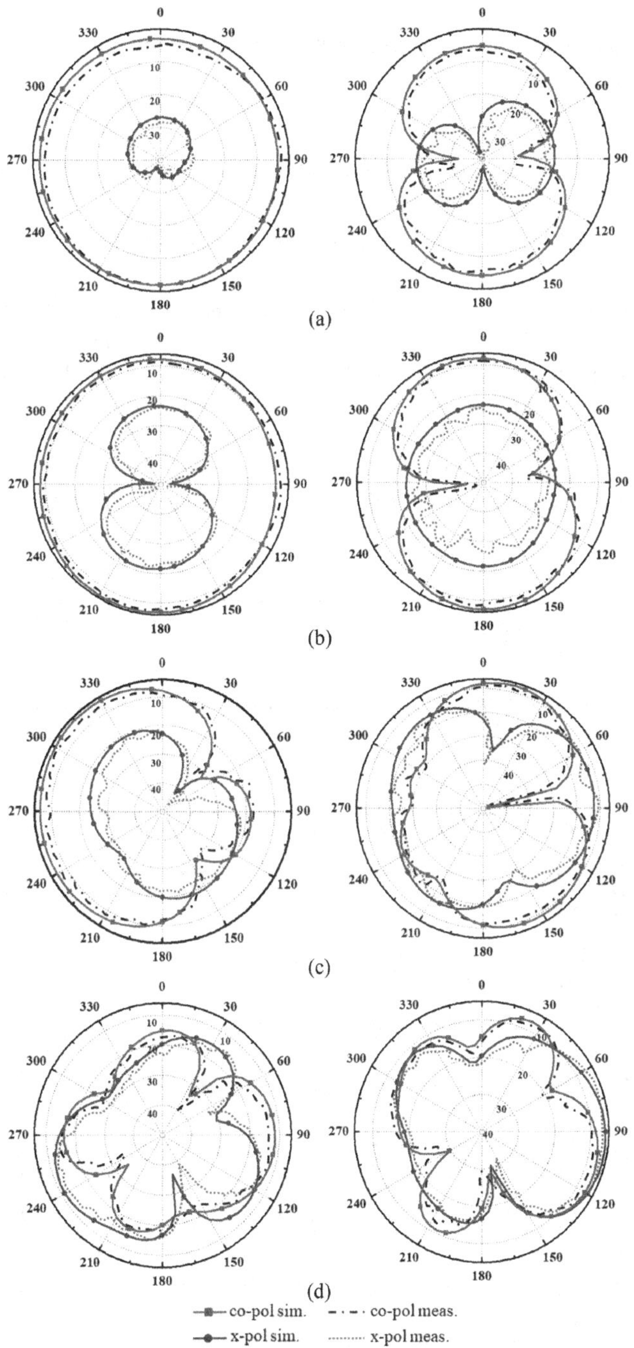

FIGURE 6.50 Radiation patterns for H-plane (x–z plane, left side) and E-plane (y–z plane, right side) at (a) 3.1 GHz, (b) 6.85 GHz, (c) 10.6 GHz, and (d) 18 GHz.

the signal, and the other one receives the signal in the far-field region. One of the important features that define the suitability of the UWB antenna is its fidelity factor (F), and it is calculated as

$$F = \max\left[\left|\frac{\int_{-\infty}^{\infty} s_i(t)s_r(t+\tau)dt}{\sqrt{\int_{-\infty}^{\infty} s_i^2(t)dt\int_{-\infty}^{\infty} s_r^2(t)dt}}\right|\right] \tag{6.13}$$

here, $s_i(t)$ and $s_r(t)$ are the input and received signals, respectively. The input and received signals of the face-to-face and side-by-side orientations are shown in Figure 6.51. From this figure, fidelity factors of face-to-face and side-by-side orientations of the antennas are calculated and they are found to be 87.32% in face-to-face configuration and 84.72% in side-by-side configuration.

The variation of τ against frequency is shown in Figure 6.52 for face-to-face and side-by-side orientations. From this figure, it is clear that the group delay fluctuation

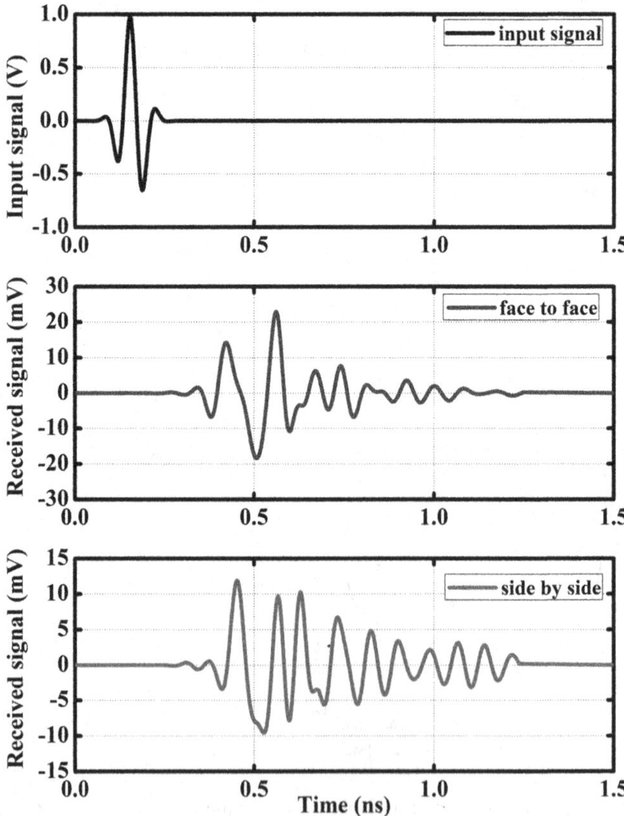

FIGURE 6.51 Input and received signals for face-to-face and side-by-side configurations.

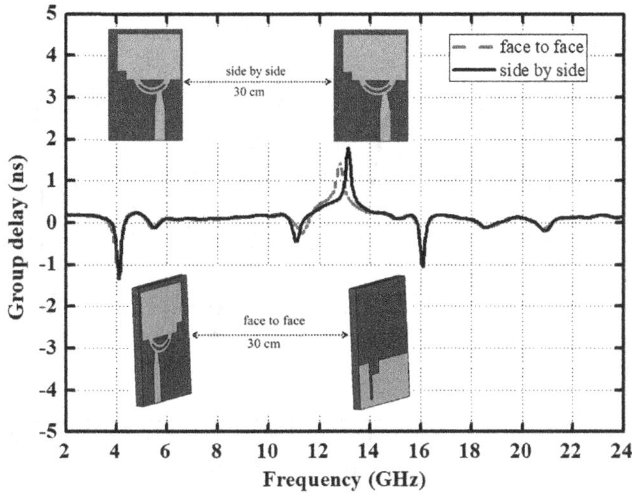

FIGURE 6.52 The group delay variation for the two configurations of the antenna.

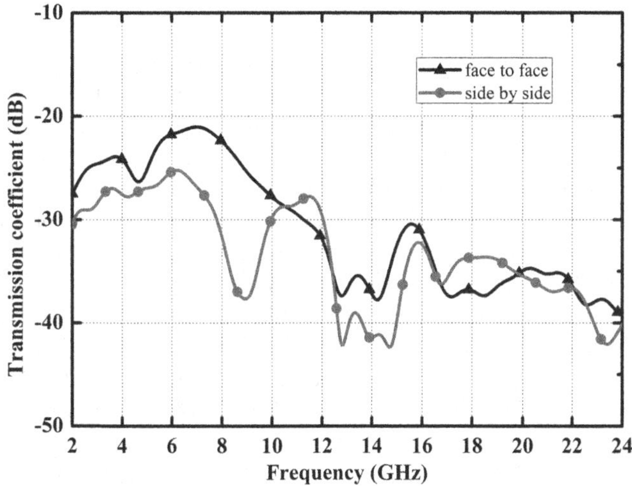

FIGURE 6.53 Variation of transmission coefficient versus frequency.

is less than ±2.0 ns over the entire operating band, which is quite good for UWB technology.

The variation of isolation (S_{21}) with frequency is illustrated in Figure 6.53. From this curve, it is noted that the variation of S_{21} value is less than −20dB for face-to-face/side-by-side configuration.

The variation of phase S_{21} with frequency for the two configurations is shown in Figure 6.54. It is observed that the phase S_{21} shows a linear and almost constant variation and less distortion with received signals for both the configurations.

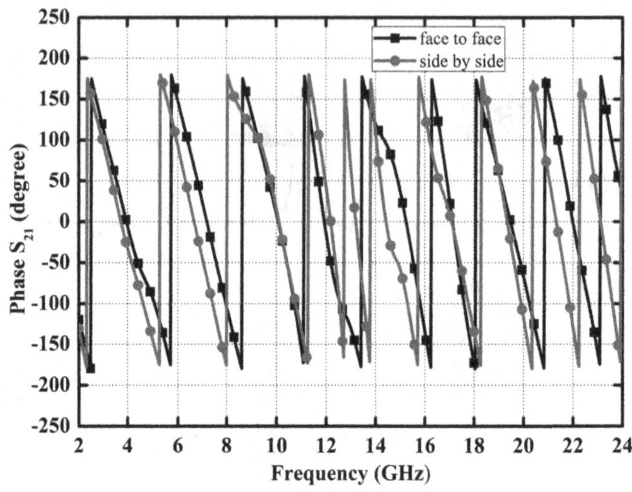

FIGURE 6.54 Variation of phase (S_{21}) with frequency.

6.6 CONCLUSIONS

In this chapter, compact and printed antennas were designed, fabricated, and measured for UWB communication systems. In the first design, a parametric study of the patch and ground plane showed that the impedance bandwidth and various resonance frequencies could be tuned by optimizing the values of L_2, L_3, L_5, and N. The maximum bandwidth of 107.35% (2.27–7.53 GHz) was obtained by optimizing the dimensions of the antenna patch and ground plane. Various resonance frequencies were formulated theoretically, which agreed well with the simulated resonance frequencies. H-plane patterns calculated at different frequencies were stable and omnidirectional, while E-plane patterns showed bidirectional radiation. In the second design, two different feeding methods were used, which were small in size, compact, and simple in geometry. The antenna with CPW line showed a bandwidth over 147.13% (2.51–16.48 GHz), while a microstrip line-fed antenna produced a bandwidth of 139.88% (2.86–16.17 GHz). The UWB characteristics were achieved by modifying the radiating patch and using the slots at suitable position in the ground plane. Both the antennas exhibited stable omnidirectional radiation patterns with acceptable gain in the whole operational band. The group delay variation for both the antennas was quite low (>0.5 ns), which makes these antennas very promising candidates for UWB communication systems and applications. A novel small-size printed monopole half-cut antenna was described in the third design. The perfect magnetic wall technique was implemented to miniaturize the antenna structure upto 50% as compared to the corresponding full design. This antenna revealed the improved bandwidth of (16.12 GHz) as compared to full design (14.2 GHz). The radiation characteristics, gain, and group delay results permit the feasibility of the antenna to cover the various wireless spectra. In the fourth design, a novel feeding method was demonstrated and successfully implemented to achieve a compact planar antenna that can operate in UWB region and even beyond that. The mirror-imaged '*P*'-shaped slot in the ground plane plays the key role in obtaining UWB response. The presented design achieved a bandwidth of 153.22% (2.94–22.2 GHz) for VSWR ≤ 2. The far-field radiation patterns, antenna realized gain, directivity, and radiation efficiency were also calculated and found quite acceptable for the entire UWB range. The time-domain analysis was performed to calculate the group delay, signal quality, and fidelity factors for face-to-face and side-by-side orientations. The various antenna designs described in this chapter cover many of the wireless communication spectrums such as GPS (1.575 GHz), DCS (1.710–1.880 GHz), PCS (1.850–1.990 GHz), UMTS (1.920–2.170 GHz), LTE 2300/LTE 2500 (2.3–2.4 GHz/2.5–2.69 GHz), Bluetooth (2.4 GHz), Wi-Fi (2.4–2.485 GHz), WLAN (2.4–2.48, 5.15–5.35, and 5.72–5.85 GHz), WiMAX (2.5–2.69, 3.40–3.69, and 5.25–5.85 GHz), ISM (5.725–5.875 GHz), mobile satellite applications (7.250–7.375 GHz), mobile applications (8.025–8.200 GHz), broadcasting satellites (12.4–12.5 GHz), defense systems (14.62–15.23 GHz), and passive sensor satellites (21.2–21.4 GHz).

REFERENCES

1. *First Report and Order. Revision of Part 15 of the Commission's Rules Regarding Ultra-Wideband Transmission Systems FCC 02–48.* Federal Communications Commission, Washington, DC, (2002).
2. Wang, F., Arslan, T.: Body-coupled monopole UWB antenna for wearable medical microwave imaging applications. *IEEE-APS Topical Conference on Antennas and Propagation in Wireless Communications (APWC)*, 146–149 (2017).
3. Mahmud, M., Islam, M. T., Samsuzzaman, M.: A high performance UWB antenna design for microwave imaging system. *Microwave Opt. Technol. Lett.* **58**(8),1824–1831 (2016).
4. Abdelhamid, M. M., Allam, A. M.: Detection of lung cancer using ultra wide band antenna. *Loughborough Antennas & Propagation Conference (LAPC)*, 1–5 (2016).
5. Qu, S.W., Li, J. L., Chen, J. X., Xue, Q.: Ultrawideband strip-loaded circular slot antenna with improved radiation patterns. *IEEE Trans. Antennas Propag.* **55** (11), 3348–3353 (2007).
6. Chen, Z. N., See, T. S. P., Qing, X.: Small printed ultrawideband antenna with reduced ground plane effect. *IEEE Trans. Antennas Propag.* **55** (2), 383–388 (2007).
7. Dissanayake, T., Esselle, K. P.: UWB performance of compact L-shaped wide slot antennas. *IEEE Trans. Antennas Propag.* **56** (4), 1183–1187 (2008).
8. Dastranj, A., Imani, A., Moghaddasi, M. N.: Printed wide-slot antenna for wideband applications. *IEEE Trans. Antennas Propag.* **56** (10), 3097–3102 (2008).
9. Dastranj, A., Abiri, H.: Bandwidth enhancement of printed E-shaped slot antennas fed by CPW and microstrip line. *IEEE Trans. Antennas Propag.* **58** (4), 1402–1407 (2010).
10. Liu, W. C., Wu, C. M., Tseng, Y. J.: Parasitically loaded CPW-fed monopole antenna for broadband operation. *IEEE Trans. Antennas Propag.* **59** (6), 2415–2419 (2011).
11. Mitra, D., Das, D., Chaudhuri, S. R. B.: Bandwidth enhancement of microstrip line and CPW-fed asymmetrical slot antennas. *Prog.Electromagn. Res. Lett.* **32**, 69–79 (2012).
12. Yoon, C., Kim, W. S., Jeong, G. T., Choi, S. H., Lee, H. C., Park, H. D.: A planer CPW-fed patch antenna on thin substrate for broadband operation of ISM-band applications. *Microwave Opt. Technol. Lett.* **54** (9), 2199–2202 (2012).
13. Fallahi, H., Atlasbaf, Z.: Study of a class of UWB CPW-fed monopole antenna with fractal elements. *IEEE Antennas Wirel. Propag. Lett.* **12**, 1484–1487 (2013).

14. Ghatak, R., Karmakar, A., Poddar, D. R.: Hexagonal boundary Sierpinski carpet fractal shaped compact ultrawideband antenna with band rejection functionality. *Int. J. Electron. Commun. (AEU).* **67**, 250–255 (2013).
15. Mandal, T., Das, S.: Design of a CPW fed simple hexagonal shape UWB antenna with WLAN and WiMAX band rejection characteristics. *J. Comput. Electron.* **14** (1), 300–308 (2015).
16. Gautam, A. K., Yadav, S., Kanaujia, B. K.: A CPW fed compact UWB microstrip antenna. *IEEE Antennas Wirel.Propag.Lett.* **12**, 151–154 (2013).
17. Shrivastava, M. K., Gautam, A. K., Kanaujia, B. K.: A novel A-shaped monopole-like slot antenna for ultrawideband applications. *Microwave Opt. Technol. Lett.* **56** (8), 1826–1829 (2014).
18. Zhan, J., Tang, Z. J., Wu, X. F., Liu, H. L.: CPW-fed printed antenna design with multi slit patches for UWB communications. *Microwave Opt. Technol. Lett.* **55** (12), 3023–3025 (2013).
19. Jose, S. M., Lethakumary, B.: CPW-fed step-shaped microstrip antenna for UWB applications. *Microwave. Opt. Technol. Lett.* **57** (3), 589–591 (2015).
20. Aissaoui, D., Abdelghani, L. M., Hacen, N. B., Denidni, T. A.: CPW-fed UWB hexagonal shaped antenna with additional fractal elements. *Microwave Opt. Technol. Lett.* **58** (10), 2370–2374 (2016).
21. Hayouni, M., Choubani, F., Vuong, T. H., David, J.: Main effects ensured by symmetric circular slots etched on the radiating patch of a compact monopole antenna on the impedance bandwidth and radiation patterns. *Wireless Pers. Commun.* **95** (4), 4243–4256 (2017).
22. Tiwari, R. N., Singh, P., Kanaujia, B. K.: Asymmetric U-shaped printed monopole antenna embedded with T-shaped strip for Bluetooth, WLAN/WiMAXapplications. *Wireless Networks.* **26** (1), 51–61 (2020).
23. Tiwari, R. N., Singh, P., Kanaujia, B. K.: Small-size scarecrow-shaped CPW and microstrip-line-fed UWB antennas. *J. Comput. Electron.* **17** (3), 1047–1055 (2018).
24. Tiwari, R. N., Singh, P., Kanaujia, B. K.: A half cut design of low profile UWB planar antenna for DCS/PCS/ WLAN applications. *Int. J. RF Microw. Comput. Aided Eng.* **29** (9), e21817 (2019).
25. Tiwari, R. N., Singh, P., Kanaujia, B. K.: A modified microstrip line fed compact UWB antenna for WiMAX/ISM/WLAN and wireless communications. *AEU-Int. J. Electron. Commun.* **104**, 58–65 (2019).

7 Circularly Polarized Printed Antennas

Dr. Ganga Prasad Pandey
Pandit Deendayal Petroleum University, Gujarat

Dr. Dinesh Kumar Singh
G L Bajaj Institute of Technology and Management, Greater Noida

CONTENTS

7.1 INTRODUCTION

Polarization of an antenna refers to the orientation of electric field distribution in space with respect to time. For optimum reception, the orientations of transmitter and receiver must match. In certain frequency bands and in certain applications, the transmitted wave keeps on rotating with time due to which fixing the receiver orientation is a cumbersome job. This phenomenon is known as the Faraday Rotation. Circularly polarized (CP) antennas are antennas with the same reception or transmission capability in all orientations. The reflected RF signals from the ground or other objects will reverse the sense of polarization; that is, right-hand circular polarization (RHCP) reflections show left-hand circular polarization (LHCP). A RHCP antenna will reject a reflected signal which is LHCP, thus reducing the multipath interferences from the reflected signals.

The second advantage is that a CP antenna is able to reduce the 'Faraday rotation' effect caused by the ionosphere. The Faraday rotation effect causes a significant signal loss (about 3 dB or more) if linearly polarized signals pass through the ionosphere.

7.2 CIRCULARLY POLARIZED STACKED ANTENNAS

The chapter includes two designs of circularly polarized antennas. The first is reconfigurable antenna with three CP bands. PIN diodes are used to make the antenna reconfigurable. The second design is a stacked antenna with two patch layers and a suspended substrate providing four CP bands.

7.2.1 A Triple-CP Band Reconfigurable Stacked Antenna

Microstrip antennas are frequently used in many wireless communication systems because of their attractive features such as planar profile, low cost, lightweight, and easy fabrication [1]. However, the types of applications of microstrip antennas are restricted by their narrow bandwidth. Accordingly, increasing the bandwidth of microstrip antennas has been a primary goal of research in the field. In fact, many broadband microstrip antenna configurations have been reported in last few decades, such as increasing the substrate thickness and decreasing its dielectric constant [2], and using appropriate feeding techniques and impedance matching method [3]. One of the popular methods to improve the bandwidth of a microstrip antenna is to create various resonant structures into one antenna by cutting slots of different shapes, such as U-shaped slot [4] and V-shape slot [5], and by adding more patches [6]. These broadband methods cause some resonance frequencies to appear near the main patch and lead to the bandwidth broadening of the antenna. A capacitive-coupled probe-fed microstrip antenna with wideband characteristics was reported in [7].

In the current wireless communication systems, circular polarization is used as one of the most common polarization types, as it is independent of transmitter and receiver orientations [8]. The circularly polarized microstrip antennas which need two orthogonal field components with a 90° phase difference were developed with the commonly used techniques with single- and dual-feed arrangements [1]. The dual-feed approach requires the use of a 90° hybrid coupler or power splitter to provide the necessary phase shift. However, this dual-feed method has a more complex geometry, larger size, and higher loss [1]. Thus, the preference is given to single-feed circularly polarized microstrip antennas. A single-feed circularly polarized operation of a square patch by truncating a pair of patch corners is widely used in the single patch [9]. Kin-Lu Wong and Jian-Yi Wu [10] presented a design that involves cutting slits in the square patch to achieve circular polarization. The circular polarization of the square microstrip antenna with four slits and a pair of truncated corners was presented in [11]. The circular polarization can also be achieved with a circular microstrip antenna by adding a tuning stub [12]. It was shown in [13] that circular polarization can be generated by embedding a cross-shaped slot at the centre of a circular patch. In [14], a dual-band circularly polarized aperture-coupled stacked microstrip antenna was presented, but the design of the aperture-coupled stacked microstrip antenna was complicated because of its multilayer structure and feeding network. The circular polarization in two distinct bands was realized by using two perpendicular ports and two power dividers to provide 90° phase shift for each band [15]. However, this structure cannot operate in both frequency bands simultaneously. The antenna presented in [16] operated at dual frequencies with circular polarization characteristics. The dual-band circular polarization radiation was achieved by inserting slits and T-shaped elements at the patch. In [17], a single-feed slotted patch structure was presented for generating circular polarization in two frequency bands. This antenna has a problem that the axial ratio bandwidth is very narrow in both the frequency bands. A microstrip patch antenna with switchable polarization was presented in [18] with a single feed. PIN diodes are used to obtain the polarization diversity characteristics of the antenna. Many studies have been reported in the literature that describes different methods for achieving triple-band circular polarization operations [19–22]. The stacked microstrip patch antenna was used to achieve triple-band circular polarization radiation [19,20]. However, dual orthogonal feed makes the antenna complex. In [21], a three-layer single-feed stacked microstrip antenna was designed to achieve triple-band circular polarization operation. A triple-band stacked design was introduced in [22], but all these designs have a narrow axial ratio bandwidth in the three frequency bands.

This chapter includes a wideband capacitive-fed microstrip antenna with reconfigurable circular polarization. The design of antenna is carried out in three stages. In the first stage, small isosceles right triangular sections are removed from diagonally opposite corners for the generation of circular polarization. In the second stage, a truncated patch was loaded with horizontal slits of unequal lengths to create dual-CP bands and PIN diodes are inserted across both the slits to generate three circularly polarized bands. CP in three distinct bands is achieved by switching PIN diodes ON and OFF on the gap of horizontal slits. Finally, a wideband antenna with triple-band CP operation is designed. This employs an inclined slot embedded on the patch with PIN diodes across the horizontal slits to achieve the broadband performance. The impedance bandwidth of the proposed antenna is 66.61% (ON state) ranging from 4.42 to 8.80 GHz and 66.02% (OFF state) in the frequency range from 4.528 GHz to 8.986 GHz with axial ratio bandwidth of 3.81%, 3.02%, and 5.49%. The bandwidth of the presented antenna is increased from 51% to 66.61% (ON state) and 66.02% (OFF state) as compared to the capacitive-coupled probe-fed microstrip antenna [7] and also generates three distinct CP bands.

7.2.2 Quad-Band CP Stacked Antennas

In the last decade, there has been a rapid growth in the field of telecommunication technologies in the C-band frequencies (4–8 GHz). The C-band contains the frequency ranges that are used for many satellite transmissions and Wi-Fi devices (5 GHz WLAN). Nearly, all the C-band communication satellites use the band of frequencies from 5.925 to 6.425 GHz for their uplinks. The C-band of frequencies from 5.15 to 5.35 GHz (HIPERLAN/1) and from 5.47 to 5.725 GHz (HIPERLAN/2) is used for IEEE 802.11a and Wi-Fi. There is limited bandwidth available in the 2.025–2.290 GHz frequency band for earth exploration-satellite services (EESS), and tracking, telemetry, and control (TT &C) due to the fact that hundreds of satellites use these bands. A new EESS (earth-to-space) allocation in the frequency range from 7 to 8 GHz is planned in near future that would allow its use for uplinks and downlinks on the same transponder, increasing efficiency and reducing satellite complexity. The integration of multiple antennas to cover the C-band frequencies can cause mutual coupling that degrades the overall system performance. Therefore, the design of single multiband circularly polarized antennas to meet the requirements of modern wireless communication that can cover C-band simultaneously can ameliorate this problem. The multiband circularly polarized microstrip antennas are a popular choice among the present microstrip antennas as they have the advantages of better mobility, less transceiver loss caused by polarization mismatch, better weather penetration, and orientation-independent receiver system as compared to linearly polarized microstrip antennas. Several designs of stacked microstrip antennas to achieve dual CP operation have recently been reported [14,15,23–29] using dual-feed branch line coupler, multilayer structure, T-shape slits, and dual-feed aperture-coupled stacked patch. However, two CP bands are merged together by optimizing the feed location and substrate thickness between the stacked patches to obtain wideband CP microstrip antennas [30]. In [31], a four-element trap-loaded inverted-L antenna array was designed for triple-band CP operation. The inverted-L antenna array makes its structure complex. Doust et al. [20] used an aperture-coupled two-patch stacked antenna to achieve triple-band CP operation, but two feed lines having 180° phase difference were used to drive the antenna. In [22], the triple-band CP radiation was achieved by inserting two pairs of narrow slots parallel to the edges of the top patch and cutting slits in the bottom patch. Three stacked patches with a slit and I-slot were used by Falade et al. [21] to achieve triple-band circular polarization. A single-feed quad-band CP stacked patch antenna was presented in [32]; the four stacked patches were used to achieve the quad-band CP. It was observed in the literature review [20–22,31,32] that multiband CP is achieved by stacked structure and each layer is responsible for generating one CP band. The novel idea employed in our design is that only two stacked patches are used to achieve quad-band CP operation, and to the best of the authors' knowledge, there is no publication having quad-band circular polarization with two patches.

In this chapter, along with the triple-band CP antenna, a lightweight quad-band circularly polarized capacitive-coupled stacked patch antenna with wideband characteristics for C-band applications is also explained, as given in [33]. The presented antenna shows broadband behaviour with impedance bandwidth of 55.6% in the frequency band of 4.97–8.49 GHz. The 3-dB axial ratio bandwidths in the four distinct bands are 0.98%, 4.275%, 0.8869%, and 1.35%. The antenna finds applications in the areas mentioned above. In the receiver, the band-pass filters remove the signals outside of the four CP bands.

7.2.3 Triple-Band Reconfigurable Antenna Design

The geometry of a reconfigurable circularly polarized capacitive-coupled probe-fed truncated corner microstrip antenna is shown in Figure 7.1a. A pair of opposite corners is truncated with equal side length of ΔL to excite two orthogonal modes with

90° phase shift that makes the antenna circularly polarized. A pair of horizontal slits of lengths L_1, L_2 and equal width w_1 with PIN diodes is embedded on the truncated patch to achieve three circularly polarized bands, as shown in Figure 7.1b. The ON condition of the PIN diodes is implemented with a through line of length 1 mm and width 0.5 mm in simulation as well as fabrication. The inclusion of a copper strip indicates the PIN diodes in ON state, while its absence indicates the OFF state of the diodes. Figure 7.1c shows the wideband antenna with triple-band circular polarization. The slot is inclined at 135° with dimensions of 8×1 mm.

The radiating patch and feed strip are placed on an RO3003 substrate with thickness $h = 1.56$ mm, dielectric constant $\varepsilon_r = 3.0$, and loss tangent $= 0.0013$, which rose in the air by g (6 mm). The SMA connector is used to connect the feed strip that capacitively couples the energy to the radiating patch. The separation between the radiating patch and feed strip is d, the feed strip length is t, and the width is s. The structure of the antenna is based on suspended capacitive-fed microstrip strip antenna. The total height of the antenna $(g + h)$ and effective dielectric constant are the key design parameters for the patch. The dimension of the radiating patch is calculated from standard design expression after making necessary corrections in the key design parameters discussed above for the suspended dielectric [1,33–35]. The impedance bandwidth may be maximized by using the design expression [7] given as

$$g \cong 0.16\lambda_0 - h\sqrt{\varepsilon_r} \tag{7.1}$$

where g is the air gap, and ε_r and h are the dielectric constant and the thickness of the substrate, respectively. Eq. (7.1) is used to predict the initial value, while the final value would be within ±10% [7] and may be obtained with simulation tools. The feed strip can be considered as a rectangular microstrip capacitor as strip dimensions are much smaller as compared to the wavelength of operation and can be represented by terminal capacitances. The dimensions (t and s) of the terminal capacitances control the reactive part of the input impedance of the antenna [7]. The optimum dimensions of the antenna are obtained via iterative processes that give broad impedance bandwidth and circularly polarized bands and are listed in Table 7.1.

The proposed antenna was fabricated in Microwave Research Laboratory of Ambedkar Institute of Advanced Communication Technology and Research (AIACTR), Delhi, INDIA. The vector network analyser (VNA) of series Agilent N5230 was used for the measurement, and Figure 7.1f shows the VSWR measurement set-up of the proposed antenna. The substrate of dimensions 5×5 cm^2 was taken for fabrication, and a white paper board is used as a support to provide an air gap.

7.2.4 Quad-Band Antenna Design

The detailed geometry of the presented antenna is shown in Figure 7.2. The structure of the antenna is basically a suspended coplanar capacitive-fed microstrip antenna. The radiating patch and feed strip are printed on an RO3003 substrate with a thickness h, which is suspended in air at a height of $g = 6$ mm above the ground plane. In the capacitive coupling method, a small feed strip is placed near the radiating

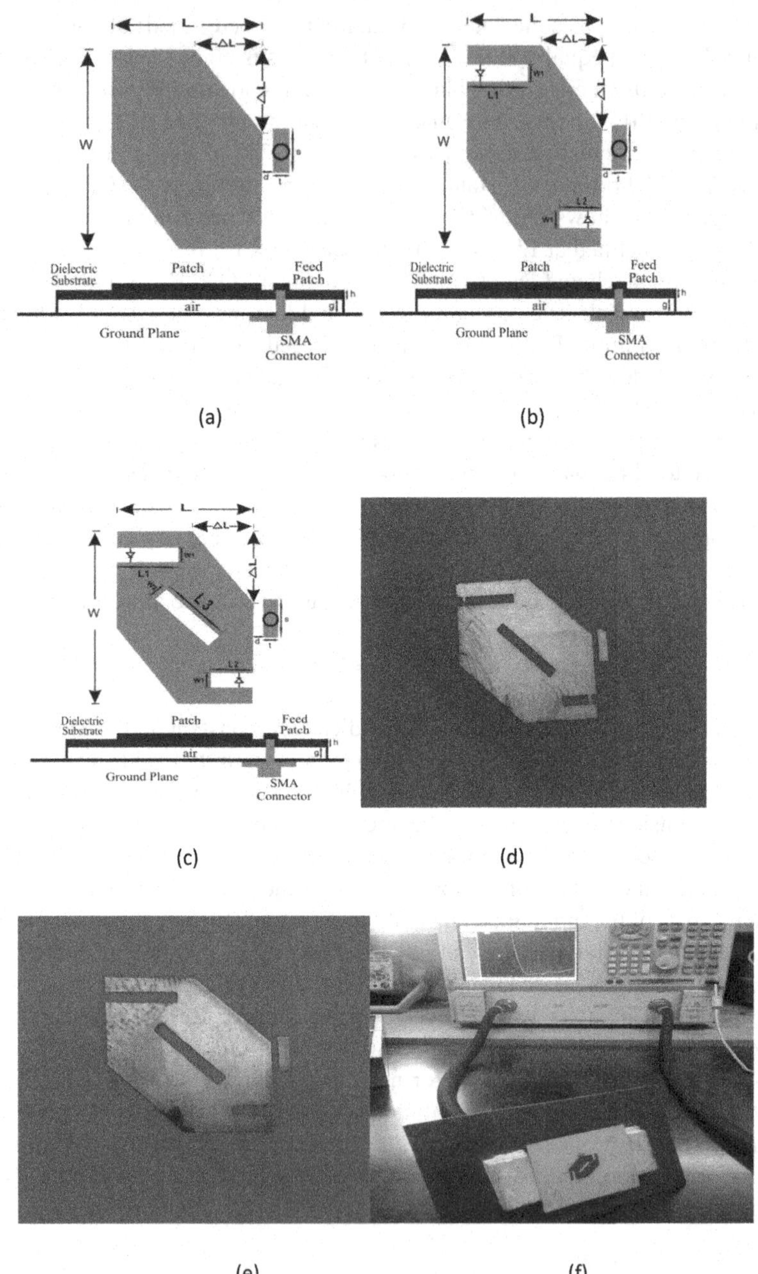

FIGURE 7.1 Geometry of a capacitive-coupled probe-fed microstrip antenna: (a) with truncated corner, (b) with PIN diodes and horizontal slits on truncated patch, (c) with PIN diodes, horizontal slits, and inclined slot on truncated patch, (d) the fabricated structure of the proposed antenna with PIN diodes ON, (e) the fabricated structure of the proposed antenna with PIN diodes OFF, (f) measurement set-up in Microwave Research Laboratory of AIACTR.

TABLE 7.1
Dimensions of Antenna Design

Parameter	L	W	s	t	D	ε_r	g
Value (mm)	15.5	16.4	3.7	1.2	0.5	3.0	6
Parameter	H	ΔL	L_1	L_2	L_3	w_1	w_2
Value (mm)	1.56	7	6.75	3.70	8.0	1.0	1.0

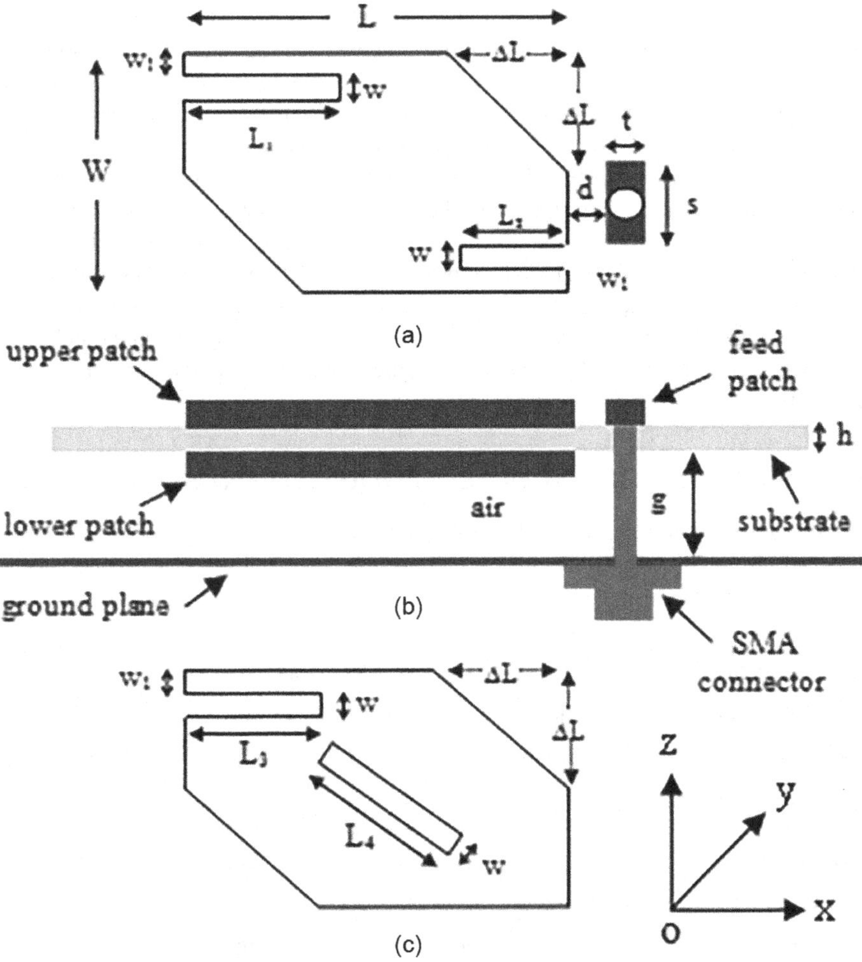

FIGURE 7.2 Geometry of the proposed stacked patch: (a) upper patch, (b) side view, and (c) lower patch.

patch. The feed strip is connected through SMA connector to capacitively couple the energy to the radiating patch. The capacitive coupling has the advantage that it can be treated as a rectangular microstrip capacitor which compensates for the reactance produced by the inductive probe. It also has the advantage that both the radiating patch and feed strip can be etched on the same substrate. The air gap enhances the impedance bandwidth without increasing the size and complexity of the antenna [35]. The air gap (g) and the separation between the feed strip and the radiating patch (d) are the key parameters for impedance matching.

The design goal is to develop an antenna having multiband CP operation. In order to achieve the goal, the design may begin with truncating a pair of opposite corners having a side length of ΔL [33] causing circular polarization. Two horizontal slits of unequal lengths (L_1 and L_2) are inserted at the opposite edges of the truncated patch. By properly optimizing the size of slits, the antenna generates dual-band CP. Further, the design employs stacking of another parasitic patch on the lower surface of the substrate, having the dimensions of $L \times W$ with the centre of the patches aligned. The opposite corners of the bottom patch are also truncated having the dimensions of $\Delta L \times \Delta L$, and a slit of length L_3 is introduced at the left upper edge. By stacking of the bottom patch, the antenna shows dual-band behaviour and generates triple CP bands. These two impedance bands are merged together to obtain the wide impedance bandwidth by optimizing the feed position at $d = 2.13$ mm. Now, a slot of length L_4 is embedded in the centre of the bottom patch along the diagonal axis of the patch. As a result, there is again enhancement in the impedance bandwidth and quad-band CP operation is obtained. The optimized dimensions of the antenna parameters at design frequency are listed in Table 7.2.

7.3 PROPERTIES OF CIRCULARLY POLARIZED ANTENNAS

For the antenna design and simulation, IE3D simulation software is used which is based on MoM. The experimental verification is carried out to authenticate the antenna results.

7.3.1 Single-Band Circularly Polarized Antennas

The antennas fabricated are shown in Figure 7.1d and e in ON and OFF states of the PIN diodes, respectively. Figure 7.3 shows the simulated VSWR of the truncated corner capacitive-coupled rectangular microstrip antenna. The truncated corner

TABLE 7.2
Dimension of Antenna Design

Parameter	(mm)	Parameter	(mm)	Parameter	(mm)	Parameter	(mm)
L	15.5	s	3.7	D	2.13	g	6
W	16.4	t	1.2	ε_r	3.0	w	1
H	1.56	L_1	6.75	L_3	5.0	w_1	1.6
ΔL	7	L_2	3.70	L_4	10		

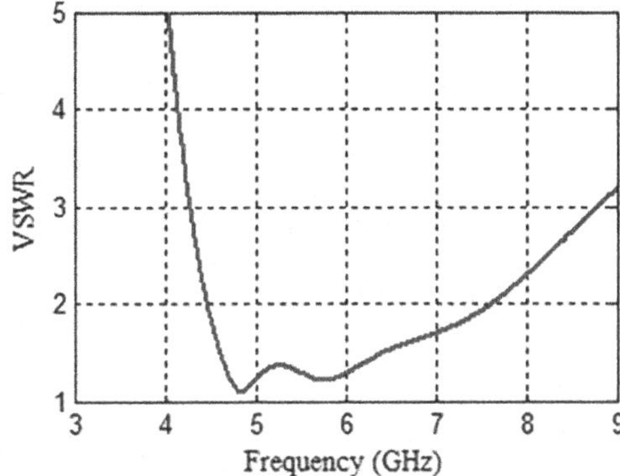

FIGURE 7.3 Variation of simulated VSWR with frequency for truncated corner antenna.

capacitive-coupled rectangular microstrip antenna achieves an impedance bandwidth of 50% (VSWR < 2).

The polarization of antenna can be changed from linear to circular by truncation. The CP band is obtained with the truncation of 7 × 7 mm at two opposite corners of the patch. Figure 7.4 shows the 3-dB axial ratio bandwidth of 11.1% corresponding to the frequency range from 5.69 to 6.36 GHz is obtained. Figure 7.5 shows the simulated gain variation against frequency. Due to truncation, the length of the electrical patch decreases, which is responsible for gain reduction with respect to reference antenna given in [7].

7.3.2 Reconfigurable Circularly Polarized Microstrip Antennas

PIN diodes are used as a switch in several microstrip antennas. GaAs PIN diodes with a forward voltage of 0.73 V and forward current of 12 mA are used for switching the antenna. An antenna with three CP bands is developed, and the structure is shown in Figure 7.1b. It is achieved by embedding slits with unequal length at the boundary of the truncated rectangular radiating patch and making the PIN diodes ON and OFF. Figure 7.6 shows the simulated VSWR of the PIN diode-loaded antenna in OFF and ON conditions of the diode. When the diode is OFF, the antenna exhibits dual-band behaviour and the impedance bandwidth is in the frequency range from 4.48 to 6.77 GHz and from 7.46 to 8.28 GHz. The corresponding impedance bandwidth with OFF state of PIN diodes is 40.71% and 10.41%. When the diode is ON, the operating frequency is in the frequency range of 4.44–7.30 GHz and 8.06–8.77 GHz. The corresponding impedance bandwidth with ON state of PIN diodes is 48.72% and 8.43%.

The variation of axial ratio with frequency is shown in Figure 7.7, and it is observed that for three frequency intervals, the axial ratio is below 3 dB, which indicates that the antenna can generate CP in three distinct bands. When the PIN diodes are OFF, they act as a simple notch in the antenna and split the single-band CP into double-band CP – one above and one below the original CP band. The two CP bands were obtained in the frequency range of 5–5.22 GHz and 6.49–6.71 GHz. The 3-dB AR bandwidth is 4.3% and 3.33%. It is seen that the

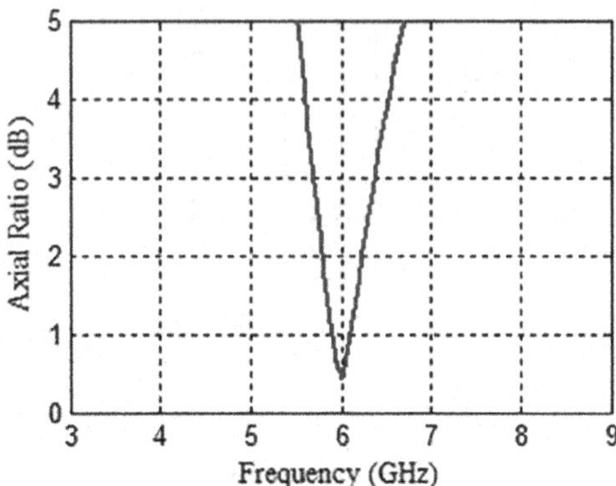

FIGURE 7.4 Variation of axial ratio with frequency for truncated corner antenna.

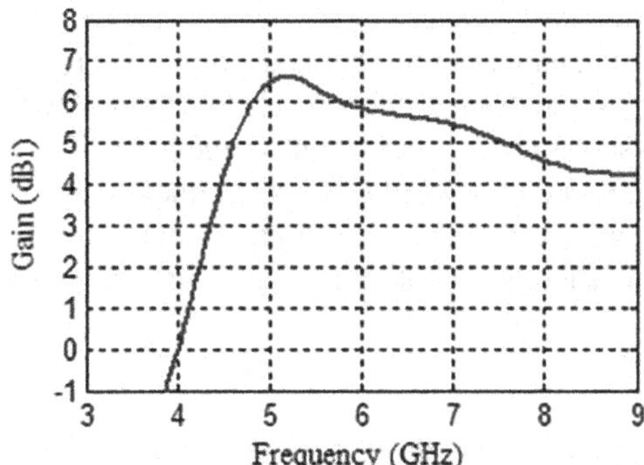

FIGURE 7.5 Variation of gain with frequency for truncated corner antenna.

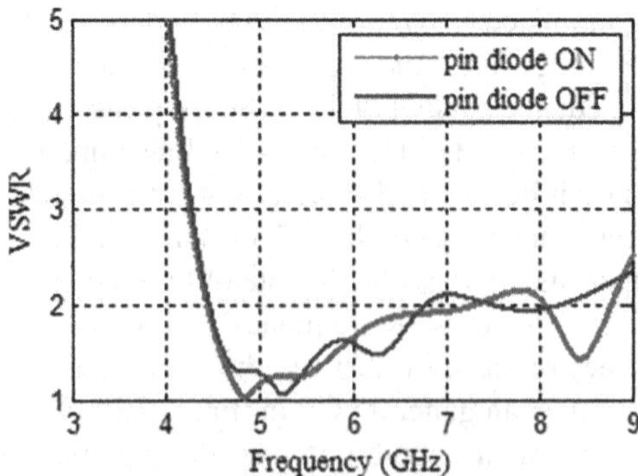

FIGURE 7.6 Variation of VSWR with frequency for PIN diode- and horizontal slit-embedded truncated patch.

centre frequency of CP operation is changed to 5.11 and 6.6 GHz from 6.025 GHz as in the antenna with truncated corners. The ON state of PIN diodes acts as an ohmic resistance and makes the gap connected and electric currents flow through the path. This effect of changed electric length of the surface current changes the resonant frequency of the two near-degenerate orthogonal modes, and the antenna gives CP at different frequencies. Figure 7.7 shows the axial ratio of a PIN diode-loaded antenna with horizontal slits also in the ON state of the diode. The 3-dB axial ratio bandwidth is 7.31% in the frequency range 5.40–5.81 GHz. It is seen from the figure that the antenna provides three CP bands by tuning the PIN diodes.

Figure 7.8 shows the simulated gain of the antenna under both conditions of the diode. It is seen that gain drops in some frequency interval. The reduction in gain occurs in the frequency range where the radiation is not in phase and the phase difference decides the gain. The reduced gain in the last CP band indicates higher-order orthogonal modes combining to produce CP.

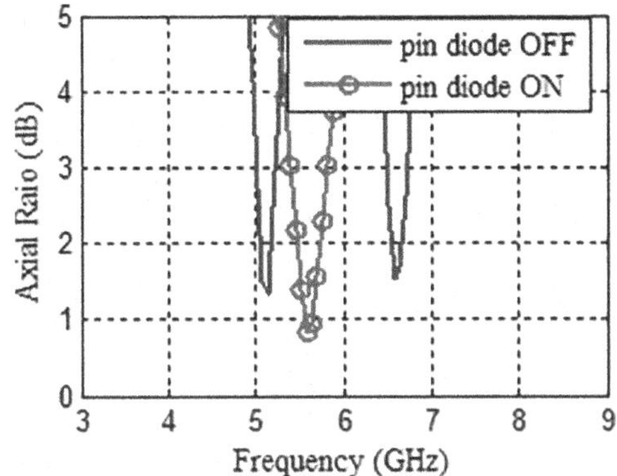

FIGURE 7.7 Axial ratio of PIN diode-loaded truncated corner antenna with horizontal slits.

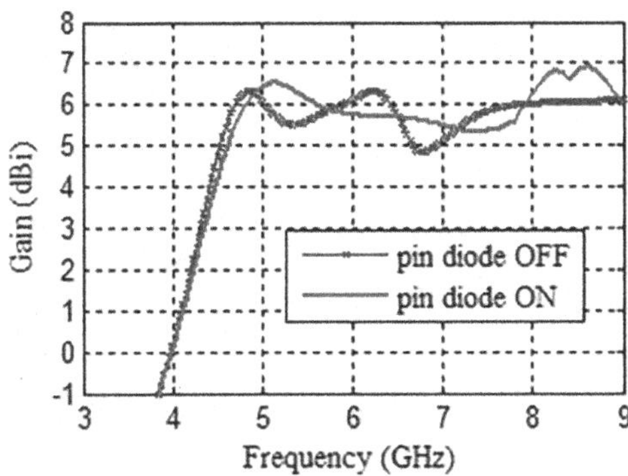

FIGURE 7.8 Gain of PIN diode-loaded truncated corner antenna with horizontal slits.

7.3.3 Impedance Bandwidth Improvement of Triple-Band CP Antennas

To increase the operational impedance bandwidth, an inclined slot was introduced, as shown in Figure 7.1c. The PIN diodes are intact at the gap between the two edges of the horizontal slits. When an inclined slot is cut inside the patch, there is a further increase in the length of the surface current path along the patch. The inclined slot and PIN diodes create two different resonances for the patch. The closeness among the resonances makes the broadband characteristics in the antenna. The broader bandwidth of the proposed antenna is due to the better control of current distribution towards the higher frequencies of the bandwidth that is achieved due to the inclined slot. Figure 7.9 shows the simulated and measured VSWR of the proposed antenna, and the simulated result shows a bandwidth of 66.61% ranging from 4.42 to 8.80 GHz with the ON state of PIN diodes, while Figure 7.10 shows the simulated and measured VSWR variations for the OFF state of the diode from 4.528 to 8.986 GHz, i.e. impedance bandwidth is 66.02%. In both the states of the diode, the proposed structure provides a better impedance bandwidth than a previous study [7]. The measured

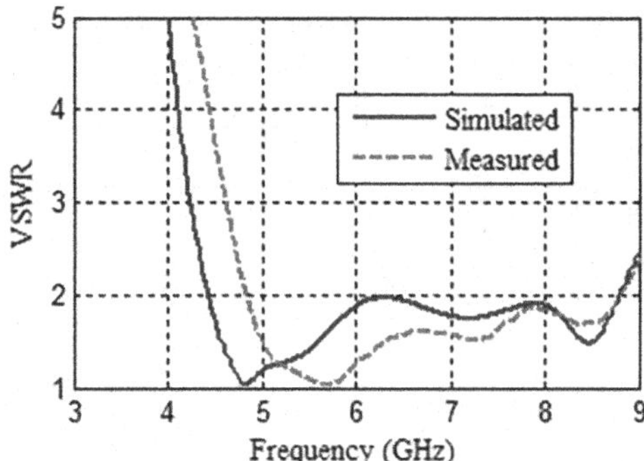

FIGURE 7.9 Measured and simulated VSWR for the proposed antenna with PIN diodes ON.

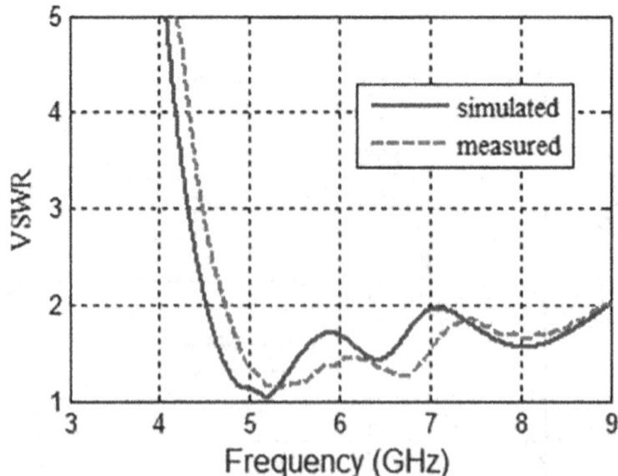

FIGURE 7.10 Measured and simulated VSWR for the proposed antenna with PIN diodes OFF.

result shown for comparison is in good agreement with the simulated result. The mismatch between the measured and simulated results existed, which may be mainly caused by fabrication imperfection.

Figure 7.11 shows the measured and simulated axial ratios of the proposed PIN diode-loaded antenna in the two states of the diode. It is clear from the figure that the antenna provides three circularly polarized bands by tuning the PIN diodes. The antenna exhibits CP in two bands with the frequency range from 4.88 to 5.07 GHz and from 6.51 to 6.71 GHz, when PIN diodes are OFF, i.e. the axial ratio bandwidth is 3.81% and 3.02%, respectively. With the ON state of the diodes, the antenna has another CP band from 5.31 to 5.61 GHz with the axial ratio bandwidth of 5.49%. The 3-dB axial ratio frequency range for all the three CP bands falls within the impedance bandwidth. Figure 7.12 depicts the measured and simulated gains with

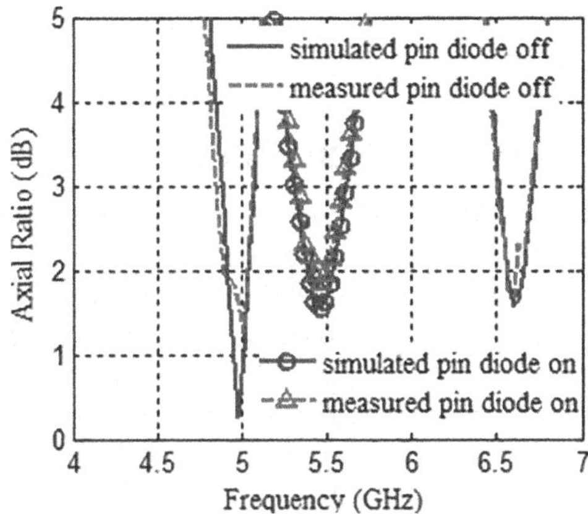

FIGURE 7.11 Measured and simulated axial ratios for the proposed antenna with PIN diodes ON and OFF.

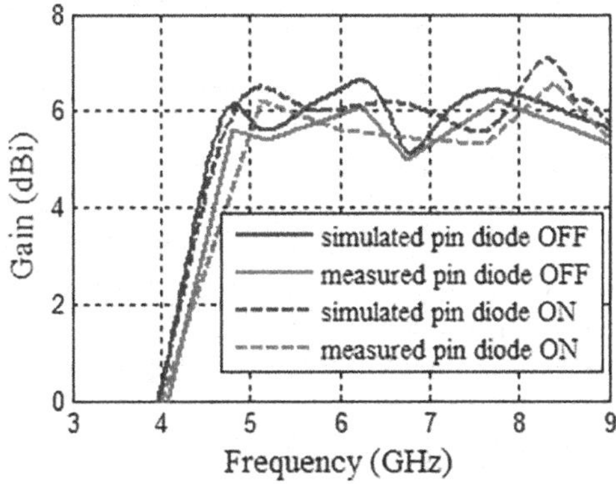

FIGURE 7.12 Measured and simulated gains for the proposed antenna.

frequency for an inclined slot-loaded microstrip antenna with PIN diodes ON and OFF. It is clear from the graph that the gain is almost constant over the CP bands.

For the reception of the signal, it is important to find the direction of field rotation in terms of LHCP wave and RHCP wave. The simulated and measured LHCP and RHCP far-field distributions in the E-plane at the centre frequencies of individual CP bands 4.98, 5.46, and 6.60 GHz are shown in Figure 7.13a–c, respectively. From these figures, it is clear that the antenna is LHCP with a considerable axial ratio beam width. A good amount of cross-polarization attenuation is obtained at all centre frequencies.

Figure 7.14a and b shows the current distribution on the radiating patch for different time frames: $t=0(0°)$, $t=T/4(90°)$, $t=2T/4(180°)$, and $t=3T/4(270°)$ at the centre frequencies of CP bands with PIN diodes OFF, while Figure 7.15 shows the same at the centre frequencies of CP band with PIN diodes ON. The surface current distribution on the radiating patch at the time frames clearly indicates the circularly polarized field radiation. The field rotates in the clockwise direction, which results in exciting a LHCP radiation.

7.3.4 Parametric Study of Quad-Band CP Antennas

The effects of dimensions of the slits, slot, and corner truncation on the reflection coefficient and axial ratio of the proposed antenna are investigated. In the investigation, one parameter is varied at a time, while others are kept constant. The effects of these geometrical parameters mentioned above on the reflection coefficient are very small and are not considered. Figure 7.16 shows the axial ratio for the change in the length of slit L_1. It can be noticed that with decreasing slit length L_1, CP band AR5 disappears and only three CP bands are obtained. Four CP bands are obtained with an increase in the slit length L_1. The effect of change in slit length L_2 is shown in Figure 7.17. With the decrease and increase in the slit length L_2, three and two CP bands are obtained, respectively. A significant effect is observed on CP bands AR5 and AR4 which disappeared with the increase in the slit length, and CP band AR5 disappears when the slit length decreased.

Figure 7.18 shows the axial ratio for the change in the length of slit L_3. It can be seen that as the L_3 decreases, CP bands AR2 and AR4 disappear and only two CP bands are obtained, while three CP bands are obtained with increasing L_3. Figure 7.19 illustrates that with decreasing slot length L_4, CP bands AR2 and AR5 disappear, while four CP bands are obtained with increasing slot length L_4. Figure 7.20 shows the axial ratio for the change in the length of corner truncation. It can be clearly seen that CP bands strongly depend on the truncation length ΔL. Only two CP bands are obtained with increasing and decreasing truncation length ΔL. The CP bands AR5 and AR4 disappear with decreasing truncation length, while CP bands AR2 and AR4 disappear with increasing truncation length. The effects on the centre frequencies (CF) and the axial ratio bandwidth (ARBW) of the different CP bands of the antenna with variation in the length of slits, slot, and corner truncation are summarized in Table 7.3. With decreasing slit length L_1, the CF of the CP bands AR2, AR3, and AR4 shifted slightly upwards, while the ARBW of CP bands AR2 and AR4 decreased and the ARBW of CP band AR3 increased. With an increase in the slit length L_1,

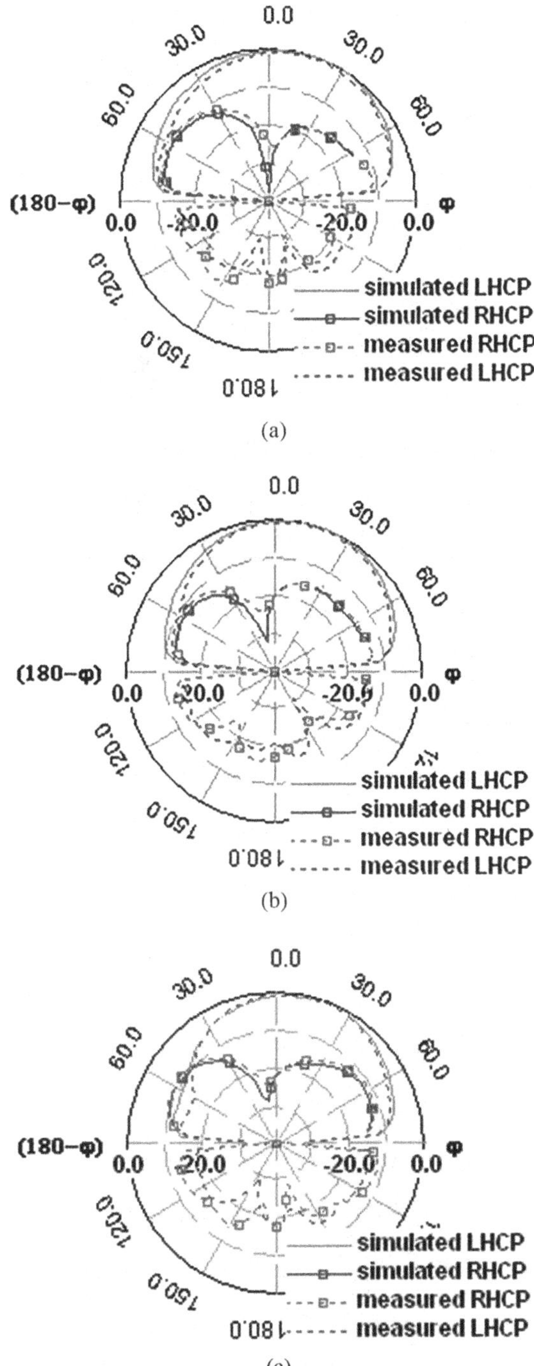

FIGURE 7.13 Simulated and measured LHCP and RHCP patterns at (a) 4.98GHz, (b) 5.46 GHz, and (c) 6.60 GHz.

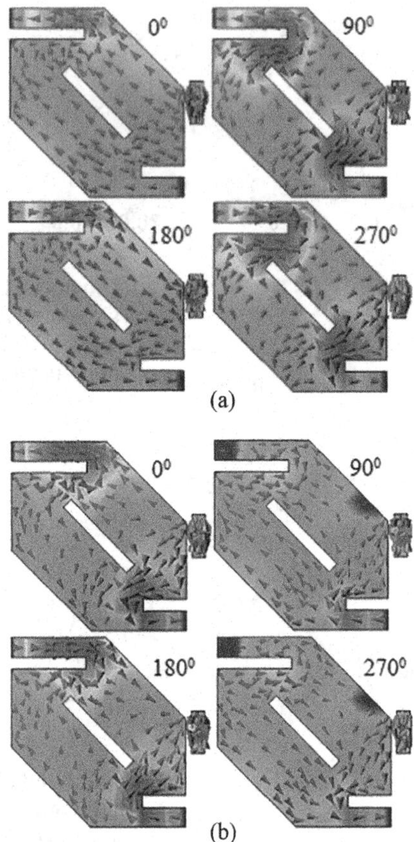

FIGURE 7.14 Surface current distribution for the proposed antenna with PIN diodes OFF at (a) 4.98 GHz and (b) 6.60 GHz.

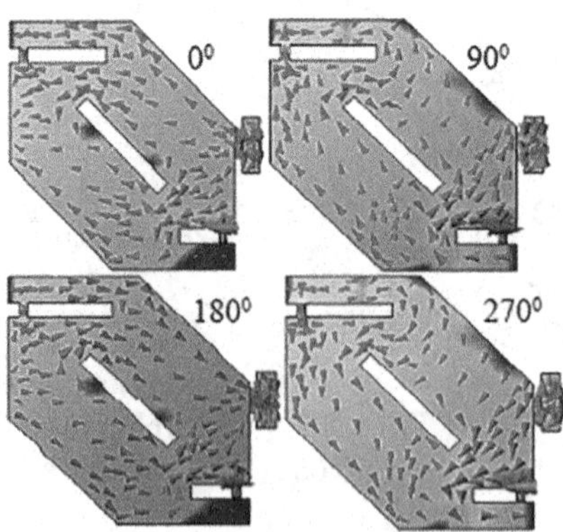

FIGURE 7.15 Surface current distribution for the proposed antenna with PIN diodes ON at 5.46 GHz.

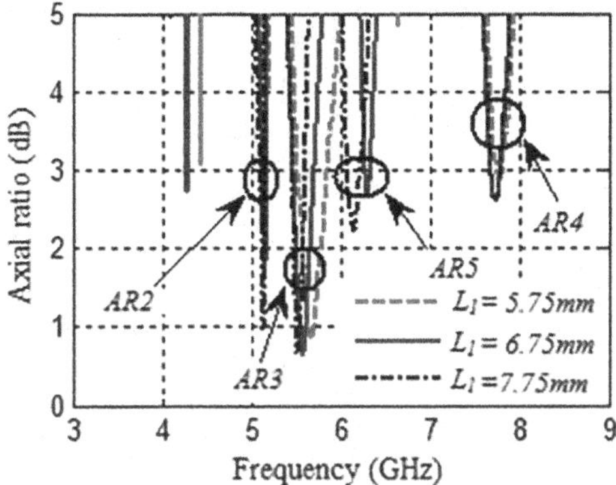

FIGURE 7.16 Variation of axial ratio for different values of slit's length L_1 in the upper patch.

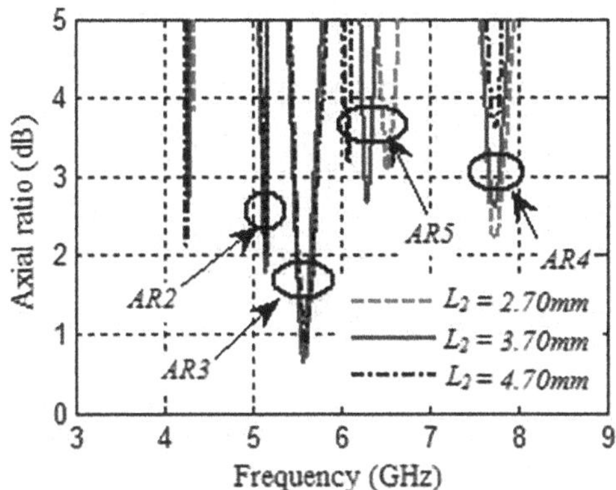

FIGURE 7.17 Variation of axial ratio for different values of slit's length L_2 in the upper patch.

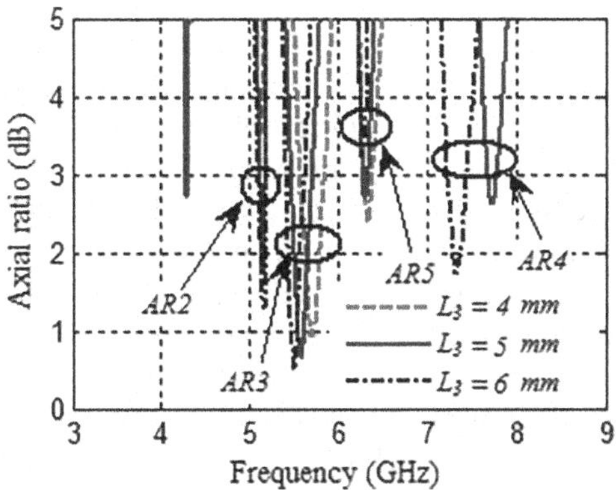

FIGURE 7.18 Variation of axial ratio for different values of slit's length L_3 in the lower patch.

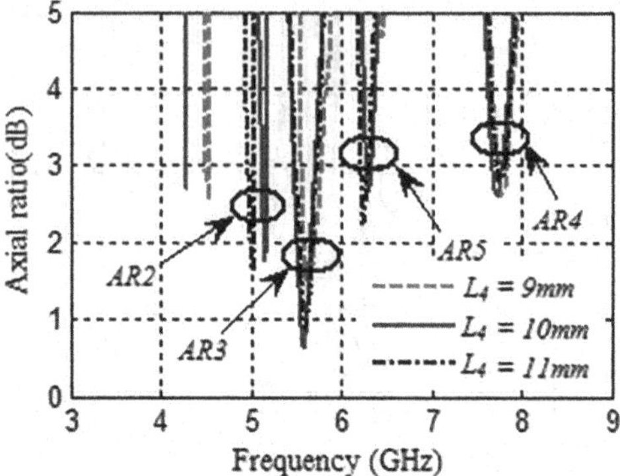

FIGURE 7.19 Variation of axial ratio for different values of slot's length L_4 in the lower patch.

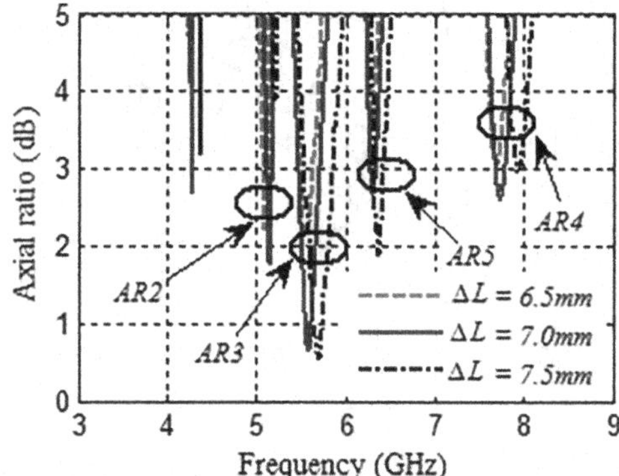

FIGURE 7.20 Variation of axial ratio for different values of corner truncation's length ΔL.

the CF of all CP bands shifted downwards. The ARBW of CP bands AR2, AR4, and AR5 increased, while the ARBW of CP band AR3 decreased. With increasing and decreasing slit length L_2, a small effect is observed on the CF and ARBW of CP bands. The CF of the CP bands AR3 and AR5 shifted upwards with a small increase in the ARBW by decreasing the slit length L_3, while the CF of CP bands AR2, AR3, and AR4 shifted slightly downwards with a small change in the ARBW by increasing the slit length L_3. A small change is observed on the CF and ARBW of the CP bands with the increase and decrease in the slot length L_4. From the table, it is also observed that the CF of CP bands AR2 and AR3 shifted downwards, while the ARBW of CP bands AR2 and AR3 decreased and increased, respectively, upon decreasing the corner truncation length ΔL. The CF of CP bands shifted upwards with increased ARBW upon increasing the corner truncation length ΔL.

By parametric studies mentioned above, the design procedure for a quad-band CP antenna is obtained and it is observed that the lengths of slits, slot, and corner

TABLE 7.3
Simulated Results of the Proposed Antenna by Varying the Length of Slits, Slot, and Corner Truncation

		AR2		AR3		AR5		AR4	
Parameter	**Value (mm)**	**CF (GHz)**	**ARBW (%)**	**CF (GHz)**	**ARBW (%)**	**CF (GHz)**	**ARBW (%)**	**CF (GHz)**	**ARBW (%)**
L_1	5.75	5.158	0.33	5.6843	5.840	NA		7.7703	0.90
	6.75	**5.1476**	**0.98**	**5.5905**	**4.275**	**6.3026**	**0.8869**	**7.7298**	**1.35**
	7.75	5.1213	1.13	5.522	2.72	6.147	2.28	7.7273	1.37
L_2	2.70	5.1459	0.88	5.6079	4.37	NA		7.7431	2.13
	3.70	**5.1476**	**0.98**	**5.5905**	**4.275**	**6.3026**	**0.8869**	**7.7298**	**1.35**
	4.70	5.151	1.12	5.6091	4.57	NA		NA	
L_3	4.0	NA		5.7036	4.71	6.3336	1.53	NA	
	5.0	**5.1476**	**0.98**	**5.5905**	**4.275**	**6.3026**	**0.8869**	**7.7298**	**1.35**
	6.0	5.1356	1.49	5.5078	3.73	NA		7.3359	2.29
L_4	9.0	NA		5.6589	3.77	NA		7.7727	1.304
	10.0	**5.1476**	**0.98**	**5.5905**	**4.275**	**6.3026**	**0.8869**	**7.7298**	**1.35**
	11.0	4.9920	1.15	5.6037	4.08	6.2488	1.44	7.7579	1.10
ΔL	6.0	5.1023	1.08	5.5587	2.64	NA		NA	
	7.0	**5.1476**	0.98	5.5905	4.275	6.3026	0.8869	7.7298	1.35
	8.0	NA		5.6934	5.52	6.3676	1.52	NA	

truncation are the critical parameters and play an important role in generating quad-band CP operation.

7.3.5 Simulated and Measured Results

The reflection coefficient and axial ratio of a slit-loaded truncated patch before stacking are shown in Figure 7.21. The impedance bandwidth of 37.83% in the frequency range of 4.65–6.82 GHz is achieved for the slit-loaded truncated patch. By properly tuning the dimensions of the slits, the antenna produces two CP bands in the frequency range 5.0–5.22 GHz and 6.49–6.71 GHz, as shown in Figure 7.21, and the 3-dB axial ratio bandwidth of 4.31% and 3.34%, respectively. Now, another rectangle patch of dimensions $L \times W$ is etched on the bottom surface of the substrate with the centre of two patches aligned and a pair of the opposite corners of the bottom patch is also removed. The truncation dimension is the same as for upper patch. But by this process, the antenna is converted to single-band CP from dual-band CP. So, to achieve the multiband CP operation again, a horizontal slit on the bottom patch is introduced, as shown in Figure 7.2c. As a result, triple-band CP radiation is achieved. Figure 7.8 shows the reflection coefficient and axial ratio of a slit-loaded stacked patch. The antenna shows dual-band response in the frequency range 4.33–6.13 GHz and 6.54–7.37 GHz with an impedance bandwidth of 34.41% and 11.93%, respectively. The corresponding CP bands of the said antenna are also characterized from Figure 7.22. From the figure, it is clear that the first three CP bands, i.e. AR1, AR2, and AR3, which are achieved by selecting the proper dimensions of the slit into the bottom patch, fall within the first band of operation. The 3-dB axial ratio bandwidths are 0.864%, 2.37%, and 3.24% in the frequency range 4.5959–4.6358 GHz (AR1), 5.5126–5.6453GHz (AR2), and 5.8471–6.0397 GHz (AR3), respectively. It is important to note that the 3-dB axial ratio bandwidth should fall within the impedance bandwidth for the proper operation of the circularly polarized microstrip antenna. It is also seen from Figure 7.22 that there is another CP band in the frequency range 7.7467–7.9017 GHz (AR4), which does not lie within the impedance bandwidth. Now, the aim here is to match the antenna impedance to cover CP band AR4. For this, a slot is embedded in the centre of the bottom patch along the diagonal axis and the

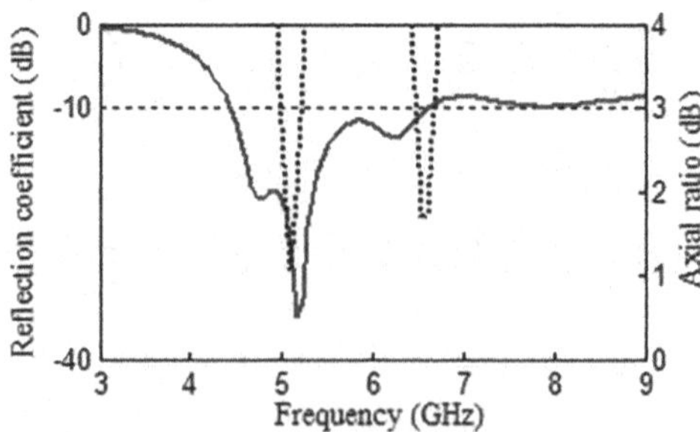

FIGURE 7.21 Reflection coefficient and axial ratio of the slit-loaded truncated patch without stacking with $d = 0.5$ mm [solid graph shows reflection coefficient and dotted graph shows axial ratio].

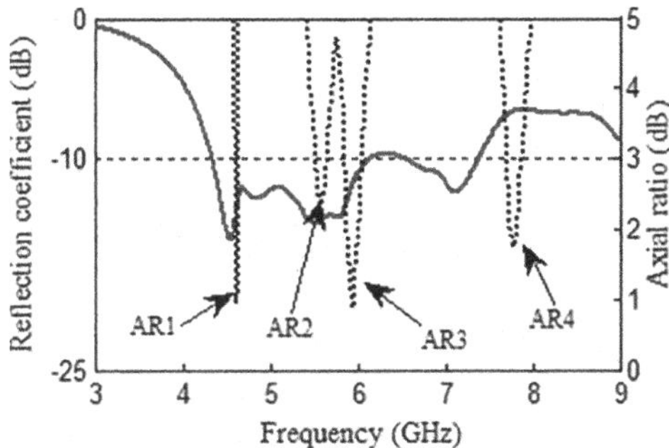

FIGURE 7.22 Reflection coefficient and axial ratio of the slit-loaded stacked truncated patch with $d=0.5$ mm [solid graph shows reflection coefficient and dotted graph shows axial ratio].

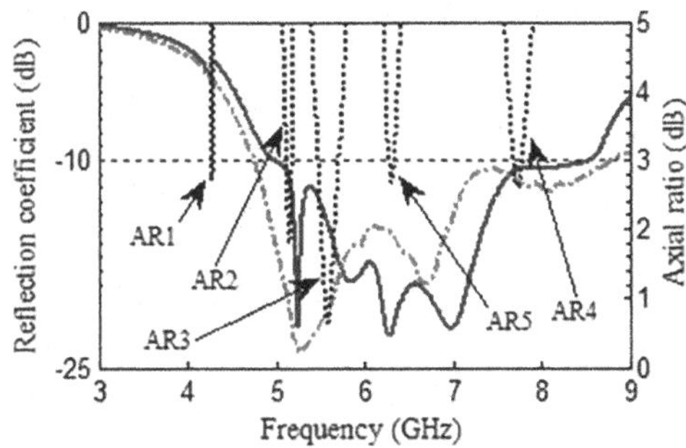

FIGURE 7.23 Reflection coefficient and axial ratio of the proposed antenna simulated reflection coefficient measured reflection coefficient axial ratio [the solid line shows simulated reflection coefficient, -. line shows measured reflection coefficient and .. line shows axial ratio].

feed location (d) is optimized at 2.13 mm. Figure 7.23 shows the simulated reflection coefficient and axial ratio at 20^0 from broadside. This proposed antenna covers the frequency band of 4.97–8.49 GHz with an impedance bandwidth of 55.6%. In this process, the lower CP band AR1 is detuned, but the old CP band AR4 falls in the band of operation and a new CP band AR5 is introduced in the band and the antenna shows quad-band CP operation. It is clear from the figure that four distinct CP bands fall within the operating band and at each CP band, the axial ratio is below 3 dB, which means that the antenna has a good CP performance. The 3-dB axial ratio bandwidths are 0.98%, 4.275%, 0.8869%, and 1.35% in the frequency band 5.1223–5.1729 GHz (AR2), 5.4710–5.7100 GHz (AR3), 6.2747–6.3306 GHz (AR5), and 7.6775–7.7822 GHz (AR4), respectively. The figure also shows the measured reflection coefficient of the proposed antenna. The measured result is in close agreement with the simulated result. Figure 7.24 illustrates the simulated and measured axial ratios of the proposed antenna for the CP bands AR2 and AR3. The simulated and measured axial ratios for the CP bands AR5 and AR4 are shown in Figure 7.25. However, to perform the

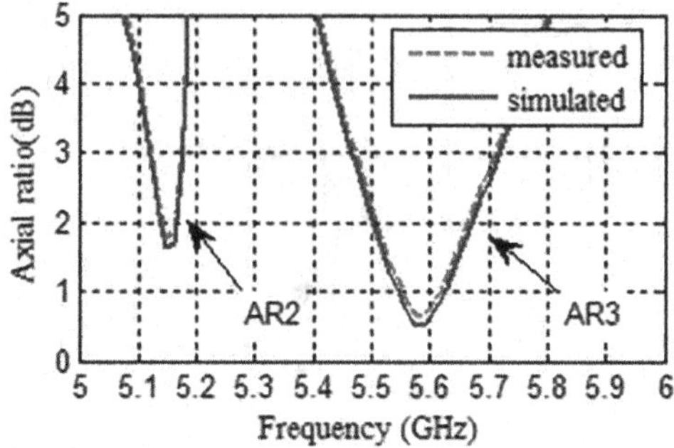

FIGURE 7.24 Simulated and measured axial ratios of the proposed antenna for CP bands AR2 and AR3.

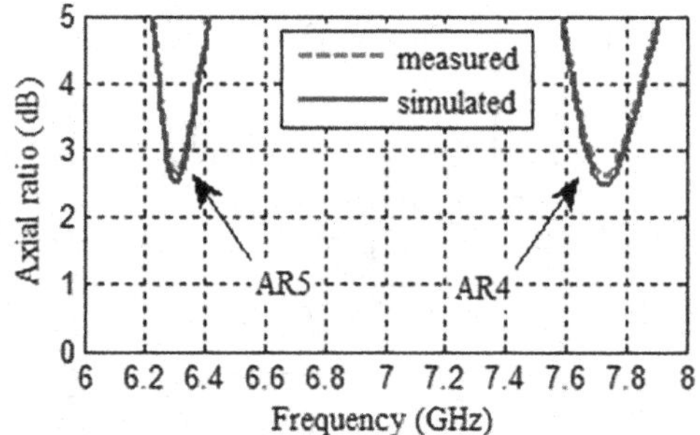

FIGURE 7.25 Simulated and measured axial ratios of the proposed antenna for CP bands AR5 and AR4.

axial ratio measurement in an anechoic chamber, the antenna under test (AUT) is referred to as source. Then, a linearly polarized antenna, usually a horn antenna, is used as the receiver antenna. A linearly polarized receiver antenna is rotated in the plane of polarization from 0° to 360°. The minimum and maximum values of power received at the given frequency are recorded. The difference between the minimum and maximum values of power was considered as the axial ratio at that frequency. This measurement process was repeated for all the frequency points, and the graph was plotted. The measured and simulated gains at 200 from the broadside are presented in Figure 7.26. For the CP bands AR2, AR3, AR5, and AR4, the antenna gain varies from 4.03 to 2.8624 dBic, 6.07 to 5.8743 dBic, 5.29 to 5.40 dBic, and 7.67 to 4.66 dBic, respectively. The gain of the antenna was measured at far-field using the substitution method in an anechoic chamber. Two calibrated horn antennas of known gain as the transmitter antenna are used to find the unknown gain of AUT. The AUT as the receiver antenna was placed on the positioner with required elevation and azimuthal coverage. The AUT was then replaced with the second horn antenna of calibrated gain, and similar measurement was repeated. Difference between the measured powers reflects the gain difference between the two receiving antennas,

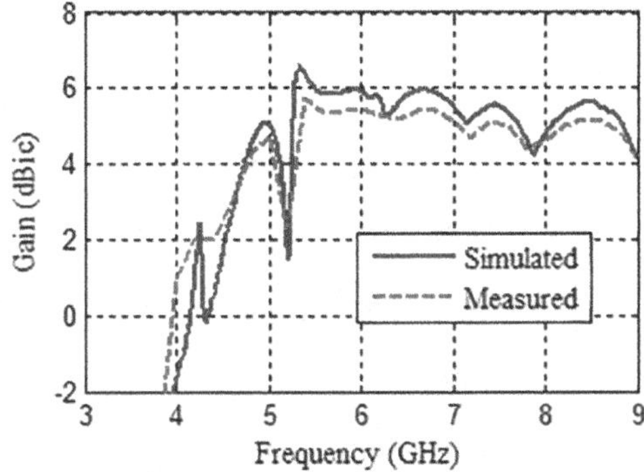

FIGURE 7.26 Simulated and measured gains of the proposed antenna.

and the absolute gain of the AUT is calculated. The whole measurement procedure was repeated by changing the distance between the transmitter and receiver antennas using the positioner movement, and the average gain was considered as the final gain value. The measured and simulated left-hand circular polarization (LHCP) and right-hand circular polarization (RHCP) radiation patterns in *xoz* plane at centre frequencies of CP bands AR2, AR3, AR5, and AR4 are presented in Figure 7.27a–d, respectively. It is observed that the LHCP is radiated for all the CP bands, and the maximum radiation is achieved at 20° from broadside. However, the squinting of the main beam can be improved to radiate at boresight by making an array of the antenna, properly setting the spacing between the antennas and adjusting the feed angle to the antenna.

7.3.6 Operating Mechanism

The operating mechanism of the antenna in the four bands of the CP operation can be explained with the help of current distribution at the two patches of the antenna. Various geometrical parameters play their role in generating different CP bands. Figure 7.28a–d shows the electric current distribution at the centre frequencies of the different CP bands for the upper patch (UP) and lower patch (LP) at four different time instants, that is, at $\omega t = 0°$, 90°, 180°, and 270°, respectively, where ω is the angular frequency and t is time. In the explanation, the path difference due to inter patch spacing is neglected. At the centre frequency of the first CP band, i.e. 5.14 GHz, the inclined slot and corner truncation are responsible for CP generation. At feed angle $\omega t = 0°$, the current vector at UP is oriented at an angle of 135°. At $\omega t = 90°$, the electric currents from the truncated corners of the LP and UP are cancelled, resulting in the radiation from the inclined slot only, which is at an angle of 45°. At $\omega t = 180°$ and 270°, the reverse radiation happens to those at $\omega t = 0°$ and $\omega t = 90°$, and the resulting radiation takes place at an angle of 315° and 225°, respectively. Hence, the E-field vector rotates clockwise as time increases, thus producing the LHCP radiation. At the centre frequency of the second CP band, i.e. 5.59 GHz, the radiation from both patches determines the direction of the resultant electric field. At $\omega t = 0°$, the resultant two vector currents make the radiation at

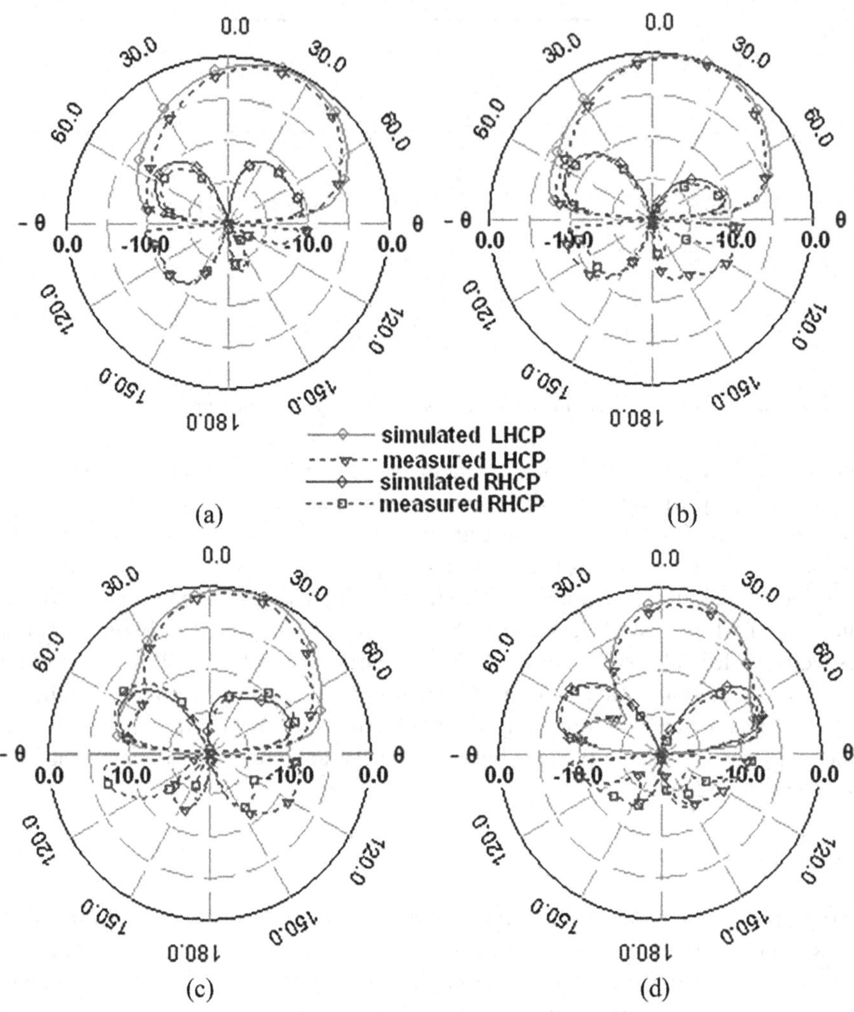

FIGURE 7.27 Simulated and measured radiation patterns of the proposed antenna in *xoz* plane: (a) at 5.14 GHz, (b) at 5.59 GHz, (c) at 6.30 GHz, and (d) at 7.729 GHz.

157.5°, while at $\omega t = 90°$, these vector currents fall at 0° and 135°, resulting in the radiation at 67.5°. Similarly, the resultant vector at $\omega t = 180°$ and $\omega t = 270°$ radiates at 337.5° and 247.5°, respectively. Hence, again, these radiation directions indicate a LHCP pattern. At the centre frequency of the third CP band, i.e. 6.30 GHz, the slit length L_1 plays an important role. The current vectors due to the truncated corners are opposite to each other cancelling the radiation. Hence, at $\omega t = 0°$, the resultant E-field vector is at an angle 180°. At $\omega t = 90°$, the horizontal E-field from LP and UP are cancelled out, resulting in only vertical E-field (90°) from the two patches. Similarly, at $\omega t = 180°$ and $\omega t = 270°$, the resulting radiation becomes at 0° and 270°, respectively, suggesting a LHCP radiation again. At the centre frequency of the fourth CP band, i.e. 7.729 GHz, the slit's lengths L_1 and L_2 on the UP are

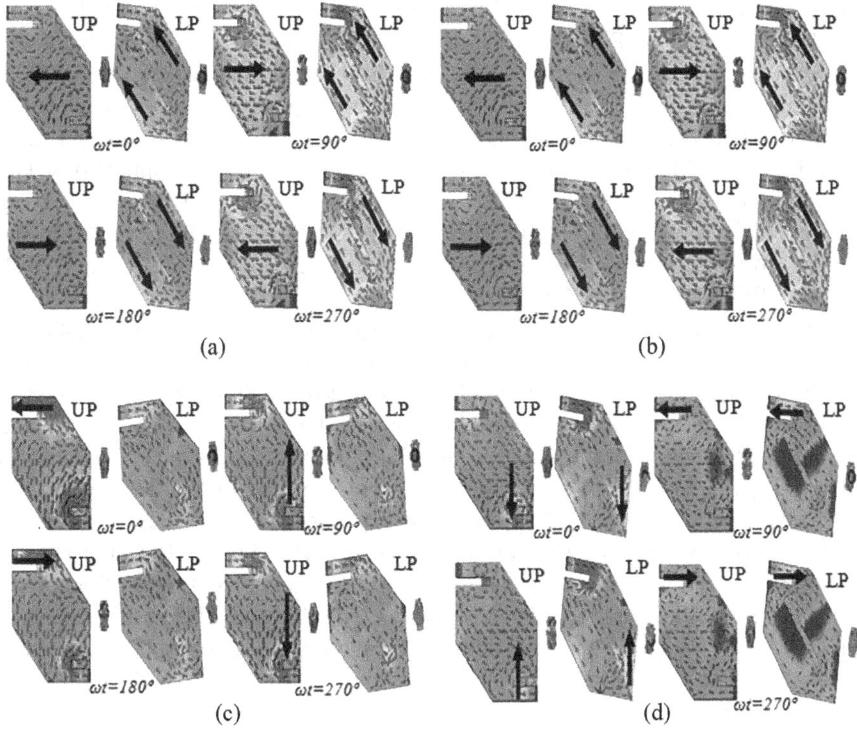

FIGURE 7.28 Simulated current densities of the proposed antenna at centre frequencies of circularly polarized bands: (a) at 5.14 GHz, (b) at 5.59 GHz, (c) at 6.30 GHz, and (d) at 7.729 GHz.

responsible for CP radiation. All the E-field vectors except vertically downward vectors at the edge of lower slit are cancelled out, and the resultant radiation occurs at an angle of 270° for $\omega t = 0°$. At $\omega t = 90°$, the upper slit radiates at 180° dominating other E-field vectors. Similarly, at $\omega t = 180°$ and $\omega t = 270°$, the resultant E-field vectors radiate at 90° and 0°, respectively, rotating the field clockwise indicating a LHCP radiation.

7.4 CONCLUSION

In this chapter, two antennas of three and four CP bands were discussed. The first antenna operated in three CP bands using PIN diodes, while the second antenna used stacking to achieve this. Both antennas were corner-truncated and suspended in air. Two horizontal slits of unequal lengths with PIN diodes on the truncated patch were cut to achieve the goal. Triple-band circular polarization was realized by switching PIN diodes across the slits ON and OFF. The broadband performance of the proposed antenna was realized by embedding an inclined slot on the patch with PIN diodes across the horizontal slits. The impedance bandwidth of the antenna increased from 51% to 66.61% (ON state) and 66.02% (OFF state) as compared to the capacitive-coupled probe-fed microstrip antenna reported earlier, and three distinct LHCP bands were also generated. An axial ratio bandwidth for the proposed antenna of 3.81%, 3.02%, and 5.49% was realized. A good LHCP performance was achieved in the three bands with respect to cross-polarization attenuation and axial ratio beam width. The results of the proposed antenna show that it is very suitable for various wireless communication system applications. The proposed antenna is useful for 5 GHz WLAN and may be used for public safety WLAN (IEEE802.11y) at 4.9 GHz, IEEE 802.11ac Wi-Fi at 5 GHz, which has the expected WLAN throughput of at least 1 GB/s and was approved in January 2014, and HIPERLAN/2(5.470–5.725 GHz).

Apart from this, the quad-band circularly polarized capacitive-coupled stacked patch antenna with wideband characteristics for C-band applications can be used as a multiband device for CP applications. Only two stacked patches were used to achieve quad-band CP operation. The antenna gave an impedance bandwidth of 55.46% and showed quad-band CP performance with axial ratio bandwidths of 0.98%, 4.275%, 0.8869%, and 1.35%. Good LHCP patterns were achieved in all the four distinct CP bands. The CP band AR2 in the frequency band 5.1223–5.1729 GHz covers HIPERLAN/1 and fixed satellite services (earth to space) (5.091–5.15 GHz). The CP band AR3 in the frequency range 5.4710–5.7100 GHz can be used for HIPERLAN/2 and amateur satellite uplink (5.650–5.670 GHz). The CP band AR5 in the frequency band 6.2747–6.3306 GHz can be used for C-band communication satellites for their uplinks (5.925–6.425 GHz).

REFERENCES

1. Garg, R.; Bhartia, P.; Bahl, I.; and Ittipiboon, A.: *Microstrip Antenna Design Handbook*. Artech House, Norwood, MA, (2001).
2. Schaubert, D.H.; Pozar, D.M.; and Adrian, A.: Effects of microstrip antenna substrate thickness and permittivity: comparision of theories and experiment. *IEEE Transactions Antennas Propagation*, 37(1989), 677–682.
3. Pues, H.F.; and Van De Capelle, A.R.: An impedance matching technique for increasing the bandwidth of microstrip antenna. *IEEE Antennas and Propagation Magazine*, 37–11(1989), 1345–1354.
4. Weigand, S.; Huff, G.H.; Pan, K.H.; and Bernhard, J.T.: Analysis and design of broadband single layer rectangular U-slot microstrip patch antenna. *IEEE Transaction on Antennas and Propagation*, 51–3(2033), 457–468.
5. Deshmukh, A.A.; and Kumar, G.: Broadband and compact V-slot loaded RMSAs. *Electron Letters*, 42–17(2006), 951–952.
6. Targonski, S.D.; Waterhouse, R.B.; and Pozar, D.M.: Wideband Aperture coupled Stacked patch antenna using thick substrate. *Electron Letters*, 32–21(1996), 1941–1942.
7. Kasabegoudar, V.G.; and Vinoy, K.J.: Coplanar capacitively coupled probe fed microstrip antenna for wideband applications. *IEEE Transaction on Antennas and Propagation*, 58–10(2010), 3131–3138.
8. Chang, F.S.; Wong, K.L.; and Chion, T.W.: Low cost broadband circularly polarized patch antenna. *IEEE Transaction on Antennas and Propagation*, 51(2003), 3006–3009.
9. Sharma, P.C.; and Gupta, K.P.: Analysis and optimized design of single feed circularly polarized microstrip antenna. *IEEE Transaction on Antennas and Propagation*, 29(1983), 949–955.

10. Wong, K.-L.; and Wu, J.-Y.: Single feed Small circularly polarized square microstrip antenna. *Electron Letters*, 33–22(1997), 1833–1834.
11. Chen, W.-S.; Wu, C.-K.; and Wong, K.-L.: Novel compact circularly polarized square microstrip antenna. *IEEE Transaction on Antennas and Propagation*, 49–3(2001), 340–342.
12. Wong, K.-L.; and Lin, Y.-F.: Circularly polarized microstrip antenna with a tuning stub. *Electron Letters*, 34–9(1998), 831–832.
13. Iwasaki, H.: A circularly polarized small-sized microstrip antenna with a cross-slot. *IEEE Transactions on Antennas Propagation*, 44–10(1996), 1399–1401.
14. Pozar, D.M.; and Duffy, S.M.: A dual band circularly polarized aperture coupled stacked microstrip antenna for global positioning satellite. *IEEE Transactions on Antennas Propagation*, 45–11(1997), 1618–1625.
15. Yu, A.; Yang, F.; and Elsherbeni, A.: A dual band circularly polarized ring antenna based on composite right and left handed metamaterials. *Progress in Electromagnetics Research*, 78(2008), 73–81.
16. Fujimoto, T.; Ayukawa, D.; Iwanaga, K.; and Taguchi, M.: Dual band circularly polarized microstrip antenna for GPS application. *IEEE Antenna Propagation Society International Symposium, S* (2008), 1–4.
17. Heidari, A.A.; Heyrani, M.; and Nakhkash, M.: A dual band circularly polarized stub loaded microstrip patch antenna for GPS applications. *Progress in Electromagnetics Research*, 92(2009), 195–208.
18. Chen, R.-H.; and Row, J.-S.: Single fed microstrip patch antenna with switchable polarization. *IEEE Transactions on Antennas Propagation*, 56–4(2008), 922–926.
19. Zhou, Y.; Chen, C.-C.; and Volkis, J.L.: Dual band proximity fed stacked patch antenna for tri-band GPS applications. *IEEE Transactions on Antennas Propagation*, 55–1(2007), 220–223.
20. Doust, E.G.; Clenet, M.; Hemmati, V.; and Wight, J.: An aperture coupled circularly polarized stacked microstrip antenna for GPS frequency bands L1, L2, L5. *IEEE Transactions Antennas and Propagation Society International Symposium*, (2008), 1–4.
21. Falade, O.P.; Rehman, M.U.; Gao, Y.; Chen, X.D.; and Parini, C.G.: Singe feed stacked circular polarized antenna for triple band operation. *IEEE Transactions on Antennas Propagation*, 60–10(2012), 4479–4484.
22. Lio, W.; Chu, Q.X.; and Du, S.: Triple band circularly polarized stacked microstrip antenna for GPS and CNSS applications. *In ICMMT Proceeding*, (2010), 252–255.
23. Yim, H.-Y.A.; Kong, C.-P.; and Cheng, K.-K.M.: Compact circularly polarized Microstrip antenna design for dual band applications. *Electron Letters*, 47–7(March 2006), 380–381.
24. Zhang, Y.-Q.; Li, X.; Yang, L.; and Gong, S.-X.: Dual band circularly polarized annular –ring microstrip antenna for GNSS application. *IEEE Transactions and Wireless Propagation Letters*, 12(2013), 615–618.
25. Lee, H.R.; Ryu, H.K.; Lim, S.; and Woo, J.M.: A miniaturized, dual – band, circularly polarized microstrip antenna for installation into satellite mobile phones. *IEEE Transactions and Wireless Propagation Letters*, 8(2009), 823–825.
26. Yuan, H.Y.; Zhang, J.Q.; Qu, S.B.; Zhou, H.; Wang, J.F.; Ma, H.; and Xu, Z.: Dual band dual polarized microstrip antenna for compass navigation satellite system. *Progress in Electromagnetics Research*, 30(2012), 213–223.
27. Yang, K.P.; and Wong, K.-L.: Dual –band circularly polarized square microstrip antenna. *IEEE Transactions on Antennas and Propagation*, 49(2001), 377–382.
28. Lai, X.-Z.; Xie, Z.-M.; and Cen, X.-L.: Design of dual circularly polarized antenna with high isolation for RFID applications. *Progress in Electromagnetics Research*, 139(2013), 25–39.

29. Yuan, H.Y.; Zhang, J.Q.; Qu, S.B.; Zhou, H.; Wang, J.F.; Ma, H.; and Xu, Z.: Dual band dual polarized microstrip antenna for compass navigation satellites system. *Progress in Electromagnetics Research*, 30(2012), 213–223.
30. Eselle, K.P.; and Verma, A.K.: Wideband circularly-polarized stacked microstrip antennas. *IEEE Antennas and Wireless Propagation Letters*, 6(2007), 21–24.
31. Rama Rao, B.; Smolinski, M.A.; Quach, C.C.; and Rosario, E.N.: Triple band GPS trap – loaded inverted L antenna array. *Microwave and Optical Technology Letters*, 38–1(July 2003), 35–37.
32. Falade, O.P.; Chen, X.; Alfadhi, Y.; and Parini, C.: Quad band circular polarised antenna. *Antenna and Propagation Conference(LAPC), Loughborough*, (2012), 1–4.
33. Kumar, G.; and Ray, K.P.: *Broadband Microstrip Antennas*. Artech House, Boston, London, (2003).
34. Singh, D.K.; Kanaujia, B.K.; Dwari, S.; Pandey, G.P.; and Kumar, S.: Reconfigurable circularly polarized capacitive coupled microstrip antenna. *International Journal of Microwave and Wireless Technologies*, 9–4(May 2017), 843–850.
35. Singh, D.K.; Kanaujia, B.K.; Dwari, S.; Pandey, G.P.; Kumar, S.: Novel quad-band circularly polarized capacitive-fed microstrip antenna for C-band applications. *Microwave and Optical Technology Letters*, 57–11(2015), 2622–2628.

8 Special Techniques of Printed Antenna

Dr. Dinesh Kumar Singh
G L Bajaj Institute of Technology and Management, Greater Noida

Dr. Ganga Prasad Pandey
Pandit Deendayal Petroleum University, Gujarat

CONTENTS

8.1 INTRODUCTION

Printed antennas are available in various shapes and most widely accepted by researchers, scientists, and engineers for various applications; however, they have bandwidth limitations. Since the inception of printed antennas, a number of techniques have been invented to enhance the bandwidth of these antennas. In this chapter, some of the special techniques of designing printed antennas for reconfiguration, polarization, feeding, etc. are discussed. This chapter is broadly divided into two parts; in the first part, two reconfigurable antennas are presented, while the second part presents a circularly polarized wideband magneto-electric dipole (MED) antenna with a defective semicircular patch for C-band applications (4–8 GHz).

The first design presents an antenna with frequency notch characteristics. This section includes two designs of reconfigurable antennas. In the first design, a novel antenna with frequency notch characteristics is presented. The antenna consists of a C-shaped Microstrip antenna with two symmetrical notches and a rectangular parasitic patch. The antenna has tunable property due to the integrated PIN diode. The second design consists of a simple radiating truncated rectangular patch with a cross-shaped slit and a ground plane embedded with L-shaped slit. The antenna produces two separate impedance bandwidths with three senses of polarization, namely, right-hand circular polarization, left-hand circular polarization, and linear polarization. The PIN diode is used to reconfigure the L-shaped slit in the ground plane. The antenna generates a dual band behavior with multiple circularly and linearly polarized bands.

The second part presents a circularly polarized wideband magneto-electric dipole (MED) antenna with a defective semicircular patch for C-band applications (4–8 GHz). In the proposed design, to get proper impedance matching and stable gain, a pair of folded vertical patches is shorted between a pair of defected semicircular patches and minimum ground plane. The defected semicircular patches work as electric dipoles, while the vertical patches work as magnetic dipoles.

8.2 C-SHAPED RECONFIGURABLE ANTENNAS

A reconfigurable antenna is useful in the changing the operating requirements to maximize the antenna performance due to its reconfigurable capabilities. The reconfigurable antenna has a capability to modify its frequency, polarization, and radiation properties in a controlled and reversible manner. Reconfigurable behavior can be obtained by modifying the antenna structure using different mechanisms such as PIN diodes, varactors, RF switches, and tunable materials. These mechanisms enable the intentional redistribution of the surface currents producing reversible modifications of the antenna properties. The reconfigurable antenna is useful in applications where multiple antennas are required. Multiple antennas can be replaced by a single reconfigurable antenna.

8.2.1 C-Shaped Antenna with Switchable Wideband Frequency Notch

The number of bands are increasing day by day with multiple high-speed services. These services need a high bandwidth and need to be separated from each other to avoid interference between channels. There are two primary design requirements for this. The first is an antenna design with frequency notch and the second is an antenna design with high cross-polarization attenuation. The notched frequency band refers to frequency response with a small stopband and cross-polarization refers to the orthonormal orientation of electric fields (vertical and horizontal). The adjacent bands in a satellite are isolated by notched frequency bands and polarization, that is, if the first channel is vertically polarized, the second will be horizontally polarized and the third will again be vertically polarized, and so on. This provides isolation among various channels. The antennas designed for multibands with a stopband in between are known as frequency notched antennas. For instance, three bands are allocated for Wireless Local Area Network (WLAN) 2.4 GHz (2,400–2,484 MHz), 5.2 GHz (5,150–5,350 MHz), and 5.8 GHz (5,725–5,825 MHz). Worldwide, the WiMAX system operates at the frequency bands of 2.5 GHz (2,500–2,690 MHz), 3.5 GHz (3,400–3,690 MHz), and 5.8 GHz (5,250–5,825 MHz). Another high-speed WLAN service of 1 Gbps speed is under development (IEEE 802.11.ac), which is to operate around 5 GHz.

Many techniques have been proposed for designing band-notched antennas. The simplest method is to etch slots on the radiating patch of the antenna, such as U-shaped slots [1], L-shaped slots [2,3], and V-shaped slots [4]. An effective way of achieving a UWB compact frequency notched antenna is the use of a split ring resonator (SRR) [5–7]. A square split ring resonator is used in [5], while in [6], a slot-type SRR has been used to design an ultrawideband notched antenna. Multiple notched frequency bands have been implemented with an SRR below feed line in [7], with simple slots in [8], and using a half-mode substrate in [9]. A frequency notched printed slot antenna has been implemented in [10], while a compact antenna is designed using fractal geometry in [11]. Instead of using the conventional fractal geometry such as Koch curves, Sierpinski triangles, and Minkowski fractals, a Koch-curve-shaped slot has been employed to achieve a compact geometry with frequency notch characteristics.

The coaxial probe feed is one of the most popular feeding techniques for electrically thick substrates, but the inductance of the probe creates impedance mismatch and causes low bandwidth. The inductance of the probe is compensated for by cutting slots on the patch, using an L-shaped strip feed [12] or by introducing a capacitive feed strip [13,14]. But these configurations create theoretical complexity in terms of analysis apart from the mechanical alignment issues while assembling and hence may increase the production cost. Capacitive feed provides a simple and ultrawide impedance matching by compensating for probe inductance with its capacitance.

A C-shaped antenna may be described as an RMSA with rectangular notch at one of the edges [15]. This is a very popular technique to achieve dual band [16] and wideband [17] characteristics.

In this design, a new technique has been presented to achieve a frequency notched antenna. A simpler capacitive coupling technique has been employed to get ultrawideband characteristics. A PIN diode has been used so that the antenna operates in ultrawide band as well as dual band with a frequency notched antenna. The gain and other radiation characteristics are normal, while the cross-polarization attenuation is very high.

8.2.2 Multiband Multipolarized Reconfigurable Circularly Polarized Monopole Antenna with a Simple Biasing Network

In recent years, circularly polarized monopole antennas have attracted a great deal of attention for the current wireless communication system, as circular polarization (CP) plays a very important role in improving the quality of the received signal [18]. These applications include the C-band communication satellite from 5.925 to 6.425 GHz for their uplink and amateur satellite operations in the frequency range of 5.830 to 5.850 GHz for downlinks. The X-band uplink frequency band uses 7.9–8.4 GHz for military communication systems. The traffic light crossing detector operates at 10.4 GHz. In Ireland, Saudi Arabia, and Canada, terrestrial communication uses the bandwidth of 10.15–10.7 GHz. The usage of multiple antennas for achieving different CP radiations will make the system complex. Hence, as per the demand of the next-generation communication systems, single multiband circularly polarized antennas can be used to reduce the complexity of the system. Therefore, designing such antennas have attracted a great deal of attention among the researchers. However, designing such antennas is challenging when the number of operating CP frequency bands increases. Generally, a monopole antenna generates linearly polarized radiation. Hence, it is difficult to radiate CP waves which were produced by two near-degenerated orthogonal resonant modes of equal amplitude with opposite phase differences. In [19], CP is generated in the triangular patch by using a Koch curve. The axial ratio bandwidth is about 1.3%. A triple proximity-fed microstrip antenna gives the CP with an axial ratio bandwidth of 0.70% for the L-band [20]. In [21], an elliptical microstrip antenna with proximity coupling is used to excite CP waves with an axial ratio bandwidth of about 0.85%. In [22], a simple circularly polarized antenna is presented. The CP operation is obtained by using an open slot having an open width at the lower side of the model. In [23], a slot antenna with straight feed is presented. An SRR-inspired structure is used to obtain the circular polarization. Wideband high gain circularly polarized antennas are proposed in [24]. The use of feed and parasitic patches makes the structure complex. Many CP monopole antennas were proposed for various applications in wireless communication [25–28]. In [25], an asymmetrically shaped radiator fed by a microstrip line and a limited ground plane is presented. In [26], the CP is achieved by loading four cylinders that are perpendicular to the substrate of the microstrip antenna near the edge of the circular patch. In [27], by using a rectangular dual-loop technology and tuning the separation between the ground plane and antenna, the monopole antenna produces a CP. In [28], an asymmetric antenna geometry is used to obtain the wideband CP. However, CP monopole antennas mentioned above focus on single CP band operation. Dual band CP has been investigated in [29–33]. The antenna presented in [29] composed of a partial ground plane and a Y-shaped radiating patch that consists of two unequal monopole arms and a modified circle. The dual CP is obtained with two unequal monopole arms and a modified circle with an axial ratio bandwidth of 3.8% and 6.8%. In [30], dual CP is obtained by embedding an inverted-L slit in the ground plane. A halved falcate-shaped dual-broadband CP printed monopole antenna is proposed in [31]. To generate dual CP, two halved falcate-shaped antennas were used to generate orthogonal modes and three stubs in the ground plane were used to give 90 degree phase difference. A novel monopole antenna with dual CP consisting of a radiating patch composed of an annular-ring linked by a square ring over the corner and a ground plane with embedded rectangular slit was proposed in [32]. In [33], the dual band CP operations are realized by using two parallel monopoles – one curved monopole and one fork-shaped monopole – and a crane-shaped strip is placed on the ground plane. A dual-feed, dual-band-stacked, CP patch antenna system is presented in [34]. The use of dual feed and stacking gives the structure complex. Triple band CP radiations are presented in [35,36]. In [35], a hexagonal slot antenna with L-shaped slits is presented with a narrow 3 dB axial ratio bandwidth of 1.7%, 3.86%, and 5.23%, while 3 dB axial ratio bandwidths of 9.8%, 4.6%, and 2.8% have been achieved in [36]. The complexity of these two designs [35,36] may hinder the integration of antennas in different applications. In [37], an inverted U-shaped radiator rotated by 45° around the horizontal axis is used to generate the triple band CP.

Recently, reconfigurable CP antennas have attracted significant attention. The PIN diodes are switched in different states to obtain the tunable property of the antenna [38]; this technique is also used to generate a reconfigurable CP microstrip antenna [39]. The authors in [40] proposed a dual feed microstrip patch antenna with frequency and polarization reconfigurability. The frequency and polarization reconfigurability are achieved by using six PIN diodes. The design is complex because of the six PIN diodes and complex biasing circuits. A frequency- and polarization-reconfigurable antenna using PIN diodes is designed in [41]. PIN diodes are commonly used as switching devices for RF and microwave application systems, as they have the advantages of low insertion loss, good isolation, and low cost [42]. It has been observed from literature reviews that mainly two considerable problems are often encountered in reconfigurable CP antenna designs. The excessive diodes will lead to a complex dc-bias network [19,43], providing independent bias for each diode by some special mechanism, such as using capacitors [44].

In this design, a very simple monopole antenna with reconfigurable multiband CP operations is presented. The antenna has overcome the abovementioned two problems of the reconfigurable CP antenna. The antenna generates right-hand circular polarization (RHCP), left-hand circular polarization (LHCP), and linear polarization (LP) in different bands with a simple biasing network without making the system complex. By removing the triangular portion at the lower edges of the rectangular radiating patch and embedding the cross-shaped slit at the right lower edge and L-shaped slit in the ground plane, the antenna provides two CP bands with three LP bands in the OFF state and three CP bands with three LP bands in the ON state of the PIN diode. The reconfigurable CP is achieved by using a PIN diode on the L-shaped slit in the ground plane. The multiband CP operation is obtained by controlling the ON/OFF state of the diode. The antenna is suitable for many applications such as traffic light crossing detectors at 10.4 GHz, C-band communication satellites for their uplink from 5.925 to 6.425 GHz, amateur satellite operations for downlinks from 5.830 to 5.850 GHz, terrestrial communication from 10.15 to 10.7 GHz, and X-band uplink frequency band from 7.9 to 8.4 GHz for military communication systems.

8.2.3 Design of C Shape Antenna with Switchable Wideband Frequency Notch

The antenna design is divided into three parts to examine the response at various phases. First, a capacitive-coupled C-shaped antenna is investigated. Two notches are cut symmetrically in the structure to introduce a new resonant frequency in its frequency response, and finally a parasitic rectangular patch is connected to the C-shaped microstrip antenna (CSMSA) via the PIN diode. The ON condition of the diode is implemented through a connection while the OFF condition is without any strip. The final antenna structure is shown in Figure 8.1. Figure 8.2a shows the fabricated antenna when the PIN diode is OFF, while Figure 8.2b shows when the diode is ON. The simulation is carried out on IE3D software and the measurements are taken using an Agilent N5230 vector network analyzer. The antenna is fabricated on an RT Duroid 6002 substrate with a dielectric constant of 2.94 and a loss tangent of 0.0012. The thickness of the substrate is 0.76 mm, which is raised by 6 mm in air. The other dimensions of the antenna are given in Table 8.1.

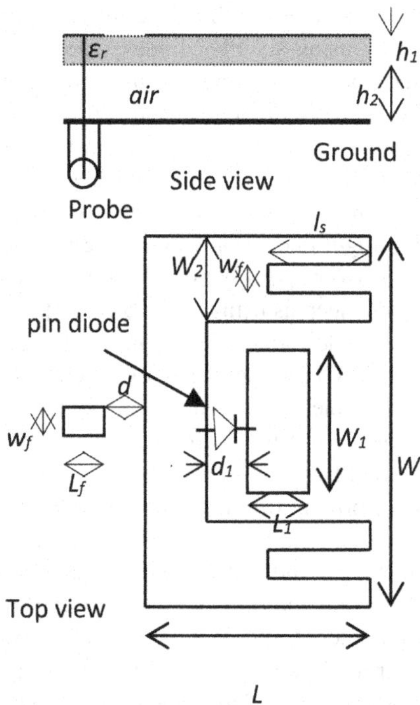

FIGURE 8.1 Schematic of the proposed antenna.

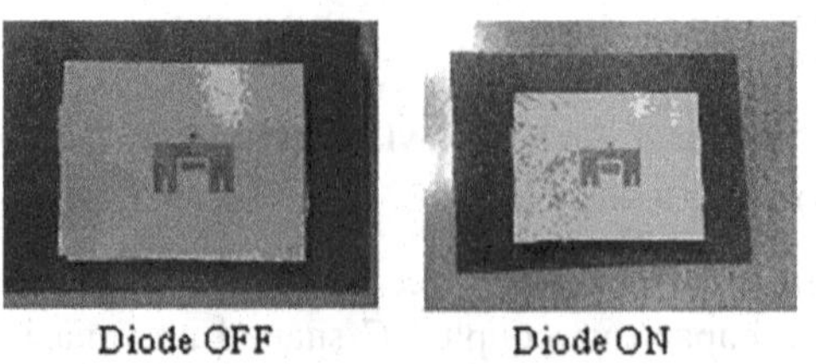

FIGURE 8.2 Fabricated antenna.

TABLE 8.1
Specification of Proposed Antenna

	Value
Substrate dielectric constant (ε_r)	3.0
Substrate height (h_1)	0.76 mm
Air height (h_2)	6 mm
Feed width (w_f)	1.2 mm
Feed length (L_f)	3.7 mm
Spacing between feed and patch (d)	0.5 mm
Main patch size ($L \times W$)	15.5 mm×25.5 mm
Parasitic patch size ($L_1 \times W_1$)	4.3 mm×7.75 mm
Notch size ($l_s \times w_s$)	4.75 mm×1 mm
Spacing between the C-shaped antenna and the rectangular patch (d_1)	1.0 mm
Width of each side wings of the C-shaped antenna (W_2)	7.875 mm

8.2.4 Multiband Multipolarized Reconfigurable Circularly Polarized Monopole Antenna with a Simple Biasing Network

The proposed reconfigurable multiband circularly polarized microstrip-fed monopole antenna is shown in Figure 8.3. The proposed CP antenna is designed and fabricated on an FR4 substrate of thickness $h=1.6$ mm and relative permittivity $\varepsilon_r=4.4$. An SMA connector is connected to a microstrip feed line of width W_1 and length L_1, which is connected to an impedance transformer of width W_2 and length L_2. The approximate value of the length of the monopole antenna radiating strip is given by the following formula [30]:

$$L_0 = \frac{\lambda g}{4} = \frac{\lambda_0}{4\sqrt{\varepsilon_{\text{eff}}}} = \frac{c}{4\sqrt{\varepsilon_{\text{eff}}}\, f_0}$$

where $\varepsilon_{\text{eff}}=(\varepsilon_r+1)/2$, c is the speed of light, λ_0 is the free space wavelength at the monopole resonant frequency f_0 and ε_{eff} is the approximated effective dielectric constant. The dimensions of the rectangular radiator of the antenna are $L_0 \times W_0$. Generally, a monopole antenna generates either vertical or horizontal linearly polarized radiations and finds difficulty to generate two orthogonal current components with an equal amplitude and 90° phase differences. CP radiations can be generated by introducing a perturbation segment into a linearly polarized antenna; the linearly polarized orthogonal modes can be converted to LHCP or RHCP. Thus, to meet the desired CP conditions, a triangular portion is cut from the left lower edge of the radiator patch to adjust for the impedance matching and an L-shaped slot is embedded in the ground plane to generate the dual CP radiation [45]. Further, to produce 90° phase differences between the currents at distinct frequencies, a triangular portion is removed from the right lower corner of the monopole radiator and a cross-shaped slit is introduced on the truncated right lower edge of the radiator patch to disturb

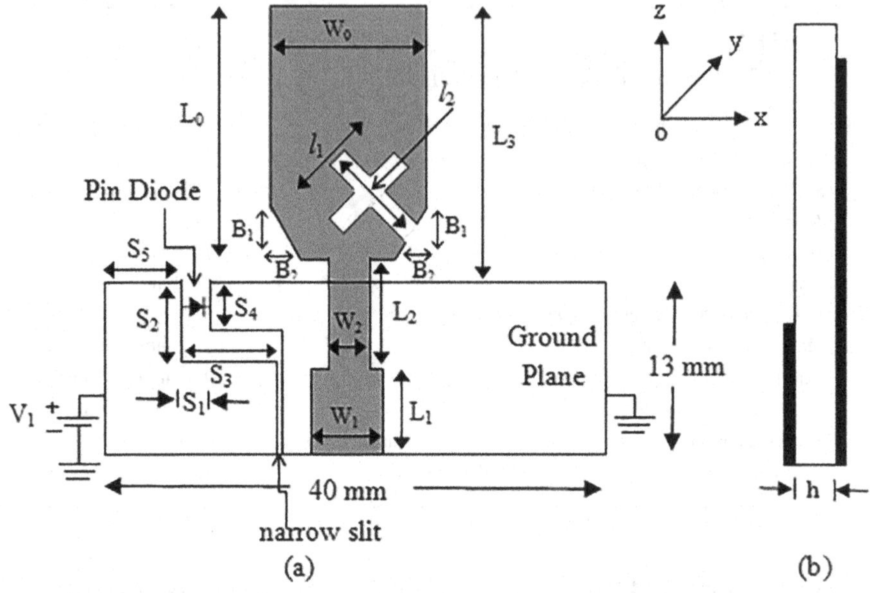

FIGURE 8.3 Schematic of the proposed antenna (a) top view (b) side view.

the surface current on the patch. The antenna exhibits a dual band behavior with two different CP radiations. To design the reconfigurable CP, a PIN diode as the switching component is used. The reconfigurable CP antenna has an L-shaped slot with a PIN diode (SMP1320-079) in the ground plane, as shown in Figure 8.3. When a positive voltage ($V_1 = +0.73$ V) is applied, a diode acts as a short circuit (ON state) with a small resistance (0.9 Ω). When a zero voltage is applied, the diode acts as an open circuit (OFF state). The geometry of the antenna varied with the ON/OFF state of the PIN diode, providing the proposed antenna with switchable CP bands in accordance with the change in geometry. A dc bias circuit is used to control the ON/OFF state of the PIN diode as shown in Figure 8.3a. A narrow slit in the ground plane is used for dc isolation. These two isolated portions of the divided ground plane are connected to each other in an ac manner by the use of capacitors. Since the bias circuit is located on the ground plane, an RF choke is not required to isolate the bias circuit from the radiation element, resulting in a reduced effect on the radiation by the bias circuit [46]. During the simulation, the ON state of the PIN diode is implemented with a copper link of length 1 mm and width 1 mm. The detailed dimensions of the proposed reconfigurable multiband CP antenna are listed in Table 8.2.

The evolution of various stages involved in the proposed antenna design is shown in Figure 8.4, while the reflection coefficients and axial ratios obtained in each stage are shown in Figures 8.5 and 8.6, respectively. The basic element used in the antenna design is a conventional rectangular patch and ground plane (Ant1). From Figure 8.5, it can be seen that in this case, three resonant modes are obtained at 7.75, 10.42, and 14.92 GHz. Ant1 exhibits the linear polarization (3-dB axial ratio > 3) as depicted in Figure 8.6. Ant2 shows that resonance frequencies shifted downward and operated at multiband frequencies due to modification of the ground plane by embedding an

TABLE 8.2
Dimensions of the Proposed Antenna

Parameters	L_0	W_0	L_1	L_2	L_3	W_1	W_2	B_1
Values (mm)	23.5	12	6.0	9.5	26	3.0	2.4	5.0
Parameters	l_1	l_2	S_1	S_2	S_3	S_4	S_5	B_2
Values (mm)	7.0	6.0	1.0	6.0	11.0	4.7	7.0	2.5

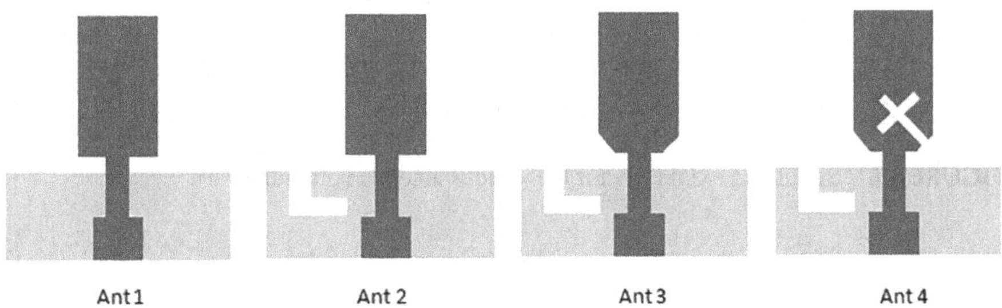

FIGURE 8.4 Antenna geometry evolution process of the proposed antenna.

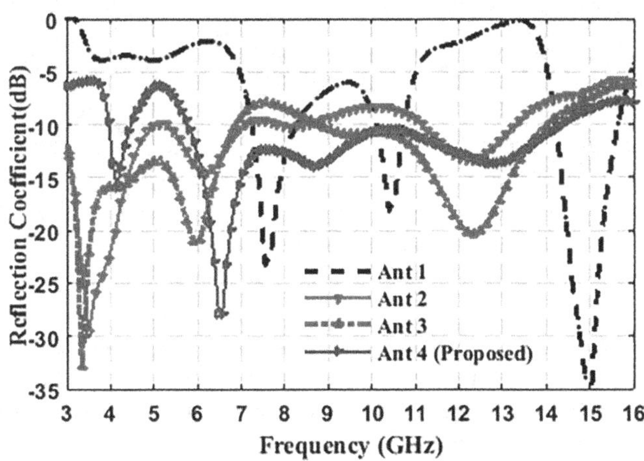

FIGURE 8.5 Simulated reflection coefficient for the various antenna configurations.

L-shaped slot in the ground plane (Ant2). The dual CP bands at the center frequencies 3.34 and 8.58 GHz are obtained as shown in Figure 8.6 (Ant2). The upper CP band is not within the operating band. Ant3 indicates that the removal of the triangular portion from the lower edges of both sides of the radiator patch provides two wideband operations and three CP operations with a small ARBW as shown in

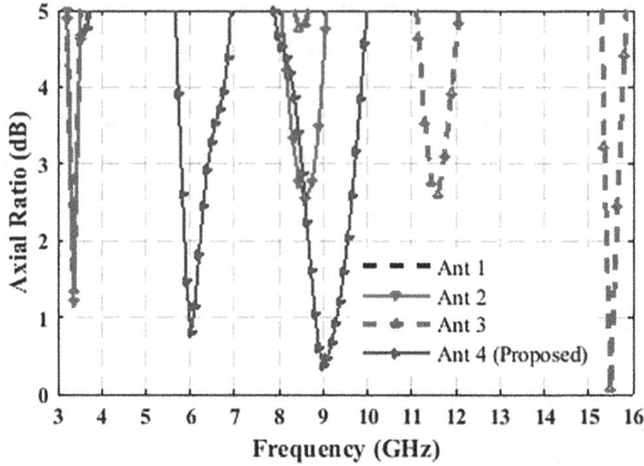

FIGURE 8.6 Simulated axial ratio for the various antenna configurations.

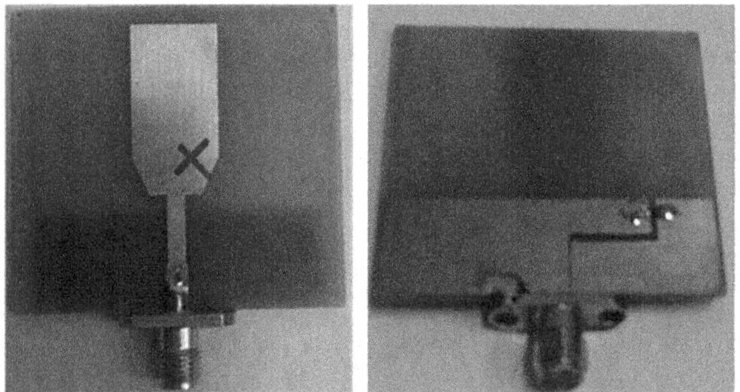

FIGURE 8.7 Fabricated prototype of the proposed antenna.

Figures 8.5 and 8.6, respectively. Finally, the radiating patch is modified by cutting a cross-shaped slot on the truncated right lower edge to obtain dual band characteristics and dual band CP behavior with a good ARBW. The proposed antenna is simulated by the Method of Moments–based IE3D simulator. Figure 8.7 shows the fabricated prototype of the antenna.

8.2.5 Characteristics of the C-Shaped Antenna with a Switchable Wideband Frequency Notch

A capacitive coupled C-shaped antenna is designed to operate in the frequency band of 5–8 GHz. The antenna exhibits two resonance frequencies. To achieve a sharper frequency response, two notches are cut symmetrically in the C-shaped microstrip

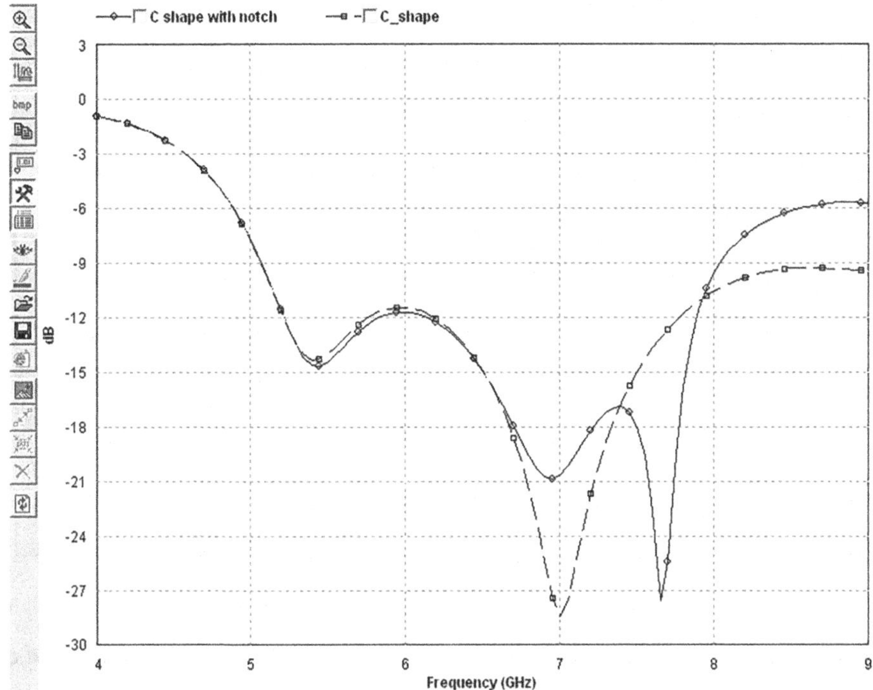

FIGURE 8.8 Variation in return loss with frequency.

antenna. Figure 8.8 shows the return loss variation of CSMSA with and without symmetrical notches. Clearly, a new resonant frequency is added at the higher side of the response.

After the cutting of notches, a parasitic patch is kept in the etched portion (mouth) of the CSMSA. The effect of putting the rectangular parasitic patch is shown in Figure 8.9. It is clear from the figure that the response remains almost unchanged after placing the parasitic rectangular patch. The slight shift in the response may be due to the mutual coupling between the structures.

After placing a parasitic element to the structure, a PIN diode (model HSMP-3860) is connected between them. The PIN diode is implemented by 1.2 mm× 1 mm through connection. The simulated and measured return loss variation when the PIN diode is OFF is shown in Figure 8.10. There is an acceptable resemblance between them. When the diode is ON, a wide frequency band around 5 GHz is notched, as shown in Figure 8.11. In practical applications, a sharp frequency notch is desirable to isolate the two bands. Here, a simple wide frequency notch is adopted to isolate the two bands. Thus, the antenna can be used both for ultrawide band and for narrow band. In fact, this notch appeared due to a shift in the lower resonant frequency towards the lower frequency spectrum and an impedance mismatch in between. This decrease in frequency was due to the enlarged current path.

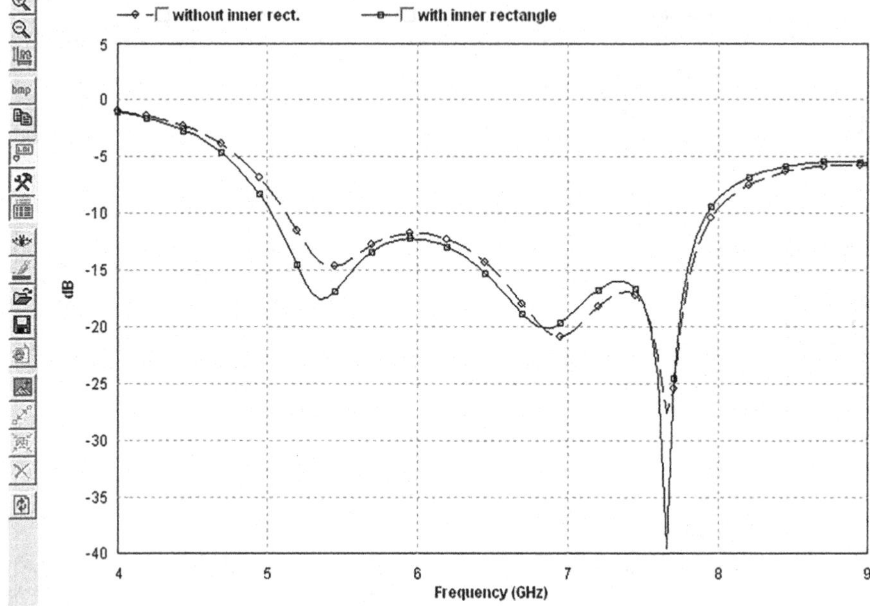

FIGURE 8.9 Variation in return loss with frequency with and without inner parasitic element.

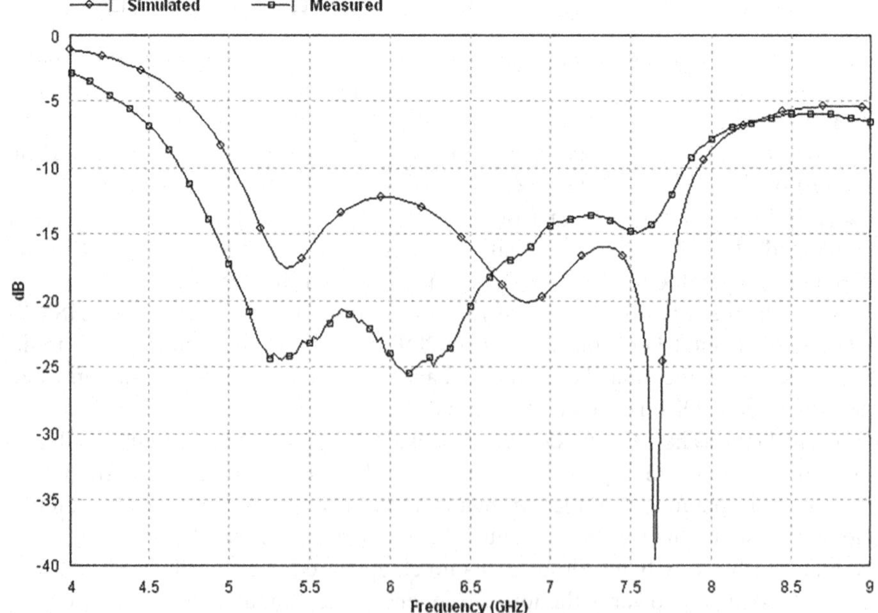

FIGURE 8.10 Simulated and measured results of tunable CSMSA when the diode is OFF.

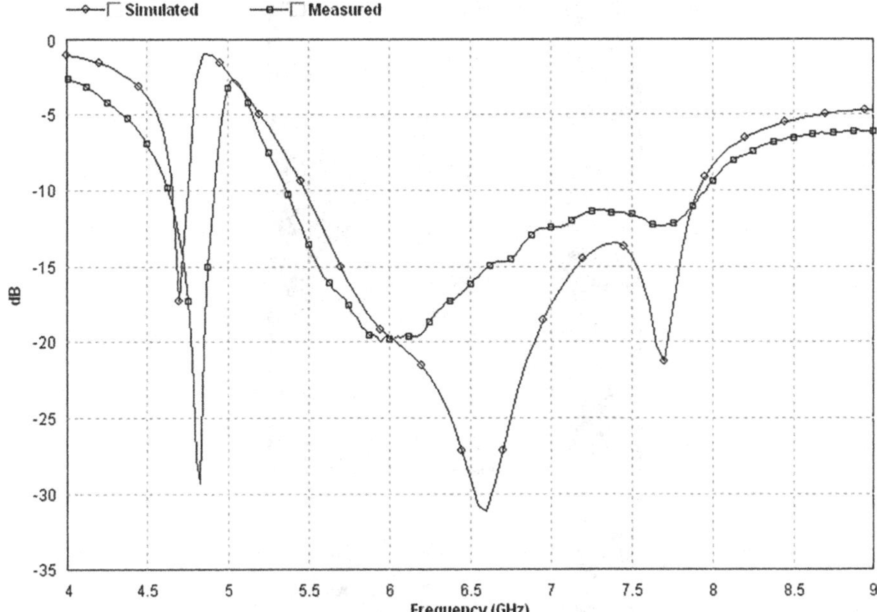

FIGURE 8.11 Return loss variation when the diode is ON.

The current distribution at old and new (shifted) resonant frequencies is shown in Figure 8.12.

To observe the behavior of the resonant frequency of the newly formed narrow band, the return loss is taken for different lengths of the parasitic rectangle, as shown in Figure 8.13. It is clear from the figure that as the length of the rectangle increases, the resonant frequency decreases, which shows the path length of current lines.

8.2.6 Other Radiation Characteristics

Figure 8.14 shows the gain of the tunable C-shaped microstrip antenna. It is observed from the figure that a better gain is obtained because of the better matching when the diode is OFF. A sharp fall in gain is observed in the stop band when the diode is ON. Figure 8.15 shows the radiation pattern at various resonant frequencies for the ON and OFF conditions of the PIN diode in the E-plane. It is observed from the figure that the radiation pattern is the same for 5.6, 6.5, and 7.7 GHz, which are in the same band of operation. When the band of operation changes (at 4.7 GHz), the main lobe of beam rotates from −10.48° to −30.3°, which is again due to the current path lines flowing in the inner rectangle. Figure 8.16 shows the current lines for 6.5 and 7.7 GHz when the diode is OFF and ON. When the diode is OFF, there is no coupling between the C-shaped antenna and the rectangle. In the ON condition, a little coupling is observed at 6.5 GHz, while being negligible at 7.7 GHz. This is why the shift in the second resonant frequency is the most (Figure 8.11).

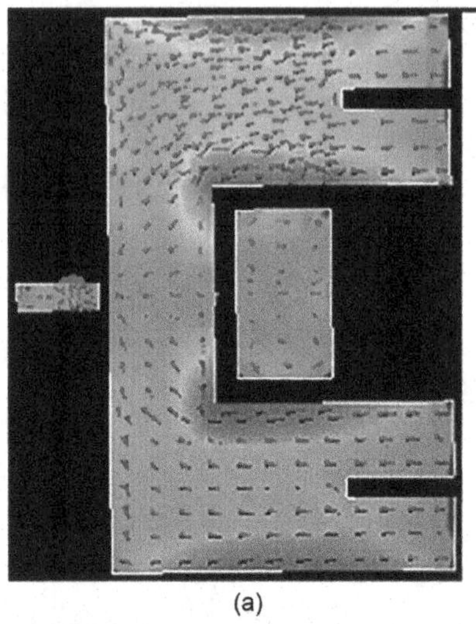

(a)

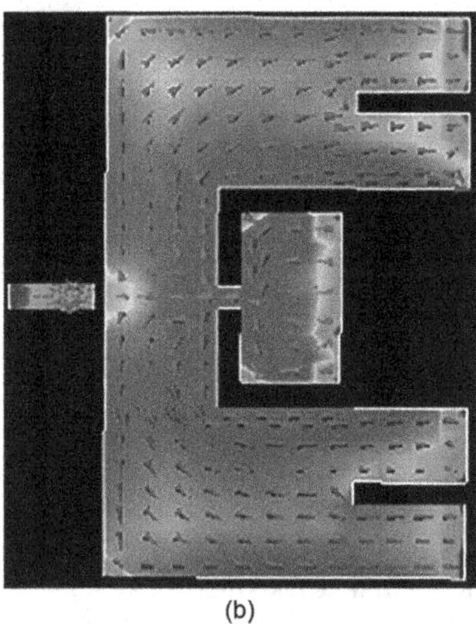

(b)

FIGURE 8.12 Current distribution (a) at 5.6 GHz when the diode OFF and (b) at 4.7 GHz when the diode ON.

8.2.7 Multiband Multipolarized Reconfigurable Circularly Polarized Monopole Antenna with a Simple Biasing Network

The measured and simulated reflection coefficients and the measured and simulated axial ratios with the OFF state of the diode are shown in Figure 8.18a and b, respectively. The proposed antenna exhibits a dual band behavior in the OFF state of the diode having an impedance bandwidth of 10.22% and 83.43% in the frequency range of 3.99–4.42 GHz and 5.84–14.20 GHz, respectively. In the OFF state of the diode, the antenna has successfully achieved two CP bands operation with a 3-dB axial ratio bandwidth of 9.83% and 13.73% in the frequency range of 5.79–6.39 GHz (CP1)

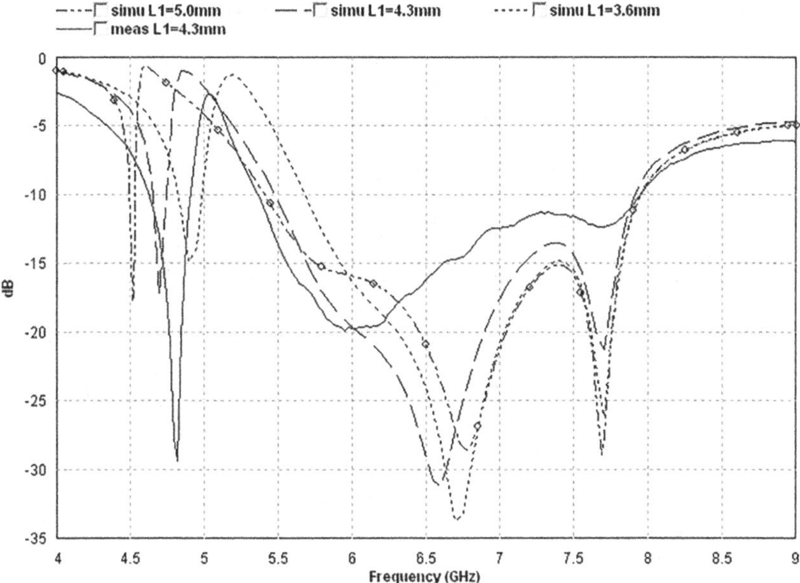

FIGURE 8.13 Return loss variation for different lengths of inner parasitic rectangle when the diode is ON.

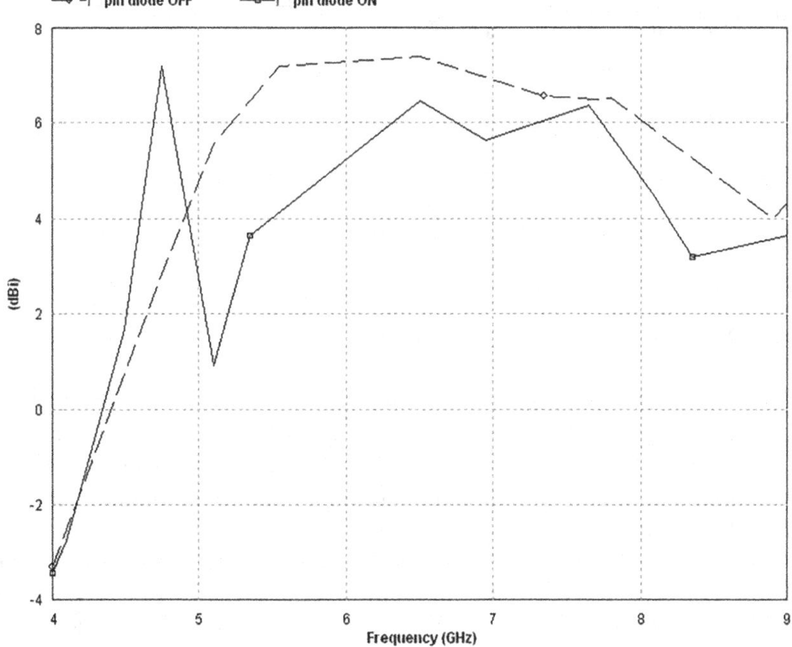

FIGURE 8.14 Variation in the gain of the tunable C-shaped microstrip antenna.

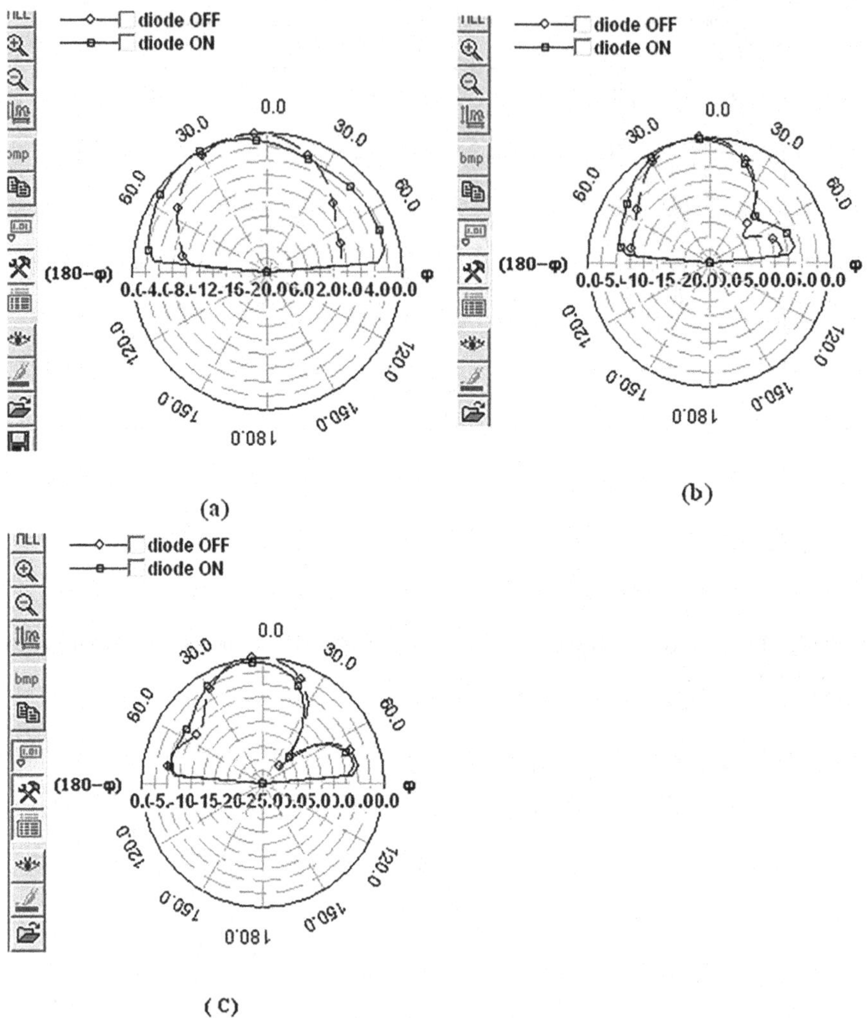

FIGURE 8.15 Radiation pattern of the tunable microstrip antenna at (a) 5.6 GHz (OFF) and 4.7 GHz (ON), (b) 6.5 GHz (both) and (c) 7.7 GHz (both).

and 8.49–9.74 GHz (CP2), respectively, and three LP bands in the frequency range of 3.99–4.42 GHz, 6.40–8.48 GHz and 9.75–14.20 GHz. The measured and simulated reflection coefficients for the ON state of the diode are shown in Figure 8.19a. Figure 8.19b illustrates the measured and simulated axial ratios in the ON state of the diode. In the ON state, the antenna also exhibits a dual band behavior with an impedance bandwidth of 49.10% and 21.31% in the frequency range of 7.20–11.04 GHz and 12.66–15.37 GHz, respectively. When the diode is set to the ON state, a switchable CP antenna is obtained. The antenna attains triple band CP operations with a 3-dB axial ratio bandwidth of 6.49%, 8.45% and 2% in the frequency range of 7.90–8.43 GHz (CP3), 10.08–10.97 GHz (CP4) and 12.87–13.13 GHz (CP5), respectively,

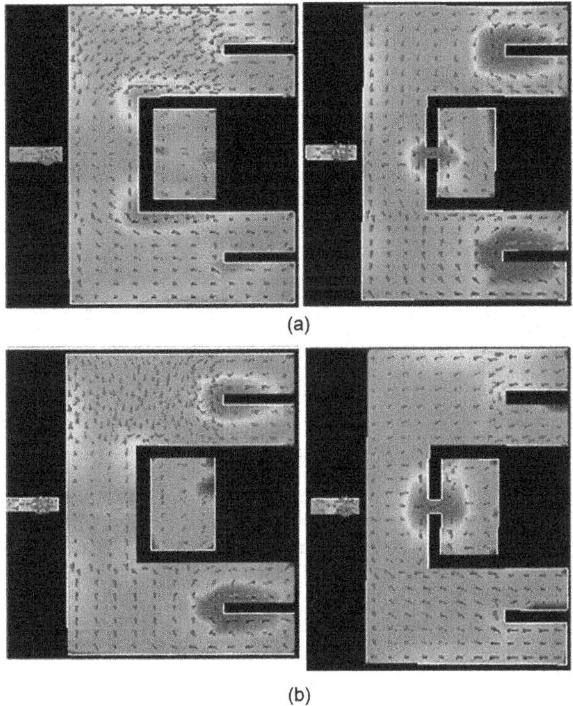

FIGURE 8.16 Current distribution at (a) 6.5 GHz and (b) 7.7 GHz.

and three LP bands in the frequency range of 7.20–7.89 GHz, 8.44–10.07 GHz and 13.34–15.37 GHz. The measured results are within reasonable agreement with simulated results. The differences between the simulated and measured results are due to the fabrication imperfections. They are also the reason that the simulation uses an open gap and a copper strip to represent the equivalent circuits of the PIN diode in the OFF and ON states, respectively. The reflection coefficient is tested by an Agilent5230A vector network analyzer, and radiation performances are measured in an anechoic chamber.

The total gain of the circularly polarized antenna can be given as [28,45,47]

$$G_T = 10\log_{10}\left(G_{\mathrm{TV}} + G_{\mathrm{TH}}\right)[\mathrm{dBic}]$$

where G_{TV} is the partial power gain with respect to linearly vertical polarization and G_{HV} is the partial power gain with respect to linearly horizontal polarization. A linearly polarized standard horn antenna with a calibrated gain is used in the measurement of G_{TV} and G_{HV}. The simulated and measured gains with the ON/OFF state of the diode are shown in Figure 8.20. The differences in the measured and simulated results of the gain are due to the loss in the PIN diode and measurement error. The peak gain of the antenna in the 3-dB axial ratio bands are 1.30 and −0.60 dBic at the center frequencies of the CP bands, respectively, for the OFF state, while 1.41, −0.41 and −0.86 dBic at the center frequencies of CP

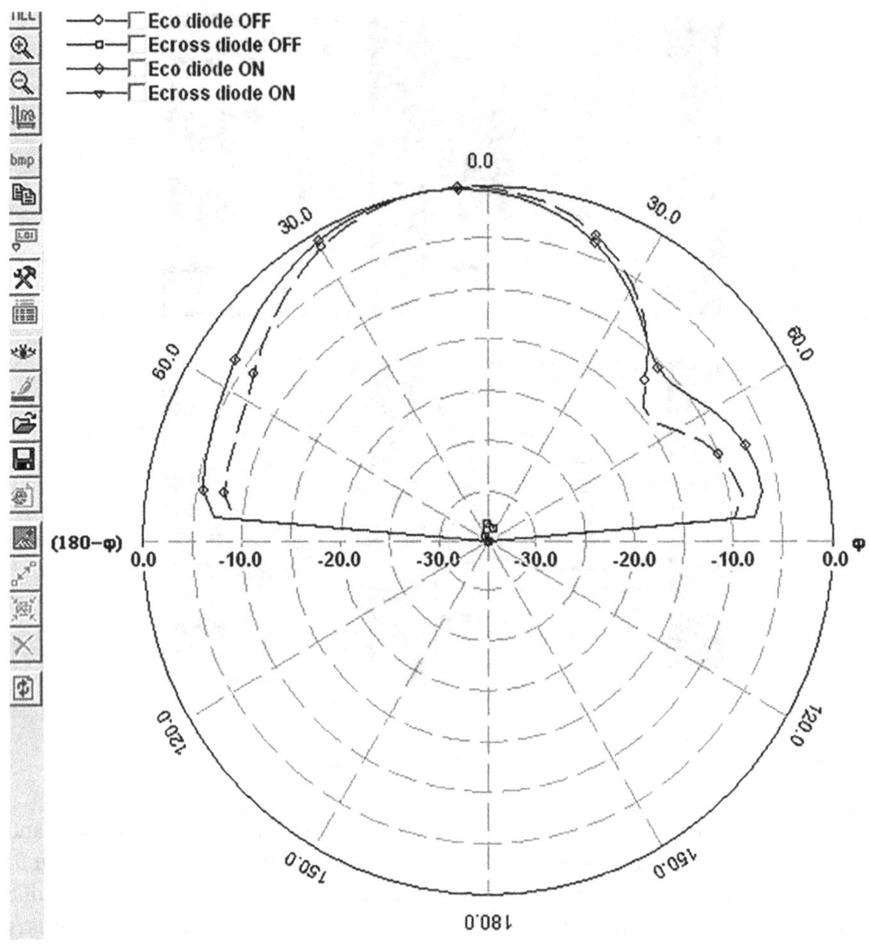

FIGURE 8.17 Copolar and cross-polar radiation pattern at 6.5 GHz.

bands, respectively, for the ON state of the diode. The simulated and measured radiation efficiencies of the proposed antenna in the ON and OFF states of the PIN diode are shown in Figure 8.21. The radiation efficiencies remain in the range of 65 ± 5%. The measured and simulated results agree well within the acceptable limit. Figure 8.22a and b plot the simulated and measured normalized RHCP and LHCP radiation patterns at the center frequencies of the CP radiations for the OFF state of the diode, while Figure 8.23a–c plot the radiation patterns at the center frequencies for the ON state of the diode in the two orthogonal planes (E-plane ($\Phi = 0°$) and H-plane ($\Phi = 90°$)). It can be seen that good LHCP and RHCP are observed at the center frequencies of the CP bands CP1 and CP2, respectively, for the OFF state of the diode. In the ON state of the diode, the RHCP, RHCP, and LHCP wave are observed at the center frequencies of the CP bands CP3, CP4, and CP5, respectively.

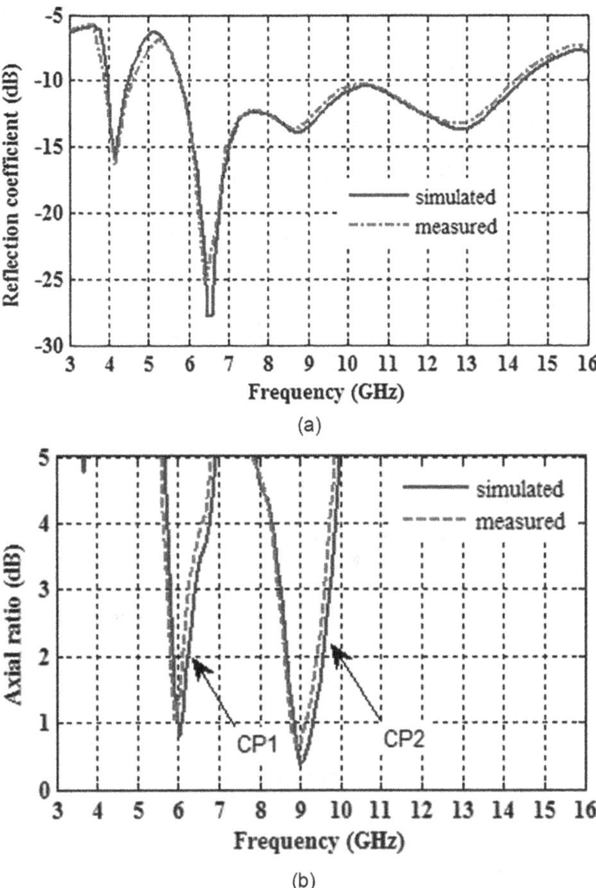

FIGURE 8.18 (a) Simulated and measured reflection coefficients of the proposed antenna in the OFF state of the diode. (b) Simulated and measured axial ratios of the proposed antenna in the OFF state of the diode.

8.2.8 Radiation Mechanism

In order to understand the operating mechanism and how CP operations are obtained, the simulated current distribution at center frequencies of different CP radiations for $t=0$, $t=\text{T}/4$, 2T/4 and 3T/4(T = time period) are shown in Figures 8.24 and 8.25 in the diode OFF and ON state, respectively. At 6.1 GHz, the resultant of the surface current is at 257.5° and clockwise rotation of surface currents at different time frames clearly indicates LHCP. The L-shaped slit in the ground plane and cross-shaped slit in the patch are responsible for generating circularly polarized waves. At 9.1 GHz, the resultant surface current is due to the patch and L-shaped embedded ground plane is about 45° at $t=0$ and rotates anticlockwise and the antenna radiates LHCP waves with time. In the diode ON condition as shown in Figure 8.25, at 8.43 GHz, the currents at the L-shaped embedded ground plane are cancelled out by each other, and the resultant current is due to the cross-shaped slit at $t=\text{T}/4$.

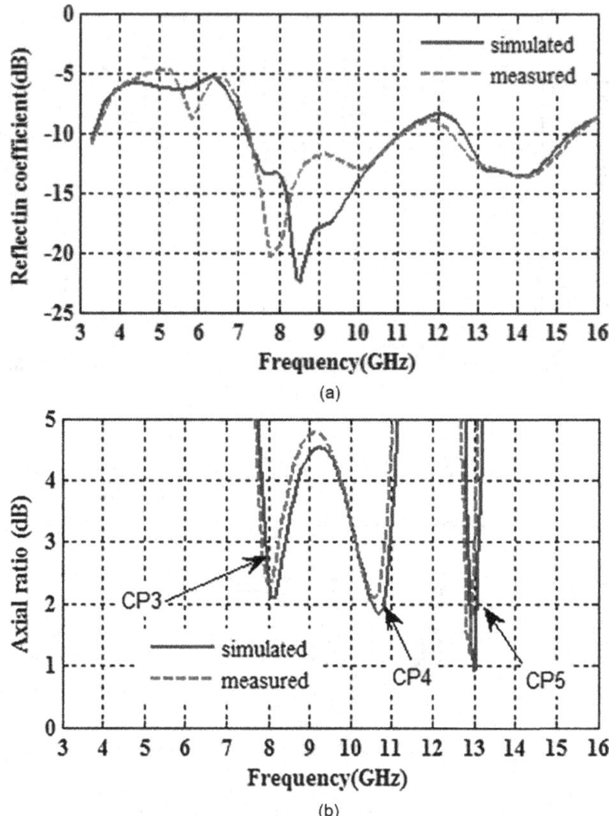

FIGURE 8.19 (a) Simulated and measured reflection coefficients of the proposed antenna in the ON state of the diode. (b) Simulated and measured axial ratios of the proposed antenna in the ON state of the diode.

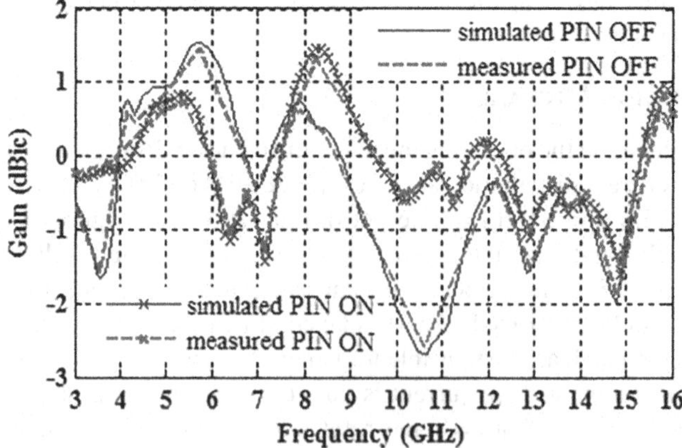

FIGURE 8.20 Simulated and measured gains of the proposed antenna in the ON and OFF states of the diode.

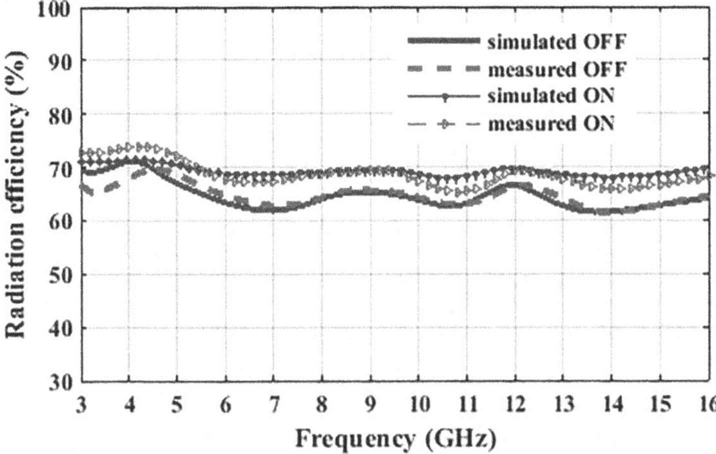

FIGURE 8.21 Simulated and measured radiation efficiencies of the proposed antenna in the ON and OFF states of the diode.

The field rotates in the anticlockwise direction, which results in exciting an RHCP radiation. At 10.525 GHz, the cross-shaped slit in the radiator patch is responsible for generating the RHCP radiation, as the surface currents rotate in anticlockwise manner. The clockwise rotation of the resulting surface current is responsible for producing the LHCP mode, at 13.00 GHz. As observed from the above figures, in the ON state of the diode, the surface currents in the L-shaped slotted ground is cancelled out and the resultant current is zero. The only CP is produced due to the asymmetric cross-shaped slit in the patch. Table 8.3 summarizes the sense of polarization of reconfigurable CP bands.

The performance parameters of the proposed antenna are compared with recently published results and are illustrated in Table 8.4.

8.2.9 Parametric Study

In order to analyze the operational performance of the presented antenna, the vital parameters are investigated to evaluate their impacts on the antenna. The parameters of cross-shaped slot and L-shaped slit are examined to find their impacts on the antenna performance and are illustrated in Figures 8.26 and 8.27, respectively. The parametric study is carried out by varying the parameters S_2, S_3, S_4, S_5, l_1, and l_2 to examine their effects on the reflection coefficients and axial ratio. It can be seen from Figure 8.26a–e, that by increasing and decreasing S_2, S_3, S_4, S_5, and l_2, little change is observed on the lower band and a very slight change is observed on the upper band of the reflection coefficient. The reflection coefficient is strongly dependent on l_2 as shown in Figure 8.26f. By decreasing the length l_2, the whole band is disturbed and reflection coefficient is above −10 dB.

The changes in CP performances by varying the parameters S_2, S_3, S_4, S_5, l_1, and l_2 are shown in Figure 8.27. The CP performances are slightly affected by the change

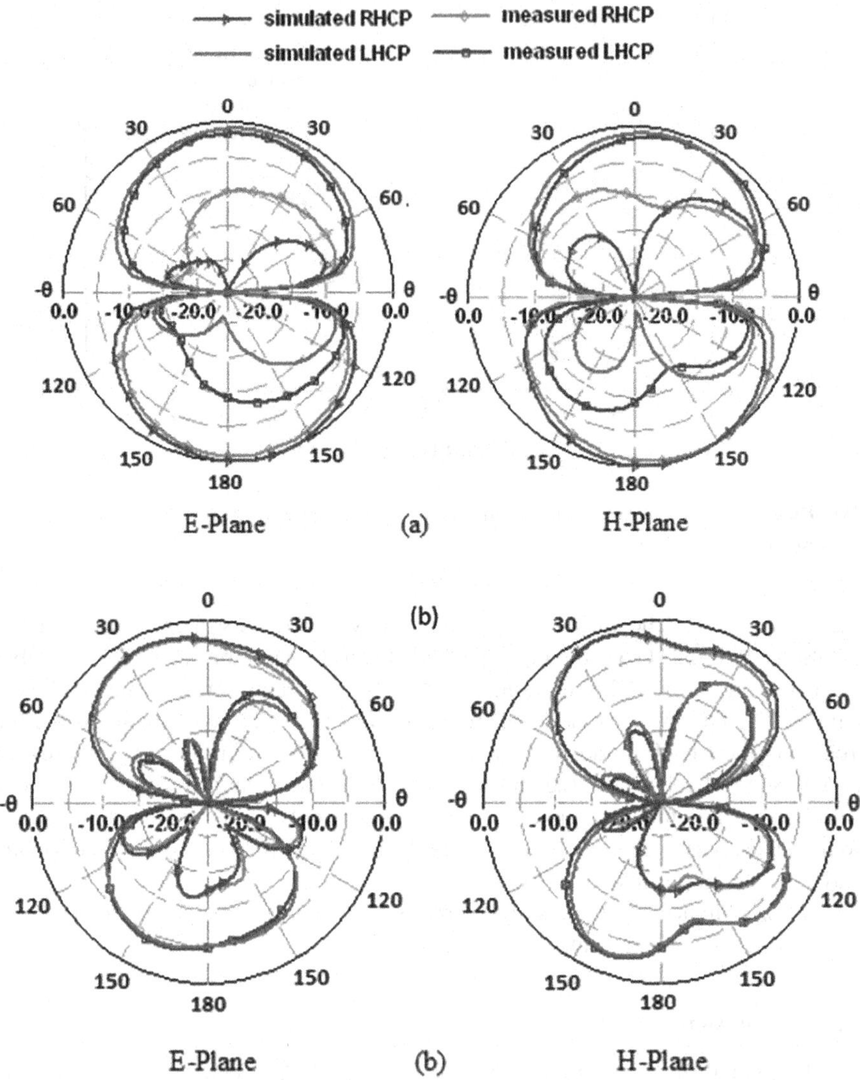

FIGURE 8.22 Simulated and measured normalized LHCP and RHCP radiation patterns at the center frequencies of the CP band when the diode is OFF (a) at 6.1 GHz and (b) at 9.1 GHz.

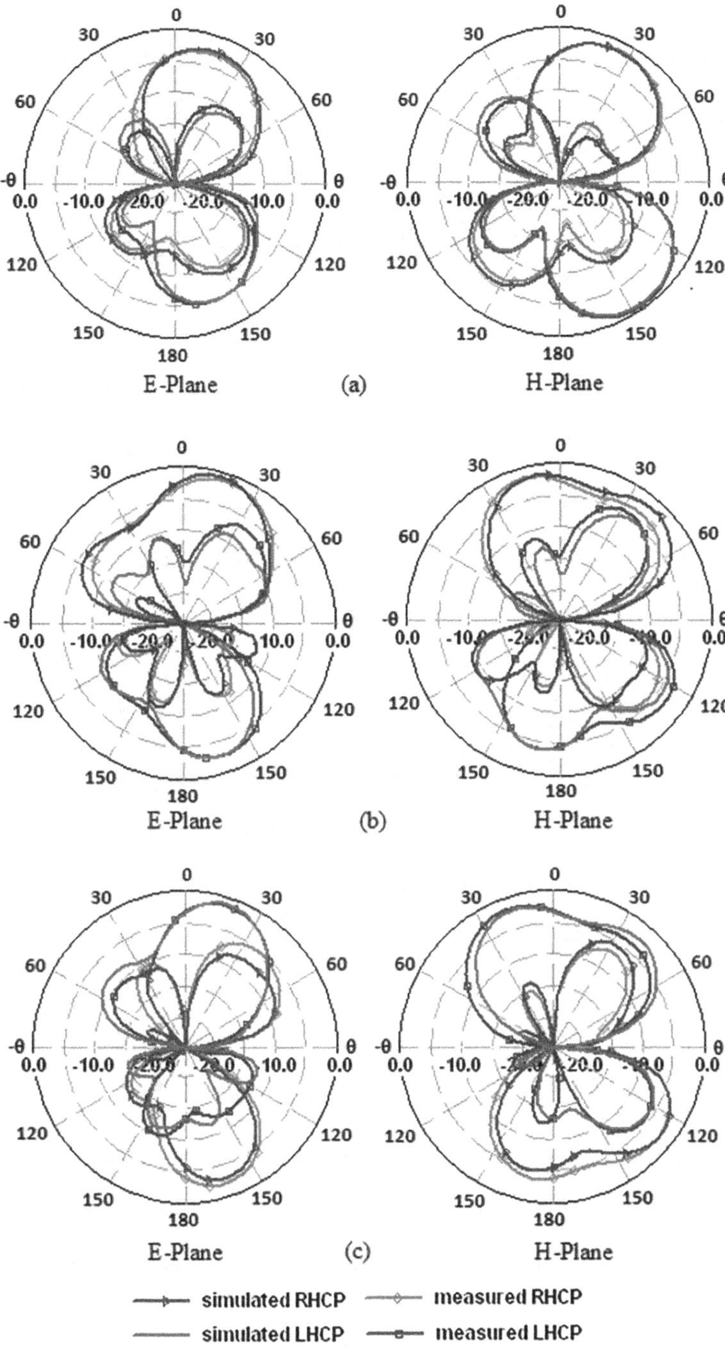

FIGURE 8.23 Simulated and measured normalized LHCP and RHCP radiation patterns at the center frequencies of CP band when the diode is ON (a) at 8.43 GHz, (b) at 10.52 GHz and (c) at 13.00GHz.

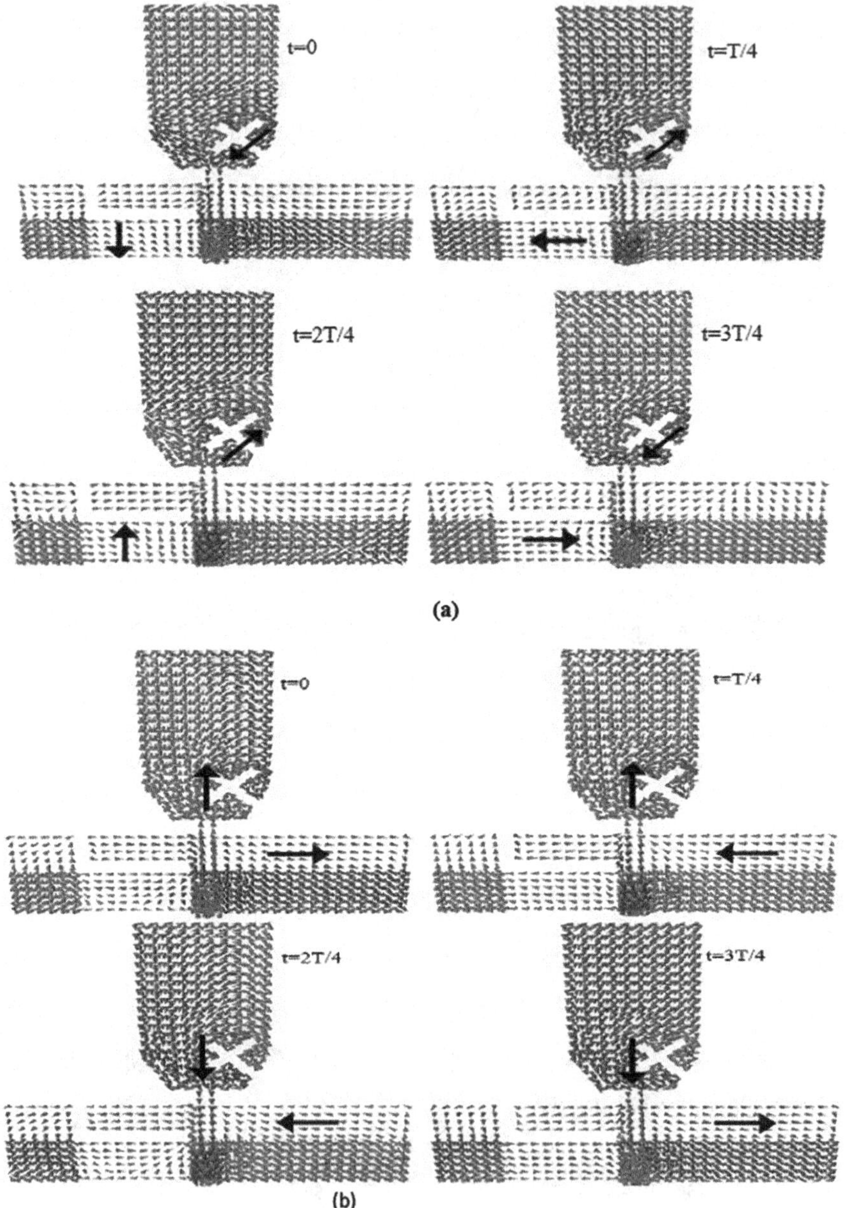

FIGURE 8.24 Surface current distribution at the center frequency of the CP bands when the diode is OFF (a) at 6.1 GHz and (b) at 9.1 GHz.

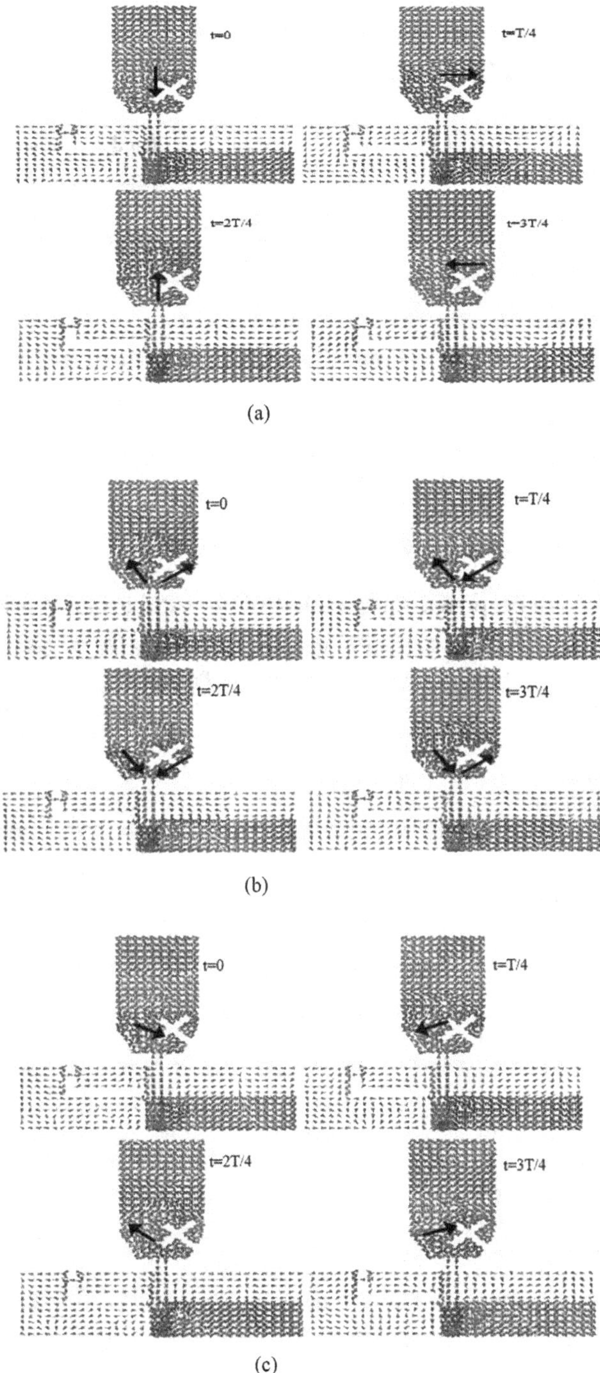

FIGURE 8.25 Surface current distribution at the center frequency of the CP bands when the diode is ON (a) at 8.43 GHz, (b) at 10.525 GHz, and (c) at 13.00 GHz.

TABLE 8.3
Summary of Antenna Polarization

Diode States		Band 1	Band 2	Band 3	Band 3	Band 5	Band 6
OFF	Senses of polarization	LHCP	RHCP	L_{P1}	L_{P2}	L_{P3}	–
	Frequency range (GHz)	5.79–6.39	8.49–9.74	3.99–4.42	6.40–8.48	9.75–14.20	–
ON	Senses of polarization	RHCP	RHCP	LHCP	L_{P1}	LP_2	L_{P3}
	Frequency range (GHz)	7.90–8.43	10.08–10.97	12.87–13.13	7.20–7.89	8.44–10.07	13.34–15.37

TABLE 8.4
Comparison of the Proposed Antenna with Other Published Work

Reference	Antenna Type	No. of Feeds	No. of Diodes	Biasing Circuit	Reconfigurable Properties	Senses of Polarization	No. of Bands
[20]	Multilayer	3	n.a.	n.a	NO	CP	01
[22]	Monopole	1	n.a.	n.a.	NO	CP	01
[24]	Multilayer with air gap	2	n.a.	n.a.	NO	CP	01
[34]	Multilayer	2	n.a.	n.a.	NO	CP	02
[38]	C-shaped microstrip antenna	1	1	Complex	Yes	LP	03
[40]	Square microstrip antenna	02	06	Complex	Yes	LP CP	02 02
[41]	Center-fed circular cavity	01	24	Complex	Yes	LP	06
[48]	Square microstrip antenna	01	04	Complex	Yes	LP CP	02 02
[46]	Square microstrip antenna	01	08	Simple	Yes	CP	02
Proposed work	**Monopole antenna**	**01**	**01**	**Simple**	**Yes**	**LP CP**	**06 05**

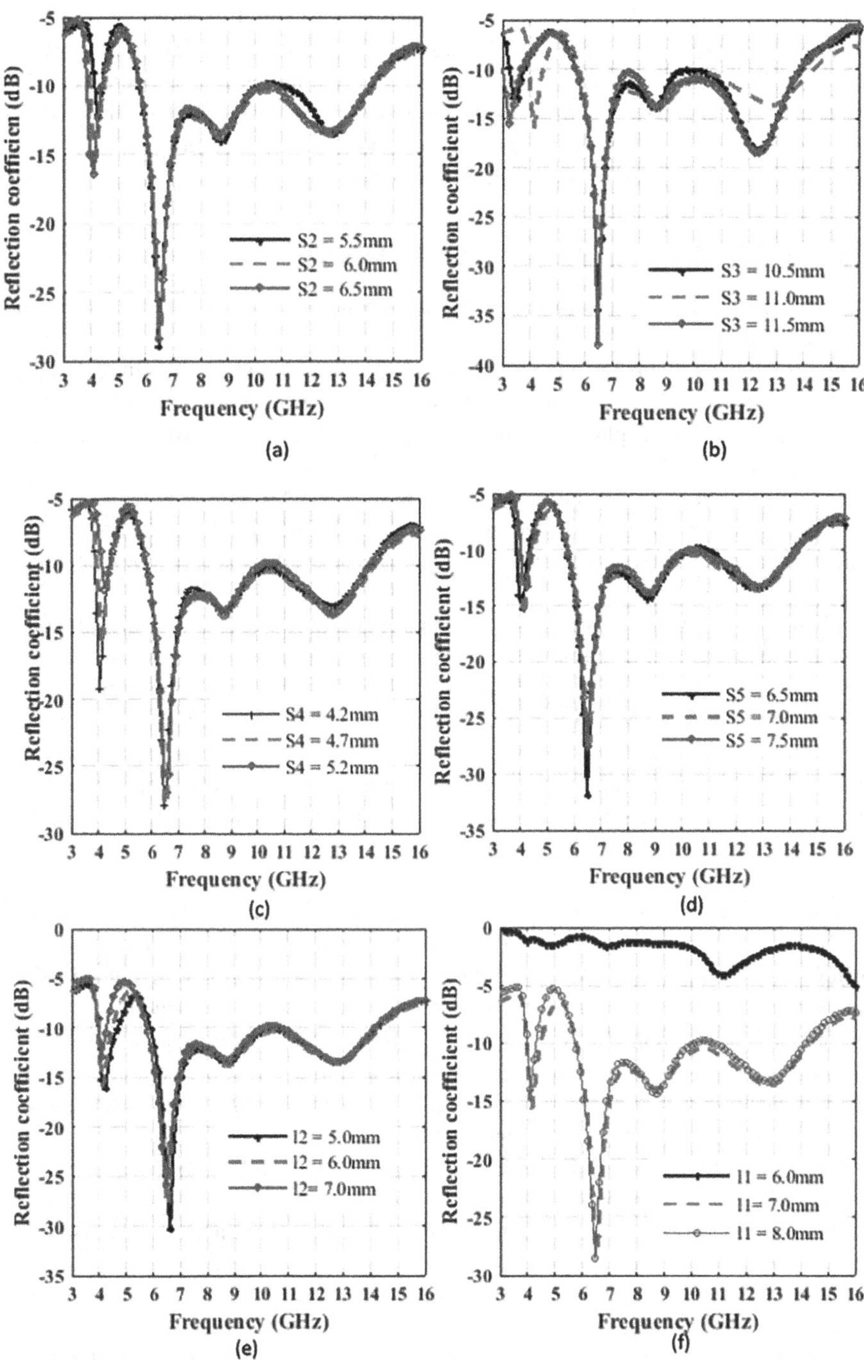

FIGURE 8.26 Parametric studies of the proposed structure: (a) variation in S_2, (b) variation in S_3, (c) variation in S_4, (d) variation in S_5, (e) variation in l_2, and (f) variation in l_1.

in the value of S_2, S_4, S_5, and l_2 as shown in Figure 8.27a–d, respectively. The CP bands strongly depend on the value of S_3 as shown in Figure 8.27e. By increasing and decreasing S_3, four CP bands with small ARBW are obtained, out of which two CP bands are not lying in the operating bands. An increase in the length l_1 has little effect on the CP operation at lower and upper bands, while a decrease in l_1 shows that the CP bands shifted upward as shown in Figure 8.27f.

8.3 MAGNETOELECTRIC DIPOLE ANTENNA

The magnetoelectric (ME) dipole antenna is a combination of an electric dipole and a magnetic dipole. The ME antenna is a complementary antenna concept that has a symmetrical radiation pattern in both E- and H-planes with a broad beam width. The vertical metallic walls act as a magnetic dipole, which is shorted with the ground plane at one end and the planer metallic electric dipole patch in another. A proximity-coupled feed is used to excite the ME dipole antenna, which performs as a combination of an electric dipole and a magnetic dipole. The ME dipole antenna exhibits various advantages, such as a broad impedance bandwidth, a stable gain, and a stable radiation pattern with low cross-polarization and back radiation levels over the operating frequencies.

Thanks to the speedy development of wireless communication systems, the focus of many researchers has been the development of advanced antennas to meet the demands of latest applications with attributes such as broadband, symmetrical radiation pattern, high gain, and compact size. Patch antennas are suitable candidates that satisfy all these attributes as well as easy fabrication; however, they encounter the problem of narrow bandwidth [49]. To achieve a broad bandwidth, Clavin et al. have presented antennas based on the complementary antenna concept, which have shown a symmetrical radiation pattern in the E- and H-planes with a broad beam width and stable gain [50,51]. In [52], the magnetoelectric dipole (MED) antenna has been presented for a femto cell base station with a rectangular ground and defected patch. The antenna has achieved an impedance bandwidth of 51.9%, but the design suffers from large size. Further, to make low-profile antennas, the height of MED antennas is decreased by implementing the folded structure as presented in [53] and the regular square-shaped ground is replaced by minimum-sized ground plane [54]. The authors in [55] have presented a low-profile antenna with a high gain of 9.2 ± 1.1 dBi and an impedance bandwidth of 28.2%. In [56], the impedance bandwidth of a broadband dual polarized antenna is enhanced by the use of an L-shaped cavity–backed ground plane with proximity coupling, but it uses a large size cavity to improve the isolation between two ports. A new feeding structure—a combination of inverted U-shaped and meandering T-shaped feeds—presented in [57] has been used to design a printed dual band MED antenna. But it has exhibited a small gain of 3 ± 0.5 dBi only.

These wideband antennas are capable of meeting the demands of the current wireless system, but incorporating the features of circular polarization (CP) will make these antennas more efficient. Therefore, CP antennas are getting more fascinating and promising within wireless systems, such as satellite communications [58], global positioning system [59], and radio frequency identification [60], because

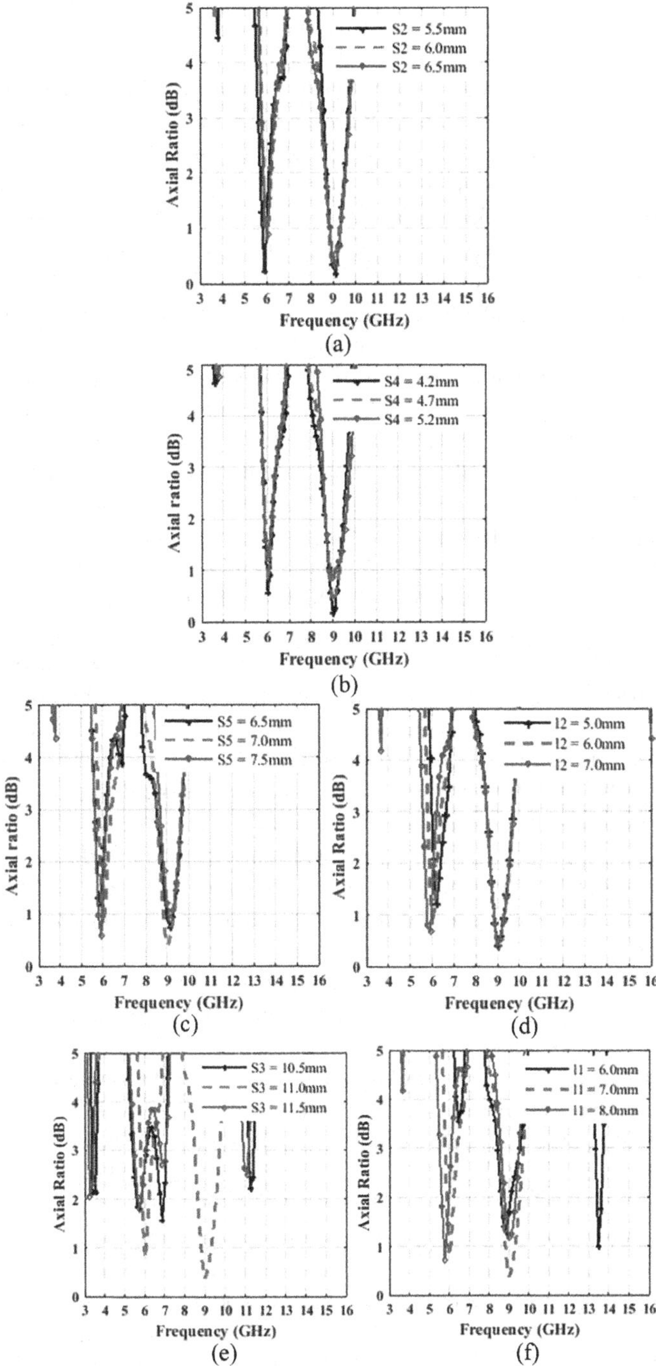

FIGURE 8.27 Parametric studies of the proposed structure: (a) variation in S_2, (b) variation in S_4, (c) variation in S_5, (d) variation in l_2, (e) variation in S_3, and (f) variation in l_1.

of their stable reception of the signal, multipath interference suppression, and additional immunity towards the polarization mismatching [61]. The fundamental operating principle of a CP antenna is the radiation of two orthogonal field components with equal amplitude and in-phase quadrature. Various techniques such as single-probe feed method [62], aperture-coupled feed method [63], and dual-fed structure method [64,65] have been used to achieve the CP in microstrip antennas. The first two methods [62,63] suffered from a narrow axial ratio (AR) bandwidth, while the later methods [64,65] have shown an AR bandwidth of more than 30%. However, these methods occupy a larger area and a complicated feeding structure; as a result, the design becomes bulky and complex. To overcome these problems, simple techniques are available to generate the CP operation, for example, by cutting the slots, inserting the slits in the patch and corner truncations [66–68]. In [69], a quad-band CP antenna with an inclined slot and corner truncations was presented for C-band applications. The antenna exhibits an impedance bandwidth of 55.46%, but its structure is complicated due to the stacked structure and exhibits very narrow CP bands. An ultra-wideband antenna with the CP characteristics is presented in [70]. To achieve the CP, the rectangular slot is truncated by two circular arcs of unequal radii. In [71], a reconfigurable circularly polarized capacitive-coupled microstrip antenna was considered to enhance the operating bandwidth of the antenna by the application of a PIN diode–based switching system, but it lacks an immediate response while the switching is applied. Another conventional approach to obtain the CP is a crossover dipole antenna in which two crossed dipoles of different lengths are selected to produce two orthogonal fields with an equal amplitude and a phase difference of 90° [72]. However, this method gives only a narrow impedance and AR bandwidth [72]. The impedance bandwidth is enhanced by 20% with the use of a crossed bowtie dipole design as proposed in [73]. The design also shows an improvement in the AR bandwidth of more than 7%. In [74], the improvement in AR bandwidth is achieved by adding a sequential rotational configuration. Further, the impedance bandwidth has been improved by using cavity-backed wide-open end cross dipole antennas, but the ease of construction is compromised [75]. However, dipole antennas with cavities [75] have a broad impedance bandwidth and good radiation patterns, but the integration of the additional cavities increases the complexity of the design and the radiation pattern on the broadside are not stable.

We herein propose a circularly polarized wideband MED antenna with a defective semicircular patch. The proposed design consists of copper-made horizontal defective semicircular patches, a pair of folded vertical patches, which work as an electric dipole and magnetic dipole, respectively, and a minimum ground plane at the bottom. A single feed and minimum ground have been chosen to maintain the simplicity of the proposed design as well as to avoid the use of a polarizer, which is mandatory for dual-feed design. The proposed antenna exhibits an impedance bandwidth of 60.37% in the frequency range of 3.71–6.91 GHz and a stable gain of 6 ± 0.5 dBic. The CP behavior has been confirmed in the proposed design in the frequency range of 3.71–4.55 GHz by embedding I-shaped slots in inverted symmetry on the semicircular patches at the appropriate positions. The easy fabrication and cheap cost may increase the popularity of the designed antenna.

8.3.1 Antenna Design

The MED antenna is a combination of an electric dipole and a magnetic dipole. The vertical metallic walls act as the magnetic dipole, which is shorted with the ground plane at one end and the horizontal metallic electric dipole patch in the other end. In the proposed MED antenna, defected semicircular shaped horizontal patches are used as electric dipoles and the folded vertical walls as magnetic dipoles. The detailed geometry of the proposed antenna is shown in Figure 8.28. It is a combination of a minimum rectangular ground of dimension $G_L \times G_W$, a horizontal pair of defected semicircular patches of radius r and two parallel vertically oriented folded walls, which are placed at an optimized distance. The vertical walls are separated by a gap G until height h_2, and the gap of $2L+G$ is maintained from height h_1 to h_2. These gaps play a very important role in stabilizing the antenna gain and reduction in the back-lobe radiation. The use of a larger ground

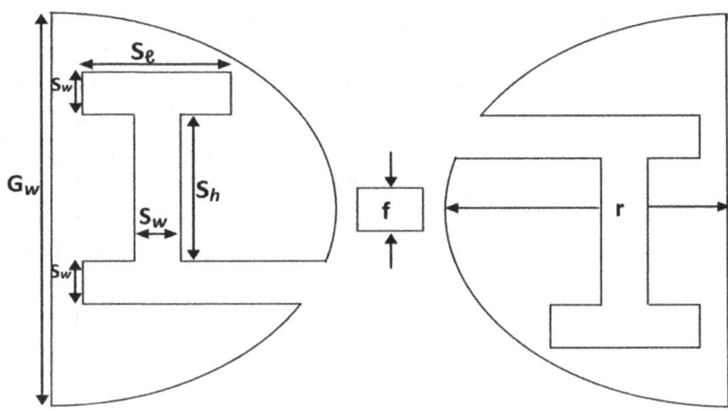

(a) Top view of the proposed antenna

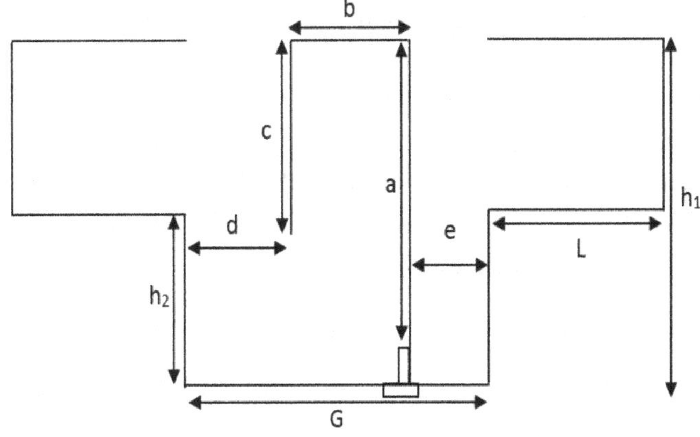

(b) Front view of the proposed antenna

FIGURE 8.28 (a) Top view of the proposed antenna. (b) Front view of the proposed antenna.

plane generally enhances the directivity with restrictions on practical issues, such as installation and bulkiness at the same time. Therefore, a minimum-sized rectangular ground has been used to maintain the compactness of the proposed design; as a result, a rectangular patch of dimension $G_L \times G_W$ is formed at height h_2 and a semicircular patch is used at height h_1, which also confirms the maximum height of the structure from the ground plane.

In order to initiate the CP operation, an I-shaped slot is cut in each semicircular patch with the help of a sharp mechanical tool. These slots have a diagonal symmetry. Due to the I-shaped slots and symmetrical folded structure, the proposed design exhibits wideband CP radiation characteristics. The gap between the extended parts of the I-shaped slots from the flat edge of the semicircular patch is 4 mm. The lower and upper parts of the I-shaped slots on the left side as well as the right side of the semicircular structures are extended until the curved edges to widen the 3-dB AR bandwidth.

To excite the antenna, the SMA connector is connected to the Γ-shaped feed line, which is divided into three segments a, b, and c with the uniform width *f*. It can be seen from Figure 8.28 that two vertical elements and one horizontal element are combined to make this type of feeding structure. The first segment (segment a) is connected to the inner conductor of the SMA connector, which is separated from one of the vertical walls by 1 mm and the other wall by 11 mm air gaps and acts as an air-filled microstrip line. The open end of this line is connected to the horizontal coupling line (segment b) of the feed line, whereas the other vertical segment of the feed line (segment c) is connected to the horizontal coupling line. The outer conductor of the connector is soldered to the ground to complete the feed loop. When the antenna is excited, the horizontal segment of the feed structure couples the energy to the semicircular defected patches and the other vertical segment is used to compensate for the inductive effect of the horizontal segment. By adjusting the size of the feed, proper impedance matching is obtained with a characteristic impedance of 50 Ω. The resonance of the antenna moves upward and downward with variation in the length *b* and height *c*. All the dimensions and parameters of the proposed design are shown in Table 8.5.

TABLE 8.5
Dimensions of and Parameters for the Proposed Antenna

Parameters	Dimension (mm)	Parameters	Dimension (mm)
G_L	12	A	19
G_W	58	B	9.5
R	29	C	12
S_ℓ	16	D	1.5
S_w	4.5	E	1.0
S_h	25	F	4.5
G	12	h_1	20
L	30	h_2	10

8.3.2 Parametric Studies

The proposed wideband circularly polarized MED antenna can operate in the frequency range of 3.71–6.91 GHz and exhibits an axial ratio bandwidth of 20.60% in the frequency range 3.71–4.55 GHz. Here, the optimization goal is to achieve the CP behavior as well as wideband operation. The parametric study in terms of its various controlling elements is carried out to obtain the optimum values. During optimization, several important parameters are investigated for optimum performance, such as the radius of the semicircular patch (*r*), the height of the semi-circular patch (h_1), the height of the metallic patch (h_2), the gap between two magnetic dipoles (G), and the width of I-shaped slots. During the investigation, while one design variable is changed, the other parameters were kept constant. In Figure 8.29, the different values of radius (*r*) are evaluated to check its impact on the reflection coefficient. It has been observed that the variation in *r* has a significant effect on the reflection coefficient variation.

The impedance bandwidth gets narrower and lower frequency point is shifted towards the lower side when *r* is increased beyond 29 mm. It can also be noted here that at $r = 31$ mm, the antenna exhibits dual band characteristics. Similarly, the bandwidth is reduced, when *r* is kept less than 29 mm, but the lower frequency point is shifted upward. However, in both circumstances, the upper cutoff frequency either remains same or decreases. Hence, it is confirmed that the optimum impedance bandwidth is achieved at the value of $r = 29$ mm.

The height of the semicircular patch (h_1) is varied, and the corresponding reflection coefficient is presented in Figure 8.30.

When h_1 is increased beyond 20 mm, the operating band gets distorted, and when h_2 is reduced below 20 mm, the impedance bandwidth gets narrower because of a significant decrease in the second and third resonance frequencies and thereby the upper cutoff frequency. When the height (h_1) is decreased below 20 mm, the whole operating

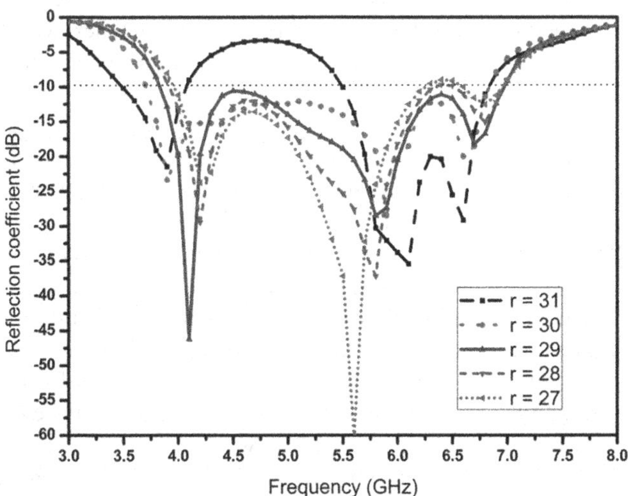

FIGURE 8.29 Variation in reflection coefficient for different values of radius (*r*).

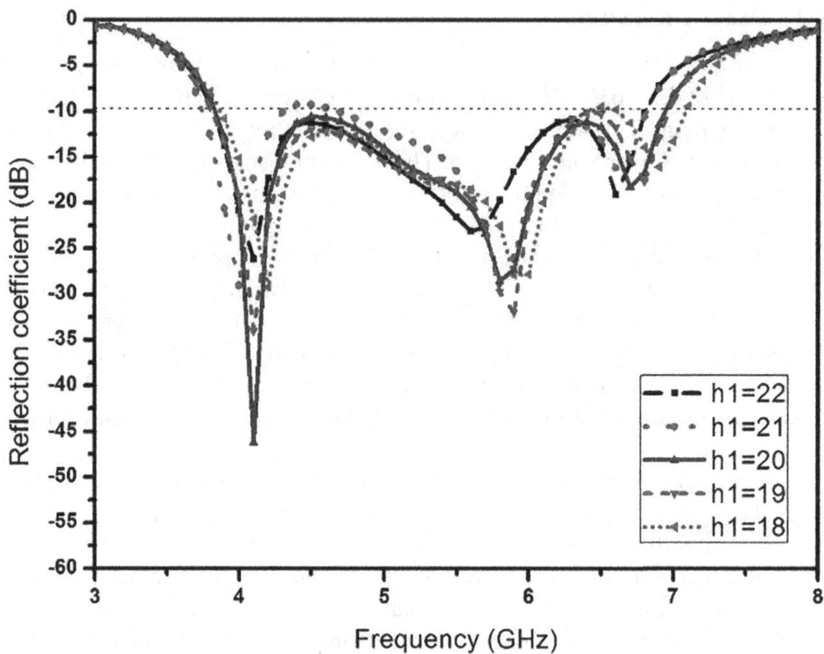

FIGURE 8.30 Variation in reflection coefficient for different values of height (h_1).

band moves upward but the resultant bandwidth decreases. Hence, the optimum value of h_1 is 20 mm. Here, it is noticeable that lower resonance frequency remains almost unchanged. Figure 8.31 shows the effect of variation in the height of the horizontal rectangular folded patch (h_2), which is varied at the intervals of 0.5 mm. Here, the lower resonance frequency is also unaffected by variation in h_2. When h_2 is increased beyond the design point of 10 mm, the second resonance frequency decreases.

Moreover, the third resonance point is mismatched and hence, the bandwidth decreases drastically. When it is decreased below the design value, the antenna behaves like a dual band antenna because of the mismatch in the operating band. Therefore, it can be projected here that the height of the folded conducting element of the antenna plays an important role in accomplishing the goal of the proposed design and the optimum value is at $h_2 = 10$ mm.

Figure 8.32 demonstrates the comparison of the reflection coefficient for different values of gap (G) between the two vertical walls.

Maximum impedance bandwidth has been achieved at $G = 12$ mm as depicted in the figure. A gradual increment in bandwidth is seen up to 12 mm, beyond which the bandwidth gets narrower and band splits into two bands. At $G = 13$ mm, the bandwidth is reduced, and hence it can be stated that the distance between two walls is also an important factor to obtain a broad bandwidth. The effect of uniform width (S_w) of the I-shaped slot on axial ratio bandwidth is explained in Figure 8.33.

The AR bandwidth has changed significantly for instantaneous values of S_w, while S_ℓ and S_h are kept constant. When $S_w \leq 4.5$ mm, the AR bandwidth gets

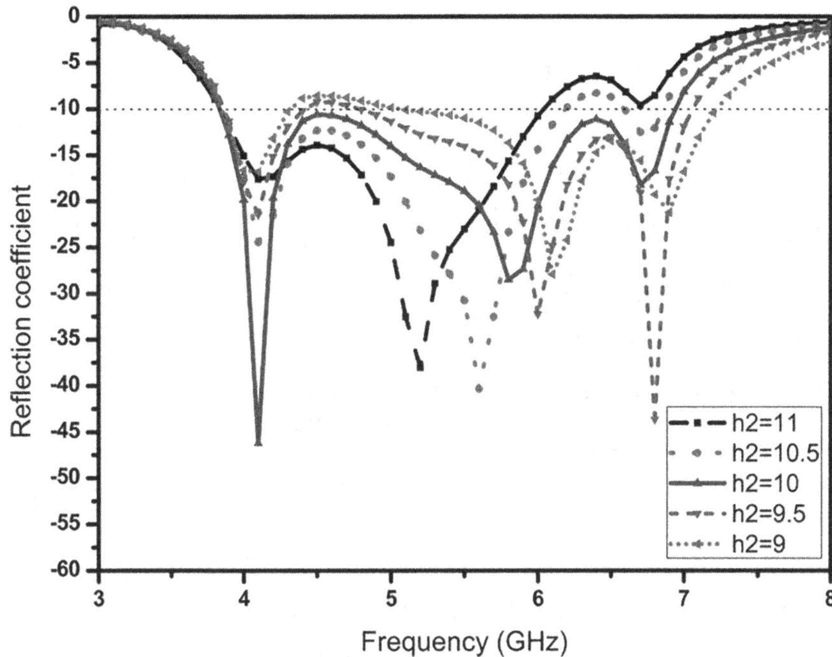

FIGURE 8.31 Variation in reflection coefficient for different values of height (h_2).

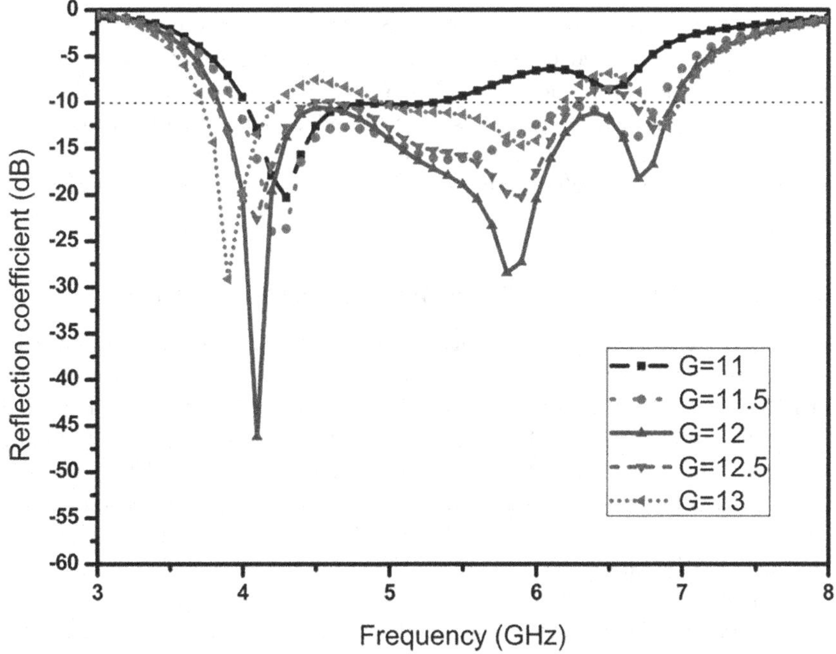

FIGURE 8.32 Variation in reflection coefficient for different values of gap (G).

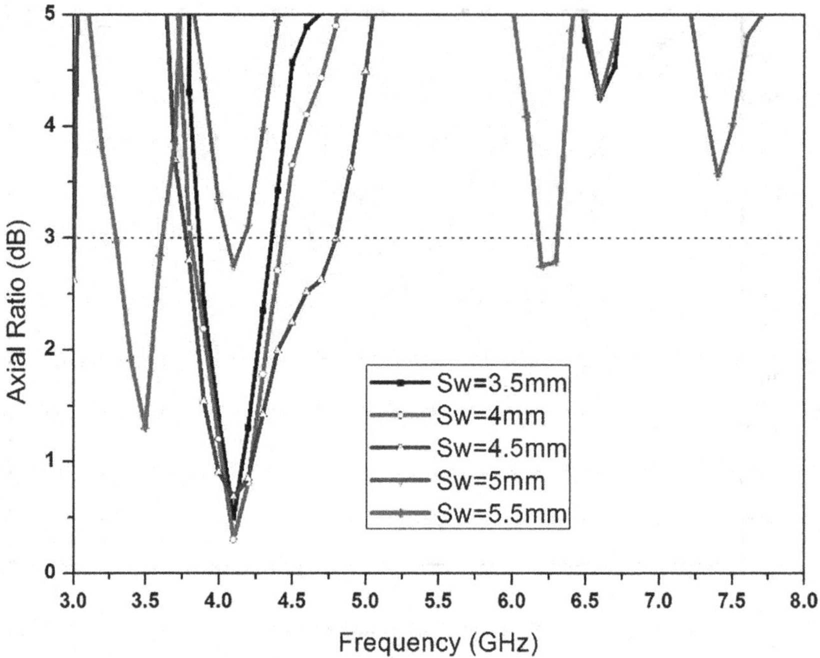

FIGURE 8.33 Variation in the axial ratio with frequency for different values of slot width (S_w).

narrower and for $S_w = 5.5$ mm, the dual CP band response is shown in the figure; out of these, one CP band does not lie in the operating frequency range. The maximum AR bandwidth is achieved when S_w is fixed at 4.5 mm throughout the I-shaped slot. Embedding I-shaped slots in an extended manner with optimized slot dimensions forced the surface currents to rotate in the counterclockwise direction. Hence, it can be concluded that the optimum AR bandwidth is achieved at $S_w = 4.5$ mm.

8.3.3 Characteristics of the Magnetoelectric Dipole Antenna

The proposed design has been simulated using an Ansys HFSS simulator and the antenna structure is fabricated by using copper sheet only. The thickness of the copper sheet used in the simulation and the fabrication are kept the same to realize the actual antenna design. The reflection coefficient has been measured in the laboratory by using the Agilent vector network analyzer (VNA) N5230. The simulated and measured values of the reflection coefficients of the proposed antenna are plotted in Figure 8.34, and it can be seen from the figure that the simulated values agree well with the experimental values in the operating frequency range.

A small deviation in the impedance bandwidth is observed because of the fabrication imperfection and the irregular surface of the metal sheet. The proposed antenna gives an impedance bandwidth of 60.37% from 3.71 to 6.91 GHz, confirming the wideband characteristics of the MED antenna. Furthermore, the antenna has achieved

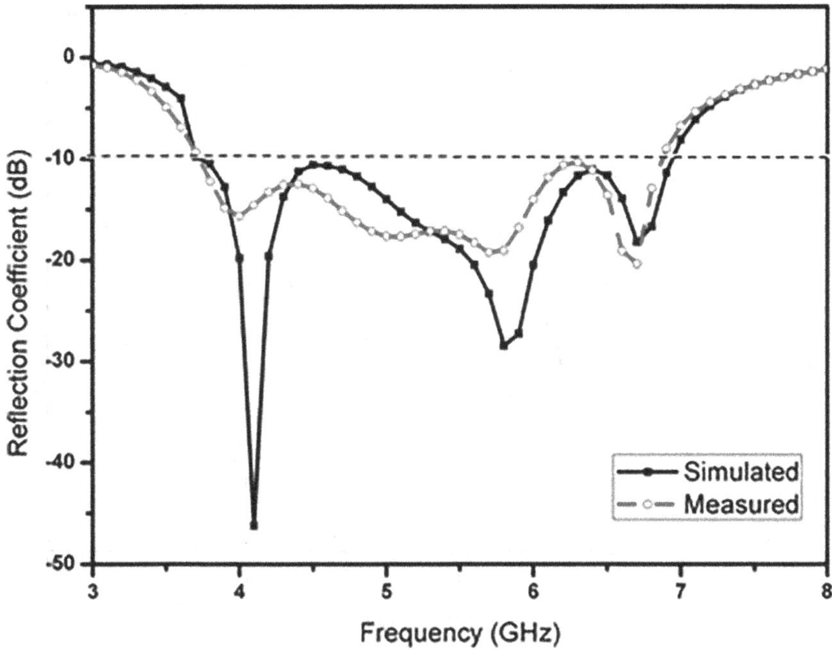

FIGURE 8.34 Measured and simulated reflection coefficients of the proposed antenna.

CP characteristics due to the insertion of I-shaped slots in an extended manner on both sides of the semicircular patches. The axial ratio is the ratio of two orthogonal components of the E-field in a given direction and is often used for the antennas in which the desired polarization is circular. The circularly polarized waves can be generated when two orthogonal E-field components with equal amplitude but with in-phase quadrature are radiated. The acceptable value of the axial ratio is less than 3dB for the patch antennas [39]. Figure 8.35 presents the variation of the measured and simulated AR values with the frequency. It is seen from the figure that a 3dB AR bandwidth of 20.60% in the frequency range of 3.71 and 4.55GHz is achieved in the simulation.

However, there is a small deviation in the measured AR bandwidth because of the inaccuracy in measurement, cutting, and finishing. The gain in the antenna along with radiation patterns were measured in an anechoic chamber by using the standard substitution method. After the analysis of outcomes, it is found that the gain is almost constant over the entire CP band as shown in Figure 8.36.

In general, errors that occur during the measurement is approx. ±0.30dB, and those further can be minimized by using precision handling processes and precisely calibrated instruments. Simulated and measured results have confirmed that the stable gain of 6±0.5 dBic is accomplished in the operating frequency range of 3.71–6.91GHz as shown in Figure 8.32. A very small variation has been seen due to errors in anechoic chamber measurement. It is confirmed from Figure 8.37 that the measured radiation efficiency matched well with the simulated result and it is more than 80% in the operating frequency range.

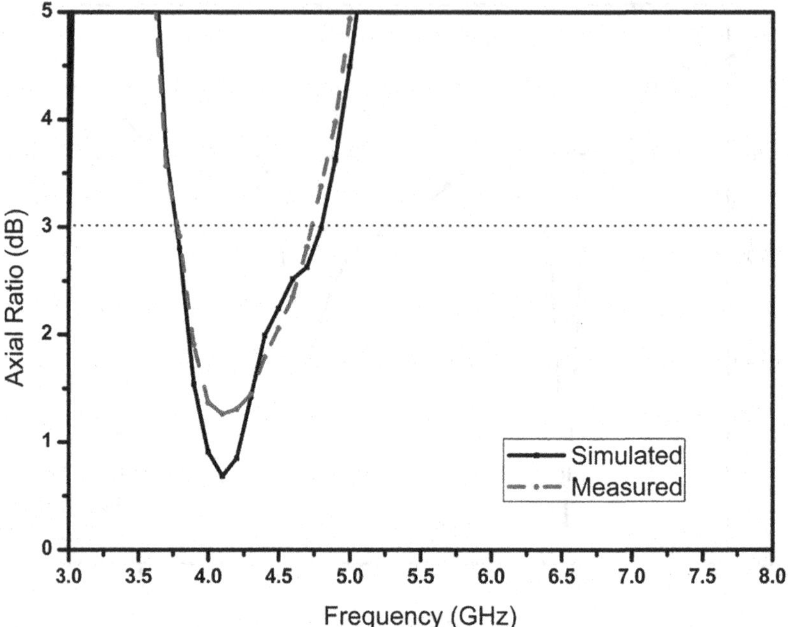

FIGURE 8.35 Measured and simulated axial ratios (dB) of the proposed antenna.

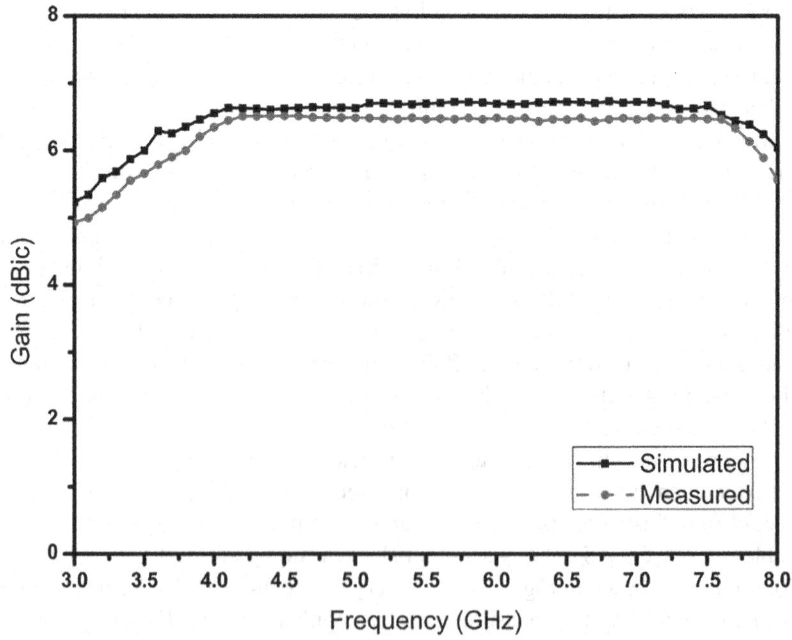

FIGURE 8.36 Measured and simulated peak gains of the proposed antenna.

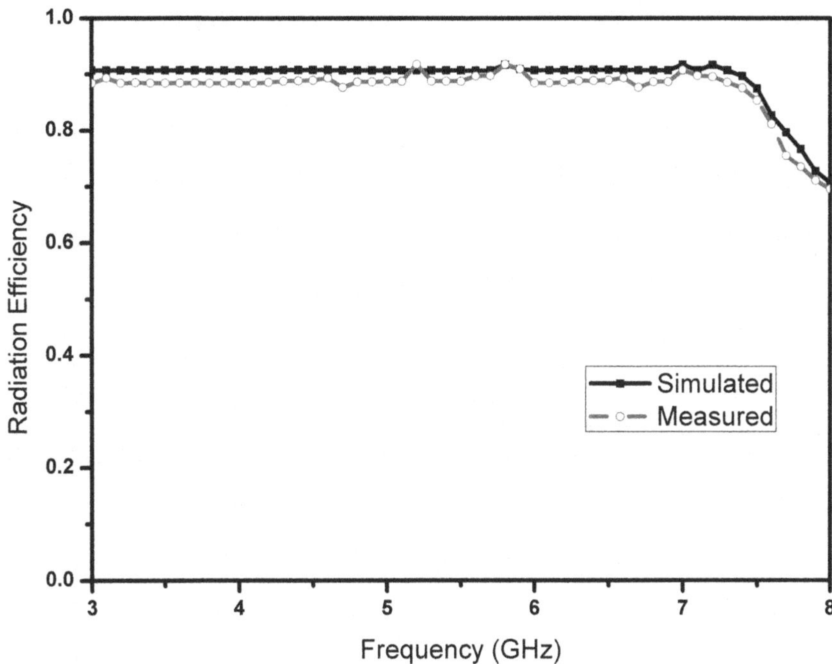

FIGURE 8.37 Measured and simulated radiation efficiencies of the proposed antenna.

To verify the CP behavior of the proposed antenna, the simulated current distributions using the simulator Ansys HFSS at different time periods are shown in Figure 8.38. Here, T is the time period of the signal at operating frequency.

Due to the symmetrical structure, the surface current distributions of the semi-circular patch of only one side has been considered for a clear visibility of the vector currents and the same pattern can be viewed on the other side also. At time $t=0$, the surface currents on the defected semi-circular surface flow along the direction of $\theta=0°$, while the direction of the surface currents has changed in the upward direction at $t=T/4$ in the direction of $\theta=90°$. Further, the direction of the vector currents is changed to $\theta=180°$ and $\theta=270°$ at time $t=T/2$ and $3T/4$, respectively. It is clear from the current distribution pattern that the direction of vector currents at $t=T/2$ and $3T/4$ are just opposite to the currents at $t=0$ and $t=T/4$, respectively. It is also confirmed that the surface currents on the electric dipole rotate in the anticlockwise direction to authenticate the right-hand circular polarization (RHCP) behavior of the antenna.

To compare the radiation patterns in the *xoz* ($\Phi=0°$) and *yoz* ($\Phi=90°$) planes, the radiation patterns obtained from simulation and measurement are plotted for the broadside direction at the center frequencies of the CP band and is shown in Figure 8.39a and b, respectively. The obtained results validate the basic characteristic of the ME dipole antenna, as the radiation patterns from simulation and measurement are unidirectional and symmetric in the E-plane and H-plane. Also, it is clearly visible from Fig 8.39 (b) that the magnitude difference between LHCP and RHCP

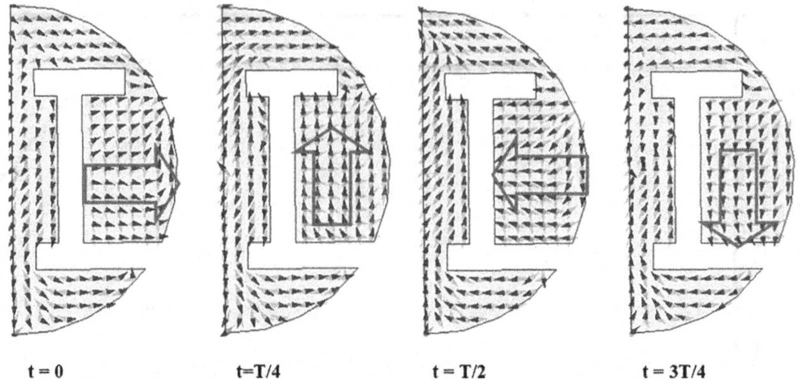

FIGURE 8.38 Current distribution showing the right-hand circular polarization for the center frequency of the CP band at 4.1 GHz at different time intervals, $t=0$, $t=T/4$, $t=T/2$ and $t=3T/4$.

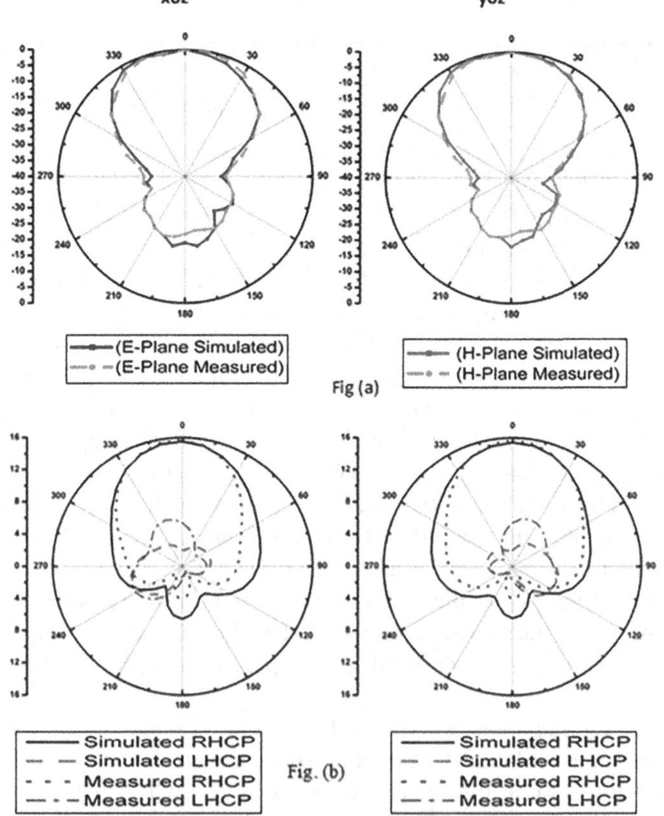

FIGURE 8.39 (a) Simulated and measured E- and H-plane radiation patterns at the centre frequency of the operating band at 5.3 GHz. (b) Simulated and measured RHCP and LHCP radiation patterns for E- and H-planes at the centre frequency of the CP band at 4.1 GHz.

is more than 3 dB, which is a remarkable point to validate the CP behavior of the antenna and confirming the RHCP waves. The experimental results agree well with the simulated results. Moreover, the radiation patterns of the *xoz*- plane and the *yoz*-plane are stable and symmetrical.

8.4 CONCLUSION

In this chapter, two designs of reconfigurable antenna have been discussed. First, a novel antenna with tunable characteristics, wide notch frequency band, and reduced cross-polarization has been designed. The experimental and simulation results match to an acceptable limit. The structure provides dual band when the diode is ON and ultrawide band when the diode is OFF. A wide frequency range around 5 GHz is notched when the diode is ON and frequency band below 5 GHz is notched when the diode is OFF. The other characteristics are normal and stable.

The second design consists of a reconfigurable multiband CP monopole antenna with a simple biasing network. The PIN diode is used to obtain multiband reconfigurable CP radiations. In the OFF state of the diode, the antenna exhibits a dual band behavior with three senses of polarization. LHCP, RHCP and three LP waves are observed. In the ON state of the diode, the antenna again exhibits a dual band behavior and produces other three CP radiations along with three LP radiations. The simple structure of the antenna makes it suitable for many applications, such as in C-band communication satellite for their uplink and amateur satellite operations (CP1), X-band uplink frequency for military communication satellites (CP3), traffic light crossing detector, and terrestrial communications (CP4).

In this part, a wideband circularly polarized MED antenna of semicircular shape embedded with diagonally symmetrical I-shaped slots has been proposed and investigated for C-band applications. In the proposed design, defected semicircular patches are placed horizontally to function as electric dipoles and two vertically oriented folded metallic walls shorted to minimum rectangular ground to act as magnetic dipoles. The Γ-shaped single feed structure is used to excite both dipoles together to achieve symmetric radiation patterns and CP waves are produced with the help of I-shaped slots. The low profile of the structure is maintained by the use of minimum ground and the folded structure. Hence, the proposed antenna is compact in design, including features of wideband behavior and circular polarization. It exhibits an impedance bandwidth of 60.37%, AR bandwidth of 20.60%, and a radiation efficiency of more than 80% within the operating frequency range. The proposed design provides CP characteristics in the frequency range of 3.71–4.55 GHz and a stable peak realized gain of 6 ± 0.5 dBic with a unidirectional radiation pattern in E- and H-planes. All these features made this antenna suitable for 5G Wi-Fi (5.15–5.875 GHz) and C band applications (3.7–4.2 GHz for downlink and 5.93–6.43 GHz for uplink).

REFERENCES

1. Vuong T. P., Ghiotto A., Duroc Y., Tedjini S. (2007). Design and characteristics of a small U-slotted planar antenna for IR-UWB. *Microwave and Optical Technology Letters*, 49, 1727–1731.

2. Lee W. S., Kim D. Z., Kim K. J., Yu J. W. (2006). Wideband planar monopole antennas with dual band-notched characteristics. *IEEE Transactions on Microwave Theory and Techniques*, 54, 2800–2806.
3. Chung K., Kim J., Choi J. (2005). Wideband microstrip-fed monopole antenna having frequency band-notch function. *IEEE Microwave Wireless Component Letters*, 15, 766–768.
4. Kim Y., Kwon D. H. (2004). CPW-fed planar ultra wideband antenna having a frequency band notch function. *Electronics Letters*, 40, 403–405.
5. Lui W. J., Cheng C.-H., Zhu H.-B. (2007). Improved frequency notched ultrawideband slot antenna using square ring resonator. *IEEE Transactions on Antennas Propagation*, 55, 2445–2450.
6. Kim J., Cho C. S., Lee J. W. (2006). 5.2 GHz notched ultra-wideband antenna using slot-type SRR. *Electronics Letters*, 42, 315–316.
7. Zhang Y., Hong W., Yu C., Kuai Z.-Q., Don Y.-D., Zhou J.-Y. (2008). Planar ultrawideband antennas with multiple notched bands based on etched slots on the patch and/or split ring resonators on the feed line. *IEEE Transaction on Antenna and Propagation*, 56, 3063–3068.
8. Ding J., Lin Z., Ying Z., He S. (2007). A compact ultra-wideband slot antenna with multiple notch frequency bands. *Microwave and Optical Technology Letters*, 49, 3056–3060.
9. Dong Y. D., Hong W., Kuai Z. Q., Yu C., Zhang Y., Zhou J. Y., Chen J.-X. (2008). Development of ultrawideband antenna with multiple band-notched characteristics using half mode substrate integrated waveguide cavity technology. *IEEE Transaction on Antenna and Propagation*, 56, 2894–2902.
10. Lui W. J., Cheng C. H., Zhu H. B. (2005). Frequency notched printed slot antenna with parasitic open-circuit stub. *Electronics Letters*, 41, 1094–1095.
11. Lui W. J., Cheng C. H., Zhu H. B. (2006). Compact frequency notched ultra-wideband fractal printed slot antenna. *IEEE Microwave and Wireless Component Letters*, 16, 224–226.
12. Pandey G. P., Kanaujia B. K., Gautam A. K., Gupta S. K. (2012). Ultra-wideband L-strip proximity coupled slot loaded circular microstrip antenna for modern communication systems. *Wireless Personal Communications*, Published on line. doi: 10.1007/s11277-012-0684-5.
13. Vandenbosch G. A. E., van de Capelle A. R. (1994). Study of the capacitively fed microstrip antenna element. *IEEE Transactions on Antennas and Propagation*, 42, 1648–1652.
14. Kasabegoudar V. G., Upadhyay D. S., Vinoy K. J. (2007). Design studies of ultra-widebandmicrostrip antennas with a small capacitive feed. *International Journal of Antennas and Propagation*, 2007, 1–8.
15. Shivnarayan, Vishwakarma B. R. (2006). Analysis of notch loaded patch for dual band operation. *Indian Journal of Radio and Space Physics*, 35, 435–442.
16. Mishra A., Singh P., Yadav N. P., Ansari J. A., Vishwakarma B. R. (2009). Compact shorted microstrip patch antenna for dual band operation. *Progress in Electromagnetic Research (PIER)*, 9, 171–182.
17. Ansari J. A., Singh P., Yadav N. P., Vishwakarma B. R. (2008). Analysis of wideband half circular disc patch antenna loaded with symmetrical notches. *Microwave and Optical Technology Letters*, 51, 1880–1883.
18. Lo Y. T., Lee S. W. (1988). *Antenna hand Book, Theory, Applications, and Design*. New York: Vav Nostrand Reinhold, ch. 21.
19. Pasumarthy N. R., Yagateela P. R. (2016). Compact single feed circularly polarized Koch island microstrip antenna. *International Journal of Electronics and Communications*, 70, 1543–1550.

20. Yohandri, Sumayanto J. T. S., Hiroaki K. (2012). A new triple proximity-fed circularly polarized microstrip antenna. *International Journal of Electronics and Communications*, 66, 395–400.
21. Baharuddin M., Victor W., Sumayanto J. T. S., Kuze H. (2011). Elliptical microstrip antenna for circularly polarized synthetic aperture radar. *International Journal of Electronics and Communications*, 65, 62–67.
22. Madhav P. T. B., Khan H., Kotamraju K. S., Jono K. (2016). Circularly polarized slotted aperture antenna with coplanar waveguide fed for broadband applications. *Journal of Engineering Science and Technology*, 11(2), 267–277.
23. Rahimi M., Maleki M., Soltani M., Arezomand S. A., Zarrabi B. F. (2016). Wide band SRR-inspired slot antenna with circular polarization for wireless application. *International Journal of Electronics and Communications (AEÜ)*, 70(September (9)), 1199–1204.
24. Zhao X., Huang Y., Li J., Zhang Q., Wen G. (2017). Wideband high gain circularly polarized UHF RFID reader microstrip antenna and array. *AEU-International Journal of Electronics and Communications*, 77(July), 76–81.
25. Rahim S. A., Danesh S., Okonkwo U. A., Sabran M., Khality M. (2012). UWB monopole antenna with circular polarization. *Microwave and Optical Technology Letters*, 54, 949–953.
26. Wu C., Han L., Yang F., Wang L., Yang P. (2012). Broad beam width circular polarization antenna: microstrip monopole antenna. *Electronics Letters*, 48, 1176–1178.
27. Wang C. J. (2011). A wideband loop-like monopole antenna with circular polarization. *Microwave and Optical Technology Letters*, 53, 2556–2560.
28. Fujimoto T. F., Jono, K. (2011). Wideband rectangular printed monopole antenna for circular polarization. *IET Microwaves, Antennas and Propagation*, 8, 649–656.
29. Wu T., Shi X. W., Li P., Bai H. (2013). Tri-band microstrip-fed monopole antenna with dual-polarization characteristics for WLAN and Wi-Max applications. *Electronics Letters*, 49, 1597–1598.
30. Jou C. F., Wu J. W., Wang C. J. (2009). Novel broadband monopole antennas with dual band circular polarization. *IEEE Transaction on Antennas and Propagation*, 57, 1027–1034.
31. Frotanpur A., Hassan H. (2011). A dual-broadband circularly polarized halved falace-shape printed monopole antenna. *International Journal of RF and Microwave Computer Aided Engineering*, 21, 636–641.
32. Ding K., Yu T., Qu D.-X., Peng C. (2013). A novel loop like monopole antenna with dual-band circular polarization. *Progress in Electromagnetics Research C*, 45, 179–190.
33. Hui C., Yung E. K. N. (2011). Dual-band circularly polarized CPW-fed slot antenna with a small frequency ratio and wide bandwidths. *IEEE Transactions on Antennas and Propagation*, 5, 1379–1384.
34. Yao Y., Liao S., Wang J., Xue K., Balfour A. E., Luo Y. (2016). A new patch antenna designed for CubeSat: dual feed, LVS dual-band stacked, and circularly polarized. *IEEE Antennas & Propagation Magazine*, 58(June (03)), 16–21.
35. Baek J. G., Hawang K. C. (2013). Triple-band unidirectional circularly polarized hexagonal slot antenna with multiple L shape slits. *IEEE Transactions on Antennas and Propagation*, 61(9), 4831–4835.
36. Park S. X., Ta I., Ziolkowiski R. W. (2013). Circularly polarized crossed dipole on an HIS for 2.4/5.2/5.8-GHz WLAN applications. *IEEE Antennas Wireless Propagation*, 12, 1464–1467.
37. Hoang V., Park H. C. (2014). Very simple 2.45/3.5/5.8-GHz triple band circularly polarized printed monopole antenna with bandwidth enhancement. *Electron Letters*, 50, 1792–1793.

38. Pandey G. P., Kanaujia B. K., Gupta S. K., Gautam A. K. (2015). A novel C shape antenna with switchable wideband frequency notch. *Wireless Personal Communications*, 80, 471–482.
39. Singh D. K., Kanaujia B. K., Dwari S., Pandey G. P., Kumar S. (2016). Reconfigurable circularly polarized capacitive coupled microstrip antenna. *International Journal of Microwave and Wireless Technologies*, 9, 843–850.
40. Anantha B., Merugu L., Rao S. (2017). A quad-polarization and frequency reconfigurable square ring slot loaded microstrip patch antenna for WLAN applications. *International Journal of Electronics and Communications (AEÜ)*, 78(August), 15–23.
41. Nguyen-Trong N., Piotrowski, A., Hall L., Fumeaux C. (2016). A frequency- and polarization-reconfigurable circular cavity antenna. *IEEE Antennas and Wireless Propagation Letters*, 16(October), 999–1002.
42. Chang K., Bahl I., Nair V. (2002). *RF and Microwave Circuit and Component Design for Wireless System*. New York: Wiley Inderscience.
43. Chenn R. H., Row J. S. (2008). Single fed microstrip patch antenna with switchable polarization. *IEEE Transactions on Antennas and Propagation*, 56, 922–926.
44. Kim B., Pan B., Kim Y. S., Papapolymerou J., Tentzeris M. M. (March 2008). A novel single feed circular microstrip antenna with reconfigurable polarization capability. *IEEE Transactions on Antennas and Propagation*, 56(3), 630–638.
45. Balanis C. A. (1997). Antenna measurements. In *Antenna Theory*, 2nd Edition. John Wiley and Sons, 839–883.
46. Yoon W. S., Baik J. W., Lee H. S., Pyo S., Han S.-M., Kim Y.-S. (2010). A reconfigurable circularly polarized microstrip antenna with a slotted ground plane. *IEEE Antennas and Wireless Propagation Letters*, 9, 1161–1164.
47. Stutzman W. L., Thiele G. A. (2012). Antenna measurements. In *Antenna Theory*, 3rd Edition. John Wiley and Sons, 559–586.
48. Sung Y. J., Jang T. U., Kim Y. S. (2004). A reconfigurable microstrip antenna for switchable polarization. *IEEE Microwave and Wireless Components Letters*, 14, 534–536.
49. James J. R., Hall P. S. (1989). *Handbook of Microstrip Antennas*. London, UK: Peter Peregrinus, ch. 4.
50. Chlavin A. (1954). A new antenna feed having equal E- and H-plane patterns. *IRE Transaction of Antennas Propagation*, 2(3), 113–119.
51. Clavin A., Huebner D. A., Kilburg F. J. (1974). An improved element for use in array antennas. *IEEE Transactions on Antennas and Propagation*, 22(4), 521–526.
52. Idayachandran G., Nakkeeran R., Anbazhagan R. (2016). Design and analysis of broadband magneto electric dipole antenna for LTE femtocell base stations. *Electronics Letters*, 52(14), 74–576.
53. Lei G., Luk K. M. (2013). A magneto-electric dipole antenna with low-profile and simple structure. *IEEE Antennas and Wireless Propagation Letters*, 12, 140–142.
54. Idayachandran G., Nakkeeran R. (2016). Compact magneto-electric dipole antenna for LTE femtocell base stations. *Electronics Letters*, 52(14), 574–576.
55. Chen D., Luk K. M. (2016). Compact low-profile magneto-electric dipole antenna. *IEEE Antennas and Wireless Propagation Letters*, 15, 1642–1644.
56. Liu Y., Chen S., Ren Y., Cheng J., Liu Q. H. (2015). A broadband proximity-coupled dual-polarized microstrip antenna with L-shape backed cavity for X-band applications. *International Journal of Electronics and Communications (AEÜ)*, 69(9), 1226–1232.
57. Feng B., Li S., An W., Hong W., Wang S., Yin S. (2014). A printed dual-wideband magneto-electric dipole antenna for WWAN/LTE applications. *International Journal of Electronics and Communications (AEÜ)*, 68(10), 926–932.
58. Arnieri E., Boccia L., Amendola G., Massa G. D. (2007). A compact high gain antenna for small satellite applications. *IEEE Transactions on Antennas and Propagation*, 55(2), 277–282.

59. Son W. I., Lim W. G. (2010). Design of compact quadruple inverted-F antenna with circular polarization for GPS receiver. *IEEE Transactions on Antennas and Propagation*, 58(5), 1503–1510.
60. Nasimuddin Z., Chen N., Qing X. (2010). Asymmetric-circular shaped slotted microstrip antennas for circular polarization and RFID applications. *IEEE Transaction Antennas Propagation*, 58(12), 3821–3828.
61. Bian, L., Guo Y. X., Ong L. C., Shi X. Q. (2006). Wideband circularly polarized patch antenna. *IEEE Transaction Antennas Propagation*, 54(9), 2682–2686.
62. Lau K. L., Luk K. M., Lee K. F. (2006). Design of a circularly-polarized vertical patch antenna. *IEEE Transaction Antennas Propagation*, 54, 1332–1335.
63. Chen H.-D., Sim C.-Y.-D., Kuo S.-H. (2012). Compact broadband dual coupling-feed circularly polarized RFID microstrip tag antenna mountable on a metallic surface. *IEEE Transaction Antennas Propagation*, 60(12), 5571–5577.
64. Guo Y.-X., Bian L., Shi X. Q. (2009). Broadband circularly polarized annular-ring microstrip antenna. *IEEE Transaction Antennas Propagation*, 57(8), 2474–2477.
65. Hu Y.-J., Ding W.-P., Cao W.-Q. (2011). Broadband circularly polarized microstrip antenna array using the sequentially rotated technique. *IEEE Antennas Wireless Propagation Letter*, 10, 1358–1361.
66. Roy J. S., Thomas M. (2008). Design of a circularly polarized microstrip antenna for WLAN. *Progress in Electromagnetics Research*, 3, 79–90.
67. Tong K. F., Wong T. P. (2007). Circularly polarized U-slot antenna. *IEEE Antennas and Propagation Society*, 55(8,), 2382–2385.
68. Hisao I. W. (1996). A circularly polarized small-size microstrip antenna with a cross-slot. *IEEE Transactions on Antennas and Propagation*, 44(10), 1399–1401.
69. Singh D. K., Kanaujia B. K., Dwari S., Pandey G. P., Kumar S. (2015). Novel quad-band circularly polarized capacitive-fed microstrip antenna for C-band applications. *Microwave and Optical Technology Letters*, 57(11), 2622–2628.
70. Ram Krishna R. V. S., Kumar R. (2013). Design of ultra-wideband trapezoidal shape slot antenna with circular polarization. *International Journal of Electronics and Communications (AEÜ)*, 67(12), 1038–1047.
71. Singh D. K., Kanaujia B. K., Dwari S., Pandey G. P., Kumar S. (2017). Reconfigurable circularly polarized capacitive coupled microstrip antenna. *International Journal of Microwave and Wireless Technologies*, 9(4), 843–850.
72. Bolster M. F. (1961). A new type of circular polarizer using crossed dipoles. *IEEE Transactions on Microwave Theory and Techniques*, 9(5), 385–388.
73. Yang D, Yang H.-C. (2012). A novel circularly polarized bowtie antenna for Inmarsat communications. *IEEE Antennas and Propagation Magazine*, 54(4), 317–325.
74. Baik J.-W., Lee K. J., Yoon W.-S., Lee T.-H., Kim Y.-S. (2008). Circularly polarized printed crossed dipole antennas with broadband axial ratio. *Electron Letter*, 44(13), 785–786.
75. Qu S. W., Li J. L., Xue Q., C. H. Chan. (2008) Wideband cavity-backed bowtie antenna with pattern improvement. *IEEE Transactions on Antennas and Propagation*, 56(12), 3850–3854.

9 Reconfigurable Printed Antennas

Dr. Deepak Gangwar
Bharati Vidyapeeth's College of Engineering, New Delhi

Dr. Sachin Kumar
Kyungpook National University, Korea

Dr. Surendra Kumar Gupta
Ambedkar Institute of Technology, Delhi

Ghanshyam Singh
Feroze Gandhi Institute of Engineering and Technology, Raebareli

Ankit Sharma
Galgotias College of Engineering and Technology, Greater Noida

CONTENTS

9.1 INTRODUCTION

Antennas are critical and vital elements for radar and communication systems. Proliferation in wireless communication and radar systems over the past 50 years results in different types of antennas. The different classes of antennas possess different built-in assets and detriments that make them reasonable for specific applications. As technology is changing day by day and a conventional antenna has fixed functionality, reconfigurable antennas are required to accommodate new systems. The reconfigurable antennas provide a better solution, as they can adapt to the changing system requirements.

Reconfigurable antennas can tune their characteristics such as frequency, pattern, and polarization, as per the demand of application [1–3]. This reconfigurability can be single, for instance, frequency diversity, pattern diversity, or polarization diversity, or it may be a combination of two or more [4–9]. It is very tedious and space-consuming to design separate antennas for multiple operations, by frequency reconfiguration, and a single antenna could be used for numerous frequency bands. Modern communication systems are multi-functional and compact. Antenna reconfigurability not only saves space and cost but also reduces the production time. To accomplish multi-functionality in present-day specialized gadgets, metamaterial components are stacked in traditional antennas.

A single reconfigurable antenna could deliver the functionality equivalent to multiple traditional single-purpose systems, thus offering significant savings in cost, volume, weight, and maintenance resources. The introduction of new kinds of functionality into antennas will result in higher or comparable performance and lower costs. The system designers need to be willing to exploit these new degrees of freedom and functionality, so that the antenna becomes a more active part of the communication link, working together with new circuits, communication protocols, and signal processing techniques. The customary techniques used for antenna characterization require two types of information: the input impedance response over frequency (typically called the frequency response) and the radiation characteristics (or radiation patterns). Generally, the frequency response is given preference, because, in the absence of a reasonable match of input impedance, severe reflections may damage the components of the transmitting system and can also result in power wastage, whereas receiving systems may require signal amplification due to the reduced sensitivity. Once the antenna frequency response is known, the examination of the radiation patterns can be done. This chapter briefly reviews both the frequency and radiation features of the antennas that can be manipulated through the reconfiguration of material and physical parameters.

9.2 DIFFERENT APPROACHES FOR ATTAINING RECONFIGURABILITY

The antenna reconfigurability can be accomplished by electrical, optical, or mechanical/physical means, or using special materials such as liquid crystals, as shown in Figure 9.1. Physically reconfigurable printed antennas are complex and have a bulky configuration. In view of this, the antenna researchers are more

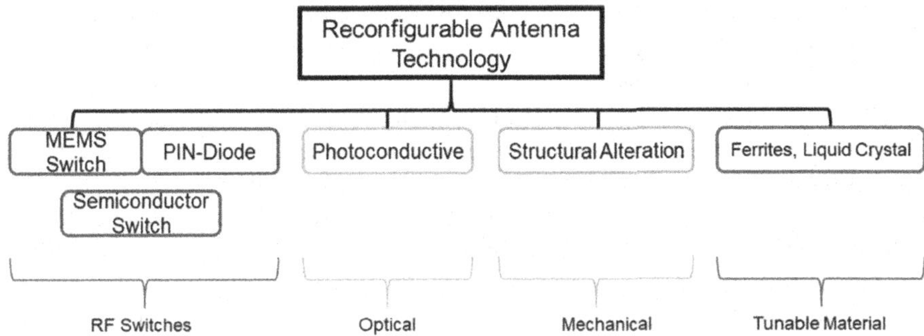

FIGURE 9.1 Different methods to attain reconfigurability.

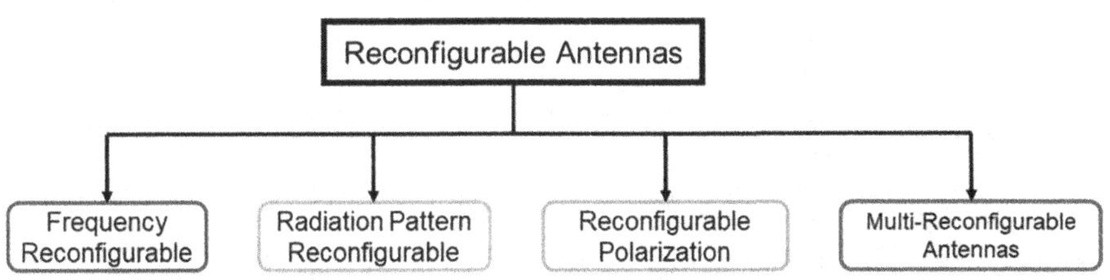

FIGURE 9.2 Types of reconfigurability in antennas.

attracted to electrically reconfigurable antennas, as such antennas are easy to fabricate, possess less volume, and are less complex. In the last few years, various mechanisms for reconfigurable antennas have been explored by several researchers. The accompanying passage distinctive reconfiguration systems give a few models and argue the advantages and disadvantages of each approach. Based on the antenna functionality, the various types of reconfigurability are illustrated in Figure 9.2.

9.2.1 Electrical Method-Based Reconfigurable Antennas

Before the evolution of semiconductor diodes, the reconfigurable antennas were constructed by using mechanical means. These mechanical ways were slow and bulky. Performances of reconfigurable antennas rely on switching mechanisms, and with the development of electronic switches, their performance improved considerably. The electronic switches are fast in response, have high power handling capacity, and consume significant power. PIN diodes, RF micro-electromechanical systems (MEMS), and varactor diodes are common to be worked as switching elements to alter the geometry of the main radiator to accomplish either type of reconfigurability. PIN diodes offer high switching speed, small volume, low cost, and low insertion loss. A DC bias is needed to operate PIN diodes.

Due to recent developments in the semiconductor fabrication, small-size MEMS received great attention to be used as switching elements in reconfigurable wireless systems. MEMS are superior to PIN diodes because of isolation, insertion loss, power consumption, and linearity. RF-MEMS functioning as a switch is based on mechanical movement. The drawback of the RF-MEMS is slow switching as compared to the PIN and varactor diodes. RF-MEMS response time is in the range of 1–200 μs, and due to their slow speed, they are not suitable for some applications. The response time of the varactor diode and the PIN diode is in the range of 1–100 ns [10]. The switches implemented by PIN diodes operate in two different modes, that is, ON and OFF, depending upon the DC biasing. While the varactor diode operates in continuous mode, its capacitance continuously varies by changing its biasing voltages. The varactor diode-based antenna provides a very wide tuning range and it needs a small current, in comparison with the PIN diode and MEMS. However, it has non-linear characteristics and needs a variable power supply. The various types of RF switch-based reconfigurable antennas are illustrated in Figure 9.3. A comparison of different switching components is provided in Table 9.1.

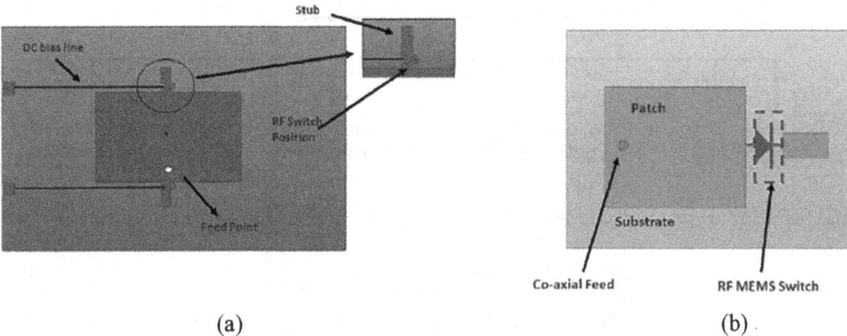

FIGURE 9.3 Methods for achieving frequency-reconfigurable antennas. (a) RF switch-based (varactor, PIN diode) antenna and (b) RF-MEMS switch-based antenna.

TABLE 9.1
A Comparison of Different Switch Components [1–10]

Reconfiguration Technique	Advantages	Disadvantages
PIN diodes	• High reliability • Extremely low cost • Common choice for reconfiguration	• High tuning speed • High DC bias in the ON state • High power handling capacity
Varactors	• Small current flow • Continuous tuning • Ease of integration	• Non-linearity • Low dynamic range • Complex bias circuitry
RF-MEMS	• High isolation and linearity • Wide impedance bandwidth • Low power losses and low noise figure	• High control voltage • Slow switching speed • Limited life cycle

9.2.1.1 RF-MEMS-Based Reconfigurable Antennas

A few RF-MEMS-based reconfigurable antennas were investigated and reported in Refs. [10–14]. A tunable dual-band slotted antenna with implanted RF-MEMS was analyzed in Ref. [11]. Huff et al. investigated a pattern-reconfigurable square spiral antenna integrated with RF-MEMS switches in Ref. [12]. Nasiri et al. [13] studied a micro-machined frequency-agile patch antenna configured by MEMS, where the driving mechanism was thermal actuation. The thermal actuators were connected to a silicon platform and on the application of voltage; they shift downward with patch antenna resulting in a downward frequency shift. The actuator-driven voltages are CMOS-compatible. The antenna operation band changes from 17.37 to 15.07 GHz in upstate–downstate on the application of bias voltage to the actuator. The achievable bandwidth is different in both the states, and a bandwidth of 1.4% is achieved in the downstate, while a bandwidth of 3.9% is achieved in the upstate. MEMS switches were integrated into an array of parasitic antennas to accomplish pattern diversity, as suggested by Petit et al. [14]. A prototype was fabricated and tested. Experimental results exhibit good impedance bandwidth and ±25° beam tilting for a 5.6-GHz resonance band.

9.2.1.2 PIN Diode-Based Reconfigurable Antennas

Various PIN diode-based reconfigurable antennas were analyzed and discussed in Ref. [15–23]. A frequency-reconfigurable antenna for WLAN and GSM applications, based on PIN diode-embedded epsilon negative transmission line was reported in Ref. [15]. The pattern-reconfigurable antennas presented in Refs. [16,17] were realized by the classical Yagi–Uda approach. In general, the PIN diodes are installed in between parasitic elements to change their electrical lengths, so that these elements function as a reflector–director or vice versa, based on the operating states of PIN diodes. A frequency-reconfigurable antenna with a customized ground plane was studied in Ref. [18]. The PIN diodes were embedded at the appropriate position on U- and L-shaped slots in the ground plane. The etching of the ground plane results in the miniaturization of the reported antenna compared to the traditional antennas. The antenna can operate at multiple frequency bands by switching the states of the PIN diodes. Khidre et al. [19] reported a single-feed wideband polarization-reconfigurable E-shaped patch antenna, where two PIN diodes were placed at the top and bottom slots. The polarization mode (right-hand circular polarization (RHCP) or left-hand circular polarization (LHCP)) depends upon the operating states of the PIN diodes. The proposed design was symmetric, and a 7% bandwidth was achieved in the 2.4-GHz band. The DC control circuit of the antenna was simple, and fabricated antenna was useful for WLAN IEEE 802.11b/g. In Ref. [20], a polarization-agile antenna with a customized ground plane and L-shaped slots etched at the four corners underneath the main radiator was suggested. The antenna polarization sense switched between RHCP and LHCP by PIN diodes installed at an appropriate location on the L-type slot. Yang et al. [21] achieved a wide axial ratio bandwidth by incorporating the staircase slots on the ground surface. The PIN diodes were implanted at staircase slots, and the DC biasing circuit was very simple, as there is no need for RF choking inductor at the ground plane. A PIN diode-based polarization-reconfigurable antenna is shown in Figure 9.4.

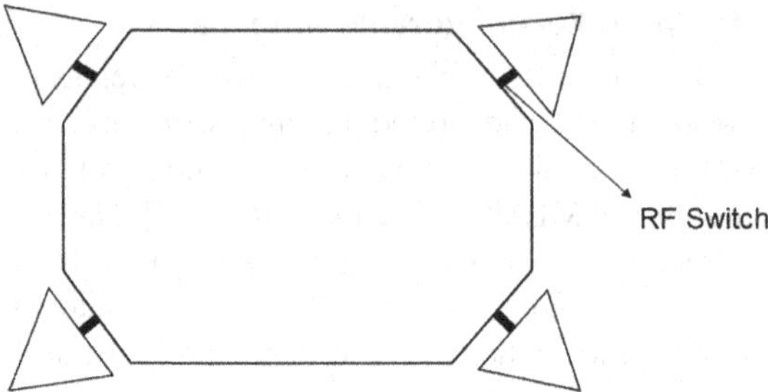

FIGURE 9.4 PIN diode-based polarization-reconfigurable antenna.

A coplanar waveguide (CPW)-fed wideband end-fire pattern diversity antenna was studied by Zhang et al. [22]. The antenna was comprised of the main radiator and two-step-shaped parasitic patches. Four PIN diodes were installed at the parasitic patches, and by changing their states, the functioning of the parasitic elements changed from the reflector to the director or vice versa. This antenna switches beam in the opposite direction by switching (ON or OFF) PIN diodes. Experimental results depicted a good back-to-front ratio. A CPW-fed compound reconfigurable antenna (pattern and frequency) was presented in Ref. [23]. The proposed antenna was comprised of closed-ring resonator (CRR), electric-inductive-capacitive (ELC) resonator, and the parasitic elements. A PIN diode D_1 was installed between ELC and CRR, and other diodes (D_2–D_5) were placed between the parasitic elements and the ground plane. When the diode D_1 operating state was changed, the antenna demonstrated frequency agility, while the states of diodes D_2–D_5 were changed to vary the electrical length of the parasitic elements. The diode D_1 was inserted between the CRR and ELC resonators for frequency reconfigurability. By controlling the electrical length, the main beam steers to accomplish end-fire pattern diversity. The fabricated antenna was planar and compact and depicts multi-band features along with pattern diversity. A PIN diode-based compound reconfigurable antenna is shown in Figure 9.5.

The frequency diversity is achieved by exciting the CRR and ELC resonators separately and collectively, and by mutual coupling between the parasitic elements and two resonators. The PIN diodes implanted between the ground plane and four parasitic elements control the electrical length of the ground plane to accomplish pattern diversity. The ground plane and parasitic elements along with PIN diodes function as a reflector–director or vice versa, depending upon the state of implanted PIN diodes. The ground plane and parasitic elements jointly steer the primary omnidirectional beam to bidirectional and unidirectional end-fire radiated beam at multiple frequency bands, depending upon the state of PIN diodes.

The surface current distribution is studied to interpret the end-fire beam reconfigurability. The simulated surface current distributions and 3D radiation patterns at 4.76 GHz, when the antenna is in state 5 and state 6, are illustrated in Figure 9. 6. It is noticed that when the diode D_2 is ON, the parasitic element at the right end of the antenna is attached to the ground plane and works as a reflector, directing the main beam toward (+x) direction. Next, when the diode D_3 is ON, the

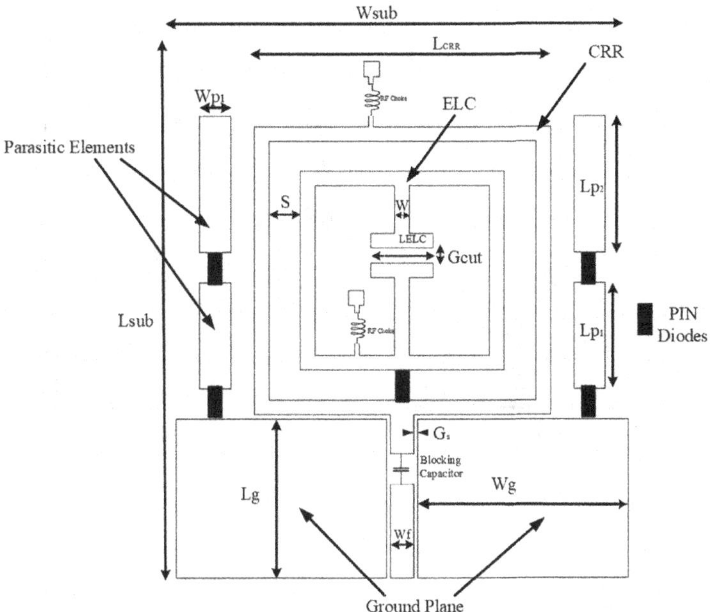

FIGURE 9.5 Geometry of the proposed reconfigurable antenna. (a) Top view and (b) side view.

parasitic element at the left end is attached to the ground plane and works as a reflector, directing the main beam toward (–*x*) direction. This is due to the non-symmetrical distribution of surface current around the y-axis, on the resonator and on the ground plane. Here, the parasitic element and ground plane jointly work as a reflector. This results in an end-fire beam in *XZ*-(H)-plane. When the parasitic elements are not connected to the ground plane, then they are not resonating. It is also clear from Figure 9.7 that current distribution is symmetrical about the *y*-axis. The maximum current is concentrated along the outer surface of CRR and at the top edge of the ground plane, while very small current lies along the periphery of ELC. This type of surface current distribution results in an omnidirectional beam in *XZ*-(H)-plane.

Another reconfigurable L-strip-fed circular microstrip antenna on thick substrate with CSRR in the ground plane has been analyzed and investigated in Ref. [24]. The antenna is analyzed using the cavity model and the circuit theoretic approach for initial design and then simulated on IE3D simulation software. The antenna is made reconfigurable with PIN diode, which makes it to work in different configurations. Two diodes were used to implement a double annular slot, one annular slot, and one split slot and CSRR in the ground plane. While other configurations of diodes provide bandwidth and radiation pattern diversity, CSRR provides size reduction of up to 13.31% along with high gain directivity and radiation efficiency. A maximum gain of 8 dBi and directivity of 8.3 dBi have been achieved in the respective band of operations. The antenna exhibits wideband along with multi-band characteristic.

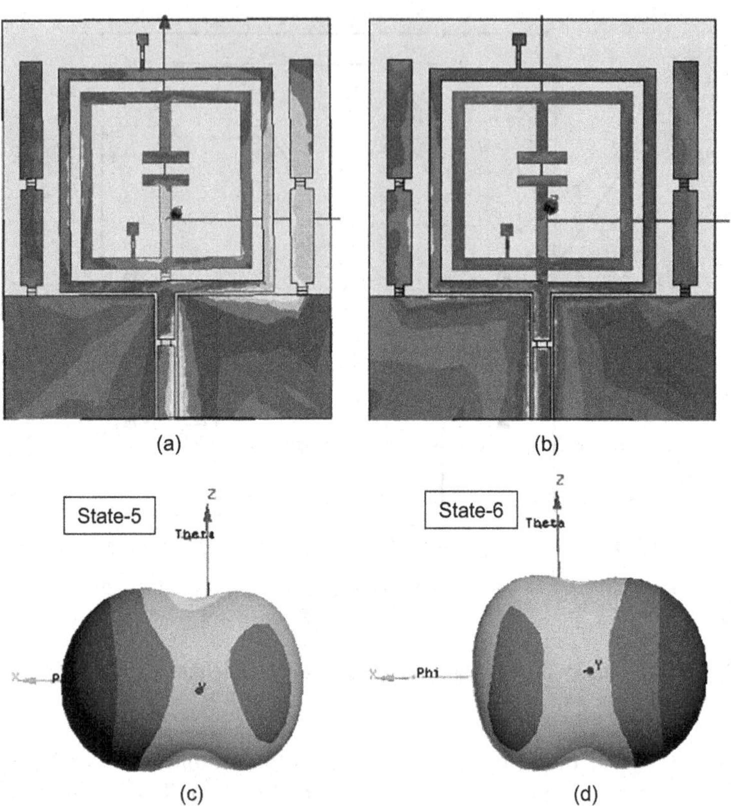

FIGURE 9.6 Simulated surface current distribution and 3D radiation pattern in end-fire mode. (a) Surface current when diode D_2 is ON, (b) surface current when diode D_3 is ON, (c) 3D radiation pattern when diode D_2 is ON (+*x*), and (d) 3D radiation pattern when diode D_3 is ON (−*x*).

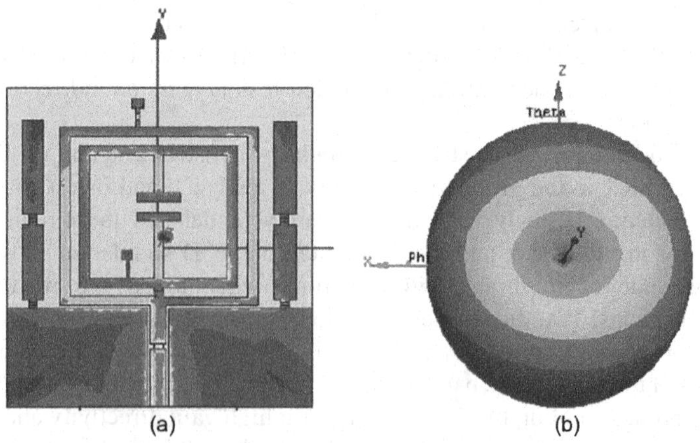

FIGURE 9.7 Simulated surface current distribution and 3D radiation pattern in omnidirectional mode. (a) Surface current and (b) 3D radiation pattern.

9.2.1.2.1 Theoretical Modeling of L-Strip-Loaded CMSA

The CSRR-loaded L-strip CMSA is shown in Figure 9.8. It consists of a CMSA on thick substrate fed with L-shaped microstrip feed. Two annular slots are cut in the ground plane, and PIN diodes are connected at diagonally opposite faces. The antenna is analyzed using the cavity model with the circuit theoretic approach. The whole analysis is divided into four parts. In the first, the L-strip-loaded CMSA has been analyzed; the second part includes characterization of CSRR; the third part takes care of integration of the two structures; and the last subsection describes the behavior of a tunable antenna. The proposed antenna fed with L-strip is analyzed using the circuit theoretic approach and the cavity model. Figure 9.9 shows equivalent circuit of an L-strip-fed circular microstrip antenna. To insure capacitive nature of feed, the length of horizontal part of L-strip under patch is kept less than quarter wavelength. The capacitance thus introduced is suppressed by the inductance arising from vertical part

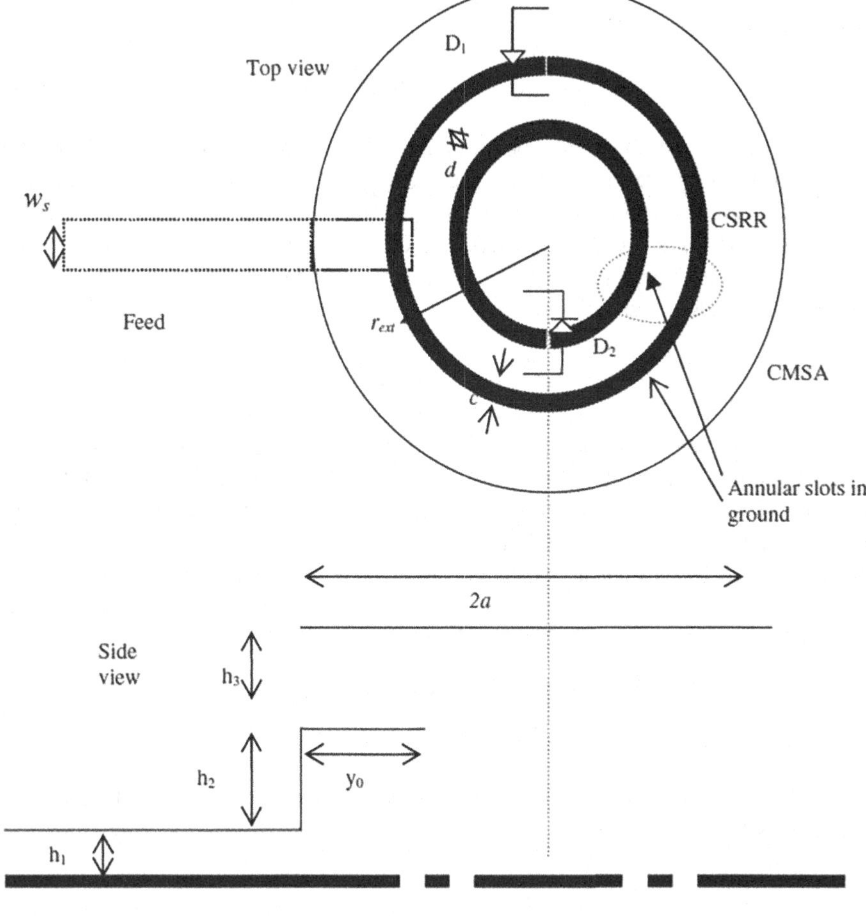

FIGURE 9.8 Schematic of the proposed antenna.

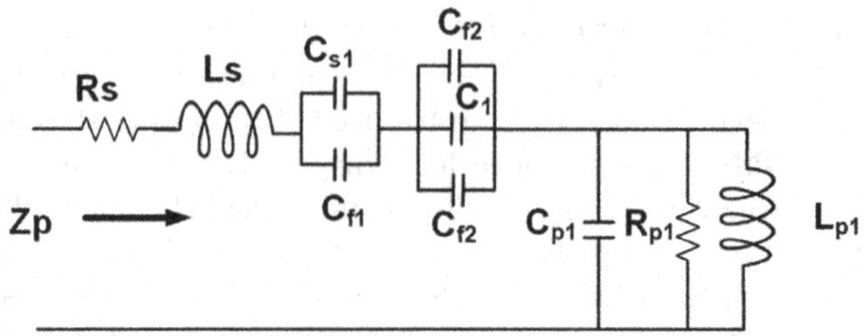

FIGURE 9.9 Equivalent circuit of an L-strip proximity-coupled CMSA.

of L-strip. Apart from these, a series resistance arises due to finite conductivity of copper used. The calculation of series resistance (R_s) and inductance (L_s) is given as [25]:

$$L_s = 0.2h_2\left[\ln\left\{2h_2/(w_s+t_s)\right\}+0.2235\left\{(w_s+t_s)/h_2\right\}+0.5\right](nH) \tag{9.1}$$

$$R_s = 4.13h_2(w_s+t_s)\sqrt{f.\rho/\rho_0} \tag{9.2}$$

where w_s is the width, t_s is the thickness, and h_2 is the height of L-strip in mm; f is the operating frequency in GHz; ρ is the specific resistance of the strip (Ω cm); and ρ_0 is the specific resistance of copper. The antenna metallization is taken as perfect except the vertical part of L-strip. Capacitance (C_{s1}) is arising due to vertical electric fields between the horizontal part of L-strip and ground plane in series with above L_s and R_s and is calculated as:

$$C_{s1s1} = \frac{\varepsilon_r\varepsilon_0 w_s y_0}{h_2+h_2} \tag{9.3}$$

where y_0 is the length of the horizontal part of L-strip and ε_r is the relative dielectric constant. There is a fringing capacitance (C_{f1}) between the open end of L-strip and ground plane, (C_{f2}) between the open end of L-strip and patch, and between the radiating edge of patch and the horizontal part of L-strip (C_{f2}). These capacitances are calculated by evaluating extended effective length of L-strip due to fringing field. The expression of extension in the length of an open-ended microstrip line is given as:

$$l_e = \frac{0.412\,h(\varepsilon_e+0.3)(w_s/h+0.264)}{(\varepsilon_e-0.258)(w_s/h+0.8)} \tag{9.4}$$

where ε_e is the effective dielectric constant of the material buried under the microstrip line and ground plane. Thus, the associated fringing capacitance is calculated as:

$$C_f = \frac{l_e\sqrt{\varepsilon_{\text{reff}}}}{cZ_0} \tag{9.5}$$

where l_e is the extension in effective length of L-strip feed, c is the velocity of light in vacuum, Z_0 is the characteristic impedance of feed, and ε_{reff} is the effective dielectric constant of the material. The fringing capacitance between the horizontal part of L-strip and ground plane (C_{f1}) is calculated by putting $h = h_1 + h_2$, and the two capacitances between patch and the horizontal part of L-strip (both C_{f2}) are calculated by putting $h = h_3$. The capacitance due to the vertical electric field between the horizontal part of L-strip and patch is calculated as:

$$C_1 = \varepsilon_r \varepsilon_0 y_0 w_s / h_3 \tag{9.6}$$

The equivalent circuit of L-strip-fed circular microstrip antenna is shown in Figure 9.2. The structure contains a series RLC resonant circuit of L-strip feed in series with a parallel RLC resonant circuit of the antenna. The resonance resistance of patch at a given feed location ρ_0 is calculated as:

$$R_p = \frac{1}{G_T} \frac{J_0^2(k\rho_0)}{J_0^2(kr)} \tag{9.7}$$

The antenna capacitance C_p and inductance L_p are calculated as [25]:

$$C_p = Q_T / \{2\pi f_{\text{res}} R_p\} \tag{9.8}$$

and

$$L_p = R_p / \{2\pi f_{\text{res}} Q_T\} \tag{9.9}$$

where Q_T is the total quality factor, G_T is the total conductance of patch of radius r incorporating radiation loss, conduction loss, and dielectric loss [26], and f_{res} is the resonant frequency [27].

9.2.1.2.2 Analysis of CSRR

The basic particle used in miniaturization is split ring, which can be electrically or magnetically driven. It is obvious that if a $\lambda/2$ transmission line is made in a closed form like ring, it has a diameter of $\lambda/2\pi$ at the given frequency. Hence, size reduction is observed. In order to reduce the size of element further, another ring may be added as edge coupled. This will add edge capacitance into the structure, thus further reducing the size [28]. The CSRR is the dual of split ring resonator (SRR) that gives very similar characteristics. There is a slight difference in the analysis process of the two structures. CSRR is taken as CPW, while SRR is taken as CPS. But under ideal conditions, it can be demonstrated that the resonant frequency of both structures is the same. The capacitance of CSRR is the capacitance of a disk of radius $r_0 - c/2$ surrounded by a ground plane at a distance of c from its edge, with average radius r_0.

Figure 9.10a shows the basic CSRR structure, and Figure 9.3b shows its equivalent circuit. The capacitance C_c is given by [28]:

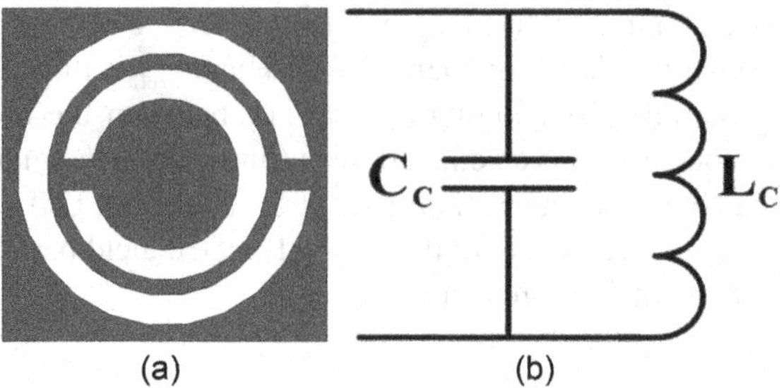

FIGURE 9.10 (a) Schematic of CSRR and (b) its equivalent.

$$C_c = \frac{\pi^3 \varepsilon_0}{c^2} \int_0^{\infty} \frac{\left[bB(kb) - aB(ka)\right]^2}{k^2} \left[\frac{1}{2} \left(1 + \frac{\varepsilon_r + \varepsilon_r^2 \tanh(kh)}{\varepsilon_r + \tanh(kh)} \right) \right] dk \tag{9.10}$$

The notations and functions used above are given in Ref. [27]. The inductance of CSRR, L_c, is calculated from the inductance corresponding to a CPW structure of length $2\pi r_0$, strip width d, and slot width c [28].

9.2.1.2.3 CSRR-Loaded L-Strip-Fed CMSA

A CSRR is etched in the ground plane of L-strip-fed CMSA. The two resonators formed so share the same field distribution, or in other words, the same L-strip is used to feed power to both resonators [24]. The equivalent circuit of CSRR comes in parallel with the resonant circuit of patch which in turn comes in series with L-strip equivalent. Simple L-strip-fed antenna is expected to resonate at two frequencies – first due to patch itself (lower) and second due to L-strip (higher). To simplify the numerical study, only TM_{11} mode is excited. By changing the y_0, higher order mode (TM_{21}) may be excited. This will complicate the problem and may divert the attention from the main work.

9.2.1.2.4 Implementation of Reconfigurable L-Strip-Fed CMSA

To see the bandwidth reconfigurability and radiation pattern diversity, a reconfigurable CSRR is implemented with the help of PIN diode and annular slot. For that, instead of making CSRR, two annular slots are made in the ground plane. One diode is connected at one end of the outer annular slot and another diode at the opposite end of the inner annular slot. This makes bandwidth of antenna tunable and radiation pattern diversified. When one of the diodes is "ON," it makes a combination of annular slot and split annular slot; when both diodes are "OFF," it becomes two annular slots; and when both diodes are "ON," it makes CSRR. In this way, three types of antennas are implemented in one.

9.2.1.2.5 Design Parameters

The antenna is made using foam material as it provides a very low value of permittivity (near to air). The three-layered structure is fed with a microstrip feed of 5 mm width, which is bent to make L-strip feed. The height of feed is 7.8 mm, and the length of the horizontal part of L-strip is 12 mm. The radius of patch is 23 mm, below which CSRR is etched. The separation between slots and the width of slots both are 0.2 mm. A PIN diode (model HSMP-3860) is taken for modeling with "ON" voltage and current of 0.73 V and 12 mA, respectively. In the simulation, the diode is implemented through metallic connection.

9.2.1.2.6 Results and Discussion

MATLAB codes were developed to analyze the behavior of the antenna. The antenna was simulated using full-wave electromagnetic simulator IE3D. Naturally, the L-strip-fed antenna provides a wide bandwidth when the resonances corresponding to antenna and L-strip feed are close to each other [30] and merged with higher order modes. But in this design, only TM_{11} mode is excited. First of all, the behavior of a simple L-strip-fed CMSA is designed and analyzed theoretically.

Figure 9.11 shows the return loss of antenna without CSRR for different horizontal length of L-strip. The graph gives a very clear idea about the resonance frequencies due to L-strip and basic patch. As patch antenna feed length increases, the associated capacitance increases, causing the reduction in the upper resonant frequency. It is also interesting to see that the strip length may be varied to tune the antenna for various applications. The behavior of antenna return loss v/s frequency loaded with CSRR is shown in Figure 9.12. It is very similar to an unloaded antenna except the shift in resonant frequency. In an unloaded antenna, the theoretically calculated upper and lower resonances are at 3.83 GHz and 1.83 GHz, which have been shifted at 3.32 and 1.77 GHz, respectively, after loading CSRR (all theoretical values). That shows that a significant reduction in size is obtained.

Now, the antenna was made tunable by putting two PIN diodes (D_1 and D_2) at the opposite sides of the two annular slots made in the ground. Three combinations are possible for this setup: first, when only diode D_1 is ON (making one of them the

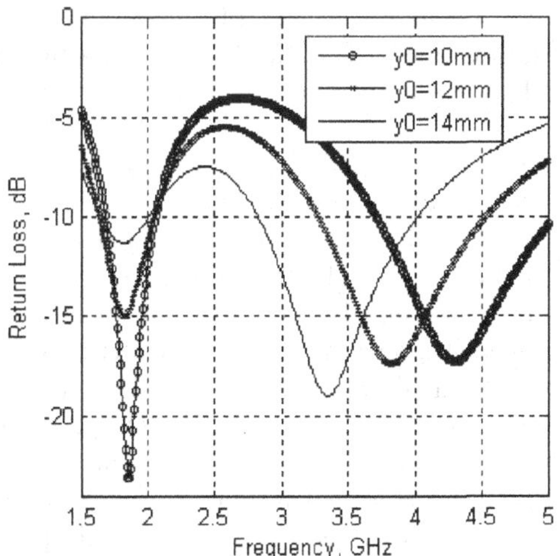

FIGURE 9.11 Variation of return loss with frequency without CSRR.

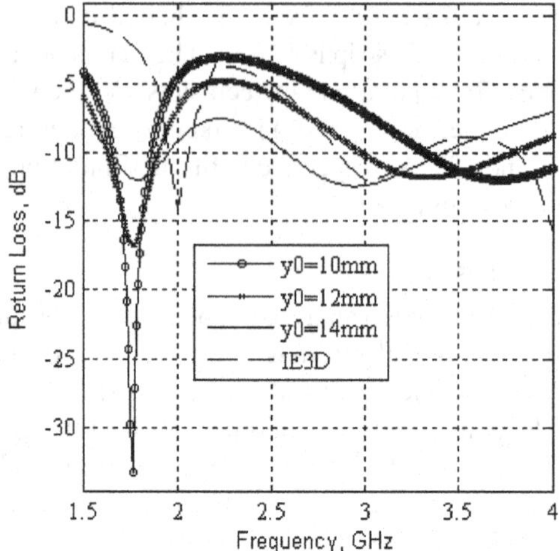

FIGURE 9.12 Variation of return loss versus frequency with CSRR.

annular slot and the other the split slot); second, when both D1 and D2 are ON (making it CSRR); and third, when both are OFF (making two annular slots in the ground plane). All the three cases have been analyzed, and return loss variations with frequency are shown in Figure 9.13 from IE3D simulation software along with a simple antenna without any ring. The figure shows how the antenna may be made tunable for different diode conditions. The basic antenna operates in two frequency bands: 2.42–2.8 and 4.13 GHz onward. The antenna with only one of the diodes ON shows the lower resonance at almost the same frequency, but upper resonance is shifted from 4.5 to 3.5 GHz. This is due to one split ring resonator formed [26]. As soon as we ON both diodes, making it CSRR-loaded, the whole band is shifted toward the lower frequency side. It is due to the extra edge capacitance introduced between the two split rings. When both diodes are OFF, an extra band is introduced around 4.4 GHz due to higher order mode. The variation of gain with respect to an isotropic antenna is shown in Figure 9.14. It is clear from the figure that gain is higher than reported earlier [20].

Figure 9.15 shows the variation of directivity with frequency for different antenna structures. A complete picture of the antenna gain and directivity is shown in tabular form in Table 9.2. From the table, it may be concluded that the antenna has very good gain and directivity in their different bands of operation.

The radiation pattern is a very important parameter in antenna design. In the present design, the antenna has a variable radiation pattern. The antenna without any ring has a radiation pattern as shown in Figure 9.16. The pattern is for two frequencies in two bands of operation, i.e., 2.5 and 4.4 GHz. The pattern is shifted clockwise by 30.8°. This again shows resemblance with the TM_{21} mode [31]. Figure 9.17 shows the radiation pattern for a CSRR-loaded antenna, i.e., when both diodes are ON. The pattern is shown for all the three bands. The back lobe of the pattern increases as the electrical length increases. But the graph at 3 GHz has no back lobe. This is due to the high gain of 8.6 dBi. Figure 9.18 shows a radiation pattern when one of the diodes

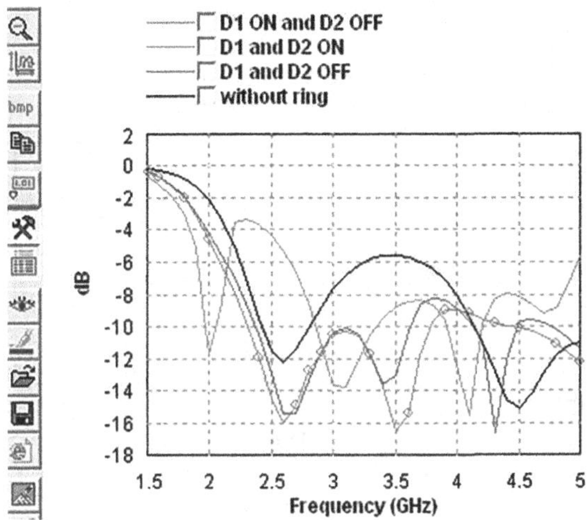

FIGURE 9.13 Return loss versus frequency of PIN diode-tuned antenna.

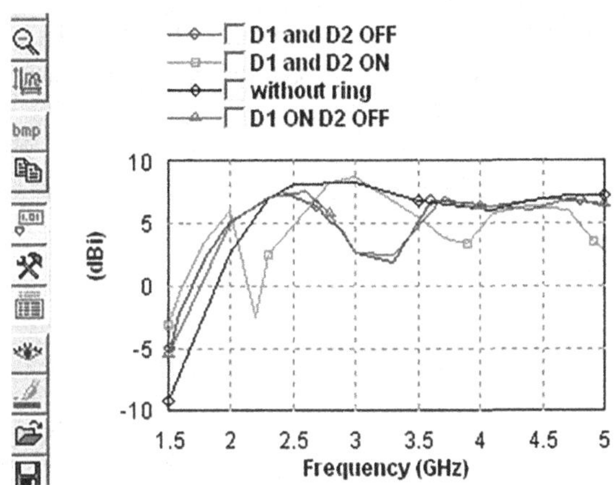

FIGURE 9.14 Variation of gain for different antennas.

is ON. The pattern shows its omnidirectional nature along with a back lobe. That is why the antenna has relatively low gain and directivity. The antenna with two rings in the ground plane has a radiation pattern as shown in Figure 9.19. The higher order mode is excited at a higher frequency, while at a lower frequency, the back lobe is more. This is due the radiation from the slots at lower frequencies.

9.2.1.3 Varactor Diode-Based Reconfigurable Antennas

Varactors have variable reactance (capacitance), so they can be used as tuning elements in the reconfigurable antennas [29]. In Ref. [30], the varactor diode was placed at the corners of the slot, which resulted in dual-band operation of the antenna, with independent tuning at both the bands. A dual-band frequency tunable square

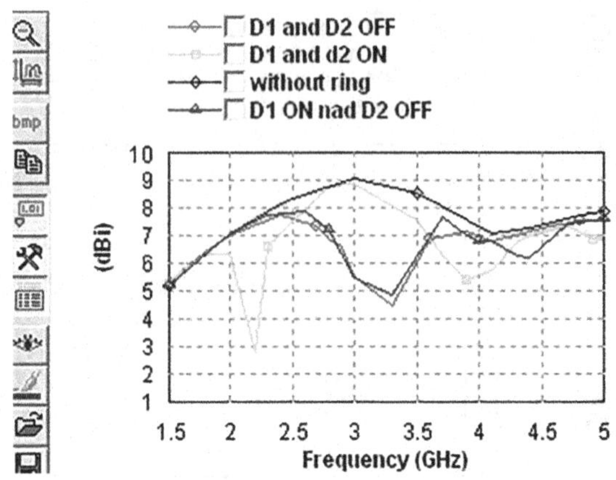

FIGURE 9.15 Variation of directivity for different antennas.

TABLE 9.2
Gain and Directivity for Different Antenna Structures

Frequency	Without Ring		D_1 and D_2 OFF		D_1 ON D_2 OFF		D_1 and D_2 ON	
Band	Gain	Directivity	Gain	Directivity	Gain	Directivity	Gain	Directivity
First band	8	8.35	7.1	7.6	4.7	5.9	8	8.35
Second band	>6	>7	>6	>7	6.2	6.4	>6	>7
Third band	NA	NA	NA	NA	>6	>7	NA	NA

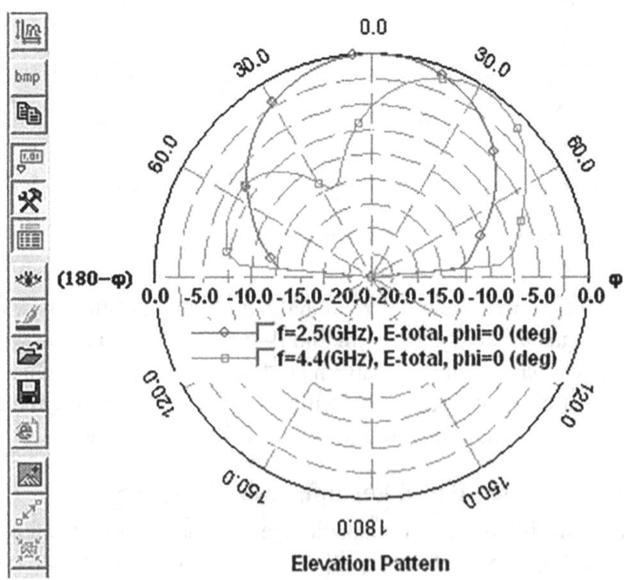

FIGURE 9.16 E-plane radiation pattern of a basic antenna without rings.

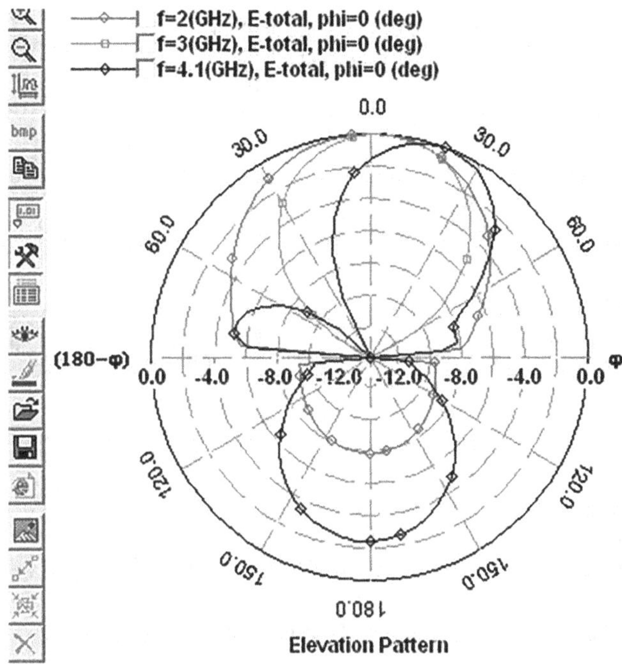

FIGURE 9.17 E-plane radiation pattern of a CSRR-loaded antenna.

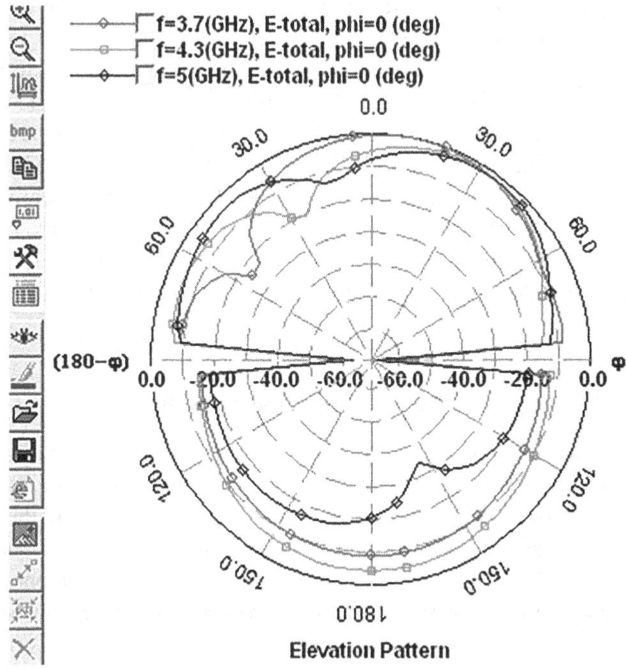

FIGURE 9.18 E-plane radiation pattern with D_1 ON and D_2 OFF.

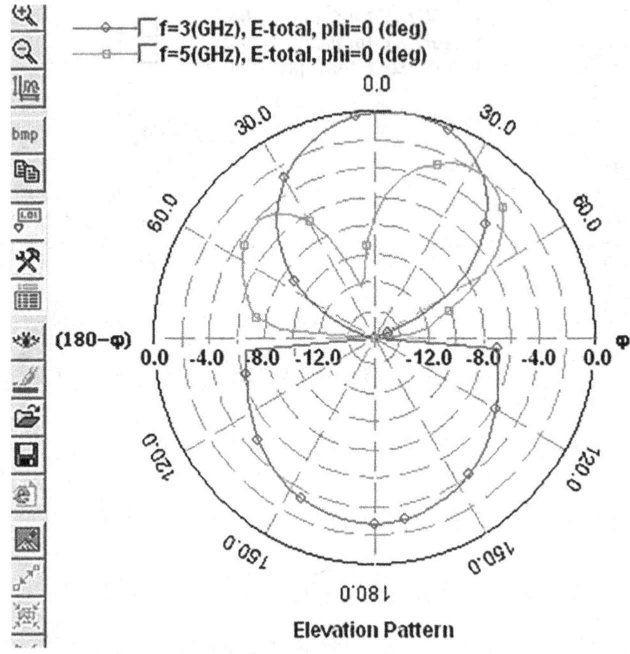

FIGURE 9.19 E-plane radiation pattern with D_1 and D_2 OFF.

patch was proposed in Ref. [31]. Varactor diodes were placed at the extended arms of the hexagonal slot, etched on the main radiator. Both bands were independently controlled by varying the varactor bias supply. The radiation characteristics of the reported antenna were constant for the entire tuning range (80.11% and 65.69%).

The authors in Ref. [32] demonstrated the optimal use of the varactor diode to find the maximum efficiency and tuning range for a frequency-agile antenna. A generalized model was presented for varactor tuning. Two prototypes were studied and tested to show the importance of tuning devices and their positions to obtain a high efficiency and wide tuning range. One- and three-varactor-diode models were investigated to obtain a wide capacitance range. The measured and simulated results were in close resemblance.

9.2.2 Optical Methods/Photoconductive Switches

Reconfigurable antennas based on the optical switching mechanism use photoconductive switches made from semiconductor materials (Si, GaAs). The states (ON and OFF) of the photoconductive/optical switches change by controlling the intensity of laser light. The integration of laser and power supply raises some issues for using optical switches in the antenna reconfiguration. A reconfigurable monopole embedded with three optoelectronic switches among four segments was investigated [33], where optical fiber was used to illuminate the switches. The effective length of the monopole depends upon the state of the switch, which results in the tuning of frequency. The proposed design with optoelectronic switches eliminates the effect of the biasing line. Panagamuwa et al. [34] reported an optically controlled compound (pattern and frequency) reconfigurable antenna. In this design, a balanced dipole was constructed on a silicon substrate. Two silicon photodiodes were embedded in dipole arms to reconfigure dipole attributes. The diode states were controlled by infrared laser conveyed through fiber-optic wires. Frequency reconfigurability was achieved when both the diodes were either ON or OFF. When only one diode was ON, then there was a shift in pattern null toward the longer side of the dipole by 50°. In the pattern shift state, there is a small variation in the resonant frequency.

An infrared LED-controlled reflectarray system was reported for Ka-band [35], where a reconfigurable miniaturized unit cell fabricated by silver inkjet printing on a silicon wafer was designed. When the wafer gets illuminated by various levels of intensities (LED), it results in the tapering of radiation pattern or variation in the gain of the reflectarray. A photoconductive switch integrated frequency and pattern-reconfigurable antenna array for mm-wave applications was reported [36]. Two photoconductive switches were installed at the slot for frequency tunability and beam steering. A planar pattern-reconfigurable antenna controlled by RF PIN photodiode and phototransistor was suggested in Ref. [37]. Optical switches were installed between the microstrip elements to control the main beam direction. The designed antenna has one omnidirectional and four directional radiation patterns. The isolation and insertion loss studies were carried out for the two switching systems, and the test results confirm the simulation results. A microstrip-fed frequency-agile patch antenna was presented in Ref. [38], where the resonant frequency was switched by changing the state of photoconductive switches through optical power. The proposed design operates with low photoconductivity and needs low optical power. Photoconductive switches controlled the impedance of branched stubs connected to the patch antenna optically.

9.2.3 Physically/Mechanically Reconfigurable Printed Antennas

The antennas based on electrical/electronic lumped elements need complex DC biasing circuitry and proper isolation between the DC supply and RF signal. There is some extra power loss in switches and biasing, and isolation circuitry affects the antenna performance. In view of these drawbacks, the antenna researchers recently investigated the physically/mechanically reconfigurable antennas [39–41]. These types of antennas have less RF power loss, better isolation, and more linearity in comparison with the antenna topologies based on RF switches. However, they possess a larger area and a complex mechanical system. A physically rotatable compound (frequency and pattern) reconfigurable antenna having two microstrip patches connected to a single feed line was studied [39]. A mechanically rotatable frequency-tuned antenna for cognitive radios (2–10 GHz) was reported in Ref. [40]. It is possible to accomplish pattern diversity by a single radiating patch using mechanical movements. As this approach is slow, the antennas based on this method are more attractive, where fast (pattern) reconfigurability is not needed. This technique is less expensive as compared to the expensive phased array-based technique. Zhu et al. [41] reported mechanically configured frequency agility in a patch antenna, without using any switching components. The proposed design consisted of a rotatable metasurface placed over a circular patch. The rectangular-loop-type unit cells were etched on the top of the metasurface substrate facing opposite the patch. As the metasurface rotated mechanically in steps 10°, 25°, 35°, 55°, and 80°, the patch resonant frequency changed to 4.77, 4.9, 5.07, 5.31, and 5.5 GHz, respectively. There is no need for extra DC biasing circuits, but they need mechanically movable parts that limit their applications.

9.2.4 Material-Based Reconfigurable Antennas

Antenna attributes can be reconfigured by varying the structure of the antenna without using any electrical or optical switches, and this technique is known as the material-based reconfiguration technique. In material-based techniques, the antenna attributes are reconfigured by changing substrate material characteristics, and ferrite or liquid crystals are used as material. If the antenna substrate material permittivity or permeability is altered, then the antenna characteristics get changed. This change in material attributes happened by modifying the orientation of the molecules through an external voltage. This change is non-linear for a liquid crystal for different applied voltage levels, while a static magnetic or electric field changes the material permeability/permittivity of a ferrite material. The use of a tunable substrate material is the latest trend to achieve reconfigurability. The tunable material-based antennas face challenges such as reliability and lack of proper modeling. In recent years, people reported many antennas with this approach [42,43]. The ferroelectric material used as an antenna substrate in Ref. [42] was tuned by applying DC bias voltage across the material. They need a large power supply to change the dielectric constant; however, they possess better directivity and narrow beamwidth.

Liu et al. reported [43] a liquid crystal antenna with a tuning range of 4%–8% at the 5-GHz frequency band. The antenna demonstrated a low efficiency because of high losses. The latest trends in material-based (use of microfluidics, graphene, and optical controls) reconfigurable antennas were reported in Refs. [44,45]. Kelley et al. [44] demonstrated a frequency-agile antenna inspired by liquid metal as the switching method. The designed edge feed frequency-reconfigurable antenna was able to switch between GPS band (1.6 GHz) and ISM band (2.4 GHz). The dielectric fluid was displaced on top of the main radiator by pressure, which makes the antenna to switch over from one state to another state. A eutectic gallium–indium (EGaIn)-based fluidic mechanism was utilized as a switch in the proposed design. A chamfered corner polarization-agile printed antenna based on metal switching was studied [45]. A pressure-actuated liquid was used as a switching mechanism, and the design was able to shift between the circular and linear polarization. The antenna design was simple, as it does not require biasing circuits. A summary of comparison between different reconfiguration techniques is given in Table 9.3.

9.3 APPLICATIONS

9.3.1 Frequency-Reconfigurable Antennas for Cognitive Radio System

Designing an antenna for cognitive radio (CR) systems has been an intriguing issue for scientists in the course of time. CR systems prove to be important for optimized

TABLE 9.3
Advantages and Disadvantages of Different Reconfiguration Techniques

Reconfiguration Technique	Advantages	Disadvantages
Electrical reconfiguration	• Ease of implementation • Low-cost	• Complex structure • Requirement of biasing systems
Optical reconfiguration	• No need to use bias lines • No intermodulation distortion	• Lossy behavior • Complex activation mechanism
Mechanical reconfiguration	• No need of active elements • No need of biasing systems	• Slow response • Requirement of power source
Smart-material-based reconfiguration	• Low profile • Lighter weight	• Low efficiency • Limited application

usage of the available frequency spectrum. Normally, a CR antenna system contains a frequency-reconfigurable ultra-wideband (UWB) antenna for sensing. A CR-enabled system can switch among different unused bands to improve spectrum utilization efficiency. A stepper motor-controlled frequency-reconfigurable antenna for CR applications was demonstrated in Ref. [40]. The reconfigurable design was composed of a UWB monopole antenna, a circular substrate with five different-shaped parasitic patches, a stepper motor, and a sensing section. The antenna structure was planar as both reconfigurable and sensing antennas were incorporated on the same substrate. A stepper motor rotated the reconfigurable antenna, and the experimental results confirm the simulation data.

A band-reconfigurable (wideband-to-narrowband) antenna for CR systems was reported in Ref. [46]. The reported design was composed of a microstrip patch, a wideband monopole, and a rectangular slot engraved on the ground surface. One PIN diode was implanted between the wideband monopole and strip line, while two PIN diodes were engraved between the slot and the small patch. When all the switches were in the ON state, the antenna operated for a wideband (1.85–7.18 GHz), and when all the switches were in the OFF state, it operated for a narrowband (5.29 GHz). A single-substrate, frequency-reconfigurable multi-band MIMO antenna for CR applications was studied [47]. The fabricated antenna sensing element was integrated with a two-element, multiple-input multiple-output (MIMO) antenna. The MIMO antenna was equipped with varactor and PIN diodes for frequency agility. The design was compact, and it used a single substrate for etching the reconfigurable antenna and the sensing antenna. The MIMO reconfigurable antenna covered a 0.72–2.55 GHz bandwidth, while the sensing antenna covered a 0.72–3.44 GHz bandwidth.

9.3.2 Pattern-Reconfigurable Antennas for the MIMO Systems

Kotwalla et al. [48] demonstrated a planar UWB frequency-reconfigurable MIMO antenna array. Four different-sized rectangular stubs along with PIN diodes were used to feed microstrip, which introduces narrowband features. The original antenna functions for 3.1–10.6 GHz, while for narrowband operation, the resonating frequencies were 4.5 and 5.93 GHz, which makes the antenna suitable for high-speed WLAN applications. PIN diodes were installed between the feed lines, and four stubs were responsible for frequency reconfiguration between the UWB and narrowband. The proposed antenna showed good isolation (15 dB) and a reasonable value of ECC. Thao et al. [49] proposed a planar and miniaturized frequency-reconfigurable antenna for MIMO systems. Two PIN diodes were installed at the ground plane between two symmetric frequency-reconfigurable elements. A slot was engraved in the ground plane to reduce mutual coupling among the elements of the antenna. The designed antenna covered 1.9, 2.3, and 2.6 GHz bands of UMTS, with their respective peak gains of 1.55, 1.81, and 0.87 dBi. Figure 9.20 depicts a schematic of the pattern-reconfigurable antenna configuration.

9.3.3 Reconfigurable Antennas for Satellite Systems

A frequency-agile ink-printed graphene material-based patch antenna was proposed for satellite communication [50]. The antenna showed switching of modes (TM_{10}, TM_{20}, and TM_{02}) by controlling the excitation of the feed location through the PIN diode. A multilayer substrate was used to overcome low graphene conductivity and to boost the antenna radiation performance. A miniaturized multiport CPW-fed beam steerable antenna for satellite K-band was demonstrated in Ref. [51]. The proposed antenna has a rectangular radiator of $5\times4\,\text{mm}^2$ fed by four independent ports. A 90° beam steering was obtained by the excitation of each individual port while wide bandwidth was achieved due to the CPW feed. This antenna was suitable for different satellite applications such as ground exploration and navigation that falls in the K-band.

A high-gain, low-profile, low-cost, PIN diode-embedded printed loop antenna capable of altering polarization was reported in Ref. [52]. The antenna has a dual loop over which four PIN diodes were placed symmetrically to switch polarization between RHCP, LHCP, and linear polarization. The polarization state depends upon

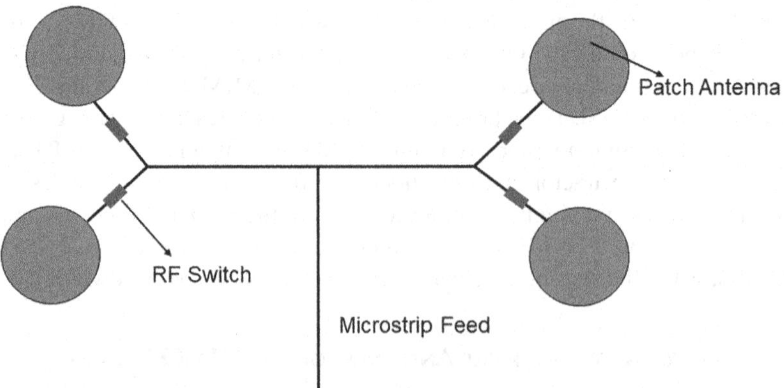

FIGURE 9.20 Pattern-reconfigurable antenna configuration.

the operating state of PIN diodes placed over the dual loop. The antenna finds applications in terrestrial and satellite communication systems.

9.4 MULTI-RECONFIGURABLE ANTENNAS: THE LATEST TREND

The multi-reconfigurable antennas can have two or more features modified independently: frequency and bandwidth; frequency and polarization; frequency and radiation patterns; polarization and radiation patterns; and frequency, polarization, and radiation patterns. There is an emerging demand for multi-service radios, which function for different spectrum standards and make the antenna operational for a single band, multi-band, or wideband. The multi-functional characteristic turns out a significant research field [48,53–60]. Recently, a number of antenna design techniques have achieved this multimode functionality:

- The switching between narrowband and wideband is achieved using a reconfigurable band-pass filter on the feed line of the antenna [53–56].
- The modified ground plane was implanted with active elements, slots, slit structures, and parasitic elements [48,57–62].
- Different antennas with narrowband/wideband operations were used for CR or MIMO communications [60].

The antenna radiation patterns or resonating frequency were reconfigured in Ref. [61]. The antenna radiation patterns were reconfigured among broadside, end-fire, and omnidirectional modes in the paper [62], where the pattern reconfiguration enhanced the system performance, suppressed noise, and saved energy by refining the signal path. To decrease the required number of antennas and to encounter the interference caused by other wireless devices/systems, frequency tuning proves to be very effective [63]. Because of this, several antenna configurations of this type were reported [64–68]. The customary approach in most designs was to change the antenna operating frequency by means of active elements, specifically varactor or PIN diodes [64–66].

The pattern reconfiguration can be realized with a modified ground surface (with slit/slot arrangements), where the current movement is varied to steer the direction of the radiation pattern [67]. It can also utilize switchable directors across the main radiator to obtain the desired patterns [68]. Such antennas are advantageous in imaging, radar, sensing, and tracking applications [69]. With frequency reconfigurability, the antenna offers tunability and utilizes the available spectrum efficiently. Polarization diversity can decrease the multipath fading and improve channel capacity. Of late, such types of reconfiguration arrangements were used frequently, and many designs have been reported [70]. Furthermore, an active electromagnetic band gap (EBG) or metasurface with switches mounted on the antenna radiator was reported [69–71]. However, the use of active elements for realizing frequency and polarization reconfigurability is the most common method reported so far [72,73].

The reconfigurable antennas offer pattern diversity and support different polarizations through a single resonating element [74]. They also can increase the capability of communication systems, avoid polarization mismatch, enhance signal power, and improve radiation coverage. The radiation pattern reconfigurability is capable to steer the radiation patterns of the antenna to fulfill the system needs [75]. It can increase channel capacity without increasing the volume of the radiator [76]. In the literature, one design was proposed [77] for three-parameter reconfiguration. In Ref. [78], an antenna design was suggested for frequency, pattern, and polarization reconfigurability, but with tuning limitations. Unlike the design proposed in Ref. [77], it is not capable of delivering simultaneous and independent reconfiguration, which is a major research gap. However, it is more difficult to design an antenna possessing flexibility in terms of polarization, pattern, and frequency. A comparative study on various reconfigurable antennas is presented in Table 9.4.

9.5 CONCLUSION

Reconfigurability allows the operation of a single element in multiple frequencies. This chapter covers various reconfigurable printed antennas designed using control circuitry, stacking, slits/slots, switches, and/or tunable materials to give additional degrees of operational freedom. The numerous advantages of reconfigurable antennas are briefed here:

- Capability to support multiple wireless standards, which minimizes cost, reduces volume requirements, simplifies integration, and offers high isolation among various wireless standards.
- Lower front-end processing offers good out-of-band rejection.
- The best contender for software-defined radio (SDR) is due to their ability to learn/adapt, and they can be easily automated by means of a field-programmable gate array (FPGA) or a microcontroller.
- Multi-functional capabilities facilitate functionality adjustment, as the mission changes, and they can act as a single element or like an array and provide narrowband/wideband operation.

The reconfigurable antenna signifies a potential applicant for future RF front ends for applications related to wireless and space. However, in both the systems, the manufacturing cost will increase by introducing tunability to the antenna. The increased cost can be linked to various parameters as follows:

- Design of the biasing network, which is used to activate/deactivate switching devices and add complexity to the configuration of the antenna.
- Consumption of power, which increases due to the integration of active modules, thus increasing the overall cost of the system.
- The generation of harmonics and intermodulation products.
- The need for fast tuning (antenna radiation features) to guarantee the correct working of the system.
- A numerical analysis of an L-strip-loaded tunable microstrip antenna. The antenna is made tunable with help of PIN diodes. The effect of CSRR and split ring resonators on various antenna parameters has been studied. The

TABLE 9.4
Comparative Study on Various Reconfigurable Antennas

Reference	Antenna Size (mm^2)	Operating Frequency (GHz)	Antenna Type	Reconfiguration Type	Components (No. of Switches)
[64]	41.1×49.5	2.15–2.38	Patch antenna	Radiation pattern reconfigurable	Choke inductor 2 capacitors 2 varactors
[65]	40×30	3.4–3.8 and 3.7–4.2	Slot antenna	Frequency and radiation pattern reconfigurable	4 diodes
[66]	130×160	1.82, 1.93, and 2.10 1.79, 1.89, and 2.07	Slot Antenna	Frequency and pattern reconfigurable	15 PIN diodes 32 capacitors
[67]	37×37 (circular disk)	2.29–2.50, 2.48–2.69, and 2.77–2.91	Microstrip antenna – mechanical	Frequency and radiation, pattern reconfigurable	PIN diode and 12 parasitic elements
[68]	88×114	1–1.6 GHz	Microstrip antenna	Frequency and polarization reconfigurable	32 varactor diodes
[70]	27×27	4.6 and 4.25	Dipole antenna	Polarization reconfigurable	4 PIN diodes
[71]	80×80	8–11.2	Slot antenna	Frequency and polarization reconfigurable	Metasurface
[72]	41.5×42	2.4–3.6	Microstrip antenna	Frequency and polarization reconfigurable	12 varactors
[73]	120×100	2.02–2.56, 2.32–2.95, 1.92–2.70	Slot antenna	Frequency and polarization reconfigurable	2 PIN diodes
[74]	41.5×42	2.56	Microstrip patch antenna	Polarization reconfigurable	PIN diode Switch
[75]	150×150	2.4	Slot antenna	Polarization reconfigurable	2 PIN diodes 3 inductors 1 capacitor 1 SMA connector
[76]	44×44 (circular disk)	2.45–2.65	Patch antenna	Polarization reconfigurable	2 PIN diodes 1 resistor
[77]	240×120	2.5	Microstrip patch antenna	Frequency and polarization reconfigurable	2 LED indicators 1 RF choke
[78]	50×60	5.2/5.8	Patch antenna	Frequency and polarization reconfigurable	3 pairs of PIN diodes

antenna shows significant amount of miniaturization with higher gain and directivity. Finally, it may be concluded that L-strip-fed antenna provided a better result.

REFERENCES

1. Bhartia, P., and I. Bahl. "A frequency agile microstrip antenna." In *1982 Antennas and Propagation Society International Symposium*, vol. 20, pp. 304–307. IEEE, 1982.
2. Haupt, R. L., and M. Lanagan. "Reconfigurable antennas." *IEEE Antennas and Propagation Magazine* 55, no. 1 (2013): 49–61.
3. Costantine, J., Y. Tawk, S. E. Barbin, and C. G. Christodoulou. "Reconfigurable antennas: Design and applications." *Proceedings of the IEEE* 103, no. 3 (2015): 424–437.
4. Weedon, W. H., W. J. Payne, and G. M. Rebeiz. "MEMS-switched reconfigurable antennas." In *IEEE Antennas and Propagation Society International Symposium. 2001 Digest. Held in conjunction with: USNC/URSI National Radio Science Meeting (Cat. No. 01CH37229)*, vol. 3, pp. 654–657. IEEE, 2001.
5. Huff, G. H., and J. T. Bernhard. "Reconfigurable antennas." In *Modern Antenna Handbook*. New York: Wiley online, 2008, pp. 369–398.
6. Yang, S., C. Zhang, H. K. Pan, A. E. Fathy, and V. K. Nair. "Frequency-reconfigurable antennas for multiradio wireless platforms." *IEEE Microwave Magazine* 10, no. 1 (2009): 66–83.
7. Lai, M.-I., T.-Y. Wu, J.-C. Hsieh, C.-H. Wang, and S.-K. Jeng. "Design of reconfigurable antennas based on an L-shaped slot and PIN diodes for compact wireless devices." *IET Microwaves, Antennas & Propagation* 3, no. 1 (2009): 47–54.
8. Sung, Y. "Investigation into the polarization of asymmetrical-feed triangular microstrip antennas and its application to reconfigurable antennas." *IEEE Transactions on Antennas and Propagation* 58, no. 4 (2009): 1039–1046.
9. Bai, Y.-Y., S. Xiao, C. Liu, X. Shuai, and B.-Z. Wang. "Design of pattern reconfigurable antennas based on a two-element dipole array model." *IEEE Transactions on Antennas and Propagation* 61, no. 9 (2013): 4867–4871.
10. Biyikli, N., Y. Damgaci, and B. A. Cetiner. "Low-voltage small-size double-arm MEMS actuator." *Electronics Letters* 45, no. 7 (2009): 354–356.
11. Topalli, K., E. Erdil, O. A. Civi, S. Demir, S. Koc, and T. Akin. "Tunable dual-frequency RF MEMS rectangular slot ring antenna." *Sensors and Actuators A: Physical* 156, no. 2 (2009): 373–380.
12. Huff, G. H., and J. T. Bernhard. "Integration of packaged RF MEMS switches with radiation pattern reconfigurable square spiral microstrip antennas." *IEEE Transactions on Antennas and Propagation* 54, no. 2 (2006): 464–469.
13. Nasiri, M., H. Mirzajani, E. Atashzaban, and H. B. Ghavifekr. "Design and simulation of a novel micromachined frequency reconfigurable microstrip patch antenna." *Wireless Personal Communications* 72, no. 1 (2013): 259–282.
14. Petit, L., L. Dussopt, and J.-M. Laheurte. "MEMS-switched parasitic-antenna array for radiation pattern diversity." *IEEE Transactions on Antennas and Propagation* 54, no. 9 (2006): 2624–2631.
15. Nasir, U., A. S. Afzal, B. Ijaz, K. S. Alimgeer, M. F. Shafique, and M. S. Khan. "A compact frequency reconfigurable CPS-like metamaterial-inspired antenna." *Microwave and Optical Technology Letters* 59, no. 3 (2017): 596–601.
16. Majid, H. A., M. K. A. Rahim, M. R. Hamid, and M. F. Ismail. "Frequency and pattern reconfigurable Yagi antenna." *Journal of Electromagnetic Waves and Applications* 26, no. 2–3 (2012): 379–389.

17. Sharma, S. K., F. Fideles, and A. Kalikond. "Planar Yagi-Uda antenna with reconfigurable radiation patterns." *Microwave and Optical Technology Letters* 55, no. 12 (2013): 2946–2952.
18. Han, L., C. Wang, X. Chen, and W. Zhang. "Compact frequency-reconfigurable slot antenna for wireless applications." *IEEE Antennas and Wireless Propagation Letters* 15 (2016): 1795–1798.
19. Khidre, A., K.-F. Lee, F. Yang, and A. Z. Elsherbeni. "Circular polarization reconfigurable wideband E-shaped patch antenna for wireless applications." *IEEE Transactions on Antennas and Propagation* 61, no. 2 (2012): 960–964.
20. Yoon, W.-S., J.-W. Baik, H.-S. Lee, S. Pyo, S.-M. Han, and Y.-S. Kim. "A reconfigurable circularly polarized microstrip antenna with a slotted ground plane." *IEEE Antennas and Wireless Propagation Letters* 9 (2010): 1161–1164.
21. Yang, Z.-X., H.-C. Yang, J.-S. Hong, and Y. Li. "Bandwidth enhancement of a polarization-reconfigurable patch antenna with stair-slots on the ground." *IEEE Antennas and Wireless Propagation Letters* 13 (2014): 579–582.
22. Zhang, G.-M., J.-S. Hong, G. Song, and B.-Z. Wang. "Design and analysis of a compact wideband pattern-reconfigurable antenna with alternate reflector and radiator." *IET Microwaves, Antennas & Propagation* 6, no. 15 (2012): 1629–1635.
23. Singh, G., B. K. Kanaujia, V. K. Pandey, D. Gangwar, and S. Kumar. "Pattern and frequency reconfigurable antenna with diode loaded ELC resonator." *International Journal of Microwave and Wireless Technologies* 12 (2020): 163–175.
24. Pandey, G. P., B. K. Kanaujia, S. K. Gupta, and A. K. Gautam. "CSRR loaded tunable L-strip fed circular microstrip antenna." *Wireless personal communications* 74, no. 2 (2014): 717–730. doi: 10.1007/s11277-013-1317-3, 2014.
25. Guha, D. "Resonant frequency of circular microstrip antennas with and without air gaps." *IEEE Transaction on Antenna and Propagation* 49 (2001): 55–59.
26. Marqués, R., F. Martin, and M. Sorolla. *Metamaterials with Negative Parameters: Theory, Design, and Microwave Applications*, vol. 183. New York: John Wiley & Sons, 2011.
27. Bahl, I., and P. Bhartia. *Microwave Solid State Circuit Design*. New York: John Wiley, 1988.
28. Garg, R., P. Bhartia, I. J. Bahl, and A. Ittipiboon. *Microstrip Antenna Design Handbook*. Norwood, OH: Artech House, 2001.
29. Sam, W. Y., and Z. Zakaria. "The investigation of the varactor diode as tuning element on reconfigurable antenna." In *IEEE 5th Asia-Pacific Conference on Antennas and Propagation (APCAP)*, pp. 13–14. IEEE, 2016.
30. Behdad, N., and K. Sarabandi. "A varactor-tuned dual-band slot antenna." *IEEE Transactions on Antennas and Propagation* 54, no. 2 (2006): 401–408.
31. Shynu, S. V., G. Augustin, C. K. Aanandan, P. Mohanan, and K. Vasudevan. "Design of compact reconfigurable dual frequency microstrip antennas using varactor diodes." *Progress in Electromagnetics Research* 60 (2006): 197–205.
32. Nguyen-Trong, N., and C. Fumeaux. "Tuning range and efficiency optimization of a frequency-reconfigurable patch antenna." *IEEE Antennas and Wireless Propagation Letters* 17, no. 1 (2017): 150–154.
33. Freeman, J. L., B. J. Lamberty, and G. S. Andrews. "Optoelectronically reconfigurable monopole antenna." *Electronics Letters* 28, no. 16 (1992): 1502–1503.
34. Panagamuwa, C. J., A. Chauraya, and J. C. Vardaxoglou. "Frequency and beam reconfigurable antenna using photoconducting switches." *IEEE Transactions on Antennas and Propagation* 54, no. 2 (2006): 449–454.
35. Alizadeh, P., C. G. Parini, and K. Z. Rajab. "Optically reconfigurable unit cell for Ka-band reflectarray antennas." *Electronics Letters* 53, no. 23 (2017): 1526–1528.
36. Da Costa, I. F., A. Cerqueira, D. H. Spadoti, L. G. da Silva, J. A. J. Ribeiro, and S. E. Barbin. "Optically controlled reconfigurable antenna array for mm-wave applications." *IEEE Antennas and Wireless Propagation Letters* 16 (2017): 2142–2145.

37. Patron, D., A. S. Daryoush, and K. R. Dandekar. "Optical control of reconfigurable antennas and application to a novel pattern-reconfigurable planar design." *Journal of Lightwave Technology* 32, no. 20 (2014): 3394–3402.
38. Pendharker, S., R. K. Shevgaonkar, and A. N. Chandorkar. "Optically controlled frequency-reconfigurable microstrip antenna with low photoconductivity." *IEEE Antennas and Wireless Propagation Letters* 13 (2014): 99–102.
39. Khan, M. S., A. Iftikhar, A. D. Capobianco, R. M. Shubair, and B. Ijaz. "Pattern and frequency reconfiguration of patch antenna using pin diodes." *Microwave and Optical Technology Letters* 59, no. 9 (2017): 2180–2185.
40. Tawk, Y., J. Costantine, K. Avery, and C. G. Christodoulou. "Implementation of a cognitive radio front-end using rotatable controlled reconfigurable antennas." *IEEE Transactions on Antennas and Propagation* 59, no. 5 (2011): 1773–1778.
41. Zhu, H. L., X. H. Liu, S. W. Cheung, and T. I. Yuk. "Frequency-reconfigurable antenna using metasurface." *IEEE Transactions on Antennas and Propagation* 62, no. 1 (2013): 80–85.
42. Long, S. A., and G. H. Huff. "A substrate integrated fluidic compensation mechanism for deformable antennas." In *2009 NASA/ESA Conference on Adaptive Hardware and Systems*, pp. 247–251. IEEE, *2009.*
43. Liu, L., and R. J. Langley. "Liquid crystal tunable microstrip patch antenna." *Electronics Letters* 44, no. 20 (2008): 1179–1180.
44. Kelley, M., C. Koo, H. McQuilken, B. Lawrence, S. Li, A. Han, and G. Huff. "Frequency reconfigurable patch antenna using liquid metal as switching mechanism." *Electronics Letters* 49, no. 22 (2013): 1370–1371.
45. Champion, M., D. Jackson, B. Cumby, and E. Belovich. "Polarization reconfigurable antennas using a liquid metal switching mechanism." In *2017 IEEE International Symposium on Antennas and Propagation & USNC/URSI National Radio Science Meeting*, pp. 415–416. IEEE, 2017.
46. Majid, H. A., M. K. A. Rahim, M. R. Hamid, and M. F. Ismail. "Reconfigurable wide to narrow band antenna for cognitive radio systems." In *2011 IEEE International RF & Microwave Conference*, pp. 285–288. IEEE, 2011.
47. Hussain, R., and M. S. Sharawi. "A cognitive radio reconfigurable MIMO and sensing antenna system." *IEEE Antennas and Wireless Propagation Letters* 14 (2014): 257–260.
48. Kotwalla, A., and Y. K. Choukiker. "Design and analysis of microstrip antenna with frequency reconfigurable in MIMO environment." In *2017 International conference of Electronics, Communication and Aerospace Technology (ICECA)*, vol. 1, pp. 354–358. IEEE, 2017.
49. Thao, H. T. P., V. T. Luan, N. C. Minh, B. Journet, and V. V. Yem. "A company frequency reconfigurable MIMO antenna with low mutual coupling for UMTS and LTE applications." In *2017 International Conference on Advanced Technologies for Communications (ATC)*, pp. 174–179. IEEE, 2017.
50. Kumar, J., B. Basu, F. A. Talukdar, and A. Nandi. "Graphene-based multimode inspired frequency reconfigurable user terminal antenna for satellite communication." *IET Communications* 12, no. 1 (2017): 67–74.
51. Sulakshana, C., K. Dahal, and L. Anjaneyulu. "Pattern reconfigurable antenna with multi-port excitation for K-band application." In *2017 IEEE International Conference on Microwaves, Antennas, Communications and Electronic Systems (COMCAS)*, pp. 1–5. IEEE, 2017.
52. Zhang, L., S. Gao, and Q. Luo. "Polarization reconfigurable loop antenna for satellite communications." In *2014 Loughborough Antennas and Propagation Conference (LAPC)*, pp. 649–652. IEEE, 2014.
53. Qin, P.-Y., F. Wei, and Y. J. Guo. "A wideband-to-narrowband tunable antenna using a reconfigurable filter." *IEEE Transactions on Antennas and Propagation* 63, no. 5 (2015): 2282–2285.

54. Kingsly, S., D. Thangarasu, M. Kanagasabai, M. G. N. Alsath, R. R. Thipparaju, S. K. Palaniswamy, and P. Sambandam. "Multiband reconfigurable filtering monopole antenna for cognitive radio applications." *IEEE Antennas and Wireless Propagation Letters* 17, no. 8 (2018): 1416–1420.
55. Sharma, S., and C. C. Tripathi. "An integrated frequency reconfigurable antenna for cognitive radio application." *Radioengineering* 26, no. 3 (2017): 746–754.
56. Bitchikh, M., W. Rili, and M. Mokhtar. "An UWB to narrow band and BI-bands reconfigurable octogonal antenna." *Progress in Electromagnetics Research* 74 (2018): 69–75.
57. Hussain, R., and M. S. Sharawi. "Integrated reconfigurable multiple-input–multiple-output antenna system with an ultra-wideband sensing antenna for cognitive radio platforms." *IET Microwaves, Antennas & Propagation* 9, no. 9 (2015): 940–947.
58. Augustin, G., B. P. Chacko, and T. A. Denidni. "Electronically reconfigurable uni-planar antenna for cognitive radio applications." *IET Microwaves, Antennas & Propagation* 8, no. 5 (2013): 367–376.
59. Nachouane, H., A. Najid, A. Tribak, and F. Riouch. "Reconfigurable and tunable filtenna for cognitive LTE femtocell base stations." *International Journal of Microwave Science and Technology* 2016 (2016). Article ID 9460823.
60. Nachouane, H., A. Najid, A. Tribak, and F. Riouch. "Dual port antenna combining sensing and communication tasks for cognitive radio." *International Journal of Electronics and Telecommunications* 62, no. 2 (2016): 121–127.
61. Purisima, M. C. L., M. Salvador, S. G. P. Agustin, and M. T. Cunanan. "Frequency and pattern reconfigurable antennas for community cellular applications." In *2016 IEEE Region 10 Conference (TENCON)*, pp. 3767–3770. IEEE, 2016.
62. Nguyen-Trong, N., L. Hall, and C. Fumeaux. "A frequency-and pattern-reconfigurable center-shorted microstrip antenna." *IEEE Antennas and Wireless Propagation Letters* 15 (2016): 1955–1958.
63. Nguyen-Trong, N., L. Hall, and C. Fumeaux. "A dual-band dual-pattern frequency-reconfigurable antenna." *Microwave and Optical Technology Letters* 59, no. 11 (2017): 2710–2715.
64. Zainarry, S. N. M., N. Nguyen-Trong, and C. Fumeaux. "A frequency-and pattern-reconfigurable two-element array antenna." *IEEE Antennas and Wireless Propagation Letters* 17, no. 4 (2018): 617–620.
65. Han, L., C. Wang, W. Zhang, R. Ma, and Q. Zeng. "Design of frequency-and pattern-reconfigurable wideband slot antenna." *International Journal of Antennas and Propagation* 2018 (2018). Article ID 3678018.
66. Majid, H. A., M. K. A. Rahim, M. R. Hamid, and M. F. Ismail. "Frequency and pattern reconfigurable slot antenna." *IEEE Transactions on Antennas and Propagation* 62, no. 10 (2014): 5339–5343.
67. Ye, M.., and P. Gao. "Back-to-back F semicircular antenna with frequency and pattern reconfigurability." *Electronics Letters* 51, no. 25 (2015): 2073–2074.
68. Li, N., W. Leng, A.-G. Wang, T.-F. Guo, and Z.-Y. Zhang. "A compact reconfigurable microstrip antenna with frequency and radiation pattern selectivity." *Microwave and Optical Technology Letters* 57, no. 12 (2015): 2848–2854.
69. Liang, B., B. Sanz-Izquierdo, E. A. Parker, and J. C. Batchelor. "A frequency and polarization reconfigurable circularly polarized antenna using active EBG structure for satellite navigation." *IEEE Transactions on Antennas and Propagation* 63, no. 1 (2014): 33–40.
70. Yi, G., C. Huang, X. Ma, W. Pan, and X. Luo. "A low profile polarization reconfigurable dipole antenna using tunable electromagnetic band-gap surface." *Microwave and Optical Technology Letters* 56, no. 6 (2014): 1281–1285.
71. Ni, C., M. S. Chen, Z. X. Zhang, and X. L. Wu. "Design of frequency-and polarization-reconfigurable antenna based on the polarization conversion metasurface." *IEEE Antennas and Wireless Propagation Letters* 17, no. 1 (2017): 78–81.

72. Nguyen-Trong, N., L. Hall, and C. Fumeaux. "A frequency-and polarization-reconfigurable stub-loaded microstrip patch antenna." *IEEE Transactions on Antennas and Propagation* 63, no. 11 (2015): 5235–5240.
73. Liu, J., J. Li, and R. Xu. "Design of very simple frequency and polarisation reconfigurable antenna with finite ground structure." *Electronics Letters* 54, no. 4 (2018): 187–188.
74. Narbudowicz, A., X. Bao, and M. J. Ammann. "Omnidirectional microstrip patch antenna with reconfigurable pattern and polarisation." *IET Microwaves, Antennas & Propagation* 8, no. 11 (2014): 872–877.
75. Gu, C., S. Gao, H. Liu, Q. Luo, T.-H. Loh, M. Sobhy, et al. "Compact smart antenna with electronic beam-switching and reconfigurable polarizations." *IEEE Transactions on Antennas and Propagation* 63, no. 12 (2015): 5325–5333.
76. Lin, W., H. Wong, and R. W. Ziolkowski. "Circularly polarized antenna with reconfigurable broadside and conical beams facilitated by a mode switchable feed network." *IEEE Transactions on Antennas and Propagation* 66, no. 2 (2017): 996–1001.
77. Rodrigo, D., and B. A. Cetiner. "Frequency, radiation pattern and polarization reconfigurable antenna using a parasitic pixel layer." *IEEE Transactions on Antennas and Propagation* 62, no. 6 (2014): 3422–3427.
78. Selvam, Y. P., L. Elumalai, M. G. N. Alsath, M. Kanagasabai, S. Subbaraj, and S. Kingsly. "Novel frequency-and pattern-reconfigurable rhombic patch antenna with switchable polarization." *IEEE Antennas and Wireless Propagation Letters* 16 (2017): 1639–1642.

10 Dielectric Resonator-Based Multiple-Input Multiple-Output (MIMO) Antennas

Gourab Das and Dr. Ravi Kumar Gangwar
Indian Institute of Technology (Indian School of Mines), Dhanbad

CONTENTS

10.1 INTRODUCTION

The rapid improvement in wireless communication in the last two decades has shaped our social and personal life in such a way that the wirelessly transmitted information is an integral part of human life. In the case of wireless communication, the information is transmitted from one point to another through unbounded media. In unbounded media, the transmission and reception of information can only be done through an antenna. Hence, the antenna acts as a backbone of a wireless communication system.

Nowadays, most of the people want small form factor wireless communication devices that have powerful performance and computing capabilities. Using a mobile phone, one can make calls, play music and movies, and also browse the Internet. In a portable computer, now people can play different games, watch high-definition videos, or access the Internet even while moving in a car or on a train. The introduction of different improved wireless technologies is the main reason for the rapid expansion of high-performance devices. The current wireless communication standards depend on three different technologies to achieve improved performance: (i) adaptive modulation and coding (AMC); (ii) orthogonal frequency-division multiple access (OFDMA); and (iii) multiple-input multiple-output (MIMO) technology. The first two technologies deal with data coding and modulation to improve data transmission over wireless channels. The last one, i.e., MIMO technology, uses multiple antennas to improve the system performance. This technology uses multiple antennas at the transmitter and receiver sides to combat multipath fading. So, the MIMO technology is currently utilized in most of the wireless standards and it will also be used in future technologies as well since it provides improved system throughput compared to other techniques.

On the other hand, wireless technology needs some special characteristics in the transmitting and the receiving antennas, such as high bandwidth, lightweight, compact structure, high radiation efficiency, high gain, flexible feeding mechanism, and diversified radiation patterns. To fulfill these demands, two different antennas are mostly used in wireless technology: (i) microstrip patch antenna and (ii) dielectric resonator antenna (DRA). But, due to high metallic and surface wave losses, the microstrip patch antenna suffers from low gain and radiation efficiency with narrow bandwidth. To overcome these drawbacks, the dielectric resonator antenna comes into the picture. So, the DRA becomes a good alternate candidate for traditional low-gain antennas (monopole, slot, dipole, and patch antenna). So, the DRA-based MIMO antenna is one of the good choices in most of the wireless standards due to its several advantages over the microstrip patch antenna. When commercial wireless components are considered, DRAs will be placed on the ground plane. As a result, reduced coupling can be expected with other electronic components.

10.2 WIRELESS COMMUNICATION SYSTEMS

In the late 19th century, the experiment of transmission and reception of electromagnetic waves was done by Heinrich Hertz. This was the first experiment in the history of wireless communication. The new era of wireless communication was begun in the 20th century by Marconi. After that, a rapid improvement in the wireless communication devices and systems was begun. To improve public and private radio communication, wireless networking, and mobile communication, the advancement of wireless technology was increased rapidly. This enables the requirement of low-cost devices with less power consumption and smaller size.

The demand for faster transmission and reception of data looks to be the limitless and main reason for the rapid improvement behind the development of modern wireless communication systems. The first mobile phone was commercialized in the early 1980s, which supported voice calls and limited data transfer in the form of text. This is the beginning of wireless communication and is also considered as the first generation (1G) in cellular terminals. This is also called the advanced mobile phone system (AMPS). After a few years later, the technology advanced more when Global System for Mobile Communications (GSM) standard was introduced, and this is called second generation (2G) in cellular and mobile terminals. This is capable of both voice and data communication, and the data rate is increased by 200 times compared to 1G. In order to increase the wireless capability and data rate, the cellular industry progressed toward third-generation (3G) and fourth-generation (4G) systems in 2006 and 2011, respectively. The requirement of high data rate and spectral efficiency explores the scope of fifth-generation (5G) standards or next-generation wireless systems. With the progression of each generation, a massive improvement in data rates is achieved with the use of new technology and standards. This allowed the data transfer rate to increase to more than 100 Mbps in cellular phones [1]. Different wireless standards such as LTE advanced or IEEE 802. 11 also deliver reliable wireless services with high data rates.

The size of the mobile phone is reduced dramatically with the advancement of antenna design. The phone used in the 1G standard was bulky with limited capabilities. The antenna was generally placed outside of the phone package. After that, the size of the phone reduced. The complexity of the phones increased in 2G and 3G phones due to the integration of microelectronic chips and compact packaging. Generally, printed antennas are used in these phones and they are embedded in the phone package. In 3.5G and 4G phones, again the size of the phone increased without much increase in weight. But at the same time, all the capabilities, as well as battery life, also increased. Such type of phones consists of a multiple-antenna system that supports multiple wireless standards with high data rates [2].

10.3 NEED OF MULTIPLE ANTENNAS

The revolution in wireless communication has taken place with the expansion of wireless standard and wireless technology. Usually, in urban environments, there is no line-of-sight (LOS) path that exists between the transmitter and the receiver. A large number of signals with different paths arise due to the reflections from buildings,

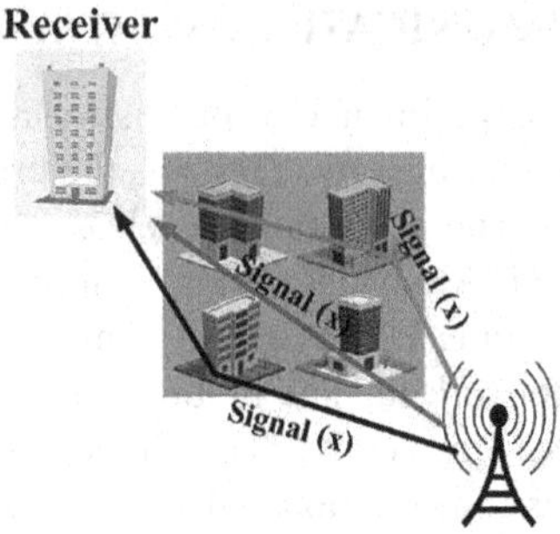

FIGURE 10.1 Schematic diagram of multipath propagation.

trees, mountains, and other reflecting surfaces. As a result, the signal is reflected through multiple paths and finally received at the receiver section. This phenomenon is also called multipath propagation. The schematic diagram of the multipath propagation phenomenon is shown in Figure 10.1. This multipath propagation creates a variety of signal paths that exist between transmitter and receiver. This gives rise to the interference of various signals, which causes a distortion of the signal, loss of data, and multipath fading.

Generally, the information that is transmitted from the transmitter to the receiver through a communication channel takes different paths to travel. The signal strength will vary at the receiver end since the distance travelled by each signal is different from each other. The reason for the loss of the signal power at the receiver end is mainly attributed to path loss and fading across the channel [3].

In case of line-of-sight (LOS) path between the transmitter and receiver, the signal strength at the receiver is calculated by using the Friis free space equation [4]

$$P_R = \frac{P_T G_T G_R \lambda^2}{(4\pi d)^2 L} \tag{10.1}$$

where P_T and P_R represent the transmitted and received powers of the signal, respectively. The power gains of the transmitting and receiving antennas are represented by G_T and G_R, respectively. d and λ are the distance between the transmitter and the receiver and the signal wavelength, respectively. L represents the system loss factor, and it is calculated by using the following equation

$$L = L_P L_S L_F \tag{10.2}$$

where L_P, L_S, and L_F represent the path loss, slow fading, and fast fading, respectively. The path loss depends on the distance between the transmitter and the receiver. It shows the difference between effective transmitted and received powers.

$$\text{Path Loss}(dB) = -10\log\frac{P_R}{P_T} = -10\log\frac{G_T G_R \lambda^2}{16\pi^2 d^2 L} \tag{10.3}$$

The fading effect is categorized into two different types: (i) large-scale fading and (ii) small-scale fading. When an obstacle is positioned between the transmitter and the receiver, interference occurs. This interference causes a significant reduction in signal strength since the data are blocked by the obstacle [5].

On the other hand, small-scale fading shows rapid fluctuations in signal strength over a very short distance and a short time as well. Here, the rapid signal fluctuations occur around a slowly varying mean [6].

As the complexity of the propagation channel increases, the number of propagation paths between the transmitter and receiver increases. The multipath signal components are combined at the receiver section to reconstruct the originally transmitted signal. For reconstructing the signal, the multipath signal components are combined either constructively or destructively [7]. As a result, either good or weak quality of the signal is reconstructed at the receiver section. The destructive interference at the communication channel produces a weak signal at the receiver.

On the other hand, the demand for high data rate over the communication channel increases day by day. The mobile network and wireless access point require more and more data to be transferred. In other words, the link capacity and spectral efficiency of the wireless link need to be improved.

To overcome the fading problem in the communication channel and to improve the reliability of the wireless link, multiple antennas can be placed at the transmitter and the receiver sections. They improve the performance of the communication channel by combating or exploiting multipath scattering in the communication channel. They are designed in such a way that they can take full advantage of the multipath to improve the data rate and spectral efficiency without any extra requirement of power.

10.4 MIMO WIRELESS COMMUNICATION

In radio, multiple-input multiple-output (MIMO) is used multiple antennas at the transmitter and receiver side to improve the communication performance. With the help of multiple antennas, the spatial characteristics of the communication channel can be controlled to improve the performance of the wireless link. The theory and concept of MIMO technology was initially introduced in the 1990s, but the commercial producers using MIMO technology were appearing in the market around 2003.

In a communication system, the multiple-antenna system exists at either the transmitter side or the receiver side, or both. The term MIMO is used when multiple antennas are utilized at both the transmitter and the receiver ends. So, there are different types of multiple-antenna systems that need different number of antennas and have different levels of complexity. Such type of system provides an optimum solution for different applications.

Single-Input Single-Output (SISO): In this radio system, the transmitter and the receiver consist of single antennas. It does not require any additional processing or diversity. The main advantage of this system is its simplicity. However, the disadvantage is that its channel performance degrades due to interference or multipath fading. Figure 10.2 shows the diagram of the SISO system. Figure 10.2 reveals that noise is introduced in the system when the signal is transmitted from transmitting antenna

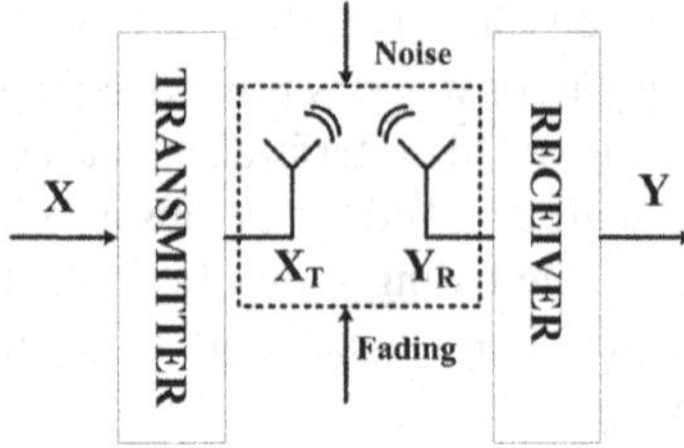

FIGURE 10.2 Block diagram of the SISO system.

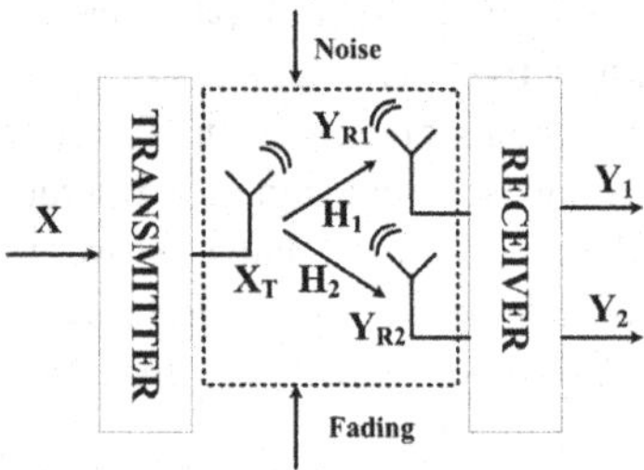

FIGURE 10.3 Block diagram of the SIMO system.

(X_T) to receiving antenna (Y_R). Also, the signal fades in this region during signal processing. The channel capacity of the proposed system is given as [8]:

$$C = B\log_2\left(1 + {}^{S}\!/_{N}\right) \tag{10.4}$$

where C and B represent the channel capacity and the bandwidth of the signal, respectively. S/N is the signal-to-noise ratio of the signal.

The channel bandwidth of the SISO system is limited by Shannon's law. It states about the theoretical maximum rate at which error-free digits can be propagated over a bandwidth-limited channel in the presence of noise.

Single-Input Multiple-Output (SIMO): SIMO system occurs when the transmitter has a single antenna and the receiver has multiple antennas. Such systems are also known as receiver diversity. When the receiver receives a signal from several independent sources, this system is required to combat the effect of fading. Figure 10.3 shows a block diagram of the SIMO system. It consists of one transmitting antenna at the transmitter side and two receiving antennas at the receiver side. (Here, two receiving antennas are considered; more than two antennas are also possible). The advantage of the SIMO system is that it is easy to implement. However, a high level of processing is required at the receiver end, since it consists of several antenna elements. The SIMO system helps to improve the diversity of the receiving antenna since it gives stronger diversity than the SISO system. The channel capacity of the SIMO system is given by following equation [8]:

$$C = M_r B\log_2\left(1 + {}^{S}\!/_{N}\right) \tag{10.5}$$

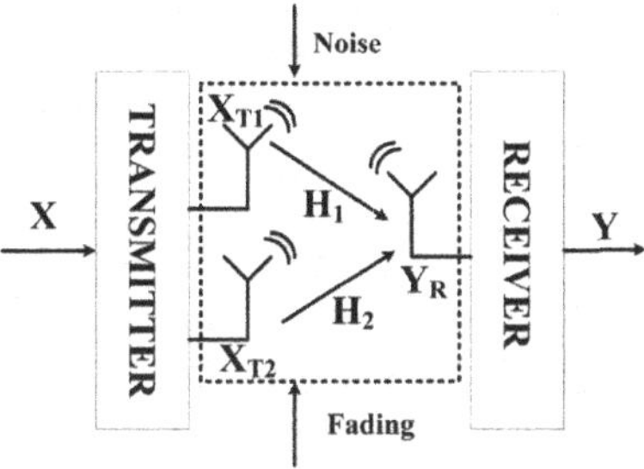

FIGURE 10.4 Block diagram of the MISO system.

where C and B are the channel capacity and the signal bandwidth, respectively. M_r represents the number of receiving antennas at the receiver side, and S/N is the signal-to-noise ratio.

Here, X is the input; Y_1 and Y_2 are the outputs; X_T is the transmitting antenna; and Y_{R1} and Y_{R2} are the receiving antennas.

Multiple-Input Single-Output (MISO): It is also named as transmit diversity. In this system, the same data are transmitted redundantly from several (more than one) transmitter antennas. The block diagram of the MISO system is shown in Figure 10.4. The receiver consists of a single antenna and is able to receive the optimum signal for further processing. The advantage of the MISO system is that it requires less space for the antennas, and a minimum level of processing is required at the receiver end. If the cell phone UEs is acting as a receiver, then it has a positive impact on phone size, cost, and battery life due to the lower processing needed. This antenna system helps to recover the original signal with lesser path loss than SISO and SIMO systems. The effect of multipath fading is also less compared to the other two techniques since it consists of multiple antennas at the transmitter end. The channel capacity is still not increased, but it is better than the SISO system. The channel capacity of the MISO system is given by using the following equation [8]:

$$C = M_t B \log_2\left(1 + {}^{S}\!/_{N}\right) \tag{10.6}$$

where C, M_t, and B represent the channel capacity, the number of antennas at the transmitter side, and the bandwidth of the signal, respectively. S/N is the signal-to-noise ratio of the transmitted signal. Here, X_1 and X_2 are the inputs; Y is the output; X_{T1} and X_{T2} are the transmitting antennas; and Y_R is the receiving antenna.

Multiple-Input Multiple-Output (MIMO): MIMO system is one of the effective radio antenna technologies that use multiple antennas at both the transmitter and the receiver sides to enable several signal paths to carry data. Since the MIMO system consists of multiple transmitting antennas, the signal can be transmitted by any of the antenna and follow any path to reach the receiving side. The path of the signal depends on the position of the antenna. That means the signal path will change if the position of the receiving antenna changes. The signal path will be changed if the antenna is moved by a small distance. The MIMO channel experience multipath fading due to the existence of multiple number of paths. Figure 10.5 shows the MIMO

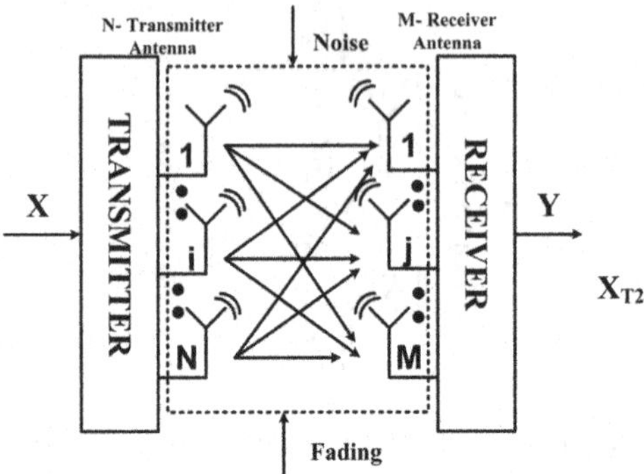

FIGURE 10.5 Block diagram of the MIMO system.

system with N-transmitting antennas at the transmitter side and M-receiving antennas at the receiver side. MIMO system offers better throughput, increased data rate, and optimized spectral efficiency compared to other techniques. In the MIMO system, multiple propagation channels are available between the transmitter and the receiver [9]. Therefore, the MIMO channel can be represented as:

$$H = \begin{bmatrix} h_{11} & \cdots & h_{1N} \\ \vdots & \ddots & \vdots \\ h_{M1} & \cdots & h_{MN} \end{bmatrix}$$

Here, h_{11}, h_{12}, etc., represent the channel coefficients between the transmitter and the receiver. When the data rate is increased for a single user, the MIMO system is called SU-MIMO (single-user MIMO), whereas if the data rate is increased for multiple users, it is called MU-MIMO (multi-user MIMO).

The relation between the input and output of the MIMO antenna system can be given by following equation [10]

$$\text{Output } y(t) = \sum_{j=1}^{N} h_{(NM)} S_{(M)}(t) \tag{10.7}$$

where $S_{(M)}(t)$ represents the received signal by the Mth antenna that is transmitted by the jth antenna.

The channel capacity of the MIMO antenna system is given in following equation [10]:

$$C = NMB \log_2 \left(1 + {}^{S}\!/_{N}\right) \tag{10.8}$$

where C, N, M, and B represent the channel capacity, the number of transmitting antennas, the number of receiving antennas, and the channel bandwidth, respectively. S/N represents the signal-to-noise ratio of the channel.

The main advantage of the MIMO system is that it provides improved performance compared to the other three techniques since it gives enhanced throughput, channel capacity, and efficiency during signal transmission by using multiple-antenna systems [11,12]. That is why the MIMO system is used in several advanced communication systems such as WLAN, WiMAX, WAM, 3G, 4G, and 5G.

10.5 MIMO TECHNIQUES

The block diagram of a MIMO system, which consists of multiple transmitting and receiving antennas, is shown in Figure 10.5. Figure 10.5 reveals that the transmitting data are passed through several antennas for transmission. Each transmitting data stream will have an identical signature. The receiving antennas are able to distinguish the multiple data streams and decode the final data. So, the MIMO techniques are able to improve the communication link by using two different methodologies: (i) spatial diversity (by combating the multipath effects) and (ii) spatial multiplexing (by exploiting the multipath effects).

Spatial Diversity: In the spatial diversity technique, the same information is sent through several independent channels to combat fading. When the same information is sent through multiple independent fading channels, the amount of fade experienced by each copy of the information will be distinct. This confirms that at least one copy of the data will suffer less fading compared to other copies. As a result, the probability of receiving transmitted data is also increased [13]. Thus, the reliability of the overall system is improved significantly. Consequently, this minimizes the co-channel interference between the channels significantly. This technique is termed "spatial diversity" in the communication system [14].

Now, consider a SISO system where data stream [1, 0, 0, 1, 0] is transmitted through a fading channel that is shown in Figure 10.6. The transmitted data stream may be lost or corrupted due to the channel variation. As a result, the receiver cannot recover the original data stream. The solution is to add independent fading channels between the transmitter and the receiver. This can be achieved by increasing the number of transmitter antennas and the receiver antennas. The SISO configuration

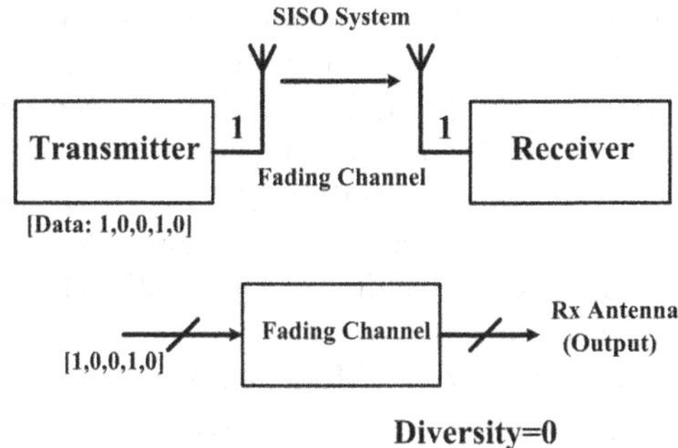

FIGURE 10.6 SISO system with fading channel configuration.

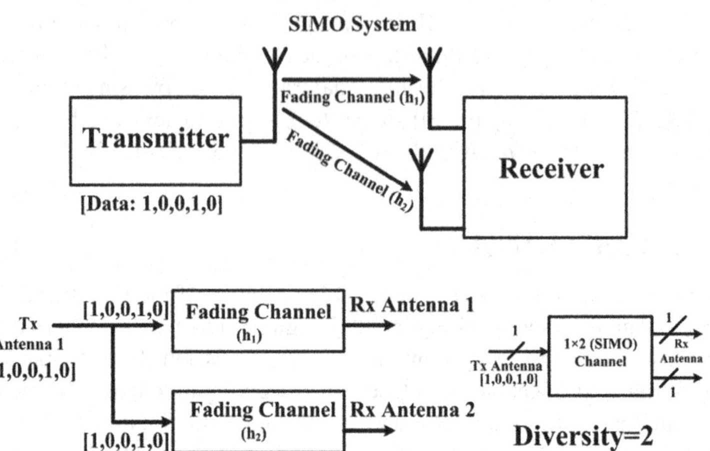

FIGURE 10.7 SIMO system with fading channel configuration.

cannot provide any diversity since no parallel link exists between the transmitter and the receiver. So, the diversity of the SISO system is zero.

Now, consider a single-input multiple-output (SIMO) antenna system that consists of one transmitting antenna and two receiving antennas. The diagram of the SIMO system configuration is shown in Figure 10.7. It reveals that two copies of the same data are propagated through two different channels, which have independent fading characteristics. If one link fails to deliver the data, the probability of delivery of the other link increases [15]. Thus, the reliability of the overall transmission increases due to the additional fading channel. Consider a system that consists of N_T transmitting antennas and N_R receiving antennas. So, the maximum number of diversity paths between the transmitter and the receiver is $N_T \times N_R$ [14].

In a similar manner, a number of diversity paths can be created by adding multiple numbers of antennas at both the transmitter and receiver sides. Figure 10.8 shows a 2×2 MIMO antenna system. In this case, the number of diversity paths is 2×2=4. The diversity can be improved more by increasing the independent channels between the transmitter and the receiver. This can be achieved by increasing the number of transmitter and receiver antennas.

Spatial Multiplexing: In the case of spatial multiplexing, each independent channel contains independent information. As a result, the data rate of the system is also increased. This can be equivalent to the orthogonal frequency-division multiplexing (OFDM) technique. In OFDM, a different part of data is carried by a different frequency sub-channel [16]. Similarly, in the case of spatial multiplexing, a number of independent sub-channels are created in the same allocated bandwidth. Thus, multiplexing gain comes into the picture with no additional power or bandwidth requirement. With reference to signal space constellation, the multiplexing gain is also called degrees of freedom. For a MIMO antenna system, the degrees of freedom are equal to min $[N_T, N_R]$, where N_T and N_R represent the number of antennas in the transmitter and the receiver, respectively [17]. In a MIMO system, the system capacity is governed by the degrees of freedom.

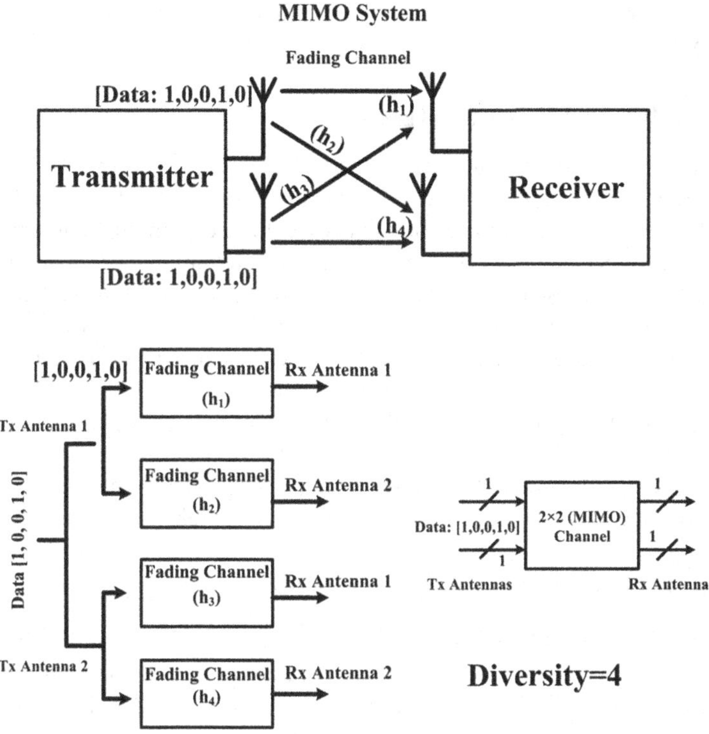

FIGURE 10.8 MIMO system with fading channel configuration.

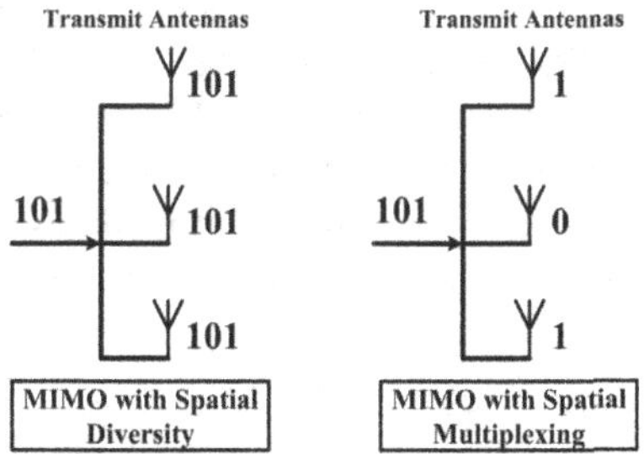

FIGURE 10.9 Difference between spatial diversity and spatial multiplexing.

Figure 10.9 shows the difference between the spatial diversity and spatial multiplexing. In a spatial diversity scheme, the same data are sent through different independent spatial channels. Here, the same data stream is transferred from three different transmit antennas. So, the diversity gain is 3 (considering 3×1 MISO system) and the multiplexing gain is 0.

In the spatial multiplexing technique, each bit of the data stream is sent across three different spatial channels. As a result, the data rate of the overall system is increased.

In this case, the diversity gain is 0, but the multiplexing gain is 3 (considering 3×3 MIMO system).

10.6 MIMO ANTENNA SYSTEMS

In the era of the modern fourth-generation (4G) wireless system, MIMO antenna systems are the key technology that led to the increase in the data rate and throughput within the limited power or bandwidth with the use of multiple antennas at the transmitter and receiver sides. To embed and integrate multiple antennas at the user terminal, the antenna size becomes a concerning factor. For useful antenna operation, the designed antennas need to operate in multiple bands. It is known to us that antenna miniaturization will affect the antenna performance in terms of its efficiency and operating bandwidth. On the other hand, since multiple antennas are placed in close proximity to each other as well as in the same ground plane, the coupling between the antennas increases. As a result, the diversity performance of the MIMO antenna system is degraded. So, the designing of the MIMO antenna systems is a very challenging task and the antenna designer needs to take care of different issues.

10.7 PERFORMANCE METRICS OF MIMO ANTENNAS

To fully characterize a MIMO antenna system, several important parameters need to be evaluated along with basic single antenna parameters such as bandwidth, gain, and radiation efficiency. The performance metrics are important for the multi-antenna system. These parameters are discussed in this section.

10.7.1 Correlation Coefficient

This parameter shows the isolation or correlation between the communication channels. It considers the radiation pattern of the radiator and shows the pattern effect when multiple antennas operate simultaneously. The envelope correlation coefficient (ECC) is obtained by squaring the correlation coefficient. The ECC of the MIMO antenna can be evaluated by using two different methods: (i) far-field method [18] and (ii) S-parameter method [19].

$$\mathrm{ECC}_F = \frac{\left|\iint_{4\pi}\left[\bar{A}_i(\theta,\phi) * \bar{A}_j(\theta,\phi)\right]d\omega\right|}{\iint_{4\pi}\left|\bar{A}_i(\theta,\phi)\right|^2 \iint_{4\pi}\left|\bar{A}_j(\theta,\phi)\right|^2} \tag{10.10}$$

where $\vec{A_i}(\theta, \phi)$ and $\vec{A_j}(\theta, \phi)$ represent the 3D far-field pattern when port-i and port-j are excited, respectively. Ω and * represent the solid angle and the Hermitian product operator.

The ECC is also calculated by using the S-parameter method when a single mode and the lossless antenna are considered. The following formula is used to calculate the ECC value by using S-parameters [18]:

$$\mathrm{ECC}_S = \frac{\left|S_{ii}^* S_{ij} + S_{ji}^* S_{jj}\right|^2}{\left(\left(1-\left(\left|S_{ii}^2\right|+\left|S_{ji}^2\right|\right)\right)\left(1-\left(\left|S_{jj}^2\right|+\left|S_{ij}^2\right|\right)\right)\right)} \tag{10.11}$$

where S_{ii} and S_{jj} represent the reflection coefficient values with port-i and port-j, respectively. Similarly, S_{ij} and S_{ji} are the isolation level between port-i and port-j, respectively. In practical applications, an ECC value of less than 0.5 is acceptable. But, for 4G wireless communication, the acceptable value of ECC is 0.3.

10.7.2 Diversity Gain

Diversity is achieved when the receiver takes multiple versions of the input signal via different propagation channels since multiple antennas are present. If the signals are not correlated to each other, the received signal at the receiver section will offer a better SNR level. As a result, a better signal reception is achieved. The effect of diversity in the communication channel is measured by the diversity gain of the antenna. It is defined as the ratio of the SNR of multiport MIMO antenna system to the SNR of single-port antenna system. The value of diversity gain is calculated with the help of ECC value [20]. The following formula is utilized to obtain the diversity gain of any MIMO antenna system [18]:

$$\mathrm{DG} = 10\sqrt{1-(\mathrm{ECC})^2} \tag{10.12}$$

So, the diversity gain and ECC are interrelated to each other. A higher value of diversity gain can be obtained by lowering the ECC values. In practical applications, the value of diversity gain closer to 10 is acceptable for improved MIMO performance.

10.7.3 Mean Effective Gain

In a predefined wireless condition, the mean effective gain (MEG) shows the antenna performance where the environmental effect is considered. The stand-alone antenna gain is not a good practice to validate the antenna performance since, in practical applications, the antenna is not used in the anechoic chamber. The designed antenna is utilized in a practical environment for a certain application. So, the environmental effect on the antenna's radiation characteristics is an important factor in evaluating its performance. To estimate the antenna performance, one way is to fabricate the designed antenna and obtain the antenna performance with respect to the standard antenna with known characteristics. But, the above-mentioned procedure is time-consuming and costly. The other method is to calculate the mean effective gain of the antenna, which is more practical and easier. It can be calculated by using the following formula [18]:

$$\mathrm{MEG}_i = 0.5\left[1-\left|S_{ii}\right|^2-\left|S_{ij}\right|^2\right] \tag{10.13}$$

$$\mathrm{MEG}_j = 0.5\left[1-\left|S_{ij}\right|^2-\left|S_{jj}\right|^2\right] \tag{10.14}$$

where MEG_i and MEG_j are the mean effective gains of port-i and port-j, respectively. For better diversity performance, the difference between MEG_i and MEG_j should be less than 3 dB.

10.7.4 Total Active Reflection Coefficient

The scattering parameter alone is not enough to properly characterize the MIMO antenna system. The total active reflection coefficient (TARC) is an important MIMO performance parameter that characterizes the MIMO antenna system. TARC is defined as [21]:

$$\text{TARC} = \sqrt{\frac{\text{Total Reflected Power}}{\text{Total Incident Power}}}$$

The TARC is evaluated by utilizing S-parameters of the antenna system. For a MIMO antenna system that consists of N-elements, the TARC is defined as [22]:

$$\Gamma_{\text{MIMO}} = \frac{\sqrt{\sum_{i=1}^{N} |b_i|^2}}{\sqrt{\sum_{i=1}^{N} |a_i|^2}} \tag{10.15}$$

where b_i and a_i represent the reflected and incident signals, respectively. These parameters can be calculated with the help of scattering parameters. The relationship between the incident and reflected signals of a MIMO antenna system can be given by following equation:

$$[b] = S[a] \tag{10.16}$$

where S represents the scattering matrix of the antenna. If the MIMO antenna consists of N-ports, then the scattering matrix becomes $N \times N$. TARC also incorporates the feeding phase of the antenna ports. So, the TARC plot is utilized to obtain the impedance bandwidth and the resonance frequency of the whole MIMO antenna system for a specific phase excitation between the ports.

For a two-port radiator, TARC is calculated by using the following relation:

$$\text{TARC} = \frac{\sqrt{\left(\left(\left|S_{11} + S_{12}e^{j\theta}\right|^2\right) + \left(\left|S_{21} + S_{22}e^{j\theta}\right|^2\right)\right)}}{\sqrt{2}} \tag{10.17}$$

Here, the input feeding phase is represented by θ. The input reflection coefficients of port-1 and port-2 are represented by S_{11} and S_{22}, respectively. On the other hand, S_{12} and S_{21} are the isolation between port-1 and port-2 associated with the MIMO antenna structure. The phase angle (θ) is swept from 0° to 180° to examine the effect of phase variation between the ports.

10.7.5 Channel Capacity Loss (CCL)

Channel capacity loss (CCL) shows the maximum data rate up to which the message can be constantly transmitted with negligible loss. For an N-port antenna system, the CCL can be evaluated by utilizing the relations given by eqs. (10.18)–(10.21) [23].

$$C_{\text{loss}} = -\log_2\left(\beta^R\right) \tag{10.18}$$

$$\beta^R = \begin{bmatrix} \beta_{11} & \cdots & \beta_{1,N} \\ \vdots & \ddots & \vdots \\ \beta_{N,1} & \cdots & \beta_{N,N} \end{bmatrix} \tag{10.19}$$

where

$$\beta_{ii} = 1 - \left(\sum\nolimits_{n=1}^{N} \left|S_{in}^{*} S_{ni}\right|\right) \tag{10.20}$$

$$\beta_{ij} = -\left(\sum\nolimits_{n=1}^{N} \left|S_{in}^{*} S_{nj}\right|\right) \tag{10.21}$$

for $i, j = 1, 2, 3, 4$ up to N.

10.8 PROBLEM IN MIMO ANTENNA SYSTEMS

In the case of MIMO antenna systems, multiple antennas are placed within a common substrate with a common ground plane. As a result, mutual coupling is established between the antenna elements within the MIMO antenna system. Mutual coupling shows the interaction of power coupling between the adjacent antenna elements within the multiport antenna system. The mutual coupling does not deal with the coupling between radiation patterns. Another term in MIMO antenna which is closely related to mutual coupling is "isolation." Isolation/mutual coupling is represented by using S-parameters. The transmission coefficient (S_{xy}) between two antenna elements (antenna-1 and antenna-2 are represented by x and y, respectively) measures this factor. If the antennas are placed at a distance that is less than $\lambda/4$, then high mutual coupling is established between the antennas. In medium- to small-sized communication devices of which the device dimension is not large, such as a router, USB dongle, and mobile phone, the antenna elements are placed very close to each other. As a result, high mutual coupling, i.e., low isolation, is established between the antenna elements. The low isolation level will affect the channel capacity as well as the efficiency of the MIMO antenna. To improve the MIMO performance of the antenna, isolation between the antenna elements should be high. Several isolation improvement techniques are available in the literature to minimize the mutual coupling effect between the antenna elements.

10.9 INTRODUCTION TO DIELECTRIC RESONATOR ANTENNAS (DRAs)

A dielectric resonator antenna (DRA) is a ceramic-based resonator that is able to radiate electromagnetic (EM) energy into space after proper excitation [24]. In the past, the ceramic resonators were used in filter and oscillator circuits; thus, it can offer a substitute to the waveguide cavity resonator. To prevent radiation, these ceramic resonators were bounded in a metallic cavity (for filter and oscillator applications). These ceramic resonators have a high quality (Q)-factor (20–50,000) and better temperature stability [25]. The history of dielectric resonators is very interesting, from circuit applications to antenna applications. In 1939, Richtinger was the first person who theoretically developed the dielectric resonator in the form of an un-metalized dielectric [26]. After that, in 1960s, two US-based scientists, Okaya and Barash examined the excitation of modal pattern in the dielectric resonators [27]. But still, the dielectric resonators are used for circuit applications rather than as an antenna element. In 1980s, S.A. Long and his research group examined the ceramic-based resonator as an antenna. After removing the metallic shielding and lowering the dielectric constant, the resonant modes and radiation characteristics were investigated [28]. In this way, they proved that the dielectric resonator can act as a radiator.

10.9.1 Characteristics of Dielectric Resonator Antennas (DRAs)

A dielectric resonator antenna supports several unique features, which makes it different from other radiating structures. Some of the exclusive characteristics are given as follows:

- The size of the dielectric resonator depends on both the operating frequency and permittivity of the dielectric material [24].
- The quality factor and operating frequency of a DRA depend on its aspect ratio (i.e., ratio of length to width/length to depth) [29].
- By exciting different modes inside the DRA, it is able to generate diversified radiation patterns [29].
- The dielectric resonator antennas show good power handling capability and temperature stability with the use of high-permittivity dielectric material [29].
- Due to the absence of metallic and surface wave losses, dielectric resonator antennas support improved gain and radiation efficiency values compared to planar radiators (microstrip/slot antennas) even at higher frequency values [24].
- Different feeding mechanisms such as microstrip line, aperture coupling, probe, and CPW line are used to excite the dielectric resonator. Due to such type of flexibility in the excitation mechanism, they are suitable for several current technologies [24].

Besides all of these several advantages, DRAs also have some disadvantages. Some of them are given below:

- As compared to planar structures, the fabrication cost of DRAs is high. This is due to the use of dielectric materials with a high permittivity value [24]
- Due to the native hardness of DRAs, they are difficult to design at some specific frequency as compared to microstrip/slot antennas [29]
- Different special shapes of DRAs, such as tetrahedron and hexagon, and minor geometrical alterations are very difficult because of its natural rigidity [29].
- Antenna designer should take care of modal patterns during antenna design since some modal fields which have high Q-factor are not utilized for radiating purposes [30].

10.9.2 Applications of DRAs

Due to low loss and improved gain and radiation efficiency, DRAs support various types of applications in the field of wireless communication. They can also be used in satellite communication and TV broadcasting. The multipurpose nature of DRAs makes them suitable for numerous applications such as:

- WiMAX systems.
- Radar applications.
- WLAN and GPS.
- Laptops/notebooks.
- MIMO wireless systems.
- Wearable antennas.
- RFID applications.
- Filter elements.
- The high efficiencies make DRAs suitable candidates for millimeter-wave arrays.

10.9.3 Basic Shapes of DRAs

The dielectric resonator antennas are available in several shapes and sizes. But, the radiation characteristics and theoretical analysis of different mode patterns are specified only for three shapes: (i) hemispherical, (ii) rectangular, and (iii) cylindrical DRAs [31,32]. This is why they are known as the basic shapes of DRAs.

Figure 10.10 displays the configuration of the hemispherical dielectric resonator antenna. The modal pattern of dielectric spheres was proposed by M. Gastine, L. Courtois, and J. J. Dormann in 1960 [33]. But, the theoretical analysis of hemispherical DRAs was investigated by Ahmed A. Kishk and S.A. Long [32,34]. There are two types of fundamental modes that are excited in the hemispherical DRAs, i.e., transverse electric (TE_{mnp}) and transverse magnetic (TM_{mnp}), where m, n, and p denote the field variation in radial, azimuthal, and elevation directions. The two common modes that are excited in hemispherical DRAs are TM_{101} and TE_{111}. The hemispherical DRAs are characterized by radius (a) and dielectric constant of material (ε_r). For a particular dielectric constant, the radius of the hemispherical dielectric resonator will determine the resonant frequency and radiation Q-factor since it supports only zero degrees of freedom.

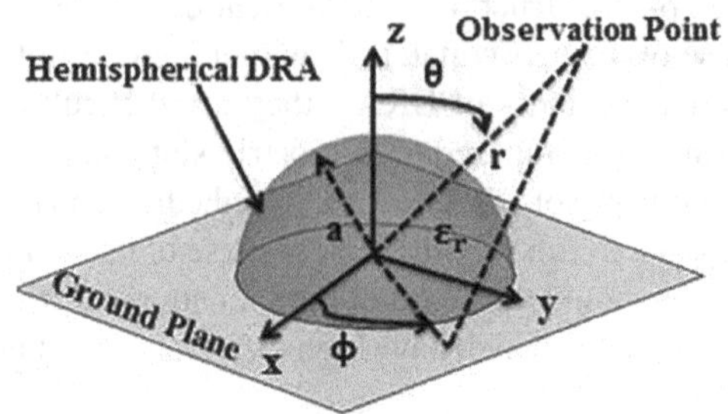

FIGURE 10.10 Configuration of the hemispherical dielectric resonator antenna [24].

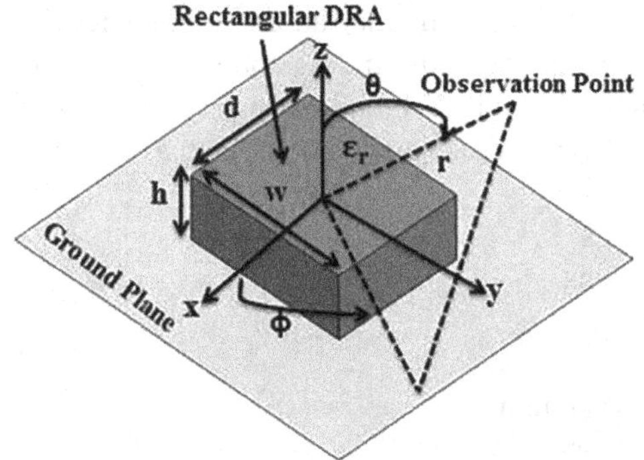

FIGURE 10.11 Geometry of the rectangular dielectric resonator antenna [24].

The geometrical layout of a rectangular DRA (rDRA) is shown in Figure 10.11. Several research groups investigated the theoretical approach to rectangular DRAs. Initially, the brief idea of rDRAs was presented by S.A. Long and his research group in 1983 [31]. After that, the possible modes and radiation characteristics of rDRAs with different feeding techniques were investigated by Mongia and Ittipiboon in 1997 [35]. The rDRAs are characterized by height (*h*), width (*w*), depth (*d*), and dielectric constant (ε_r), as shown in Figure 10.11. The rDRAs only support the non-confined mode, i.e., TE_{mnp} (*m*, *n*, and *p* show the field variation in *x*-, *y*-, and *z*-directions), since they are not a body of revolution. The rDRAs provide two degrees of freedom (by choosing the ratio of length to width and that of depth to width). Hence, the antenna designer easily achieves the desired antenna profile and resonant peak for a given dielectric constant.

The geometrical layout of a cylindrical DRA (cDRA) is shown in Figure 10.12. The experimental study on cDRAs was given by S.A. Long and his research group in 1983 for the first time [28]. The cDRAs are characterized by radius (*a*), height (*h*), and dielectric constant (ε_r), as shown in Figure 10.12. The cDRAs support three different types of mode patterns, i.e., TE_{mnp}, TM_{mnp}, and hybrid mode, where *m*, *n*, and *p* denote the field variation in azimuthal, radial, and axial directions. The bandwidth and quality factor of cDRAs can be controlled by the ratio of "*H*/*a*" for a given material, since they provide one degree of freedom for the antenna designers.

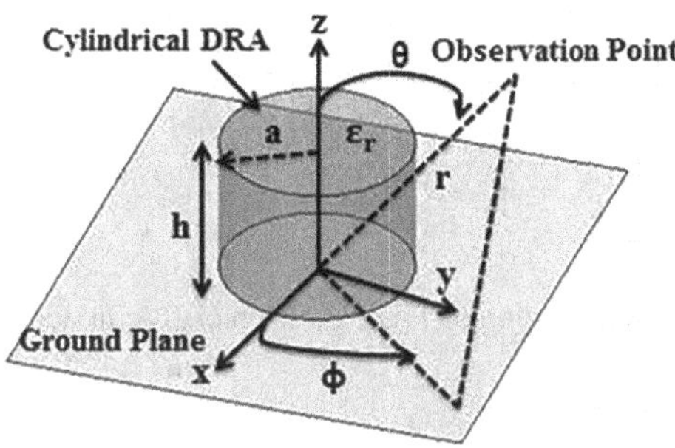

FIGURE 10.12 Configuration of the cylindrical dielectric resonator antenna [24].

The cylindrical DRA (cDRA) is the most popular shape due to two reasons: (i) It can support three different radiating modes and (ii) it is easily available in the commercial market. In the next subsection, different radiating modal patterns and their radiation characteristics are discussed.

10.9.4 Cylindrical Dielectric Resonator Antennas

In 1983, S.A. Long, M.W. McAllister, and L.C. Shen performed the experimental study on cDRAs for the first time. They also performed an investigation on the radiation characteristics of cDRAs. This work gave a new era in the field of DRA research. The benefits of a cylindrical dielectric resonator as an antenna have already been discussed in brief. In this section, these advantages are explained in detail. The modes in the cDRAs are categorized into three types: (i) TE mode ($E_z=0$), (ii) TM mode ($H_z=0$), and (iii) hybrid mode (both E_z- and H_z-components are present). The hybrid mode can further be divided into two different parts, i.e., HE (E_z-component is dominant) and EH (H_z-component is dominant). For TE and TM modes, there is no azimuthal dependence, since they are axially symmetric, whereas azimuth dependence exists in the case of HE modes. Three different indices (m, n, and p) are added to the subscripts for each mode, which show the variation of field in azimuthal, radial, and axial directions [36]. The subscripts of the modes show the field variations in the azimuthal ($m=\phi$), radial ($n=r$), and axial ($p=z$) directions. Now, the modes in cDRAs are specified as TE_{0np}, TM_{0np}, HEM_{mnp}, and EH_{mnp}. The span of δ is between 0 and 1 in the cylindrical coordinate system. The modes in the cDRAs are divided into two parts: (i) radiating modes and (ii) non-radiating modes. Some radiating modes in cDRAs are $TE_{01\delta}$, $TM_{01\delta}$, $HEM_{11\delta}$, $HEM_{12\delta}$, and $HEM_{11\delta+1}$. On the other hand, some of the non-radiating modes are $TE_{01\delta+1}$, $HEM_{21\delta}$, and $HEM_{22\delta}$ [37,38]. Some of the radiating modes are discussed below.

$TE_{01\delta}$ mode: The resonant frequency of $TE_{01\delta}$ mode is computed by utilizing the following formula [35]:

$$f_{r,\,TE_{01\delta}} = \frac{2.327 v_0}{2\pi d\sqrt{\varepsilon_r + 1}}\left[1.0 + 0.2123\left(\frac{a}{H}\right) - 0.00898\left(\frac{a}{H}\right)^2\right] \tag{10.22}$$

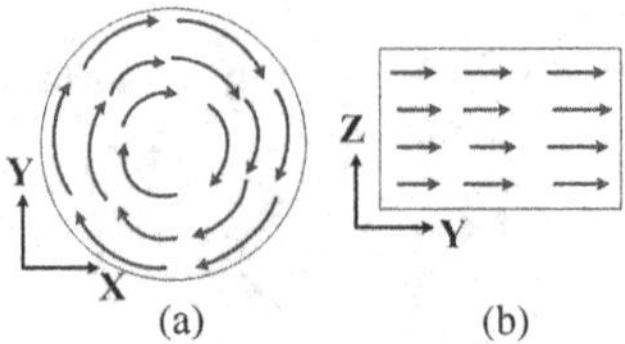

FIGURE 10.13 E-field distribution of $TE_{01\delta}$ mode on cDRA: (a) top view; (b) side view.

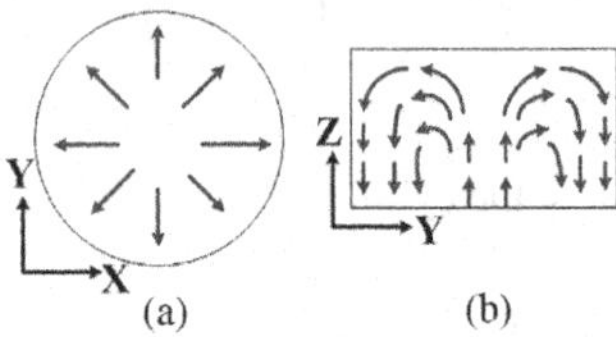

FIGURE 10.14 E-field distribution of $TM_{01\delta}$ mode on cDRA: (a) top view; (b) side view.

where v_0 is the velocity of light. "a," "h," and "ε_r" are the radius, height, and permittivity of the cylindrical DRA, respectively. Figure 10.13 displays the E-field distribution in the cDRA for $TE_{01\delta}$ mode [39]. Figure 10.13 reveals that the E-field distributions are very comparable to the magnetic dipole. The tilted radiation pattern is obtained in case of $TE_{01\delta}$ mode, since the maximum direction of radiation arises at an angle away from the broadside direction (θ=0°; ϕ=0°) [35].

$TM_{01\delta}$ mode: The resonant frequency of $TM_{01\delta}$ mode in the cDRAs is realized by utilizing the following empirical formula [35]:

$$f_{r,\,\mathrm{TM}01\delta} = \frac{v_0\sqrt{3.83^2 + \left(\dfrac{\pi a}{2H}\right)^2}}{2\pi d\sqrt{\varepsilon_r + 2}} \tag{10.23}$$

The different variables used in the above equation are the same as those given in eq. (10.23). Figure 10.14 displays the field distribution in the cylindrical dielectric resonators for $TM_{01\delta}$ mode [39]. Figure 10.14 reveals that this mode is excited toward the end-fire direction (null at broadside direction). So, such type of modes shows the monopole type of radiation pattern [35].

a. **$HEM_{11\delta}$ mode**: The resonant frequency of $HEM_{11\delta}$ mode in the cDRAs is given by using the following equation [35]:

$$f_{r,\,\mathrm{HE}11\delta} = \frac{6.321 v_0}{2\pi a\sqrt{\varepsilon_r + 2}}\left[0.27 + 0.36\left(\frac{a}{2H}\right) + 0.02\left(\frac{a}{2H}\right)^2\right] \tag{10.24}$$

In the above equation, the meaning of the variables is the same as in eq. (10.22). Equations (10.22)–(10.24) are valid only if the aspect ratio of the DRA lies in the range $0.5 < (H/a) < 5$. Figure 10.15 shows the E-field distribution for $HEM_{11\delta}$ mode inside the cylindrical resonator [39]. The $HEM_{11\delta}$ mode is excited toward the

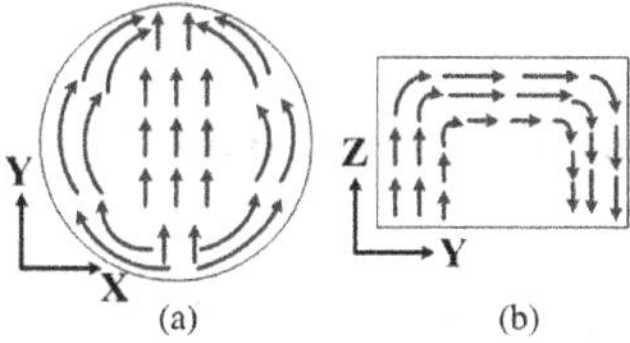

FIGURE 10.15 E-field distribution of $HEM_{11\delta}$mode on cDRA: (a) top view; (b) side view.

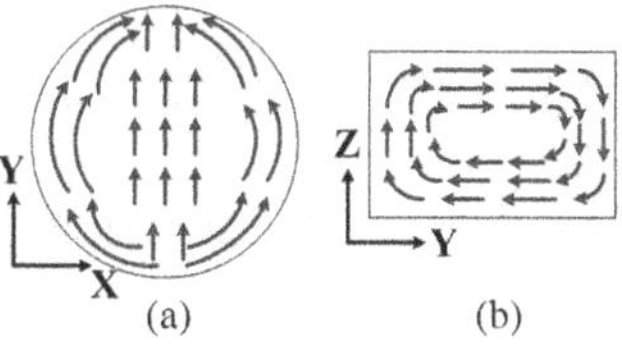

FIGURE 10.16 E-field distribution of $HEM_{11\delta+1}$ mode on cDRA: (a) top view; (b) side view.

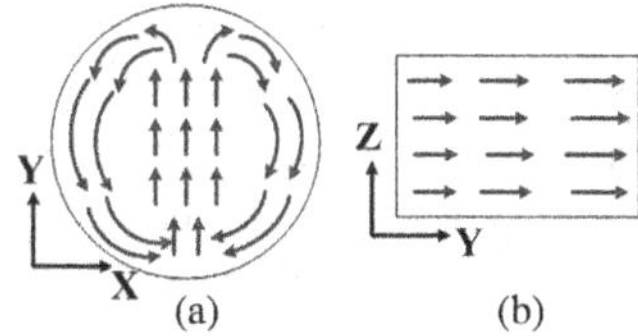

FIGURE 10.17 E-field distribution of $HEM_{12\delta}$ mode on cDRA: (a) top view; (b) side view.

broadside direction ($\theta=0°$; $\phi=0°$) since the E-field line of $HEM_{11\delta}$ mode is similar to a horizontally placed magnetic dipole. This mode is regularly utilized in DRAs since this mode is very easy to excite and has favorable radiation characteristics [35].

$HEM_{11\chi+1}$ mode: It is the next higher-order mode of $HEM_{11\delta}$ mode. The field variation of this mode is shown in Figure 10.16 [40]. Its field distribution is also similar to a horizontally placed magnetic dipole. Like $HEM_{11\delta}$ mode, this mode also radiates in the broadside direction [40].

b. **$HEM_{12\chi}$ mode**: The E-field distribution of $HEM_{12\delta}$ mode is shown in Figure 10.17. Its field distribution is the same as that of a horizontally placed electric dipole. This mode was investigated by *D.* Guha and his research group. The exclusive feature of this mode is that it can radiate in broadside direction with higher gain values than $HEM_{11\delta}$ mode [41]. To calculate the resonant frequency of $HEM_{12\delta}$ mode, no formula is available in the literature. But, the resonant frequency is anticipated with the help of resonant frequency of $HEM_{11\delta}$ and the aspect ratio (*H/a*) of cDRAs, which is given in Table 10.1 [38].

10.9.5 Feeding Mechanisms

The feeding mechanism plays a significant role in the case of dielectric resonator antennas since the aforementioned modal patterns are created with the help of

TABLE 10.1
Ratio of Resonant Frequency of $HEM_{11\delta}$ and $HEM_{12\delta}$ Corresponding to Aspect Ratio

Aspect Ratio (*H*/*a*)	Ratio of Resonant Frequency of $HEM_{11\delta}$ and $HEM_{12\delta}$ [38]
0.5	1.6
0.6	1.64
0.7	1.68
0.8	1.72
0.9	1.8
1.0	1.85

feeding techniques. Also, the dielectric resonator antennas are very flexible with different feeding techniques. Several feeding mechanisms such as probe feeding, microstrip line feeding, aperture coupling, coplanar feeding, and dielectric image guide coupling are generally used to excite the dielectric resonators [24].

For lower-frequency applications, coaxial probe feeding is a very popular technique to excite DRAs. Different types of modes can be excited in the DRAs by simply changing the location of the coaxial probe feed. If the probe is placed at the center of cDRAs, then $TM_{01\delta}$ mode is excited. On the other hand, $HEM_{11\delta}$ mode is generated in the cDRAs when the probe is placed adjacent to the cDRAs. The main advantage of the coaxial probe feed is that it does not require any additional matching network [24].

Another important feeding mechanism of DRAs is microstrip line coupling. This coupling mechanism is able to excite $HEM_{11\delta}$ mode in DRAs, since it acts as a horizontal magnetic dipole. It provides the advantage of easy fabrication [42].

One of the most common and popular feeding mechanisms in the case of DRAs is aperture coupling. The aperture is etched from the ground plane upon which the DRA is placed. The most important feature of aperture coupling is that the spurious radiation from the feed is less since the feed network is placed below the ground plane. The slot dimension should be electrically small. As a result, the amount of spilling radiation beneath the ground plane can be minimized. This type of coupling behaves as a horizontally placed magnetic dipole. So, $HEM_{11\delta}$ mode is easily excited in cDRAs with this type of excitation [24,38]. Coplanar waveguide (CPW) is one of the important excitation mechanisms in the DRA. The excitation mechanism of CPW is similar to the microstrip line. The control over impedance matching can be realized by adding stubs or loops at the edge of the line. Different types of modes can be easily excited by changing the location of CPW. For example, $HEM_{11\delta}$ mode is excited in the cDRAs when a coplanar loop is placed at the edge. On the other hand, centrally positioned loop excites $TM_{01\delta}$ mode in the cDRAs [43].

Dielectric image guide feeding techniques are generally used in millimeter-wave frequencies for DRA excitation since this feeding technique does not suffer from conductor losses. Generally, such type of feeding techniques is utilized in series-fed DRA arrays [44].

10.10 MIMO DIELECTRIC RESONATOR ANTENNAS

In most of the wireless standards, the MIMO antenna system is generally used and it will be used in future technologies as well. Currently, 802.11n WLAN standards utilize a MIMO antenna system for wireless access points. Generally, wireless access point boxes are noticed in buildings, shops, and other populated places and usually occupy large volume and space. The MIMO antenna used in the access point box is also large and sometimes almost twice the size of the access point box. Typically, the planar MIMO antenna is generally preferred in such a system. But, the main problem of the metallic printed antennas is their low efficiency and gain when a commercial substrate is used.

On the other hand, a dielectric resonator antenna (DRA) offers improved radiation efficiency and is easily integrable with other electronic components. It can also cover a wide frequency band by changing the size or altering the dielectric constant of the dielectric material used. Also, DRAs can be also utilized to cover a wide bandwidth or generate multiband characteristics. With this viewpoint, the dielectric resonator antennas are being used in the MIMO antenna systems due to their several advantages compared to microstrip patch antennas. The literature is not very rich in the case of MIMO DRAs. Recently, some DR-based MIMO antennas have been reported in the literature, and additional research should be done to achieve optimum MIMO performance. In this chapter, several isolation mechanisms are discussed to achieve improved performance in MIMO DRAs. In the next section, some isolation improvement techniques are discussed in the case of DR-based MIMO antennas.

10.11 MIMO DRA EXAMPLES

Similar to microstrip patch antennas, a multiport dielectric resonator is also used with several isolation enhancement mechanisms to achieve improved MIMO performance. High isolation also provides excellent diversity performance. In the open literature, several methods are used by different researchers in case of MIMO DRAs for obtaining a better isolation value: (i) generation of orthogonal mode; (ii) excitation of degenerated modes; (iii) introduction of the defected ground plane; (iv) use of decoupling structures; (v) introduction of meta-surface/frequency-selective surface/EBG between two DRAs; and (vi) separation of radiation patterns.

10.11.1 Generation of Orthogonal Mode

This method is a very popular technique to improve isolation in MIMO DRAs. Generally, the researchers use this technique to reduce the mutual coupling. The main advantage of this technique is that it does not require any extra decoupling network. As a result, the radiation characteristics of the antenna in terms of gain or radiation efficiency are not deteriorated. A dual-polarized MIMO DRA with two feeding ports was presented [45]. The rectangular DRA was excited with the help of a microstrip line-based conformal strip line. The two feeding ports were placed orthogonally to excite orthogonal modes in the DRA. Guo and Luk investigated a dual-polarized dual-port MIMO DRA. A cylindrical DR was excited with the help of a rectangular slot. The two slots were placed orthogonally and generated decoupled orthogonal modes. As a result, more than 30 dB of isolation was established between the ports [46]. Tang et al. proposed a dual-port dual-polarized MIMO DRA with low cross-polarization characteristics. At the horizontal port, an H-shaped slot-coupled feed was used to excite the rectangular DRA. On the other hand, at the vertical port, a balanced rectangular slot was used. The antenna covered a bandwidth of 440 MHz with a center frequency of 5.7 GHz [47]. Thamae and Wu proposed a cDRA with improved diversity performance. They discussed two antenna structures. In the first structure, the cDRA was excited by two orthogonally placed coaxial probes. In the second structure, three coaxial probes placed at 120° from each other were used to excite the cDRA [48]. A multiband MIMO DRA for mobile handsets was discussed [49]. The antenna operated at three bands, covering DVB-H, Wi-Fi, and WiMAX bands. Two different excitation mechanisms were placed orthogonally to generate orthogonal modes. The minimum isolation between the ports was 5 dB. A CPW-fed dual-polarized MIMO DRA was proposed with the excitation of even and odd mode generation. At port-1, CPW transmission line was used to excite TE^{x}_{111} mode. On the other hand, an L-shaped microstrip line was used at port-2 to generate TE^{Y}_{111} mode in the DRA. With this technique, more than 20 dB of isolation was established between the ports [50]. Thamae et al. introduced the concept of orthogonal mode generation in the MIMO DRA for isolation enhancement. They used two coaxial probes to excite the cylindrical DRA. In this antenna structure, the authors did not reveal the exact isolation value between the two ports. They did channel analysis for ensuring the proper MIMO applications [51]. Sun and Leung proposed a dual-band dual-polarized MIMO cDRA for DCS and WLAN applications. Two different excitation mechanisms, i.e., strip feed and slot feed, were used to generate $HE_{11\delta}$ and HE_{113} modes in the cDRA. The port isolation of higher than 36 dB was achieved with the generation of orthogonal modes [52]. A reduced-size MIMO DRA for 4G applications was discussed by Nasir et al. The rectangular DRA was fed by two orthogonally placed similar feed lines. The size reduction was achieved by placing a rectangular patch at the top of the DRA [53]. Messaoudene et al. proposed a dual-band MIMO DRA for LTE terminals. A single cDRA was excited by two orthogonal ports, and an improved isolation between the ports was achieved [54]. Jamaluddin and his team presented an F-shaped DRA for 4G applications. The antenna operated in between 2.3 and 3.14 GHz and achieved 33 dB of isolation. Several antenna orientations were discussed, such as parallel, horizontal, orthogonal, and mirrored parallel. The orthogonal orientation provided improved diversity performance among others [55]. Khan et al. proposed a dual-band MIMO DRA for LTE applications. The antenna covered LTE band 11 (1.43–1.5 GHz) and LTE band 7 (2.5–2.69 GHz). The DRA was excited by a coaxial probe and a rectangular slot in orthogonal orientation. As a result, isolation values of 23 and 32 dB were established in the first band and second band, respectively [56]. The same research group also proposed another dual-band MIMO DRA for WLAN and WiMAX applications. The proposed structure operated at 3.5 and 5.25 GHz bands. Two different feeding mechanisms were placed orthogonally to generate orthogonal modes to achieve high isolation [57]. Sharma and Biswas proposed a wideband two-element MIMO DRA. The wideband characteristic was established with the help of a mushroom-shaped DRA. This

DRA was excited with the help of a conformal trapezoidal patch. The two antenna elements were placed orthogonally to exploit polarization diversity [58]. Akhtar and his research group presented a SIW-fed MIMO DRA for mm-wave future 5G applications. The proposed structure consisted of four DRs, and each DR was fed by substrate integrated waveguide (SIW). The impedance bandwidth of the proposed antenna was 6.92% (26.64–28.55 GHz), and within the operating band, 30 dB of isolation was achieved [59]. In 2017, Gangwar and his research group presented a dual-port aperture-coupled MIMO cDRA for WiMAX applications [60]. Figure 10.18 shows the proposed dual-port MIMO antenna. The feeding structures are oriented in such a manner that orthogonal modes are generated in the cDRA. In the same year, the same research group proposed another cDRA-based MIMO antenna with high port isolation. Figure 10.19 shows the proposed dual-feed MIMO cDRA with high port isolation. Port-1 consists of Wilkinson power divider to excite the cDRA; on the other hand, a CPW-fed conformal strip line is used at port-2. These two feedings are oriented in such a manner that two orthogonal modes $\left(HE^{x}_{11\delta}\text{ and }HE^{y}_{11\delta}\right)$ are excited about the cDRA. As a result, more than 30 dB of isolation is produced between port-1 and port-2, which is shown in Figure 10.20 [61].

A circularly polarized (CP) MIMO DRA was proposed for WLAN applications [62]. The CP characteristic was achieved by stepping two rectangular DR along with square DR. The proposed MIMO antenna structure is displayed in Figure 10.21. The port coupling was reduced with the generation of orthogonal modes. The proposed structure achieves 21.51% impedance bandwidth and 13.23% axial ratio bandwidth.

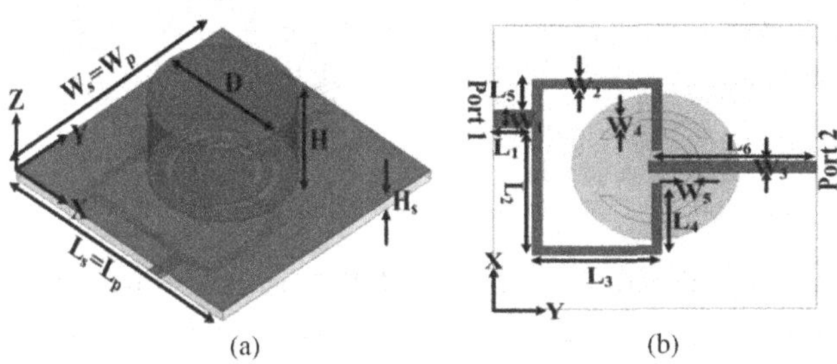

FIGURE 10.18 Dual-port MIMO antenna with orthogonal mode: (a) 3D view; (b) back view [60].

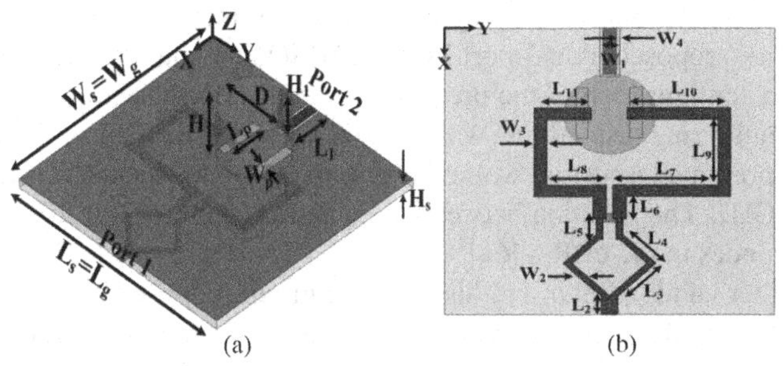

FIGURE 10.19 Dual-feed MIMO antenna with high isolation: (a) top view; (b) back view [61].

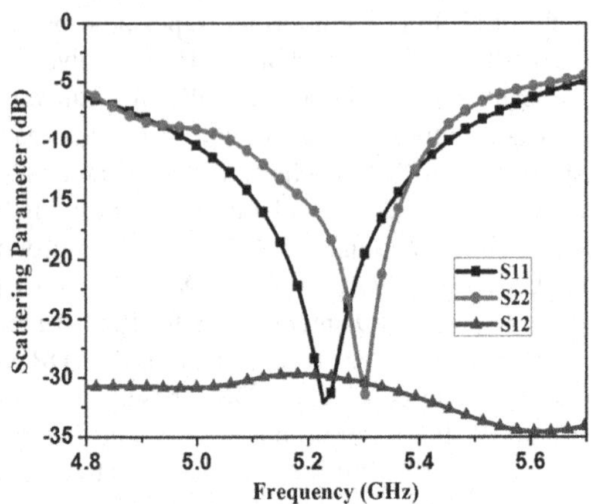

FIGURE 10.20 Scattering parameters of the proposed antenna.

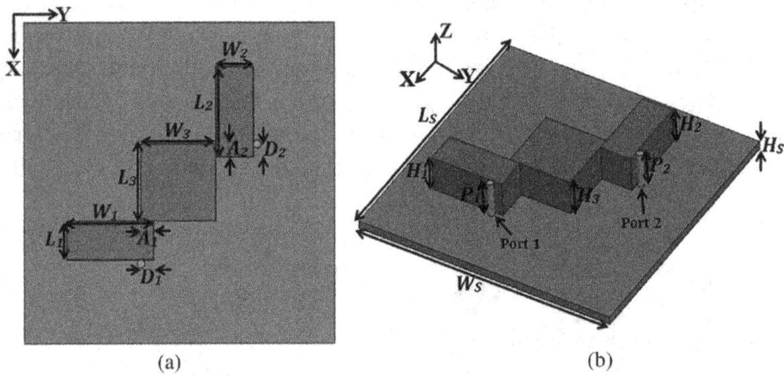

(a) (b)

FIGURE 10.21 Circularly polarized DRA-based MIMO antenna: (a) top view; (b) 3D view [62].

In the same year, Sharma et al. proposed an A-shaped DRA with the help of a conformal strip line for wideband MIMO applications. The wideband characteristic was achieved by combining two modes, i.e., TM_{101} and TM_{103}. The proposed structure offered a bandwidth of 59.7% covering 3.24–6.0 GHz band. More than 20 dB of isolation was achieved by employing orthogonal antenna element arrangements [63]. Das et al. proposed a dual-port hybrid MIMO antenna with dual-band characteristics. The configuration of the presented antenna is displayed in Figure 10.22. A modified annular ring printed line was utilized to excite $HE_{11\delta}$ and $TE_{01\delta}$ mode in the cDRA. The presented structure worked over two frequency bands, i.e., 1.75–2.4 GHz and 3.5–5.5 GHz. The isolation between the antenna elements was 20 dB by exciting orthogonal modes in the cDRA [64].

On the other hand, Akhtar and his research group proposed an equilateral triangular DRA for MIMO antennas with dual polarization characteristics. Two antenna designs were discussed. In the first design, an equilateral triangular DRA was fed by a conformal strip line. In this case, the antenna was slant-polarized along $\phi = 45°$

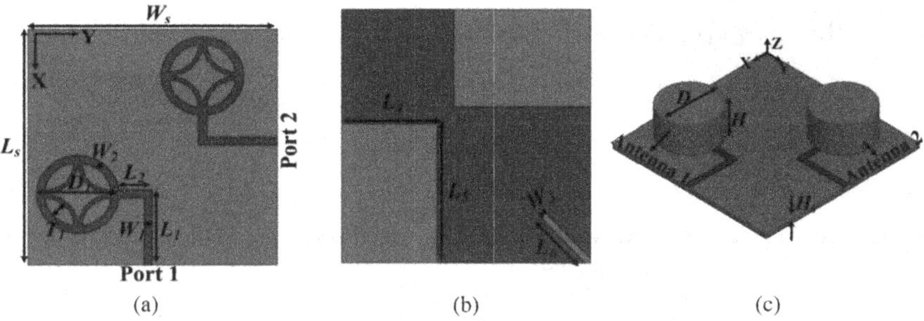

FIGURE 10.22 Dual-port hybrid MIMO antenna: (a) top view; (b) bottom view; and (c) 3D view [64].

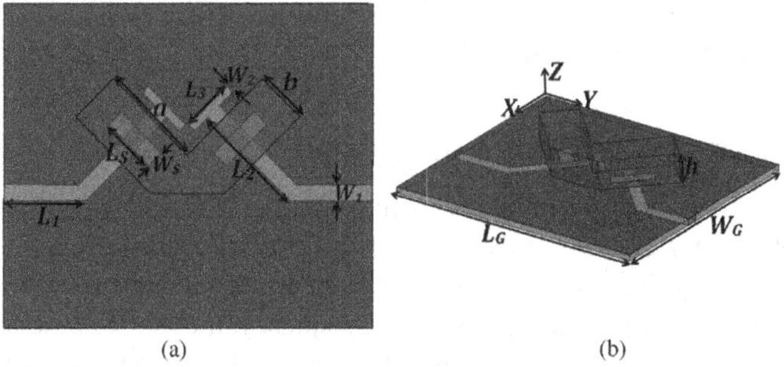

FIGURE 10.23 Truncated V-shaped dual-port MIMO antenna: (a) top view; (b) 3D view [67].

direction. When two ports were excited simultaneously with the same phase, the horizontal E-field was cancelled out. So, in-phase excitation generated only a vertically polarized field. Similarly, the out-of-phase excitation only produced horizontal E-field in the DRA. The impedance bandwidth of this antenna was 8.2% (5.27–5.72 GHz). The second design was the extension of the first antenna design. The feeding circuit consisted of a two-stage 180° hybrid coupler. The in-phase excitation was given by port-1, and on the other hand, out-of-phase excitation was given by port-2. As a result, more than 20 dB of isolation was established between the ports [65]. Pahadsingh and Sahu proposed a four-port MIMO DRA for cognitive radio platforms. The proposed design showed a dual-state operation, that is UWB and narrowband operations. The feeding lines for UWB operation and narrowband operation were orthogonal to each other. The isolation was more than 15 dB between all the ports [66]. A corner-truncated V-shaped dual-port MIMO antenna was proposed by Sahu et al. The advantage of this structure was that it can generate circular polarization also [67]. The proposed antenna structure is displayed in Figure 10.23.

Yaduvanshi and his research team proposed a super-wideband MIMO DRA. They introduced a half-split inverted-frustum-shaped DRA for UWB operation. The DRA was excited by the conformal strip line. The wideband isolation was achieved by placing two DR elements orthogonally [68]. Gangwar and his research team investigated a dual-band MIMO antenna for WLAN and WiMAX applications. To achieve dual-band operation, a dual-segment DRA was approached and each segmented DRA was

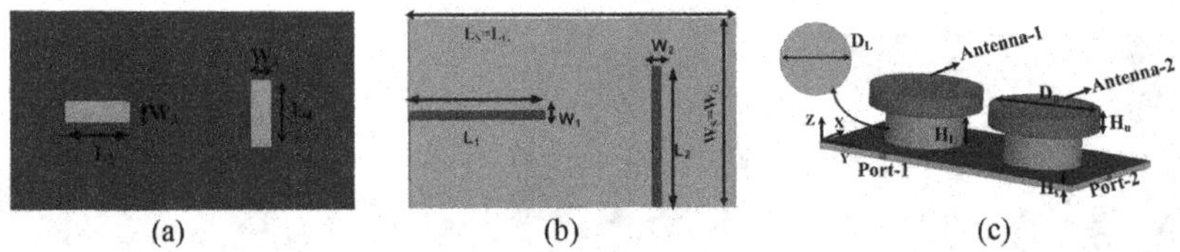

FIGURE 10.24 Dual-port MIMO antenna with multi-segmented DR (a) and (b) feeding structure; (c) the proposed antenna [69].

excited through a rectangular aperture. The dual-band was generated with the generation of $HE_{11\delta}$ and $HE_{12\delta}$ in the DRA. The feeding and the proposed antenna structure are displayed in Figure 10.24. Figure 10.24 reveals that the multi-segmented DR was placed orthogonally and achieved more than 20 dB of isolation [69].

Table 10.2 shows the comparison of the published DRA-based MIMO antennas with orthogonal mode generation.

10.11.2 Excitation of Degenerated Modes

This is one of the popular methods to improve the diversity performance of the MIMO DRAs. Degenerated modes mean two different modes excited in the same resonant frequency. Each of the modal patterns has a specific radiation pattern. If two different modes are excited at the same frequency, then the radiation patterns of the two modes are different. As a result, the isolation between the antenna elements is improved due to the pattern diversity. The excitation of degenerated modes in the DRAs is not an easy task. Several constraints are required to excite degenerated modes in the DRAs. Yan and Bernhard proposed a MIMO DRA for LTE femtocell base stations. Boundary perturbation was induced in the structure to excite degenerated modes. Two modes, i.e., TE and HE modes, of the DRAs were resonated at the same frequency and shared a common bandwidth. The isolation between the ports was more than 40 dB [70]. In 2012, Fumeaux and his research team proposed an omnidirectional MIMO cDRA with dual polarization. Two different modes, i.e., $TE_{01\delta}$ and $TM_{01\delta}$, were excited in the same frequency. A shielded metallic cavity and vias were used in the proposed structure to excite the degenerated modes [71]. Fang et al. proposed a three-port polarization diversity-based MIMO antenna for WLAN applications. The proposed structure used $TM_{01\delta}$ and $HE_{12\delta+1}$ modes in the cDRAs for degenerated mode excitation. The covariance matrix adaption evaluation strategy was used to generate such modes in the cDRA. Two pairs of slots were also used in the structure to excite $HE^{x}_{12\delta+1}$ and $HE^{y}_{12\delta+1}$ modes in the cDRA. The proposed structure showed more than 24 dB of isolation in the operating band [72]. A three-port MIMO DRA with mutually decoupled and degenerated modes was presented for X-band applications [73]. Two modes were excited in such a way that there was a low spatial overlapping of their field magnitudes, and the third mode was imposed orthogonally to the other two modes. These three modes were $TE^{x}_{\delta 32}$, $TE^{x}_{\delta 21}$, and $TE^{y}_{2\delta 1}$. This technique provided more than 20 dB of isolation between the ports. In 2019, Yang et al. proposed a dual-port cDRA with pattern diversity. The authors excited $HE_{11\delta}$ and $TM_{01\delta}$ modes of the HE and TM fundamental mode family. A cDRA was

TABLE 10.2
Published DRA-Based MIMO Antennas with Orthogonal Mode Generation

Operating Frequency Band (GHz)	Isolation Level (dB)	No. of Ports	No. of DRs	DRA Shape	Reference
2.19–2.248	20	2	1	Rectangular	[45]
1.9–2.2	35	2	1	Cylindrical	[46]
5.4–6.0	34	2	1	Rectangular	[47]
4.02–4.13	14.4	3	1	Cylindrical	[48]
0.79–0.862, 2.4–2.48, and 3.1–3.6	6	2	1	Rectangular	[49]
3.6–3.9	25	2	1	Rectangular	[50]
1.71–1.88	2.4-2.48	2	1	Cylindrical	[52]
1.7–1.9	18	2	1	Rectangular	[53]
1.82–2.0 and 2.5–2.73		2	1	Cylindrical	[54]
2.3–3.31	33	2	2	F-shaped	[55]
1.43–1.5 and 2.5–2.69	23	2	1	Rectangular	[56]
3.4–3.7 and 5.15–5.35	46	2	1	Plus-shaped	[57]
5.08–9.5	20	2	2	Mushroom-shaped	[58]
26.64–28.55	27	4	4	Cylindrical	[59]
3.1–3.68	25	2	1	Cylindrical	[60]
4.9–5.5	30	2	1	Cylindrical	[61]
5.02–6.23	18	2	1	Stepped rectangular	[62]
3.24–6.0	20	2	2	A-shaped	[63]
1.75–2.5 and 3.5–5.45	20	2	2	Cylindrical	[64]
5.27–5.65	22	2	1	Triangular	[65]
1.7–10.6	15	4	2	Cylindrical	[66]
4.89–5.42	14	2	1	Truncated V-shaped	[67]
7–34.6	20	2	2	Half-split inverted-frustum-shaped	[68]
3.3–3.8 and 5.0–5.7	20	2	2	Dual-segment cylindrical	[69]

loaded by another centrally loaded DRA with a higher dielectric constant to realize degenerated modes. The $HE_{11\delta}$ modes were excited by a rectangular cross-slot, and $TM_{01\delta}$ modes were realized by a coaxial probe. The proposed structure delivered more than 30dB of isolation between the ports [74]. In the same year, Yaduvanshi and his research team proposed a four-port MIMO DRA with pattern diversity characteristics. The antenna consisted of four DR elements. Out of four DR elements, two DRs were epsilon-shaped and the other two were cylindrical-shaped. These antenna elements were placed in such a manner that they resonated in the same frequency band. The antenna was excited with the $HE_{11\delta}$ and $TM_{01\delta}$ modes in the epsilon-shaped and cylindrical-shaped DRs, respectively. The isolation exceeded more than 15dB between the ports [75]. Yang and Leung proposed a compact dual-port pattern diversity-based MIMO DRA. The pattern diversity was achieved with the generation

TABLE 10.3
Published MIMO DRAs with degenerated modes

Operating Frequency Band	Isolation Level	No. of Ports	No. of DRs	DRA Shape	Reference
0.695–0.705	40	2	1	Split-cylindrical	[70]
3.78–4.07	15	2	1	Ring	[71]
2.4–2.5	24	2	1	Cylindrical	[72]
9.12–10	20	3	1	Rectangular	[73]
2.33–2.53	30	2	1	Cylindrical	[74]
3.64–4.24	15	4	4	Epsilon-shaped and cylindrical	[75]
2.4–2.5	30	2	1	Cylindrical	[76]

of $HE_{11\delta}$ and $TM_{01\delta}$ modes in the cDRA. A meander line-loaded annular slot was used to generate $TM_{01\delta}$ mode, and its resonant frequency was also lowered. To excite $HE_{11\delta}$, a pair of differential strips were utilized. The overlapping impedance bandwidth of these two modes was 5.7%, and within the overlapping band, the isolation was more than 30dB [76]. These published research articles are summarized in Table 10.3. Table 10.3 shows the performance comparison of various MIMO DRAs that utilized degenerated modes for isolation improvement.

10.11.3 Introduction of the Defected Ground Plane

Another technique for isolation improvement in the MIMO DRAs is the introduction of defected ground structure (DGS) between the antenna elements. In this case, a portion of the ground plane was etched between the antenna elements. Khan et al. proposed a dual-band MIMO DRA for WLAN/WiMAX applications. The dual-band was generated with the excitation of two modes, i.e., $TE_{1\delta 1}$ and $TE_{2\delta 1}$. The L-shaped DRA was excited with the help of two coaxial probes. A rectangular DGS slot was etched from the ground plane to improve port isolation between the antenna elements [77]. Nasir et al. proposed a MIMO DRA for 4G applications. The rectangular DRA was housed in the substrate and fed by two microstrip lines to excite TE modes in the DRA. The mutual coupling between the ports was reduced by etching two slits in the ground plane. The isolation was better than 20dB after etching the slits [78]. Das et al. proposed a novel circular ring-shaped defected ground structure to improve the diversity performance of the antenna. The top and bottom geometries of the proposed antenna are shown in Figure 10.25. The cDRA was exited with the help of a triangular patch antenna. The antenna operated between 3.7 and 7.25GHz, and the isolation was more than 17dB within the band [79].

Trivedi and Pujara proposed a tree-shaped fractal DRA with a defected ground structure for wideband MIMO applications. Two C-shaped periodic DGS were used to reduce the mutual coupling between the two closely spaced DRAs. The isolation of more than 15dB was obtained in the area of interest [80]. The same author proposed a two-element Maltese-shaped MIMO DRA for UWB applications. The DRA

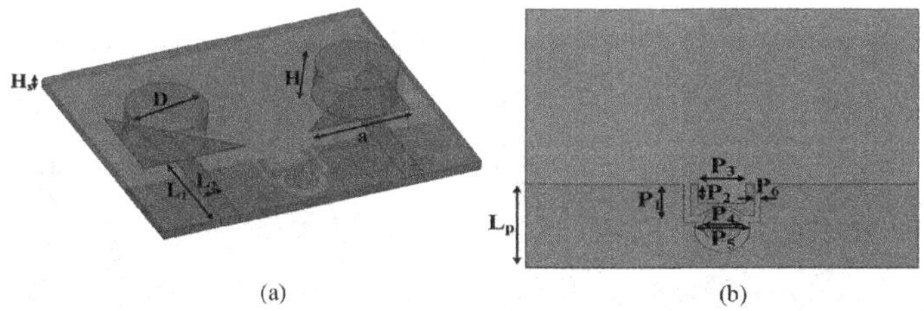

FIGURE 10.25 MIMO cDRA with DGS: (a) top view; (b) bottom view [79].

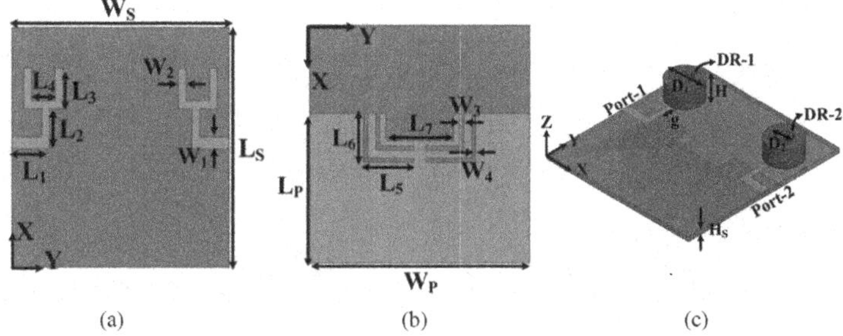

FIGURE 10.26 MIMO ring DRA with defected ground structure: (a) feeding structure; (b) bottom view; and (c) 3D view [83].

was fed utilizing a trapezoidal-shaped conformal strip line. Two C-shaped DGS were inserted between the elements, resulting in 18 dB of isolation in the entire operating band [81]. In a similar manner, a rectangular DGS was inserted between the hemispherical DRAs to improve MIMO performance. The length of the slot was optimized to enhance isolation [82].

Gangwar and his research group also proposed a dual-element hybrid ring DRA-based MIMO antenna for wideband applications. Wideband characteristics were achieved by exciting both the U-shaped printed lines along with ring DRA. Rectangular and L-shaped defects were created in the ground plane to achieve wideband isolation [83]. The presented antenna structure is displayed in Figure 10.26. The impedance bandwidth of the proposed structure was more than 60%, and isolation values were more than 15 dB within the operating band.

Sahu et al. proposed an L-shaped circularly polarized MIMO DRA for WLAN applications. The feed position was optimized in such a manner that it excited orthogonal modes inside the DRA, resulting in circular polarization in the structure. The configuration of the proposed antenna structure is shown in Figure 10.27. A pair of DGS was etched between the DR elements to improve port as well as field isolation between the antenna elements, as shown in Figure 10.27 [84].

A triple-port MIMO cDRA was presented for WLAN applications [85]. Port-1 consisted of a two-element cDRA array. Each element of the cDRA was excited by port-2 and port-3 with the help of a coaxial probe. The configuration of the triple-port MIMO antenna is shown in Figure 10.28. The isolation between the ports was

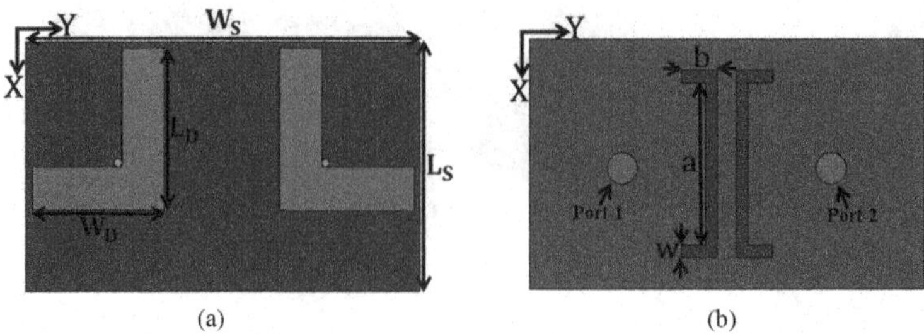

FIGURE 10.27 Circularly polarized L-shaped MIMO DRA: (a) top view; (b) bottom view [84].

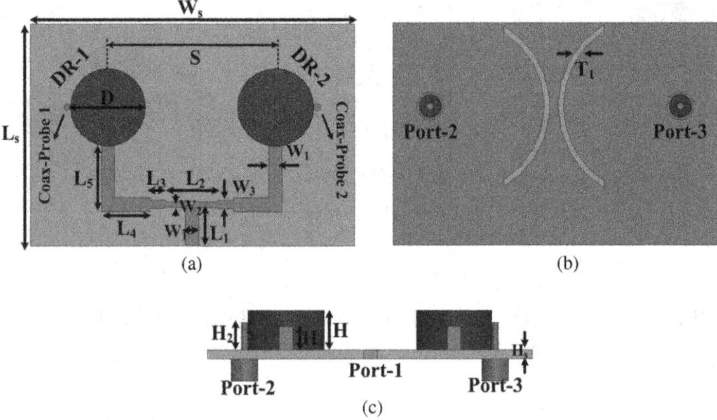

FIGURE 10.28 Triple-port MIMO antenna system: (a) top view; (b) bottom view; and (c) side view [85].

improved by two C-shaped DGS as shown in Figure 10.28. As a result, the isolation between the ports was 20 dB. Das et al. proposed a circularly polarized dual-port dual-sense MIMO antenna for WLAN applications. Figure 10.29 shows the geometry of the presented antenna. It reveals that modified circular apertures were used to excite each cDRA. The proposed structure was able to generate LHCP and RHCP depending on the selection of the port. To improve the isolation between the ports, DGS slits were utilized. As a result, the isolation between the ports was more than 25 dB within the operating band [86]. A dual-band dual-polarized MIMO ring dielectric resonator antenna was proposed for MIMO communication system [87]. A modified plus-shaped aperture was used to excite $HE^x_{11\delta}$ and $HE^y_{11\delta}$ modes inside the ring DRA. Two circular-shaped DGS were placed between the antenna elements, resulting in more than 20 dB of isolation between the DR elements. Biswas and Chakraborty proposed a complementary meander line-inspired MIMO DRA for dual-band applications. The structure consisted of an "I"-shaped DRA. Two complementary spiral meander lines and two circular slots were etched from the ground plane to improve port isolation [88]. Table 10.4 shows several published MIMO DRAs with DGS for isolation improvement. Covered band, isolation level, and the number of ports and DRs are included in Table 10.4 for comparison.

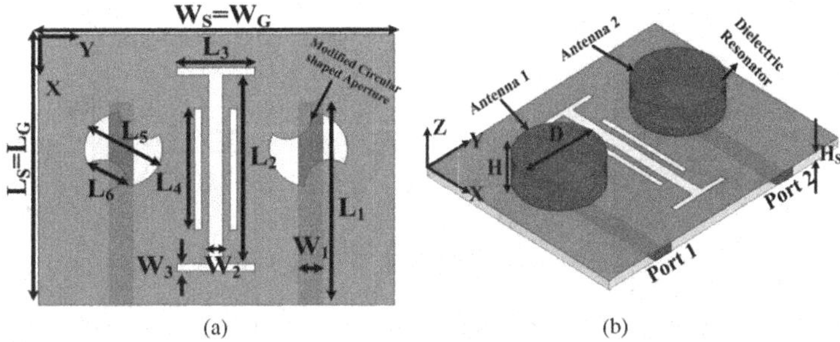

FIGURE 10.29 Dual-port circularly polarized MIMO antenna: (a) top view; (b) 3D view [86].

TABLE 10.4
Published MIMO DRAs with Defected Ground Structure

Operating Frequency Band	Isolation Level	No. of Ports	No. of DRs	DRA Shape	Reference
3.4–3.7 and 5.15–5.35	15	2	1	L-shaped	[77]
2.56–2.64	20	2	1	Rectangular	[78]
3.7–7.25	17	2	2	Cylindrical	[79]
3.6–12.6	18	2	2	Tree-shaped fractal DRA	[80]
3.95–10.4	15	2	2	Maltese-shaped	[81]
2.4–2.5 and 4.15–4.25	18	2	2	Hemispherical	[82]
3–7	18	2	2	Ring	[83]
5.2–6.08	20	2	2	L-shaped	[84]
5.0–6.0	20	3	2	Cylindrical	[85]
5.25–6.0	25	2	2	Cylindrical	[86]
2.3–2.9 and 3.4–4.0	20	2	2	Ring	[87]
3.46–5.37 and 5.89–6.49	18.5	2	2	I-shaped	[88]

10.11.4 Use of Decoupling Structures

This is one of the important techniques to improve isolation between the MIMO DRAs. This technique requires an extra matching circuit or antenna elements along with antenna elements. In 2016, Sharma et al. proposed a dual-polarized triple-band hybrid MIMO DRA. The folded microstrip feeding structure was used to excite the cDRA. The proposed antenna structure is displayed in Figure 10.30. To improve isolation, a metallic strip was used, which acts as an electromagnetic reflector [89].

Similarly, in the next year, Sahu et al. proposed a dual-polarized triple-band MIMO hybrid antenna for WLAN/WiMAX applications. The cDRA was excited by a modified Y-shaped printed line. The presented antenna structure is shown in Figure 10.31. Two metallic strips were placed to reduce mutual coupling. Metallic strips acted as a reflector and enhanced the isolation between the ports, which is shown in Figure 10.31 [90].

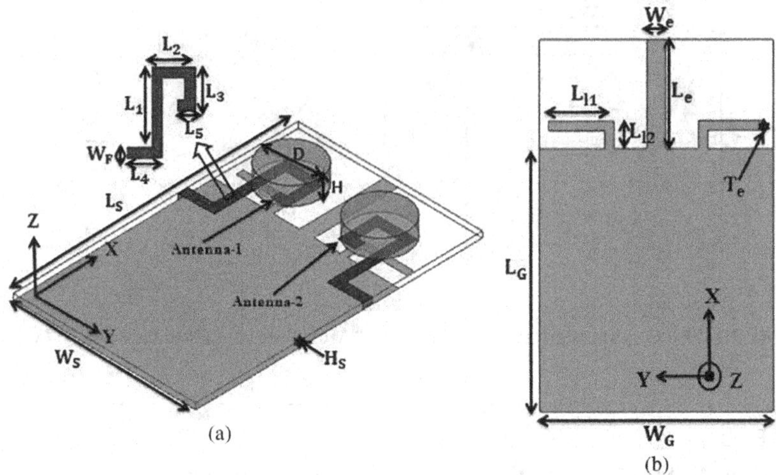

FIGURE 10.30 Dual-port MIMO cDRA with folded microstrip line: (a) top view; (b) bottom view [89].

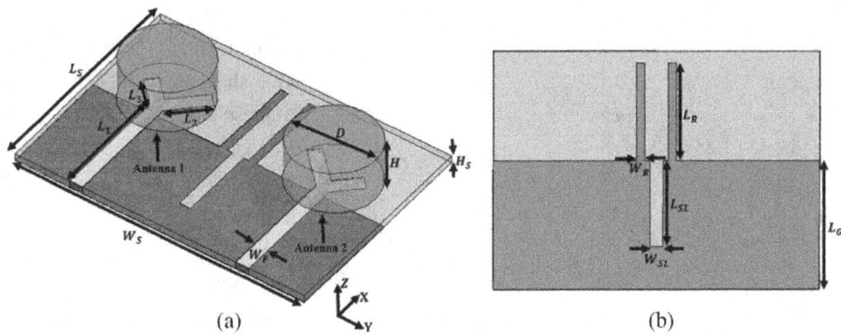

FIGURE 10.31 Dual-polarized MIMO hybrid antenna: (a) top view; (b) bottom view [90].

A dual-polarized triple-band hybrid MIMO antenna was presented by Sahu et al. The configuration of the antenna structure is shown in Figure 10.32. The rectangular DRA was excited with the help of a modified U-shaped microstrip line. To improve isolation between the ports, a metallic strip was used at the center of the ground plane [91]. Sharawi et al. proposed a dual-band DRA-based MIMO antenna for wireless access points. Two groups of DRA were chosen with different sizes. One group of cDRA covered 2.45 GHz band, and the other group operated at 5.8 GHz band. To improve isolation, a reflector element was used to tilt the antenna beam in the opposite direction. As a result, the isolation was improved between the DR elements [92]. A triple-band dual-port hybrid MIMO cDRA was presented by Sharma et al. The geometrical layout of the proposed geometry is shown in Figure 10.33. It reveals that an annular ring along with T-shaped printed line was used to excite each cDRA. To improve the isolation between the ports, two metallic strips were used. As a result, the isolation between the ports exceeded more than 20 dB [93].

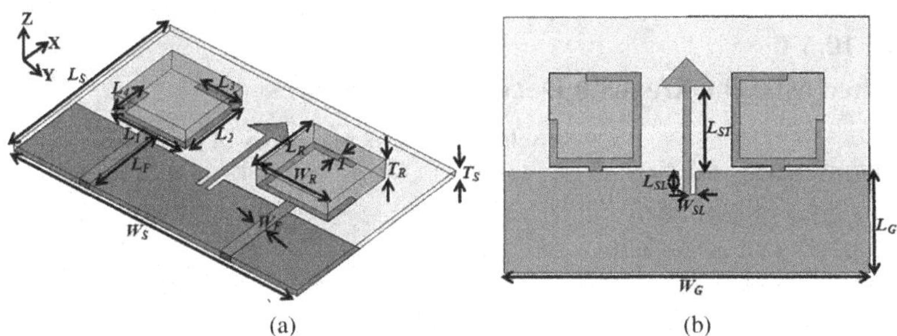

FIGURE 10.32 Geometry of the triple-band hybrid MIMO antenna: (a) 3D view; (b) bottom view [91].

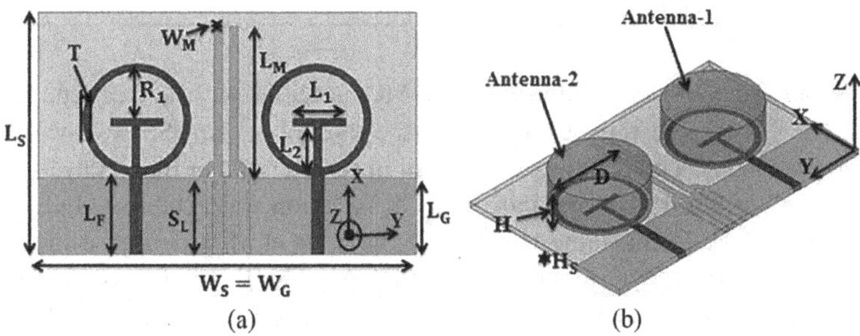

FIGURE 10.33 Geometrical layout of the triple-band MIMO antenna: (a) feeding orientation; (b) the proposed geometry [93].

Zheng and his research team used metallic vias in the DRA to reduce mutual coupling in the mm-wave MIMO DRA. The vias were placed strategically in such a manner that the vias could potentially affect the field distribution and reduce the field coupling. This method can improve the isolation of E-plane and H-plane MIMO antennas [94]. A MIMO DRA for mm-wave 5G applications was discussed [95]. Two rectangular DRAs were excited by the microstrip-fed slot line. To improve isolation, the metal strip was placed on the upper portion of the DRA where strong E-field was generated. As a result, the field coupling was suppressed and the isolation between the antenna elements was improved. More than 12 dB of isolation was achieved with this technique. Table 10.5 shows several published MIMO DRAs with a decoupling structure. This table includes operating band, isolation level, and the number of ports/DRs for comparison purposes.

10.11.5 Meta-Surface/Frequency-Selective Surface/EBG between Two DRAs

This is a new concept introduced in the DRA domain to improve MIMO performance. Al-Hasan et al. proposed an EBG structure for mutual coupling reduction in MIMO antennas. The EBG structure delivered a wide band gap region around 60 GHz and suppressed field correlation between the two DRAs [96]. Denidni and his research

TABLE 10.5
Published MIMO DRAs with Decoupling Structures

Operating Frequency Band	Isolation Level	No. of Ports	No. of DRs	DRA Shape	Reference
2.24–2.38, 2.5–3.26, and 4.88–7	15	2	2	Cylindrical	[89]
2.21–3.13, 3.40–3.92, and 5.3–6.10	20	2	2	Cylindrical	[90]
2.37–2.86, 3.18–3.84, and 4.92–5.73	16	2	2	Rectangular	[91]
2.3–2.5 and 5.7–6	15	8	8	Cylindrical	[92]
1.5–2.55, 3.21–4.0, and 4.59–5.98	25	2	2	Cylindrical	[93]
25–27	35	2	2	Rectangular	[94]
27.25–28.59	25	2	2	Rectangular	[95]

group proposed a 60-GHz mm-wave MIMO antenna with a frequency-selective surface (FSS) wall. To reduce the free space radiation, the FSS wall was inserted between the DRAs. The FSS was optimized in such a manner that it covered a wide-band from 57 to 63 GHz. More than 30 dB of isolation was established after placing the FSS wall [97]. In the same manner, Dadgopour et al. proposed a meta-surface shield for isolation improvement in MIMO DRAs. The DR elements were placed in the H-plane. The meta-surface shield consisted of SRR cells and was placed along the E-plane. The meta-surface shield delivered band-stop functionality within the operating bandwidth and is able to provide improved field isolation [98]. Denidni and his research team proposed a meta-surface orthogonalize wall for mutual coupling reduction in MIMO DRAs. By incorporating the meta-surface wall, the TE modes of the antenna became orthogonal to each other. The mutual coupling was reduced by 16 dB when the meta-surface wall was placed [99]. The same research team also introduced a metamaterial polarization-rotating wall for mutual coupling reduction in mm-wave MIMO antennas. The meta-surface wall consisted of a 1×7 unit cell and was placed along the E-plane. The same concept was also used here, i.e., using the meta-surface wall. The TE modes of the antennas became orthogonal to each other. The isolation was improved by 16 dB after placing the wall [100]. Rezapour et al. proposed double-slit SRRs for isolation enhancement in rectangular dielectric resonator antennas. The SRR structure supported epsilon-negative characteristics, and it was placed in between two E-coupled rectangular DRAs. The proposed method provided more than 11 dB of isolation [101]. The previously described MIMO DRAs with meta-surface/FSSs/EBGs are tabulated in Table 10.6. Several factors such as the operating band, mutual coupling level, and the number of ports are included in the table for comparison purposes.

10.11.6 Separation of Radiation Patterns

Separation of radiation patterns is one of the important techniques to improve the isolation between the ports in the DRA-based MIMO antenna systems. In this

TABLE 10.6
Published MIMO DRAs with Meta-surface/FSSs/EBGs

Operating Frequency Band (GHz)	Isolation Level	No. of Ports	No. of DRs	DRA Shape	Reference
54–65	20	2	2	Cylindrical	[96]
57–63	30	2	2	Cylindrical	[97]
56.6–64.8	46.5	2	2	Cylindrical	[98]
57–64	20	2	2	Cylindrical	[99]
57–64	23	2	2	Cylindrical	[100]
2.604–2.64	30	2	2	Rectangular	[101]

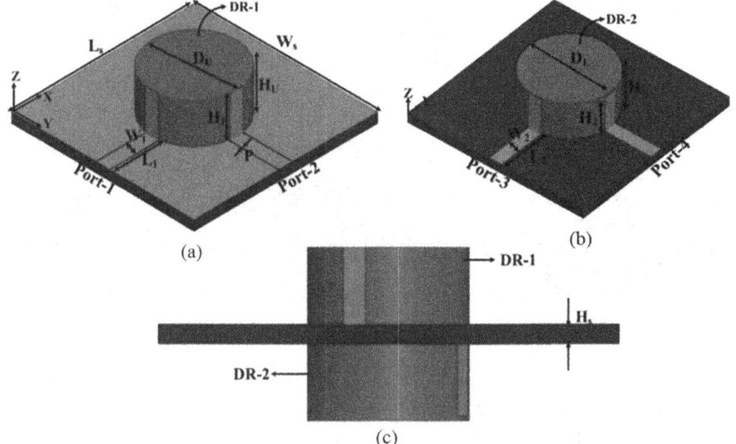

FIGURE 10.34 Four-port cylindrical MIMO antenna with bidirectional diversity: (a) top view; (b) bottom view; and (c) side view [102].

technique, the radiation beams of different antenna elements are separated spatially to improve the field correlation between the antenna elements. The advantage of this technique is that it is able to improve both the ECC and mutual coupling values. Very few research articles are available based on this technique.

Das et al. proposed a back-to-back four-port MIMO cDRA with bidirectional pattern diversity [102]. The configuration of the proposed antenna is displayed in Figure 10.34. Figure 10.34 reveals that one cDRA was placed on the top side of the substrate and the other cDRA was placed on the bottom side of the substrate. Each cDRA was excited by two ports. The top and bottom cDRAs radiated in broadside direction and in the direction opposite to broadside direction, respectively. In this way, the isolation and ECC values can be improved significantly. The same concept also presented with the help of rectangular DRAs [103]. Figure 10.35 shows the configuration of the four-port MIMO rDRA with bidirectional diversity.

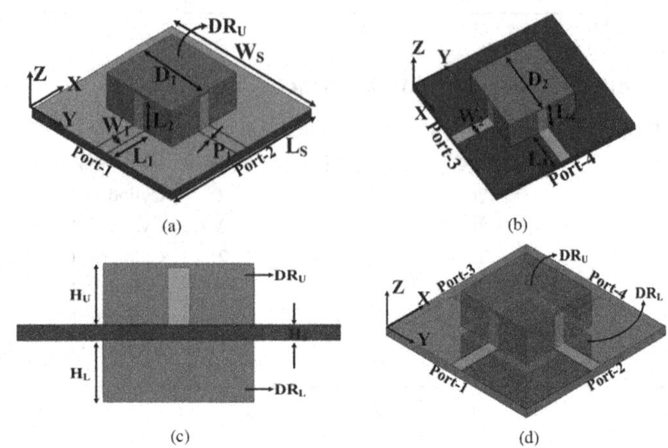

FIGURE 10.35 Four-port rectangular MIMO antenna with bidirectional diversity: (a) top view; (b) bottom view; (c) side view; and (d) 3D view [103].

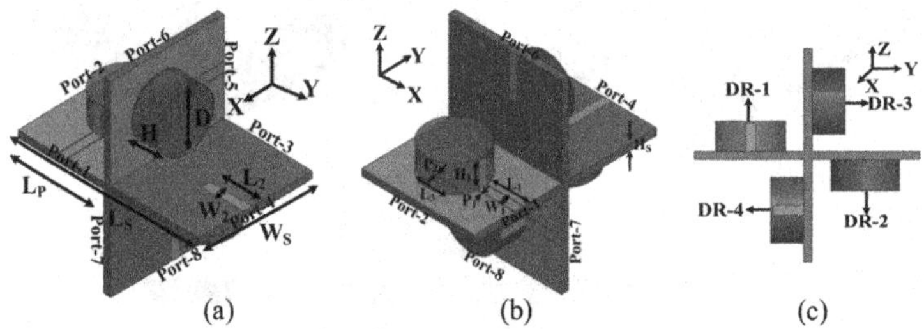

FIGURE 10.36 Eight-port MIMO antenna with multi-directional diversity: (a) and (b) 3D view; (c) side view [104].

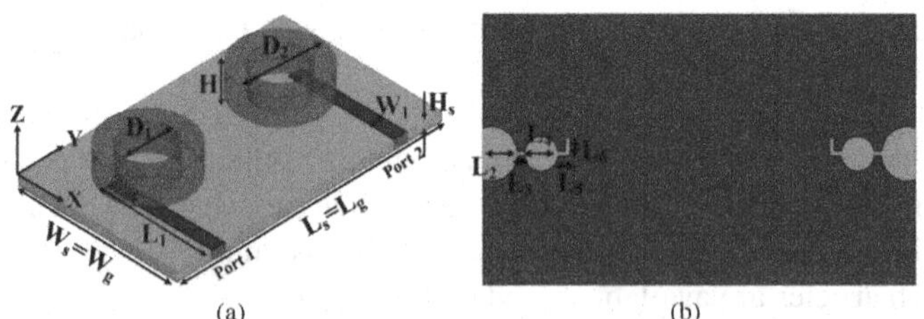

FIGURE 10.37 Wideband MIMO ring DRA: (a) top view; (b) bottom view [105].

In the same manner, four cDRAs were used to achieve multi-directional pattern diversity. The configuration of the proposed structure is shown in Figure 10.36. The four cDRAs were placed strategically in a different direction to isolate the antenna beam in a different direction. In this way, the isolation and ECC values were enhanced significantly [104]. A wideband MIMO antenna with a complementary radiation pattern was investigated by Das et al. The geometry of the proposed wideband antenna is displayed in Figure 10.37. A microstrip line was used to excite both the slots and

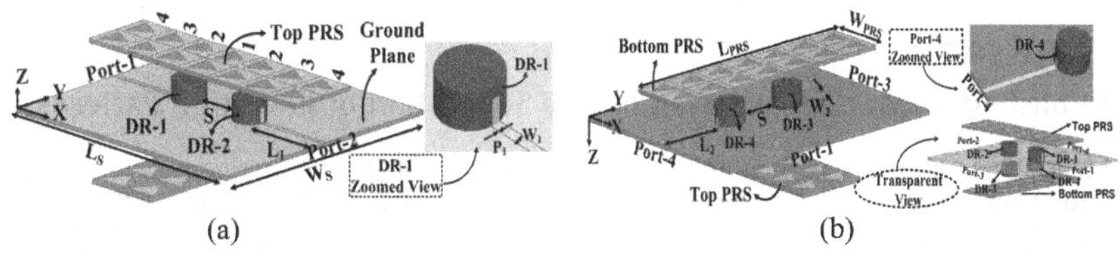

FIGURE 10.38 Four-port MIMO antenna with PRS: (a) top view; (b) bottom view [106].

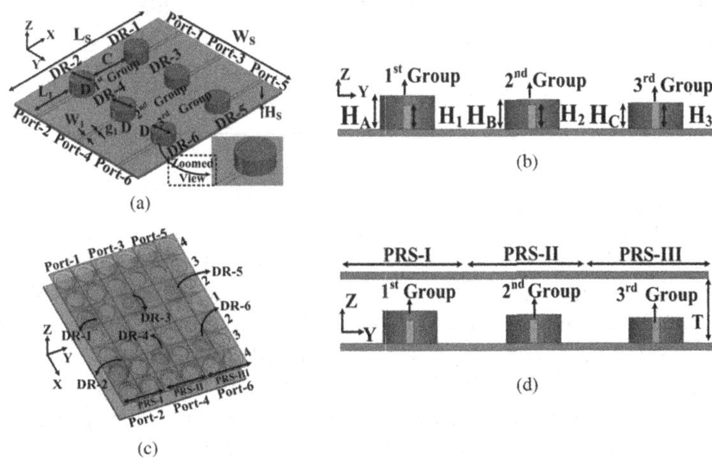

FIGURE 10.39 Six-port MIMO antenna without and with PRS: (a) top view (without PRS); (b) side view (without PRS); (c) top view (with PRS); and (d) side view (with PRS) [107].

the ring DRA. The isolation between the ports was improved by generating complementary radiation pattern between the ports [105].

Partially reflective surface (PRS) was also used to decorrelate the radiation patterns of the MIMO antennas. Gangwar and his research team proposed the use of two PRS to isolate the antenna beam in the different direction. The configuration of the proposed structure is displayed in Figure 10.38. Two cDRAs were placed on the top of the substrate, and another two cDRAs were positioned at the bottom part of the substrate. To decorrelate the radiation pattern, two PRS were placed at both sides of the substrate. The PRS tilted the antenna beam from broadside direction to tilted direction. As a result, the isolation was improved by 10–12 dB and more than 75% improvement in ECC values was realized [106].

In a similar manner, three different groups of cDRAs with different heights were placed at the top side of the substrate. Three PRS acted as a superstrate and were placed at the top side of each group of DRAs. The geometry of the proposed structure is displayed in Figure 10.39. Each PRS isolates the antenna beam in a different direction. In this way, the isolation and ECC values were improved [107]. Table 10.7 shows a comparison between different published research articles based on the separation of radiation patterns. Several factors such as covered frequency band, isolation level, and the number of ports and DRs are included in this table.

TABLE 10.7
Published MIMO DRAs based on the Separation of Radiation Patterns

Operating Frequency Band (GHz)	Isolation Level	No. of Ports	No. of DRs	DR Shape	Reference
5.4–6.0	18	4	2	Cylindrical	[102]
5.00–5.52	18	4	2	Rectangular	[103]
5.6–5.9	20	8	4	Cylindrical	[104]
3.4–8.2	20	2	2	Ring	[105]
5.15–5.35	23	4	4	Cylindrical	[106]
5.15–5.35, 5.45–5.65, and 5.7–5.9	25	6	6	Cylindrical	[107]

10.12 CONCLUSION

DRA-based MIMO antenna systems are one of the most emerging topics and can be utilized in various applications such as routers, laptops, and tablet PCs. In this chapter, initially, the basics of MIMO, MIMO techniques, and advantages of MIMO over the SISO system were discussed. Several MIMO performance metrics were also presented. After that, a detailed investigation on DRAs, their advantages over microstrip patch antennas, and different modal configurations (for cylindrical DRAs) were studied. Also, in this chapter, DRA-based MIMO antenna systems were discussed for various wireless applications with several antenna examples. Different isolation improvement techniques were investigated. In each category, a separate table was presented for comparing the several MIMO antennas which have recently been reported in the literature.

REFERENCES

1. Rappaport, T., Heath, R., Daniels, R. and Murdock, J. (2014), *Millimeter Wave Wireless Communications, ser. Prentice Hall Communications Engineering and Emerging Technologies Series from Ted Rappaport*, New York: Pearson Education.
2. Garg, V. (2007), *Wireless Communication and Networking*. San Francisco, CA: Morgan Kaufmana.
3. Molisch, A. F. (2007), *Wireless Communications*. Hoboken, NJ: John Wiley Sons.
4. Jakes, W. C. and Cox, D. C. (1994), *Microwave Mobile Communications*. Hoboken, NJ: Wiley-IEEE Press.
5. Hata, M. and Nagatsu, T. (1980), "Mobile location using signal strength measurements in a cellular system," *IEEE Transactions on Vehicular Technology*, vol. 29 (2), pp. 245–252.
6. Erceg, V., Greenstein, L. J., Tjandra, S. Y., Parko, S. R., Gupta, A., Kulic, B., Julius, A. A. and Bianchi, R. (1999), "An empirically based path loss model for wireless channels in suburban environments," *IEEE Journal on Selected Areas in Communications*, vol. 17(7), pp. 1205–1211.
7. Driessen, P. F. and Foschini, G. J. (1999), "On the capacity formula for multiple-input-multiple output wireless channels: A geometric interpretation," *IEEE Transactions on Communications*, vol. 47 (2), pp. 173–176.

8. Shannon, C. E. (1948), "A mathematical theory of communication," *The Bell System Technical Journal*, vol. 27, pp. 379–423.
9. Jensen, M. A. and Wallace, J. W. (2004), "A review of antennas and propagation for MIMO wireless communications," *IEEE Transactions on Antennas and Propagation*, vol. 25, pp. 2810–2824.
10. Rappaort, T. S. (2002), *Wireless Communications: Principles and Practice*. New York: Pearson Education.
11. Bolcskei, H. (2002), *Multiple-Input Multiple-Output (Mimo) Wireless Systems*, J. Gibson, ed., CRC Press Florida, USA
12. Paulraj, A., Nabar, R. and Gore, D. (2003), *Introduction to Space-Time Wireless Communications*. Cambridge: Cambridge University Press.
13. Telatar, I. E. (1999), "Capacity of multi-antenna Gaussian channels," *European Transactions on Telecommunication*, vol. 10, pp. 585–595.
14. Foschini, G. J. (1996), "Layered space-time architecture for wireless communication in a fading environment when using multi-element antennas," *Bell Labs Technology Journal*, vol. 1(2), pp. 41–59.
15. Foschini, G., Golden, G., Valenzuela, R., and Wolniansky, P. (1999), "Simplified processing for high spectal efficiency wireless communication employing multi-element arrays," *IEEE Journal of Selected Areas Communication*, vol. 17, pp. 1841–1852.
16. Alamouti, S. M. (1998), "A simple transmit diversity technique for wireless communications," *IEEE Journal on Selected Areas in Communications*, vol. 16(8), pp. 1451–1458.
17. Tarokh, V., Seshadri, N. and Calderbank, A. (1998), "Space-Time Codes for High Data Rate Wireless Communication: Performance Criterion and Code Construction," IEEE Transaction on Information Theory, vol. 44 (2), pp.744–765.
18. Sharawi, M. S. (2014), *Printed MIMO Antenna Engineering*. Norwood, MA: Artech House.
19. Blanch, S., Romeu, J. and Corbella, I. (2003), "Exact representation of antenna system diversity performance from input parameter description," *IET Electronics Letter*, vol. 39(9), pp. 707–708.
20. Vaughan, R. and Andersen, J. B. (2003), *Channels, Propagation and Antennas for Mobile Communication*. Hertfordshire, UK: IET.
21. Chae, S. H., Oh, S. K. and Park, S. O (2007), "Analysis of mutual coupling, correlations and TARC in WiBro MIMO array antenna," *IEEE Antennas and Wireless Propagation Letters*, Vol. 6, pp. 122–125.
22. Sarkar, D. and Srivastava, K. V. (2017), "A compact four-element MIMO / diversity antenna," *IEEE Antennas & Wireless Propagation Letter*, vol. 16, pp. 2469–2472.
23. Choukiker, Y. K., Sharma, S. K. and Behra, S. K. (2014), "Hybrid fractal shape planer monopole antenna covering multiband wireless communications with MIMO implementation for handheld mobile devices," *IEEE Transactions on Antennas and Propagation*, vol. 62, pp. 1483–1488.
24. Petosa, A. (2007), *Dielectric Resonator Antenna Handbook*. Boston, UK: Artech house.
25. Pozar, D. M. (2012), *Microwave Engineering*. Hoboken, NJ: John Wiley & Sons.
26. Richtinger, R. D. (1939), "Dielectric resonators," *Journal of Applied Physics*, vol. 10, pp. 391–398.
27. Okaya, A. and Barash, L. F. (1962), "The dielectric microwave resonators," *Proceedings of the IRE*, vol. 50, pp. 2081–2092.
28. Long, S. A., McAllister, M. W. and Shen, L. C. (1983), "The resonant cylindrical dielectric cavity antenna," *IEEE Transaction on Antennas and Propagation*, vol. 31, pp. 406–412.
29. Luk, K. M. and Leung, K. W. (2003), *Dielectric Resonator Antenna*. Baldock, England: Research Studies Press Ltd.

30. Guha, D., Gupta, P. and Kumar, C. (2015), "Dual band cylindrical dielectric resonator antenna employing $HE_{11\delta}$ and $HE_{12\delta}$ mode excited by new composite structure," *IEEE Transactions on Antennas and Propagation*, vol. 63, pp. 433–438.
31. McAllister, M. W. and Long, S. A. (1983), "Rectangular dielectric resonator antenna," *IET Electronics Letter*, vol. 45, pp. 1348–1356.
32. McAllister, M. W. and Long, S. A. (1984), "Resonant hemispherical dielectric resonator antenna," *IET Electronics Letter*, vol. 20, pp. 657–659.
33. Gastine, M., Courtois, L. and Dormann, J. J. (1967), "Electromagnetics resonances of free dielectric spheres," *IEEE Transactions on Microwave Theory and Technique*, vol. 15, pp. 694–700.
34. Kishk, A. A., Zhou, G. and Glisson, A. W. (1994), "Analysis of dielectric resonator antennas with emphasis on hemispherical structures," *IEEE Antennas and Propagation Magazine*, vol. 36, pp. 20–31.
35. Mongia, R. K. and Ittipiboon, A. (1997), "Theoretical and experimental investigations on rectangular dielectric resonator antennas," *IEEE Transaction on Antennas and Propagation*, vol. 31, pp. 406–412.
36. Mongia, R. K. and Bhartia, P. (1994), "Dielectric resonator antennas-a review and general design relations for resonant frequency and bandwidth," *International Journal of Microwave and Millimeter-wave Computer Aided Engineering*, vol. 4, pp. 230–247.
37. Kajfez, D. and Guillon, P., Eds., *Dielectric Resonators.* Norwood, MA: Artech House, 1986.
38. Guha, D., Gajera, H. and Kumar, C. (2015), "Cross-polarized radiation in a cylindrical dielectric resonator antenna: Identification of source, experimental proof, and its suppression," *IEEE Transactions on Antennas and Propagation*, vol. 63, pp. 1863–1867.
39. Kajfez, D., Glisson, A. W. and James, J. (1984), "Computed modal field distributions for isolated dielectric resonators," *IEEE Transaction on Microwave Theory and Techniques*, vol. 32, pp. 1609–1616.
40. Guo, L, Leung, K. W. and Pan, Y. M. (2015), "Compact unidirectional ring dielectric resonator antennas with lateral radiation," *IEEE Transactions on Antennas and Propagation*, vol. 63, pp. 5334–5342
41. Guha, D., Banerjee, A., Kumar, C. and Antar, Y. M. M. (2014), "New technique to excite higher-order radiating mode in a cylindrical dielectric resonator antenna," *IEEE Antennas and Wireless Propagation Letters*, vol. 13, pp. 15–18
42. Luk, K. M., Lee, M. T., Leung, K. W. and Yung, E. K. N. (1999), "Techniques for improving coupling between microstrip line and dielectric resonator antenna," *IET Electronics Letters*, vol. 35, pp. 357–358.
43. Kranenberg, A., Long, S. A. and Williams, J. T. (1991), "Coplanar waveguide excitation of dielectric resonator antennas," *IEEE Transaction on Antennas and Propagation*, vol. 39, pp. 119–122.
44. Wyville, M., Petosa, A. and Wight, J. S. (2005), "DIG feed for DRA arrays," *IEEE Antennas and Propagation Symposium Digest AP-S*, vol. 2A, pp. 176–179.
45. Huang, C. Y., Chiou, T. W. and Wong, K. L. (2001), "Dual-polarized dielectric resonator antennas," *Microwave and Optical Technology Letters*, vol. 31, pp. 222–223.
46. Guo, Y. X and Luk, K. M (2003), "Dual polarized dielectric resonator antennas," *IEEE Transaction on Antennas and Propagation*, vol. 51, pp. 1120–1123.
47. Tang, X. R., Zhong, S. S., Kuang, L. B. and Sun, Z. (2009), "Dual-polarised dielectric resonator antenna with high isolation and low cross-polarisation," *IET Electronics Letters*, vol. 45(14), pp. 719–720.
48. Thamae, L. Z. and Wu, Z. (2010), "Diversity performance of multiport dielectric resonator antennas," *IET Microwaves, Antennas & Propagation*, vol. 4(11), pp. 1735–1745.
49. Huitema, L., Koubeissi, M., Mouhamadou, M., Arnaud, E., Decroze, C. and Monediere, T. (2011), "Compact and multiband dielectric resonator antenna with pattern diversity

for multi standard mobile handheld devices," *IEEE Transactions on Antennas and Propagation*, vol. 59, pp. 4201–4208.
50. Gao, Y., Feng, Z. and Zhang, L. (2011), "Compact CPW-fed dielectric resonator antenna with dual polarization," *IEEE Antennas and Wireless Propagation Letters*, vol. 10, pp. 544–547.
51. Thamae, L. Z. and Wu, Z. (2012), "Dielectric resonator-based multiple-input multiple-output antennas and channel characteristic analysis," *IET Microwaves, Antennas & Propagation*, vol. 6 (9), pp. 1084–1089.
52. Sun, Y. X. and Leung, K. W. (2013), "Dual-band and wideband dual-polarized cylindrical dielectric resonator antennas," *IEEE Antennas & Wireless Propagation Letters*, vol. 12, pp. 384–387.
53. Nasir, J., Jamaluddin, M. H., Khalily, M., Kamarudin, M. R., Ullah, I. and Selvaraju, R. (2015), "A reduced size dual port MIMO DRA with high isolation for 4G applications," *International Journal of Microwave and Millimeter-wave Computer Aided Engineering*, vol. 25(6), pp. 495–501.
54. Messaodene, I., Denidni and T. A., Benghalia, A. (2015), "CDR antenna with dual band 1.9/2.7 GHz for MIMO LTE terminals" *Microwave and Optical Technology Letters*, vol. 57, pp. 2388–2391.
55. Roslan, S. F., Kamarudin, M. R., Khalily, M. and Jamaluddin, M. H. (2015), "An MIMO F-shaped dielectric resonator antenna for 4G applications," *Microwave and Optical Technology Letters*, vol. 57(12), pp. 2931–2936.
56. Khan, A. A., Khan, R., Aqeel, S., Kazim, J. R., Saleem, J. and Owais, M. K. (2016), "Dual band MIMO Rectangualar Dielectric Resonator Antenna with high port isolation for LTE applications," *Microwave and optical Technology Letters*, vol. 59, pp. 44–49.
57. Khan, A. A., Khan, R., Aqeel, S., Nasir, J., Saleem, J. and Owais, O. (2016), "Design of a dual-band MIMO dielectric resonator antenna with high port isolation for WiMAX and WLAN applications," *International Journal of RF and Microwave Computer Aided Engineering*, vol. 27, pp. 1–10.
58. Sharma, A. and Biswas, A. (2017), "Wideband multiple-input–multiple-output dielectric resonator antenna," *IET Microwaves, Antennas and Propagation*, vol. 11(4), pp. 496–502.
59. Sharma, A., Sarkar, A., Adhikary, M., Biswas, A. and Akhtar, M. J. (2017), "SIW fed MIMO DRA for future 5G applications," *2017 IEEE International Symposium on Antennas and Propagation & USNC/URSI National Radio Science Meeting*. IEEE, pp. 1763–1764.
60. Das, G., Sharma, A. and Gangwar, R. K. (2017), "Dual port aperture coupled MIMO cylindrical dielectric resonator antenna with high isolation for WiMAX application," *International Journal of RF and Microwave Computer-Aided Engineering*, vol. 27, pp. 1–8. doi: 10.1002/mmce.21107.
61. Das, G., Sharma, A. and Gangwar, R. K. (2017), "Dual feed MIMO cylindrical dielectric resonator antenna with high isolation," *Microwave Optical Technology Letter*, vol. 59, pp. 1686–1692.
62. Sahu, N. K., Das, G. and Gangwar, R. K. (2018), "Dielectric resonator based wide band circularly polarized MIMO antenna with pattern diversity for WLAN applications," *Microwave and Optical Technology Letters*, vol. 60, pp. 2855–2862.
63. Sharma, A., Sarkar, A., Biswas, A. and Akhtar, M. J. (2018), "A-shaped wideband dielectric resonator antenna for wireless communication systems and its MIMO implementation," *International Journal of RF and Microwave Computer Aided Engineering*, vol. 28, pp. 1–9.
64. Das, G., Sharma, A. and Gangwar, R. K. (2018), "Dielectric resonator based two element MIMO antenna system with dual band characteristics," *IET Microwaves Antennas & Propagation*, vol. 12, pp. 734–741.

65. Sharma, A., Sarkar, A., Biswas, A. and Akhtar, M. J. (2018), "Equilateral triangular dielectric resonator based co-radiator MIMO antennas with dual polarization," *IET Microwaves, Antennas and Propagation*, vol. 12(14), pp. 2161–2166.
66. Pahadsingh, S. and Sahu, S. (2018), "Four port MIMO integrated antenna system with DRA for cognitive radio platforms," *AEU - International Journal of Electronics and Communication*, vol. 92, pp. 98–110.
67. Sahu, N. K., Gangwar, R. K. and Kumari, P. (2018), "Dielectric resonator based circularly polarized MIMO antenna for WLAN applications," *3rd International Conference on Microwave and Photonics (ICMAP 2018)*, 9–11 February, Dhanbad.
68. Gotra, S., Varshney, G., Pandey, V. S. and Yaduvanshi, R. S. (2019), "Super-wideband multi-input–multi-output dielectric resonator antenna," *IET Microwaves & Antennas Propagation*, vol. 14, pp. 21–27.
69. Kumari, T., Das, G., Gangwar, R. K., and Suman, K. K. (2019), "Dielectric resonator based two-port dual band Antenna for MIMO applications," *International Journal of RF and Microwave Computer-Aided Engineering*. doi: 10.1002/mmce.21985.
70. Yan, J. B. and Bernhard, J. T. (2012), "Design of a MIMO dielectric resonator antenna for LTE femtocell base stations," *IEEE Transaction on Antennas and Propagation*, vol. 60, pp. 438–444.
71. Zou, L., Abbott, D. and Fumeaux, C. (2012), "Omnidirectional cylindrical dielectric resonator antenna with dual polarization," *IEEE Antennas and Wireless Propagation Letters*, vol. 11, pp. 515–518.
72. Fang, X. S., Leung, K. W. and Luk, K. M. (2014), "Theory and experiment of three-port polarization-diversity cylindrical dielectric resonator antenna," *IEEE Transactions on Antennas and Propagation*, vol. 62(10), pp. 4945–4951.
73. Abdalrazik, A., El-Hameed, A. S. A. and Abdel-Rahman, A. B. (2017), "A three-port MIMO dielectric resonator antenna using decoupled modes," *IEEE Antennas & Wireless Propagation Letters*, vol. 16, pp. 3104–3107.
74. Yang, N., Leung, K. W. and Wu, N. (2019), "Pattern-diversity cylindrical dielectric resonator antenna using fundamental modes of different mode families," *IEEE Transaction on Antennas and Propagation*, vol. 67(11), pp. 6778–6788.
75. Varshney, G., Gotra, S., Chaturvedi, S., Pandey, V. S. and Yaduvanshi, R. S. (2019), "Compact four-port MIMO dielectric resonator antenna with pattern diversity," *IEEE Transaction on Antennas and Propagation*, vol. 13 (12), pp. 2193–2198.
76. Yang, N., Leung and K. W. (2019), "Compact Cylindrical Pattern-Diversity Dielectric Resonator Antenna," *IEEE Antennas & Wireless Propagation Letters*, pp. 1–5. doi: 10.1109/LAWP.2019.2951633.
77. Khan, A. A., Jamaluddin, M. H., Aqeel, S., Nasir, J., Kazim, J. U. R. and Owais, O. (2017), "Dual-band MIMO dielectric resonator antenna for WiMAX/WLAN applications," *IET Microwaves, Antennas & Propagation*, vol. 11(1), pp. 113–120.
78. Nasir, J., Jamaluddin, M. H., Khalily, M., Kamarudin, M. R. and Ullah, I. (2016), "Design of an MIMO dielectric resonator antenna for 4G applications," *Wireless Personal Communications*, vol. 88, (3), pp. 525–536.
79. Das, G., Sahu, N. K., Sharma, A. and Gangwar, R. K. (2017), "Wideband MIMO hybrid cylindrical dielectric resonator antenna with improved diversity performance," *2017 IEEE Conference on Antenna Measurements & Applications (CAMA)*, 04–06 December, Ibaraki, Tsukuba, Japan.
80. Trivedi, K. and Pujara, D. (2017), "Mutual coupling reduction in wideband tree shaped fractal dielectric resonator antenna array using defected ground structure for MIMO applications," *Microwave and Optical Technology Letter*, vol. 59, pp. 2735–2742.
81. Trivedi, K. and Pujara, D. (2018), "Mutual coupling reduction in UWB modified maltese shaped DRA array for MIMO applications," *2018 48th European Microwave Conference (EuMC)*. IEEE, pp. 1117–1120.

82. Divya, G., Jagadeesh Babu, K. and Madhu, R. (2018), "Enhancement of isolation in dual-band hemispherical DRA using DGS for MIMO systems," *2018 Conference on Emerging Devices and Smart Systems (ICEDSS)*. IEEE, pp. 188–191.
83. Kumari, T., Das, G., Sharma, A. and Gangwar, R. K. (2019), "Design approach for dual element hybrid MIMO antenna arrangement for wideband applications," *International Journal of RF and Microwave Computer-Aided Engineering*, vol. 29, pp. 1–8.
84. Sahu, N. K., Das, G. and Gangwar, R. K. (2018), "L-shaped dielectric resonator based circularly polarized multi-input-multi-output (MIMO) antenna for wireless local area network (WLAN) applications," *International Journal of RF and Microwave Computer-Aided Engineering*, vol. 28. doi: 10.1002/mmce.21426.
85. Das, G., Sharma, A., Gangwar, R. K. and Sharawi, M. S. (2018), "Triple-port, two-mode based two element cylindrical dielectric resonator antenna for MIMO applications," *Microwave and Optical Technology Letters*, Vol. 60, pp. 1566–1573.
86. Das, G., Sharma, A. and Gangwar, R. K. (2018), "Dielectric resonator based circularly polarized MIMO antenna with polarization diversity," *Microwave and Optical Technology Letters*, Vol. 60, pp. 685–693.
87. Bharti, G., Kumar, D., Gautam, A. K. and Sharma, A. (2019), "Two-port ring-shaped dielectric resonator-based diversity radiator with dual-band and dual-polarized features," *Microwave & Optical Technology Letters*, pp. 1–8. doi: 10.1002/mop.32053.
88. Biswas, A. K. and Chakraborty, U. (2019), "Complementary meander-line-inspired dielectric resonator multiple-input-multiple-output antenna for dual-band applications," *International Journal of RF and Microwave Computer Aided Engineering*, vol. 29, pp. 1–12.
89. Sharma, A., Das, G. and Gangwar, R. K. (November 2016), "Dual polarized triple band hybrid MIMO cylindrical dielectric resonator antenna for LTE2500/WLAN/WiMAX applications," *International Journal of RF and Microwave Computer-Aided Engineering*, Vol. 26, pp. 763–772.
90. Sahu, N. K., Das, G. and Gangwar, R. K. (2108), "Dual polarized triple-band dielectric resonator based hybrid MIMO antenna for WLAN/WiMAX applications," *Microwave and Optical Technology Letter*, vol. 60(4), pp. 1033–1041.
91. Sahu, N. K., Sharma, A., Das, G. and Gangwar, R. K. (2017), "Design of a dual-polarized triple-band hybrid MIMO antenna for WLAN/WiMAX applications," *2017 IEEE Conference on Antenna Measurements & Applications (CAMA)*, 04–06 December, AIST, Ibaraki, Tsukuba, Japan.
92. Sharawi, M. S., Podilchak, S. K., Khan, M. U. and Antar, Y. M. (2017), "Dual-frequency DRA-based MIMO antenna system for wireless access points," *IET Microwaves, Antennas & Propagation*, vol. 11, pp. 1174–1182.
93. Sharma, A., Das, G. and Gangwar, R. K. (2018), "Design and analysis of triband dual-port dielectric resonator based hybrid antenna for WLAN/WiMAX applications," *IET Microwaves, Antennas & Propagation*, vol. 12, pp.986–992.
94. Pan, Y. M., Qin, X., Sun, Y. X. and Zheng, S. Y. (2019), "A simple decoupling method for 5G millimeter-wave MIMO dielectric resonator antennas," *IEEE Transactions on Antennas Propagation*, vol. 67, pp. 2224–2234.
95. Zhang, Y., Deng, J. Y., Li, M. J., Sun, D. and Guo, L. X. (2019), "A MIMO dielectric resonator antenna with improved isolation for 5G mm-wave applications," *IEEE Antennas Wireless Propagation Letter*, vol. 18, pp. 747–751.
96. Al-Hasan, M. J., Denidni, T. A. and Sebak, A. R. (2013), "Millimeter-wave EBG-based aperture-coupled dielectric resonator antenna," *IEEE Transactions on Antennas and Propagation*, vol. 61, pp. 4354–4357.
97. Karimian, R., Kesavan, A., Nedil, M. and Denidni, T. A. (2017), "Low-mutual-coupling 60-GHz MIMO antenna system with frequency selective surface wall," *IEEE Antennas & Wireless Propagation Letters*, vol. 16, pp. 373–376.

98. Dadgarpour, A., Zarghooni, B., Virdee, B. S., Denidni, T. A. and Kishk, A. A. (2017), "Mutual coupling reduction in dielectric resonator antennas using meta-surface shield for 60-GHz MIMO systems," *IEEE Antennas and Wireless Propagation Letters*, vol. 16, pp. 477–480.
99. Farahani, M., Akbari, M., Nedil, M. and Denidni, T. A. (2017), "Mutual coupling reduction in dielectric resonator MIMO antenna arrays using metasurface orthogonalize wall," *2017 11th European Conference on Antennas and Propagation (EUCAP)*. IEEE, pp. 985–987.
100. Farahani, M., Pourahmadazar, J., Akbari, M., Nedil, M., Sebak, A. R. and Denidni, T. A. (2017), "Mutual coupling reduction in millimeter-wave MIMO antenna array using a metamaterial polarization-rotator wall," *IEEE Antennas & Wireless Propagation Letters*, vol. 16, pp. 2324–2327.
101. Rezapour, M., Rashed-Mohassel, J., Keshtkar, A. and Naser-Moghadasi, M. (2019), "Isolation enhancement of rectangular dielectric resonator antennas using wideband double slit complementary split ring resonators," *International Journal of Microwave and Millimeter-wave Computer Aided Engineering*, vol. 29(7), pp. 1–11.
102. Das, G., Sharma, A., Gangwar, R. K. and Sharawi, M. S. (2018), "Compact back to back DRA based four port MIMO antenna system with bi-directional diversity," *IET Electronics Letters*, vol. 54, pp. 884–886.
103. Das, G, Sahu, N. K. and Gangwar, R. K. (2018), "Pattern diversity based double sided dielectric resonator antenna for MIMO applications," *IEEE Indian Conference on Antennas & Propagation (InCAP)*, 16–19 December, Hyderabad, India.
104. Das, G, Sahu, N. K., Sharma, A., Gangwar, R. K. and Sharawi, M. S. (2019), "Dielectric resonator based 4-element 8-port MIMO antenna with multi-directional pattern diversity," *IET Microwaves Antennas & Propagation*, vol. 13, pp. 16–22.
105. Das, G., Sharma, A. and Gangwar, R. K. (2018), "Wideband self complementary hybrid ring dielectric resonator antenna for MIMO applications," *IET Microwaves Antennas & Propagation*, vol. 12, pp. 108–114.
106. Das, G., Sahu, N. K., Sharma, A., Gangwar, R. K. and Sharawi, M. S. (2019), "FSS based spatially decoupled back to back four port MIMO DRA with multidirectional pattern diversity," *IEEE Antennas & Wireless Propagation Letter*, Vol. 18, pp. 1552–1555.
107. Das, G., Sharma, A., Gangwar, R. K. and Sharawi, M. S. (2019), "Performance improvement of multi-band MIMO dielectric resonator antenna system with a partially reflecting surface," *IEEE Antennas & Wireless Propagation Letter*, vol. 18, pp. 2105–2109.

11 Advances in Patch Antenna Design Using EBG Structures

Ekta Thakur, Dr. Naveen Jaglan, Prof. Samir Dev Gupta
Jaypee University of Information Technology, Solan

Prof. Binod Kumar Kanaujia
Jawaharlal Nehru University, New Delhi

CONTENTS

11.1 INTRODUCTION

High-performance applications in wireless communication systems require an advanced form of electromagnetic materials. The development of "metamaterials" with unique features has recently gained great attention from the researchers [1]. Metamaterials are used in many fields such as optics, nanoscience, material science, and antenna engineering. These materials have special characteristics that do not exist in naturally occurring materials. Hence, designing metamaterials with unique characteristics and using them in several antenna applications is an interesting concept for researchers. The concept of photonic crystals was introduced by Yablonovitch in solid-state physics [2]. The photonic crystals with a forbidden band gap are used in optics and solid-state physics. Therefore, the term photonic band gap (PBG) of the optics is used as electromagnetic band gap in the microwave domain. EBG structures are classified as a special type of metamaterial and can be defined as periodic or non-periodic structures that prevent or help the transmission of electromagnetic waves in a specific band of frequency [3,4]. EBG structures have many names in the literature, including left-handed materials, soft and hard surfaces, double-negative materials, negative refractive index materials, high-impedance surfaces (HIS), magnetomaterials, and artificial magnetic conductors (AMC) [5]. They are categorized into three groups: (i) three-dimensional volumetric structures, for example woodpile structure, (ii) two-dimensional planar structures, for example mushroom-like EBG structure, and (iii) one-dimensional transmission lines, such as holes in the ground plane. The EBG unit cell has a single band gap; however, a periodic arrangement of EBG structures can have multiple band gaps. In addition to the band gap feature, EBG structures also have some other important characteristics such as AMC and HIS [6]. For example, for both TE and TM polarizations, the two-dimensional EBG structures show a high-impedance surface. And when a wave strikes the EBG surface, 0° reflection phase is obtained and the EBG surface behaves as an AMC. These special features of EBG structures lead to a broad range of applications in microwave and antenna engineering [7,8].

In this chapter, we discuss the recent advancements in patch antenna design using EBG structures. These structures help improve the gain and bandwidth of the patch antennas. They are also used to reduce mutual coupling and to obtain band-rejection characteristics in ultra-wideband (UWB) antennas. A number of techniques to improve the gain and bandwidth, to achieve multi-band characteristics, to reduce mutual coupling, and to obtain multiple band-notch characteristics using compact EBG structures are also discussed later in this chapter. Finally, some real-life applications of EBG structure-integrated patch antennas, such as RFID and wearable electronics, are summarized.

11.2 EBG STRUCTURES AND THEIR PROPERTIES

An EBG structure is made up of metallic patches, ground plane, dielectric materials, and vias that connect patches to the ground plane [6]. From inductance and capacitance, the notch resonance frequency of an EBG cell can be determined using eq. (11.1):

$$f_s = \frac{1}{2\pi\sqrt{LC}} \tag{11.1}$$

where f_s represents the operating frequency of the EBG structure. This operating frequency can be varied by varying the inductance and capacitance. The values of inductor L and capacitor C are evaluated using the given formula [6–8]

$$L = \mu_0 h \tag{11.2}$$

$$C = \frac{W\varepsilon_0(1+\varepsilon_0)}{\pi}\cosh\frac{(2W+g)}{g} \tag{11.3}$$

where W, g, h, μ_0, and ε_0 are the width of the patch, the gap between two EBG structures, the substrate height, the permeability and permittivity of free space, respectively. The EBG structures' band gap can be evaluated by three methods, i.e., reflection phase, dispersion diagram, and the transmission characteristics calculated by suspended line method. Reflection phase characteristics of an EBG unit cell are used to predict the electromagnetic nature of the surface. For perfect electric conductor (PEC) and perfect magnetic conductor (PMC) ground planes, the reflection phase is 180° and 0°, respectively [9]. However, the PMC does not occur in the environment. For an EBG ground plane, the reflection phase varies from +180° to −180° with increasing frequency. The frequency range between +90° and −90° generally overlaps the band gap of an EBG structure [10]. Figure 11.1a indicates the reflection phase. The dispersion diagram, which is calculated using Eigenmode solver, is used to obtain the band gap of periodic EBG structures. The plot of phase constant and resonant frequency is referred to as the dispersion diagram [11]. Figure 11.1b illustrates the dispersion diagram. In the transmission characteristics method, the band gap is calculated by replacing the PEC ground plane with the EBG array. Figure 11.1c indicates the transmission loss. Transmission losses less than −15 dB are generally considered as the band gap of the EBG structure [12]. The transmission characteristics and reflection phase are enough for calculating the band gap of an EBG structure. From the transmission characteristics and reflection phase, the AMC at 0° and the surface wave band gap of the EBG structure can be easily recognized [13]. The dispersion diagram consumes more memory and time, but gives more information on band gap.

11.3 EBG STRUCTURES IN PATCH ANTENNA DESIGN

For wireless communication application, patch antenna are mostly preferred due to its low profile and low cost with high performance. The patch antennas have major benefits such as ease of installation, low cost, low profile, and integration with printed circuits. Some of the drawbacks of patch antennas are low efficiency, low power, high Q, surface wave excitation, and narrow bandwidth [15]. The unique features of the EBG structure are found useful to overcome the drawbacks of a patch antenna. The EBG structure helps in gain and bandwidth improvement [16–19]. In general, an EBG patch antenna produces a plane radiation profile, fewer side lobes, and a good antenna efficiency compared to a normal patch antenna. The EBG structure also ensures low interference to the adjacent elements and acts as a shielding material between the antenna and the communication system user. Further, the EBG structure has also been used in MIMO [20–22] systems and array antennas to alleviate the mutual coupling [23] effect. The application of EBG structures in patch antennas is a smart research area; however, some problems arise as two different structures are combined together to achieve enhanced performance. Some

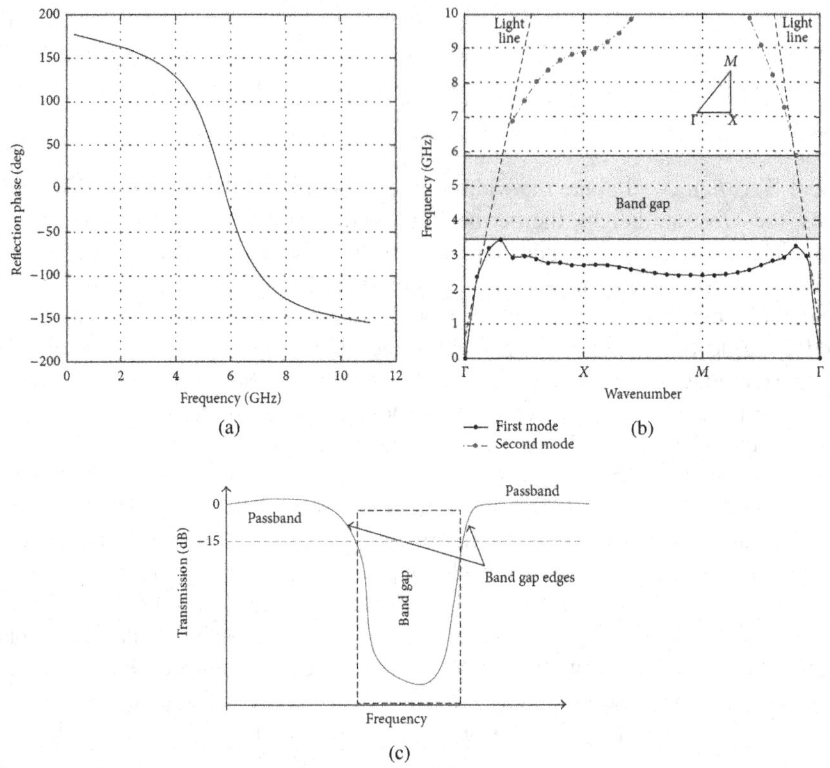

FIGURE 11.1 Different methods for calculating the band gap: (a) reflection phase methods (b), dispersion diagram methods, and (c) transmission characteristics methods [14].

new designs and their applications to enhance the performance of patch antennas are discussed in this chapter.

11.3.1 Bandwidth Improvement in Patch Antennas Using EBG Structures

For wireless communication systems, the mostly preferred antenna is the patch antenna. Generally, the conventional patch antenna has narrow bandwidth because of the PEC material in the ground plane. To achieve a wide bandwidth, the researchers have done intensive research and suggested a number of EBG structures. A wide band of 3–35 GHz is achieved by placing a periodic structure of circular and square EBG patch around the half-circular monopole antenna [24]. By inserting the conventional mushroom-type EBG structures on either side of the feed line, the impedance bandwidth is increased by 0.1 GHz as compared to the monopole UWB antenna. A good improvement in bandwidth is obtained by etching dummy EBG patterns on the feed line [25] as shown in Figure 11.2.

The EBG array is inserted on a substrate with a height of 0.4 mm and a dielectric constant of 2.33. The patch size is 8 mm × 6.3 mm, resulting in an operating frequency

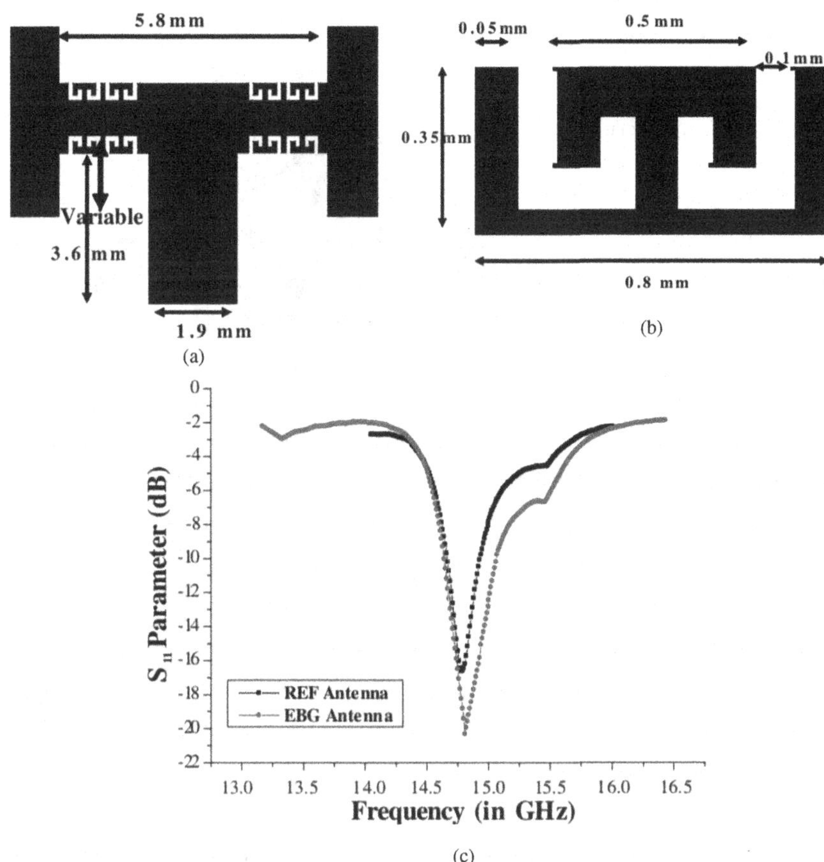

FIGURE 11.2 (a) Magnified view of feed line and (b) magnified view of EBG pattern [25] (c) S11 parameters Vs frequency.

of 14.8 GHz. Eight rows of dummy EBG structures are etched on the patch feed line with a gap of 5.8 mm in between two patches. The test data show the improvement in the bandwidth of over 0.381 GHz. In [26], Fabry–Perot (FP) antennas were integrated with EBG metamaterials for getting 4.92% impedance bandwidth and 10.7 dB gain simultaneously [27]. From Figure 11.3, it can be seen that using a trapezoidal ground plane in combination with a uniplanar EBG structure [28] increases the bandwidth and also improves the radiation characteristics of a monopole antenna. By changing the width of an EBG structure, one can vary its operating frequency. The total volume of the antenna is $100 \times 75 \times 0.762$ mm^3 including eight EBG cells. A conventional mushroom-type EBG structure was modified by inserting multiple vias to increase the band gap for noise suppression [29,30]. As shown in Figure 11.4, four vias are optimized to achieve a wider band gap and it is termed as ground surface perturbation lattice (GSPL) [31]. Figure 11.5 illustrates a lotus flower patch attached to a wide transmission line with an EBG ground plane to increase the bandwidth of the antenna. A slanted ground plane with an EBG structure was used to enhance the

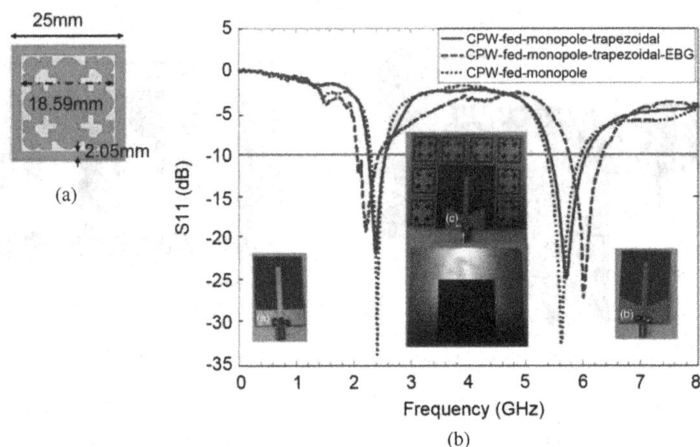

FIGURE 11.3 (a) Dual-band EBG structure; (b) return loss measurements [27].

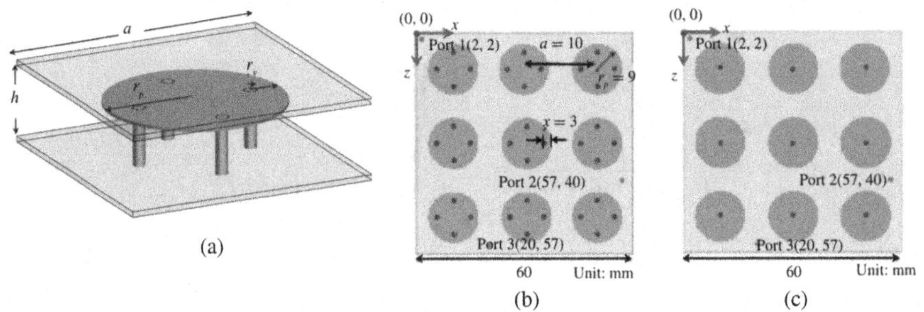

FIGURE 11.4 (a) Side view of GSPL structure; (b) GSPL structure. (c) Two-dimensional EBG structure [31].

bandwidth of the elliptical and rectangular monopole antennas [32]. In the literature, to enhance the bandwidth, a multilayer EBG structure [33] was also used. Monopole antennas of different shapes such as elliptical [34] and semi-circular [35] were suggested for enhancing the bandwidth. Further, the bandwidth of an UWB antenna can be improved by using EBG structures with a monopole antenna. Thus, in Table 11.1, some of the EBG structure approaches are described for bandwidth improvement.

11.3.2 Gain Improvement Using EBG Structures

With the arrival of new wireless communication services, the demand for ultra-wideband with low cost and compact size is increasing tremendously. The main drawbacks of printed antennas are low gain and low radiation efficiency because of the propagation of the surface wave. If an antenna is placed over a PEC ground plane such as copper or gold, it starts radiating into space. Additionally, it also produces currents that move along the sheet. These currents are also called surface waves and propagate to an edge of the antenna surface and cause multi-path interference.

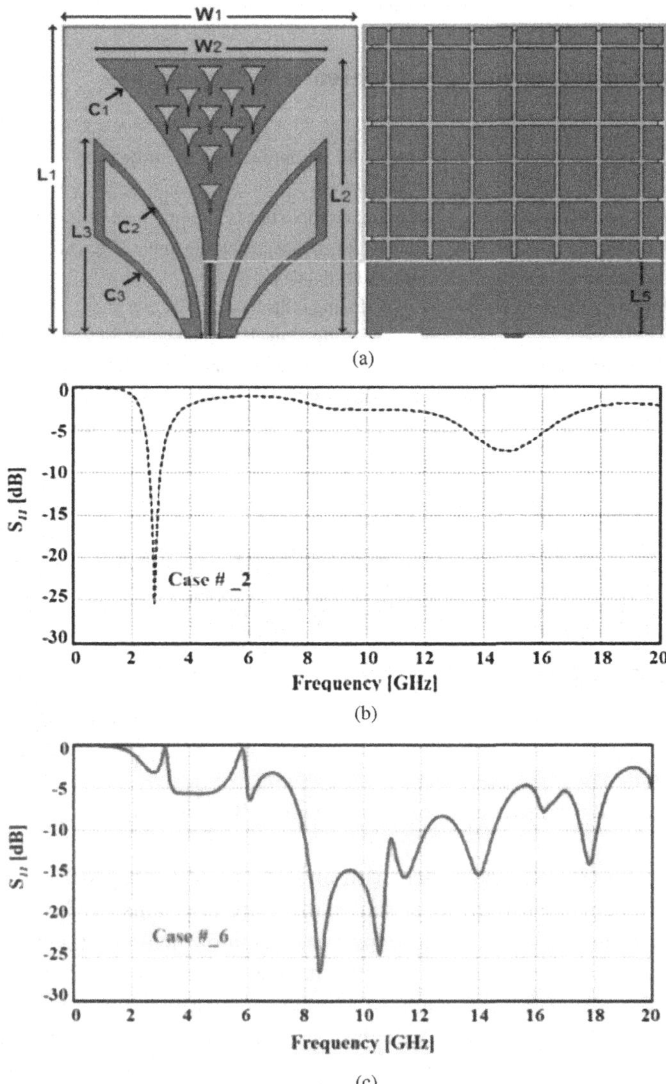

FIGURE 11.5 (a) Lotus-shaped antenna with an EBG ground plane, (b) $|S_{11}|$ of lotus-shaped antenna with partial ground plane, (c) $|S_{11}|$ of lotus-shaped antenna with EBG ground plane [32].

A number of approaches to increase the gain of patch antennas using EBG structures have been proposed. In these approaches, the EBG structure is designed in a way that its band gap and antenna resonant frequency band overlap. As a result, the surface waves cannot propagate along the substrate and the amount of radiating power increases. In order to improve gain, the EBG structure can be placed in two ways – the first is by placing the EBG structure around the patch antenna, which is [36] termed simply as the EBG structure, and the second is by replacing the ground plane with the EBG structure which is called AMC [37]. A double-rhomboid bow

TABLE 11.1
Bandwidth Enhancement Using Different EBG Approaches

S. No.	EBG Approach	Reason	Reference
1.	EBG pattern on the feed line	Bandwidth is improved due to impedance matching	[25]
2.	Multilayer EBG structure	Merging multiple bands into one wide band gap	[30,36]
3.	Symmetric placement of EBG structure around the patch	Bandwidth enhancement due to the different resonant frequencies	[28]
4.	Multiple-via EBG structure	By increasing inductance, reflection phase behavior gives much wider bandwidth	[32]
5.	EBG ground plane	To couple between the patch and EBG-AMC resonance frequencies	[24,27,33,34]

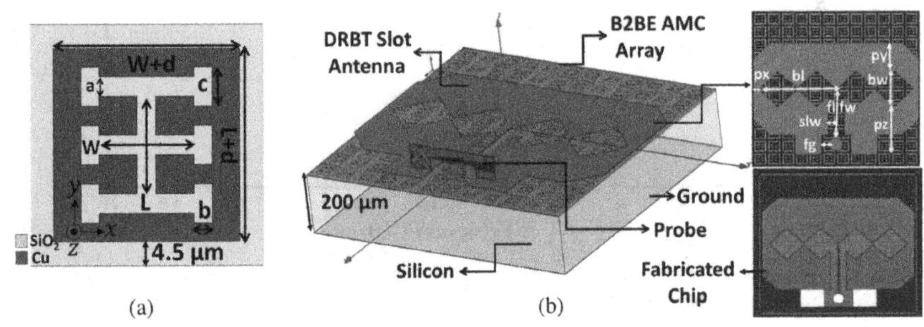

FIGURE 11.6 (a) EBG structure; (b) antenna with AMC array [38].

tie-slot antenna [38] was presented with an end-to-end E-shaped EBG surface for gain improvement at the W-band, as shown in Figure 11.6. This EBG surface behaves as a reflective surface that helps to improve the gain of the on-chip antenna. To prevent the losses due to waves entering into the lossy silicon substrate, an AMC array surface is placed beneath the patch. In another interesting study, a low-temperature co-fired ceramic (LTCC) patch antenna's gain was improved using a Sievenpiper EBG structure [39]. This LTCC was designed to resonate at 60 GHz on a DuPont™ Greentape™ 9K7 ($\varepsilon_r \sim 7.0$) of 5 mm thickness.

The two-element LTCC patch array was combined together with the Sievenpiper EBG structure to eliminate surface waves. As a result, around 4 dBi of gain enhancement and 8 dB of reduction in side lobe level were obtained, as shown in Figure 11.7. The design of eight-element MIMO antennas for 5G applications such as smartwatch and dongle [35] is presented in Figure 11.8. To improve gain and efficiency, an EBG surface was used as a ground plane. The upper layer of the substrate has eight MIMO antennas [36–40], whereas the bottom layer is composed of an EBG ground plane. The gain and antenna efficiency obtained were 8.732 dB and 92.7% at the resonant frequency, respectively.

Another study was performed [40] by changing the vias' positions in different patterns in the EBG ground surface, as presented in Figure 11.9a–d, and it was found

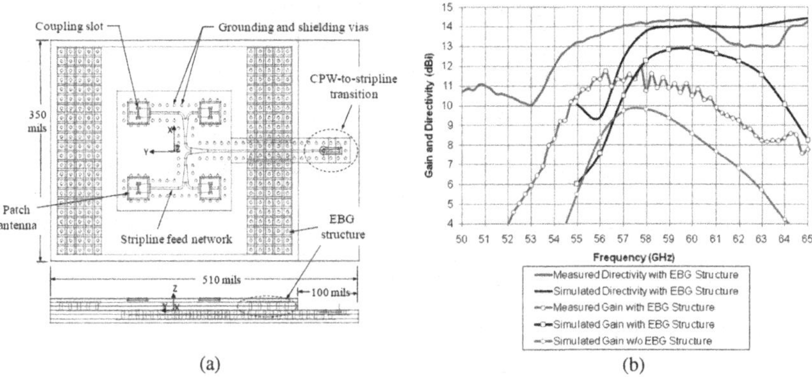

FIGURE 11.7 (a) 2×2 patch array; (b) gain and directivity versus frequency of the 2×2 patch array [39].

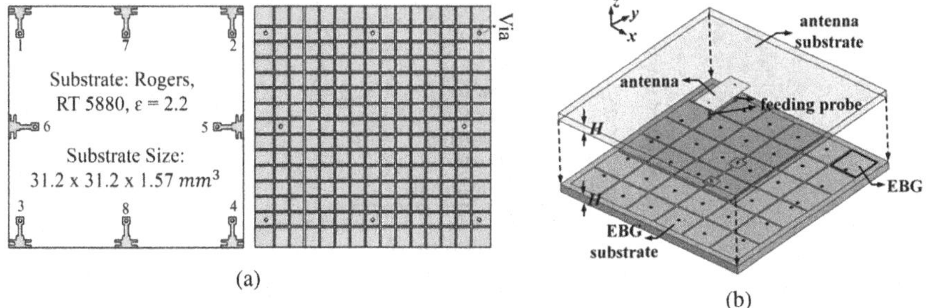

FIGURE 11.8 (a) Top and bottom views of an eight-element MIMO antenna with EBG ground plane [35]; (b) 3D view of cylindrically projected EBG planes [40].

that the antenna gain and efficiency were improved by 14.1 dB and 107.2%, respectively. Besides improving gain and directivity, the EBG superstrate also obtains dual-band dual-polarization [41–43] and suppresses the grating lobes in array antennas. Different EBG structures [42–48] were discussed based on the position of the EBG structure, such as below the patch and around the patch, to improve the overall performance of the antenna. In [49], a slotted EBG structure was used to improve the gain and to reduce the radar cross section (RCS) of a patch antenna. This slotted EBG structure was made of arrays of mushroom-type structure with rectangular slots on the patch. From the tested results, it was observed that the gain improved by 2.5 dB and RCS reduced to 4.3 dB as compared to the conventional patch antenna. Thus, few EBG structure approaches are discussed in Table 11.2 to improve patch antenna gain.

11.3.3 Mutual Coupling Reduction Using EBG Structures

Mutual coupling between antenna elements is due to the propagation of surface waves, and it affects the overall performance of the antenna. In case of a patch antenna, the E-plane coupling is stronger than the H-plane coupling. Several

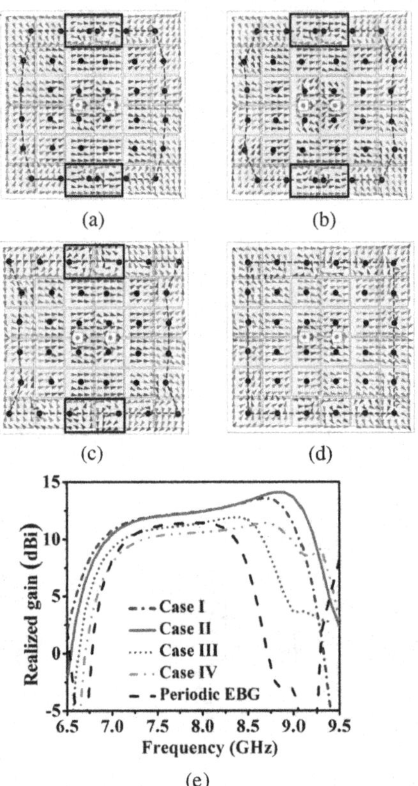

FIGURE 11.9 (a–d) Different patterns of vias in EBG ground plane; (e) realized gain of different EBG planes [40].

TABLE 11.2
Gain Enhancement Using Different EBG Structure Approaches

S. No.	EBG Structure Approach	Reason	Reference
1.	EBG structure placed around the patch	Surface wave suppression results in improved gain	[38,42,44,47,48]
2.	EBG structure ground plane	AMC property is utilized as a reflector	[35,40,49]
3.	Superstrate EBG structure layer above the patch antenna	Multiple reflections in cavity result in performance enhancement	[50,51]

methods were discussed and applied to reduce the mutual coupling between the MIMO system and antenna array, which include defected ground plane, decoupling strips, and neutralization line. For the MIMO system, it is desirable to have a mutual coupling of less than 15 dB, the envelope correlation coefficient (ECC) should be less than 0.5, and the total active reflection coefficient (TARC) should be less than 0 dB. The ECC is used to measure the correlation between radiation patterns of

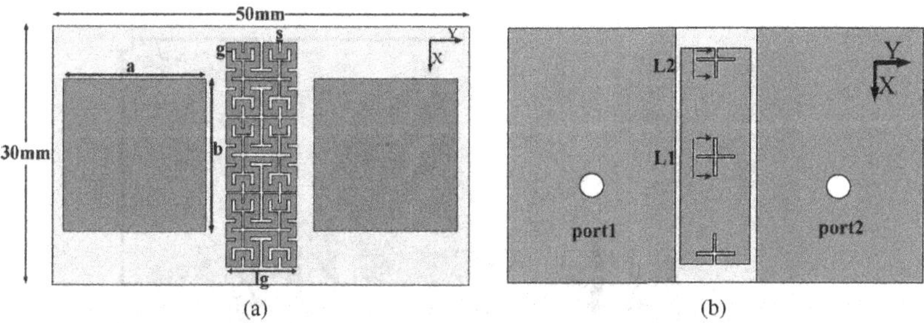

FIGURE 11.10 (a) Top view and (b) back view of patch antenna array [52].

MIMO antennas. The TARC is similar to return loss, but also considers the effect of mutual coupling. In [52], the mutual coupling was reduced by using two fractal uniplanar compact EBG structures and three cross-slots in between the patch arrays, as shown in Figure 11.10. By placing the two rows of fractal uniplanar compact EBG structure on the top layer, the mutual coupling was reduced 13 dB. Moreover, etching three cross-slots on the ground plane further improves the reduction in the mutual coupling. In [53], the conventional uniplanar compact EBG structure was modified by inserting rectangular slots. The modified EBG is shown in Figure 11.11a. Two-element MIMO antennas [53] were designed in a way that one element is positioned in front of the other. The introduction of the modified EBG array amid the two patches eliminated the propagation of surface waves. A parallel connection of L and C with series L and C connections formed a narrow band-stop filter. Because of the parallel inductance connections, the overall inductance decreased, and due to the parallel capacitance connections, the overall capacitance increased. These parallel connections aided in reducing the isolation by increasing the quality factor. The relation, given in eq. (11.4),

$$Q = \eta\sqrt{\frac{C}{L}} \tag{11.4}$$

Another investigation accomplished in [54] considered the planar compact EBG structure. To achieve better mutual coupling reduction, the planar EBG structure is placed in three different arrangements as shown in Figure.11.12. The planar compact EBG structure was fabricated using an FR4 substrate with a height of 1.6 mm and an overall dimension of 5.05×6.52 mm^2. The operating frequency of the antenna was nearly 5.6 GHz, which is at the WLAN band. In [56], a split EBG structure was placed amid the patch elements and the mutual coupling was decreased by 28 and 45 dB at 3.49 and 4.788 GHz, respectively. The split EBG structure is illustrated in Figure 11.13. In [55], miniaturized two-layer EBG structures were studied for decreasing the mutual coupling between UWB monopoles. These EBG structures contained slits in the ground plane, and as the number of slits increased, the electromagnetic coupling reduced. Figure 11.14 shows the variation in the mutual coupling with the number of slits in the ground plane. Table 11.3 shows the different EBG structures discussed in the literature to reduce the mutual coupling..

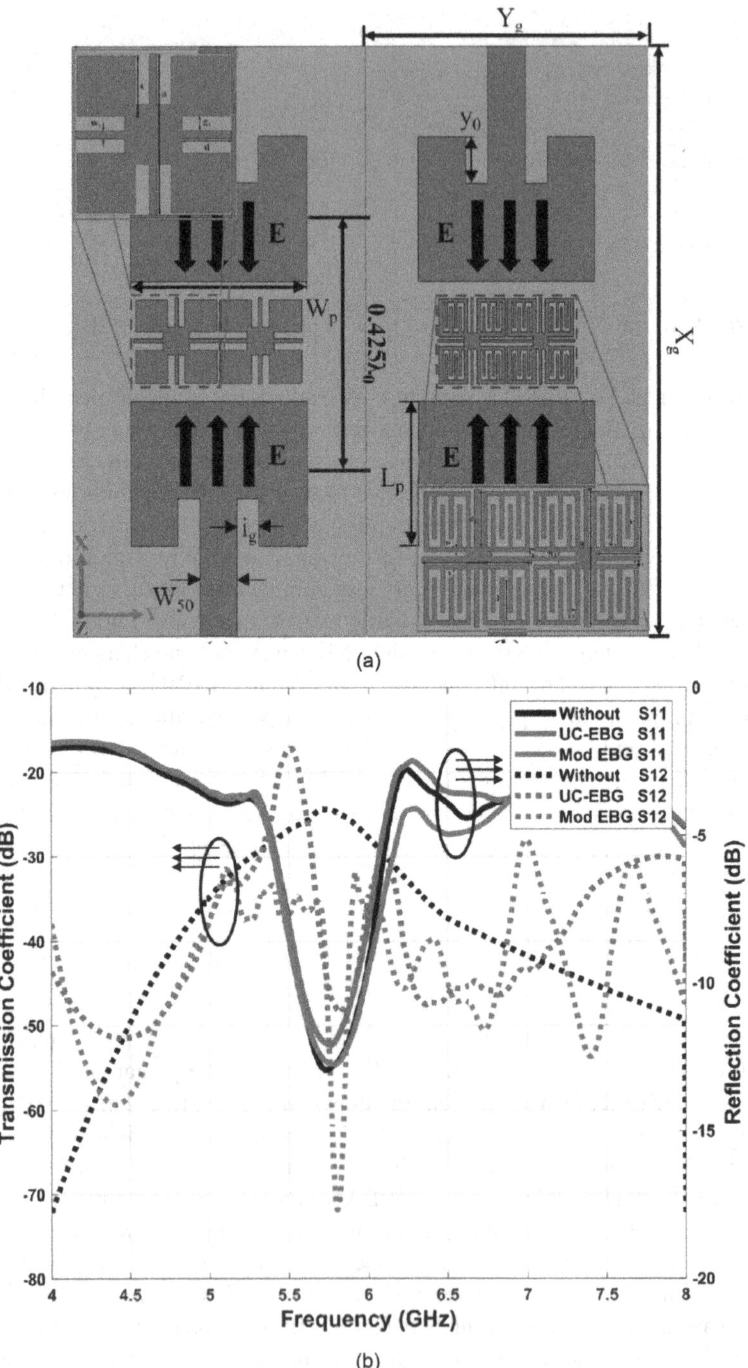

FIGURE 11.11 Two-element UWB MIMO antenna with inset feed: (a) UC-EBG structure and the proposed modified EBG structure; (b) *S*-parameters of three different cases [53].

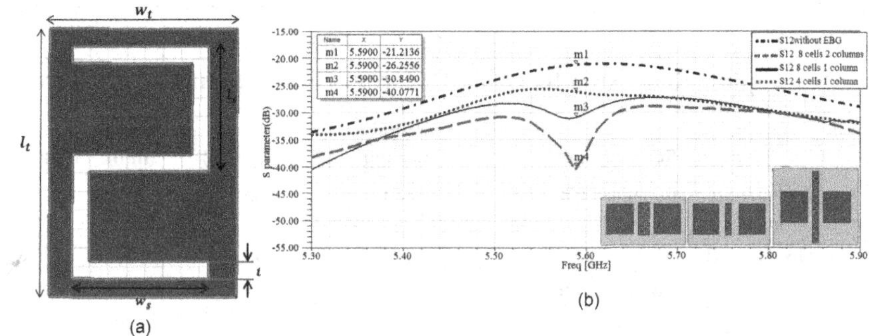

FIGURE 11.12 (a) Planar compact EBG structure; (b) mutual coupling of EBG structure array in three different arrangements [54].

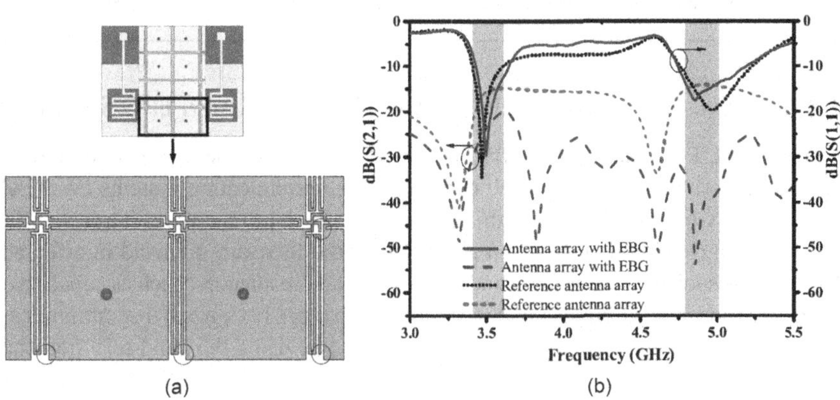

FIGURE 11.13 (a) Two-element meander line antenna; (b) reflection coefficient and mutual coupling with and without split EBG structure [56].

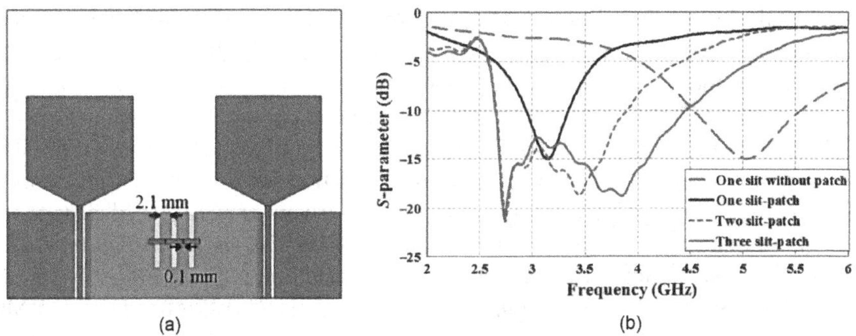

FIGURE 11.14 (a) Planar monopole MIMO antenna; (b) variation of mutual coupling [55].

TABLE 11.3
Mutual Coupling Reduction by Using Different EBG Structures

S No.	Type of EBG Structure	Mutual Coupling	Dielectric Constant (ε_r)	Height (mm)	Reference
1.	Fractal uniplanar compact EBG structure	−37 dB	2.65	1	[52]
2.	Modified EBG structure	−70 dB	4.4	1.6	[53]
3.	Planar compact EBG structure	−28 dB	4.8	1.6	[54]
4.	Double-layer EBG structure	−22 dB	4.5	1.55	[55]
5.	Split EBG structure	−44 dB	4.4	1.2	[56]
6.	Uniplanar compact EBG structure	−28 dB	10.2	1.27	[59]
7.	EM band gap metamaterial	−37 dB	4.3	1.6	[60]
8.	Uniconductor EBG structure	−46 dB	4.4	1.6	[61]
9.	Tunable double-layer EBG structure	−45 dB	4.5	1.55	[62]

11.3.4 Band-Notch Operation in Patch Antennas Using EBG Structures

In April 2002, FCC unbound the frequency range that lies from 3.1 to 10.6 GHz and this band is termed as UWB. Some narrowband communication systems (WiMAX, WLAN, and X-band) also work within this range, which produces interference. Many different design methods have been presented in the literature to avoid interference. The methods such as etching slots and using stubs and resonators, such as capacitively loaded loop and electric ring, disturb the radiation pattern because of discontinuities in the radiating elements. This problem can be solved by using the band-rejection characteristics of EBG structures. By inserting the EBG structure near the feed line of the UWB antenna, the interference of narrowband communication can be discarded [57]. In the literature [58], various designs of EBG structures have been presented to obtain band-notch characteristics. A swastika-type EBG structure [63] was used to achieve single-band-notch characteristics with a band gap of 7.5–11.1 GHz. In [64], by placing the EBG structures near feed line, three notches at WiMAX (3.5 GHz) and WLAN (5.2/5.8 GHz) bands are achieved. Mushroom-type EBG structures with vias at center [65] and edge-located mushroom-type EBG structures [65–68] are some structures presented by the researchers. A uniplanar EBG structure and two mushroom-type EBG structures [70] were placed near the feed line of the monopole antenna to have triple-band-rejection characteristics. A hexagonal-shaped c-slot mushroom-type EBG structure [66] was used to achieve band-notch function for X-band satellite communication systems (6.7–7.7 GHz). In [71], four EBG structures were placed close to the feed line to attain dual-band-notch characteristics [72], as shown in Figure 11.15a. The patch antenna and EBG structure were fabricated on a cheap FR4 substrate of dimensions $58 \times 45 \times 1.6$ mm^3. Figure 11.15b indicates the VSWR of an UWB MIMO antenna. In another interesting study [74], a circular monopole antenna [75–77] with two mushroom-type EBG structures was used to obtain band-notch characteristics for WLAN and WiMAX. Moreover, 34% of compactness was also achieved by etching an L-shaped slot on the EBG surface.

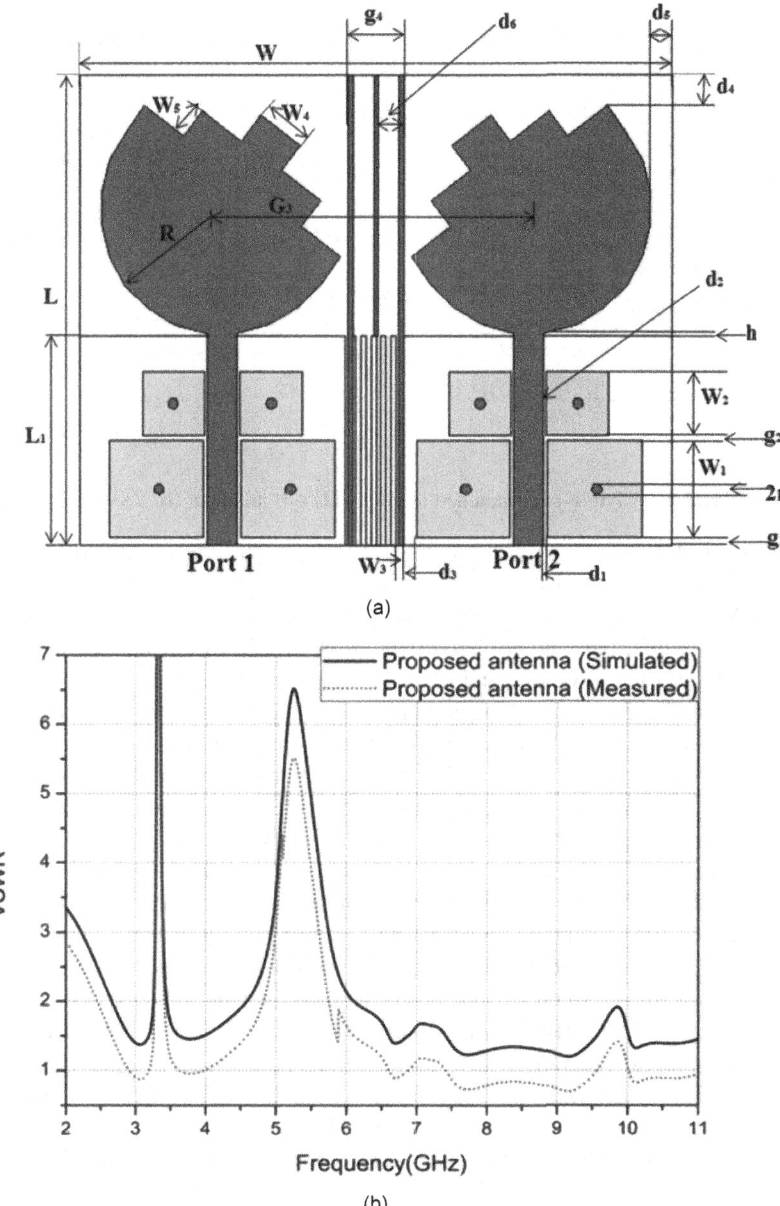

FIGURE 11.15 (a) Band-notched UWB MIMO antenna; (b) VSWR plot [71].

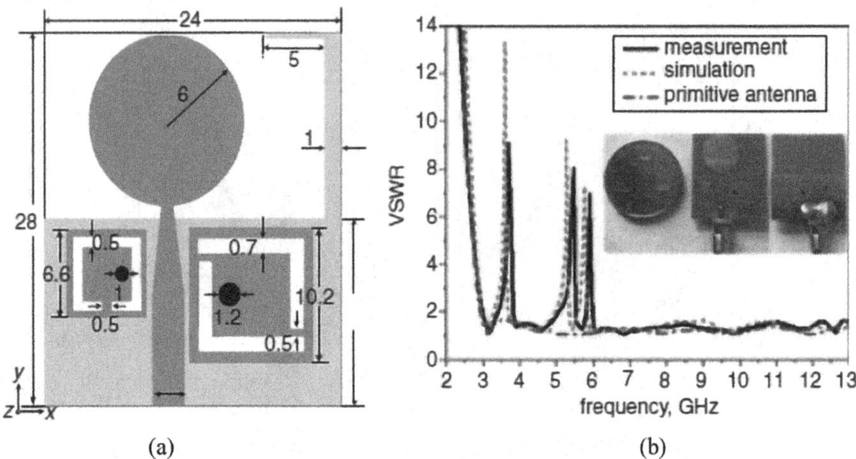

FIGURE 11.16 (a) Triple-band-notched monopole UWB antenna; (b) VSWR plot [74].

Figure 11.16a shows a circular monopole antenna with two modified EBG structures. This modified EBG structure contains two L-shaped slots with an edge-located via (ELV) that achieve two notched bands. So, in this, a single EBG structure is used to obtain dual-band-notch characteristics. A dual-band-notched MIMO antenna obtains notches in WiMAX band (3.3–3.6 GHz) and WLAN band (5–6 GHz), as presented in Figure 11.16b. A triple-band-notched UWB MIMO antenna [70] was realized by using three EBG structures. Figure 11.17 indicates the band-notched antenna. The designed antenna avoids the interference from WiMAX band ranging from 3.3 to 3.6 GHz, WLAN band ranging from 5 to 6 GHz, and the X-band for satellite communication systems ranging from 7.2 to 8.4 GHz. Another modified EBG structure connected with the feed line was used to reject the interference from narrowband communication systems [73]. This modified EBG structure also rejects are for WiMAX, WLAN, and X-band satellite communication system.

This single EBG structure rejects are for WiMAX, WLAN, and X-band satellite communication systems. Figure 11.18a displays the band-notched UWB antenna. The antenna and EBG structure are designed using a FR4 substrate with a height of 1 mm, a dielectric constant of 4.4, and an overall dimension of $30.5 \times 26 \times 1$ mm^3 and 8×5.95 mm^2, respectively. A pentagonal printed UWB monopole antenna having three notched bands is shown in Figure 11.19a. Two slots are inserted in the EBG structure to attain multiple band rejection. The dimension of the EBG structure is 9.4×4.5 mm^2. Figure 11.19b indicates the VSWR of the band-notched UWB antenna. Thus, Table 11.4 shows several cases of different EBG structures for achieving band-notch characteristics.

11.3.5 Dual-Band and Multi-Band Characteristics Using EBG Structures

In the previous section, numerous EBG structures were discussed to improve bandwidth and gain, to reduce mutual coupling, and to obtain band-notch characteristics.

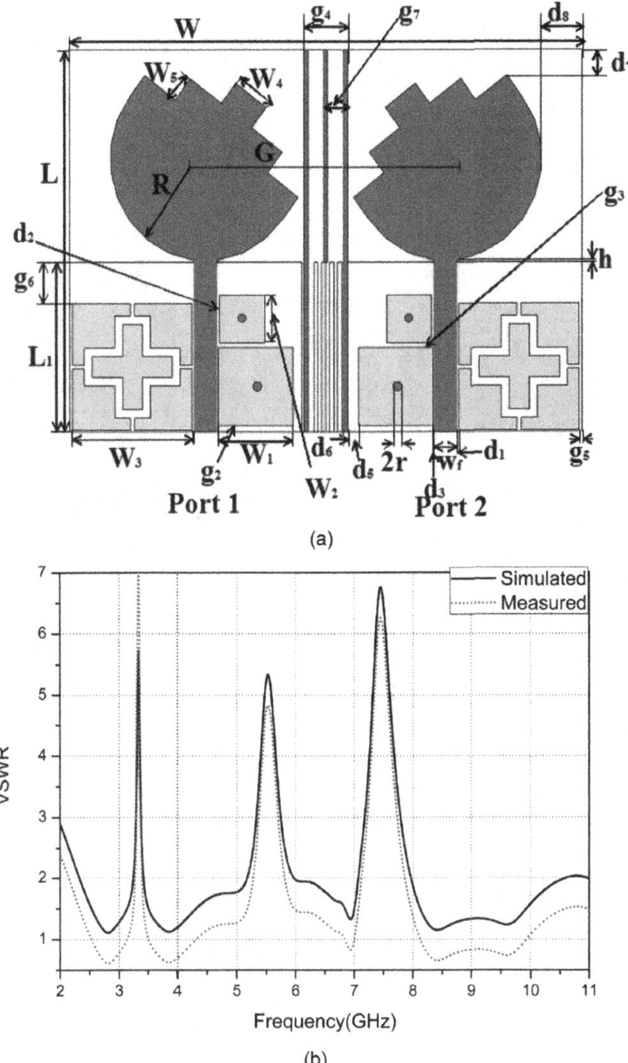

FIGURE 11.17 (a) Triple-band-notched UWB MIMO antenna; (b) VSWR plot [70].

The motivation of this section is to study the idea of multi-band operation in patch antennas using EBG structures. The resonant frequency of EBG structures can be calculated using a distributed lumped network. Using this distributed lumped network, the researchers have designed numerous EBG structures to obtain dual-band and multi-band characteristics in patch antennas.

In [79], dual-band characteristics were obtained by placing pinwheel-shaped slot EBG structure periodically around the antenna. The antenna combined with periodical EBG structures resonated at 4.9 and 5.4 GHz. The tested result also showed a bandwidth improvement of 41% and 25.4% at low frequency and high frequency,

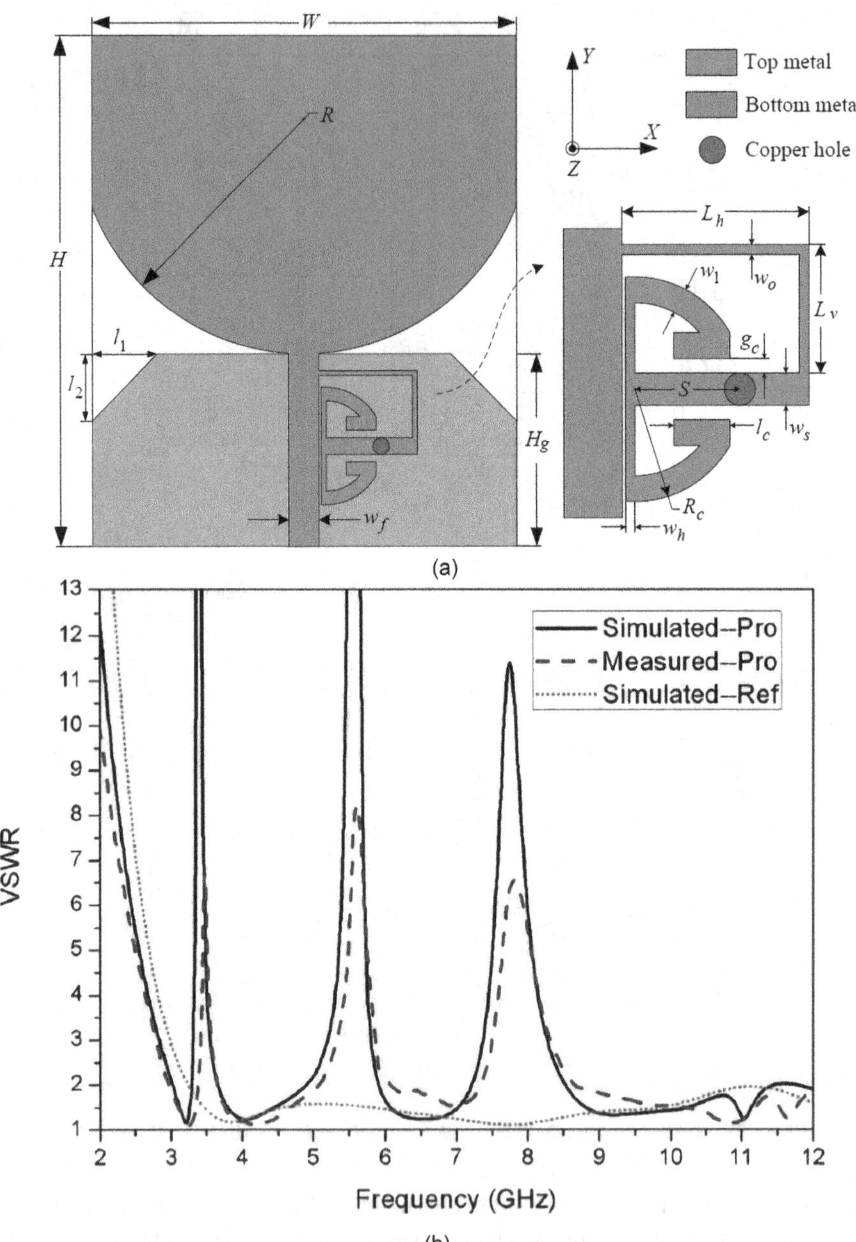

FIGURE 11.18 (a) Band-notched UWB antenna; (b) VSWR of triple-band-notched UWB MIMO antenna [73].

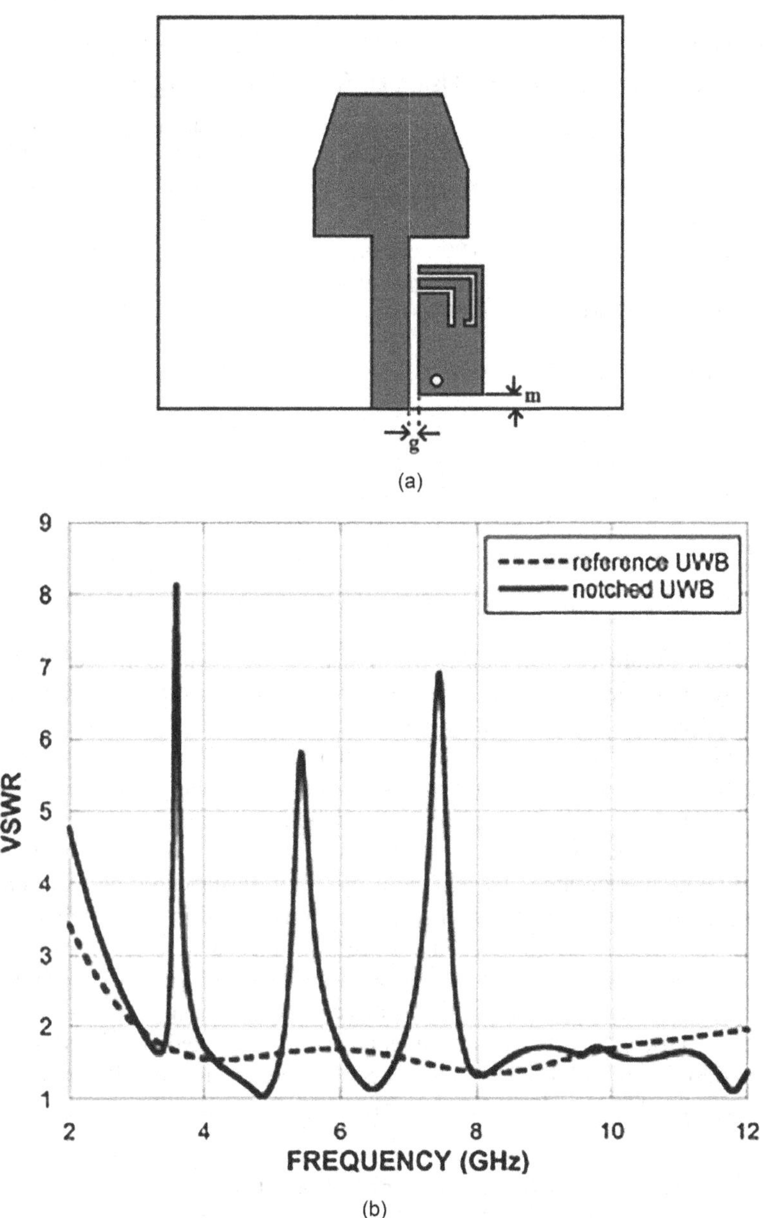

FIGURE 11.19 (a) Pentagonal UWB monopole antenna; (b) VSWR plot [75].

TABLE 11.4
Band-Notch Characteristics by Using Different EBG Structures

S. No.	No. of Notches	Type of EBG Structure	Notched Band (GHz)	Size (mm²)	Reference
1.	Single-notched band	Mushroom EBG structures	5.36–6.34	6.25×6.25	[22]
2.	Triple-notched band	Mushroom EBG structures and two split ring resonators	6.7–7.7	3.9×3.9	[66]
3.	Dual-notched band	Mushroom EBG structures	3.3–3.6 5–6	9.25×9.25 6.1×6.1	[76]
4.	Dual-notched band	DG-CEBG	3.3–3.6 5–6	5×5 3×3	[77]
5.	Dual-notched band	Uniplanar EBG structures with a π-shaped slot	3.45–3.9	5.2×5.2	[78]
6.	Triple-notched band	Uniplanar EBG structure Mushroom EBG structure	3.3–3.6 5–6 7.1–7.9	15×15 9.25×9.25 5.6×5.6	[79]

respectively. Figure 11.20 illustrates the H-shaped MPA with pinwheel-shaped slot EBG structures and the VSWR of three different cases. Another study performed in [80] presented a polarization-dependent EBG structure, as presented in Figure 11.21. This EBG structure behaves as a reflector of a dual-band dipole antenna. This reflector transforms the linearly polarized wave into a circularly polarized wave [80]. The tested antenna attained impedance bandwidths and axial ratio bandwidths of 13.4% and 3.2% and 2.4% and 3.5%, respectively. Another dual-polarized dual-band patch antenna was designed by loading a modified mushroom-type EBG unit cell [81], as presented in Figure 11.22. A square slot was etched from the radiating patch antenna, and the modified mushroom unit cell was placed at the slot to attain triple-band characteristics [82]. A new interesting study was performed in which the PEC ground plane was replaced by an EBG surface to obtain dual-band characteristics. The dual-polarized dual-band patch antenna and its simulated and tested results are presented in Figure 11.23. Thus, Table 11.5 lists several cases of EBG structures for obtaining multi-band/dual-band characteristics of patch antennas.

11.3.6 A Low-Profile MPA Using EBG Structures

In wireless communication systems, the dimension of the patch antenna depends on the resonating frequency. In this condition, it is a challenge to design the smallest possible antenna at low frequency. If the total thickness of the antenna is less than one-tenth of the operating wavelength, then the antenna is called a low-profile antenna.

The mutual coupling effect of the nearby ground plane degrades the overall performance. In [81], the performance of an MPA was examined by placing three different ground planes, namely PEC, PMC, and EBG, with the same height. When PEC

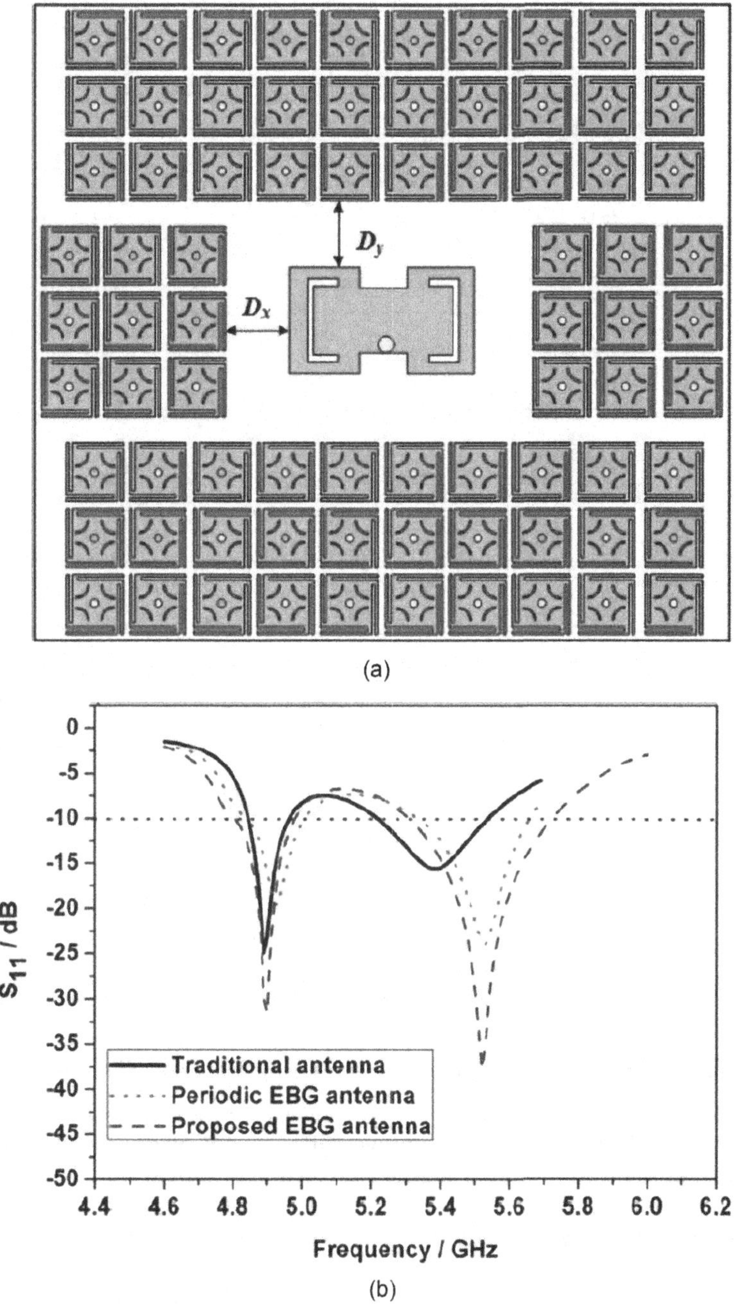

FIGURE 11.20 (a) Patch antenna with coaxial feed; (b) VSWR plot of microstrip antenna [79].

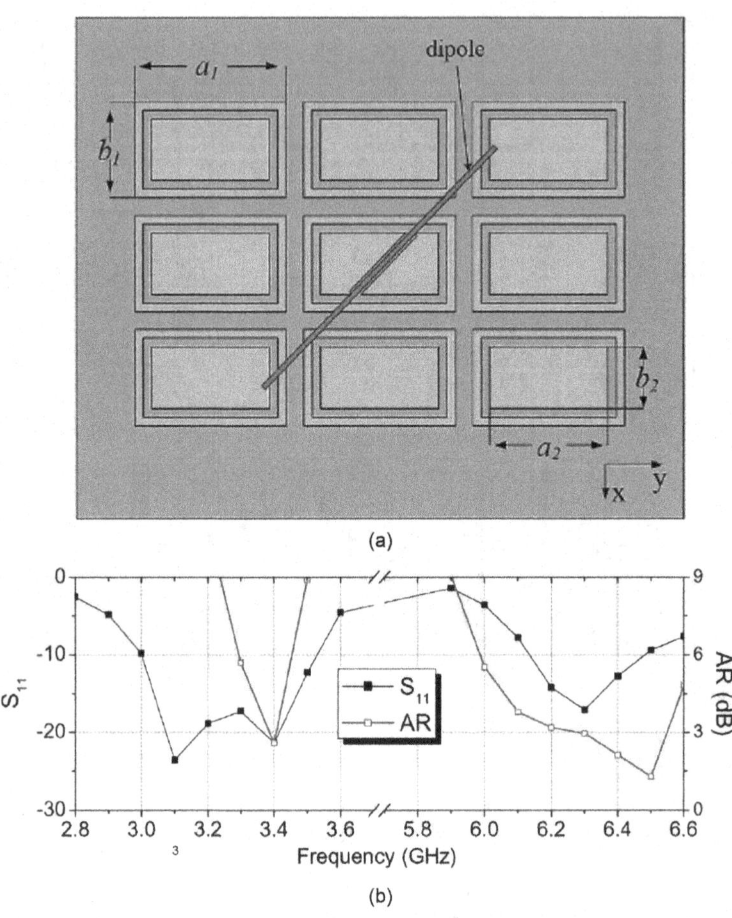

FIGURE 11.21 (a) Circularly polarized antenna; (b) S_{11} parameter and axial ratio plot [80].

TABLE 11.5

Multi-Band/Dual-Band Characteristics by Using Different EBG Structures

S. No.	Type of EBG Structure	Resonating Band without EBG Structure (GHz)	Resonating Band with EBG Structure (GHz)	Reference
1.	Dual-band EBG structure	2.265–2.563 and 5.434–6.061	2.062–2.453 and 5.765–6.343	[33]
2.	Pinwheel-shaped EBG structure	4.85–4.96 and 5.20–5.52	4.82–4.97 and 5.31–5.72	[79]
3.	Modified mushroom-like unit cell	2.21–2.29 and 2.45–2.5	2.21–2.29, 2.34–2.39 and 2.5–2.55	[81]
4.	Spiral conductor EBG structure	2.12–3.35 and 6.01–7.16	2.12–2.98, 5.24–5.80 and 6.05–7.77	[82]
5.	Uniplanar EBG structure	1.7–5.63 and 9.9–11.36	1.5–5.63 and 9.52–13.06	[83]

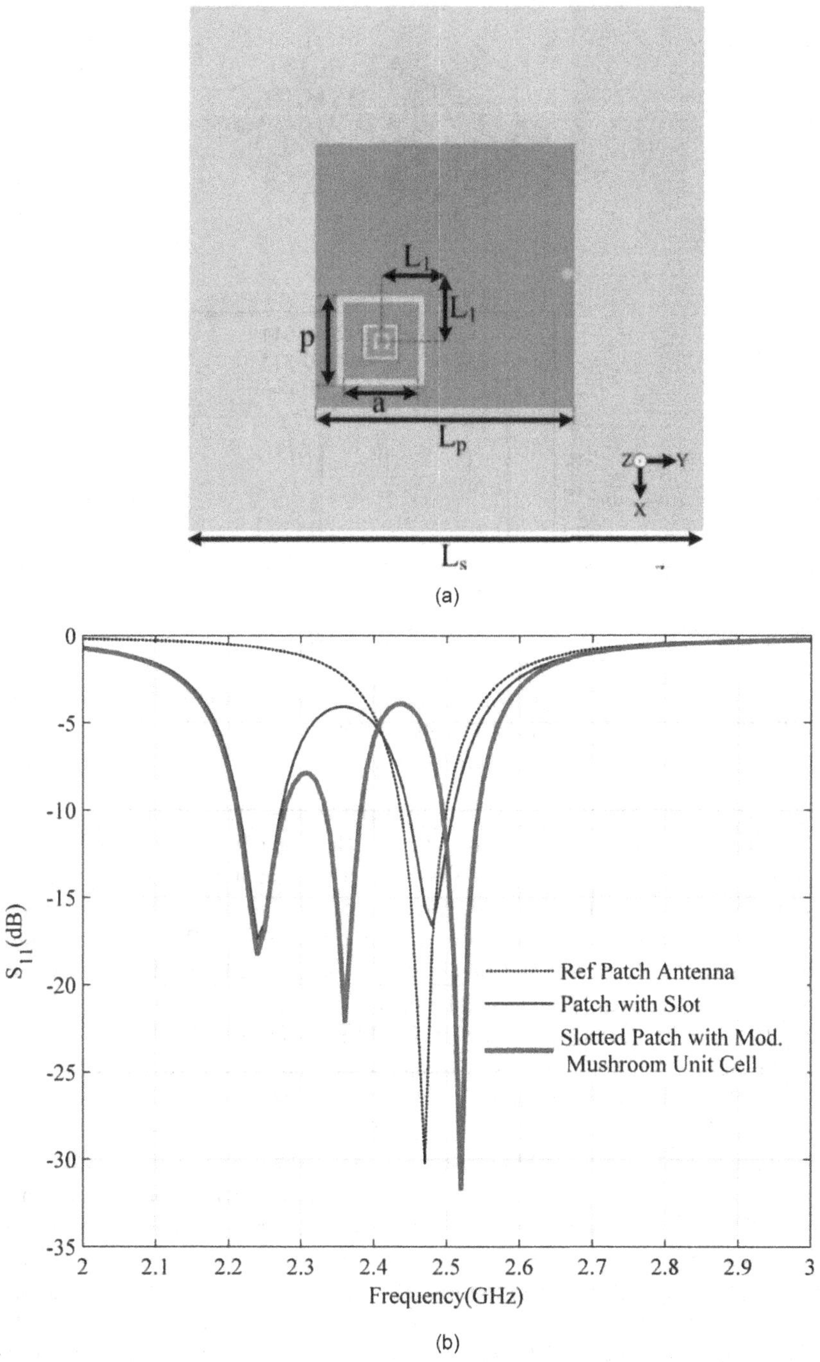

FIGURE 11.22 (a) Dual-polarized dual-band PA; (b) return loss of dual-polarized dual-band PA [81].

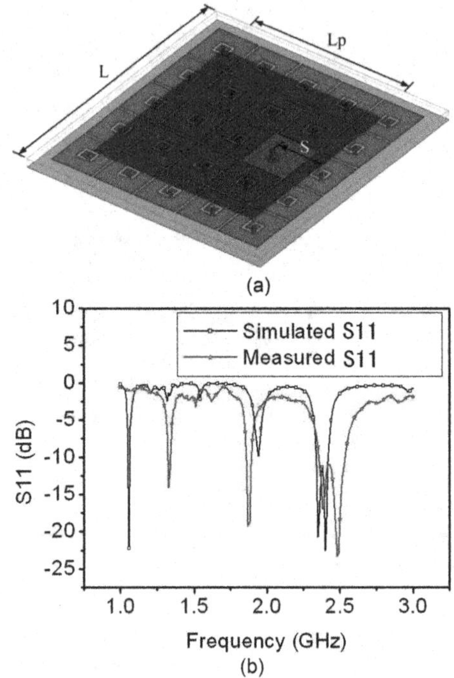

FIGURE 11.23 (a) Multi-band patch antenna; (b) return loss plot [82].

materials such as copper and gold are used as the ground plane, the return loss is –3.7 dB. The reason is that the PEC surface has a reflection phase of 180° and the image current has the opposite direction to that of the printed antenna. The reversal image current cancels the radiation from the printed antenna, resulting in poor return loss. When the ground plane is a PMC, the reflection phase is 0°. But the strong mutual coupling between the patch and image current results in a mismatch between the 50-Ω transmission lines [82]. Only if a proper impedance transformer is used, a good return loss can be obtained. Moreover, the PMC is an ideal surface and does not occur in the environment. However, when the ground plane is replaced by an EBG surface, the reflection phase varies from –180° to +180° and results in constructive interference of image current and radiation from the printed antenna [83–85]. The best return loss of –30 dB is obtained by the printed antenna over the EBG ground plane. From the above discussion, it is observed that the EBG ground plane has the desired features to design low-profile antennas. In [86], a dual-band, low-profile antenna with mushroom-like structures loaded with circular slots was described and is shown in Figure 11.24a. From Figure 24b, we can observe that the first resonating frequency shifts toward the right in case of EBG-CS that leads to compact size at low frequency. By using an EBG-CS superstrate, a low-profile antenna was achieved. In another study, a printed slot antenna placed above an AMC plane was used to obtain a low-profile, wideband antenna [87]. The radiating slots are shown in Figure 11.25a, with three unequal arms that are etched for widening the impedance bandwidth.

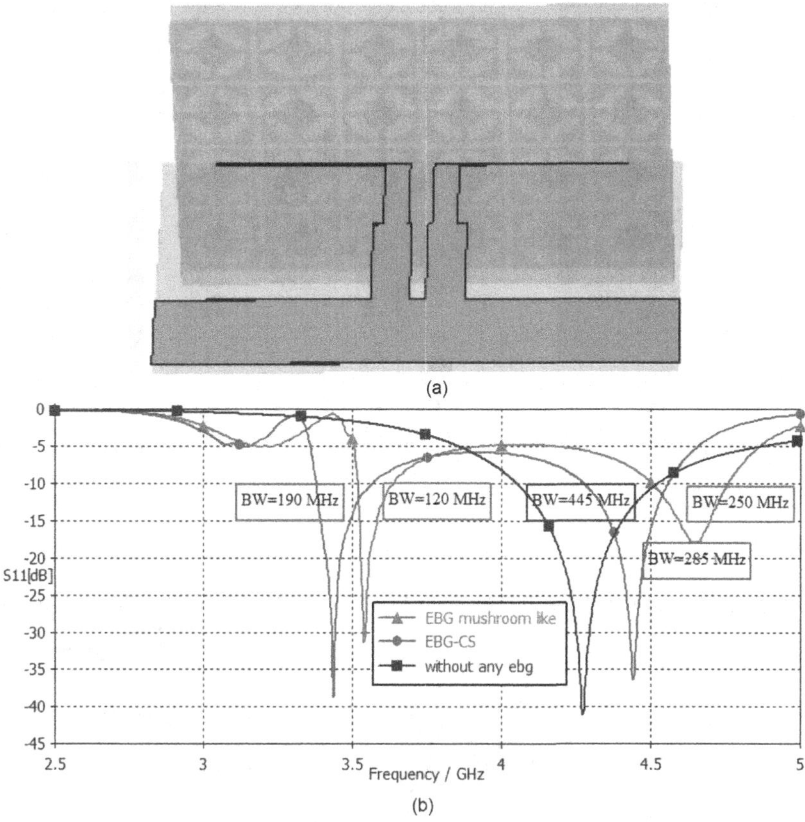

FIGURE 11.24 (a) Dual-band antenna with mushroom-like structures loaded with circular slots; (b) return loss of three different cases [86].

By placing an AMC [88,89] array as the ground plane, the radiation pattern was improved. Using an AMC with a printed antenna resulted in 62.82% compactness in size, a bandwidth improvement of 41%, and a perfect impedance matching. In [90–92] EBG structures that can generate adjustable bandgaps. The EBG structures can be placed as AMC in modern electronic products and potentially cover multiple frequency bands of wireless communications. Similarly, in [93], an AMC array was placed as a ground plane underneath a dual-wideband circularly polarized [80,94] antenna to achieve a low-profile antenna.

Two barbed-shape and bow tie dipoles printed on FR4 substrates are presented in Figure 11.26. To obtain an antenna with a higher broadside gain and a low profile, the ground plane was replaced by a square-shaped cavity plane. Finally, the cavity was modified to a pyramid-shaped cavity and it was found out that the inclination angle should be around 45° to further improve the gain, particularly in the high-frequency band. Figure 11.27a–c illustrates the high-gain reconfigurable antenna and a comparison of three different ground planes.

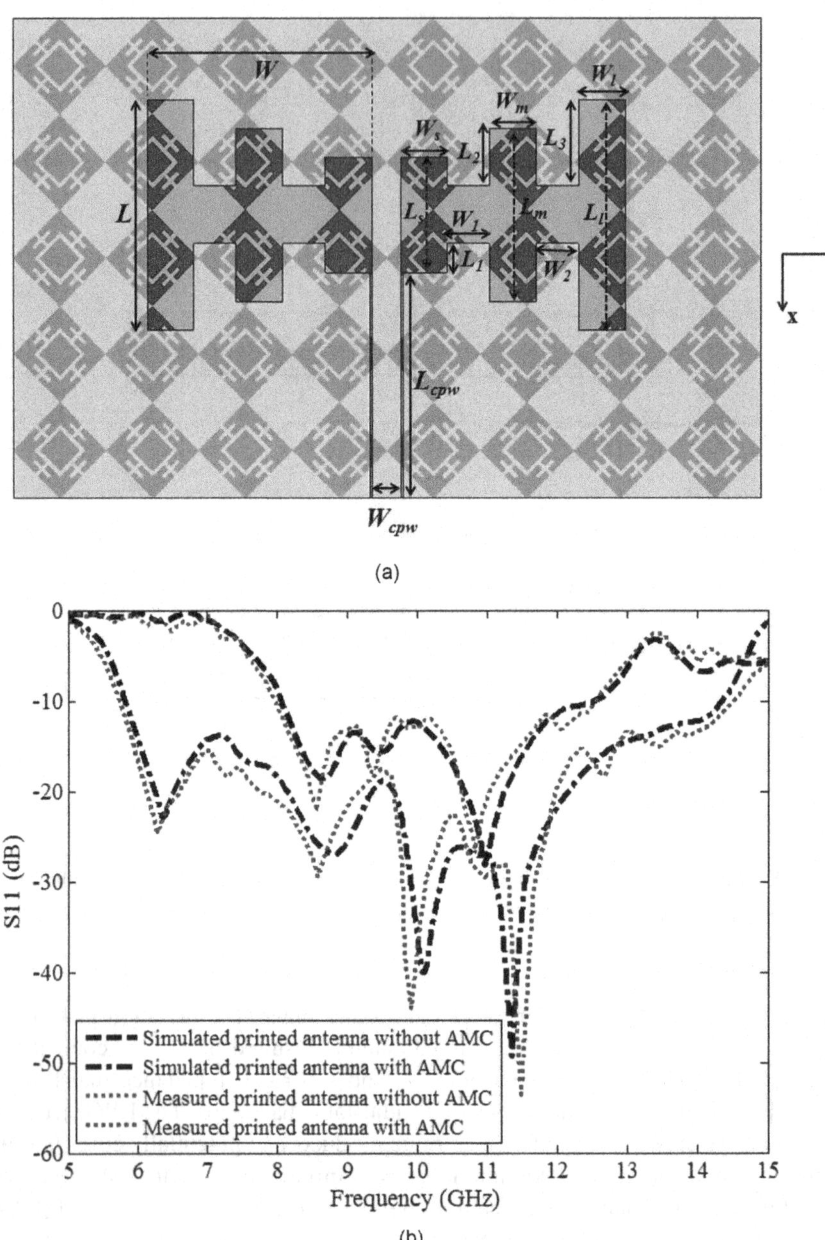

FIGURE 11.25 (a) Slot antenna with an AMC array; (b) return loss of slot antenna with an AMC array [110].

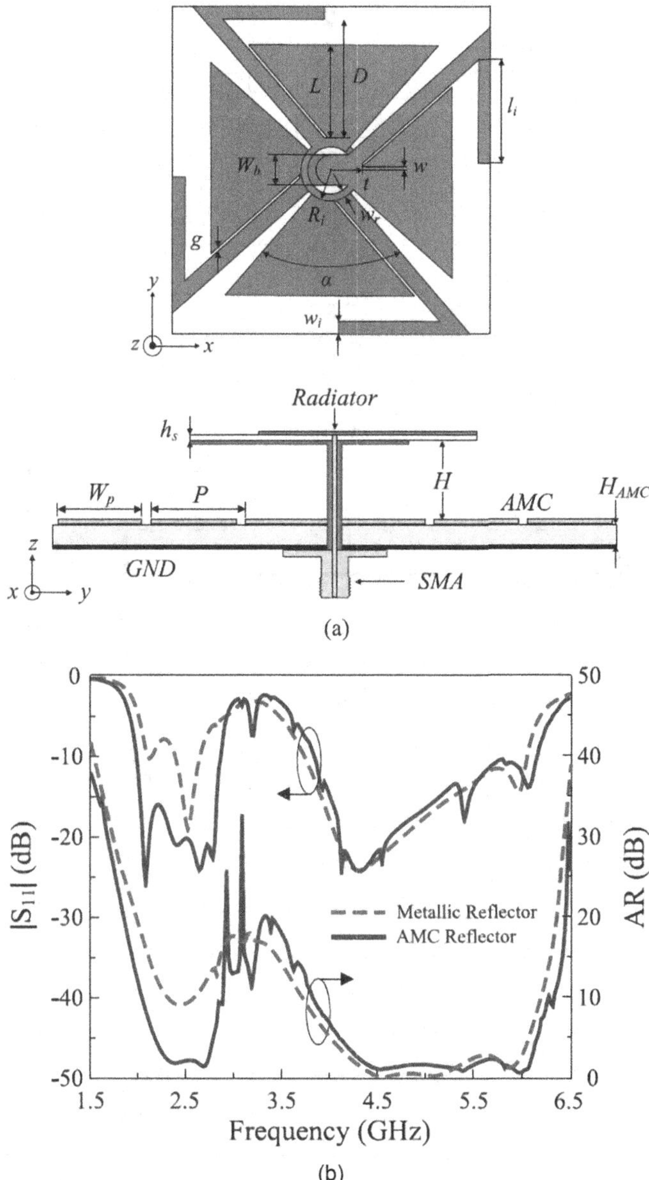

FIGURE 11.26 (a) Dual-wideband circularly polarized antenna; (b) return loss of dual-wideband circularly polarized antenna with an AMC and metallic reflector [93].

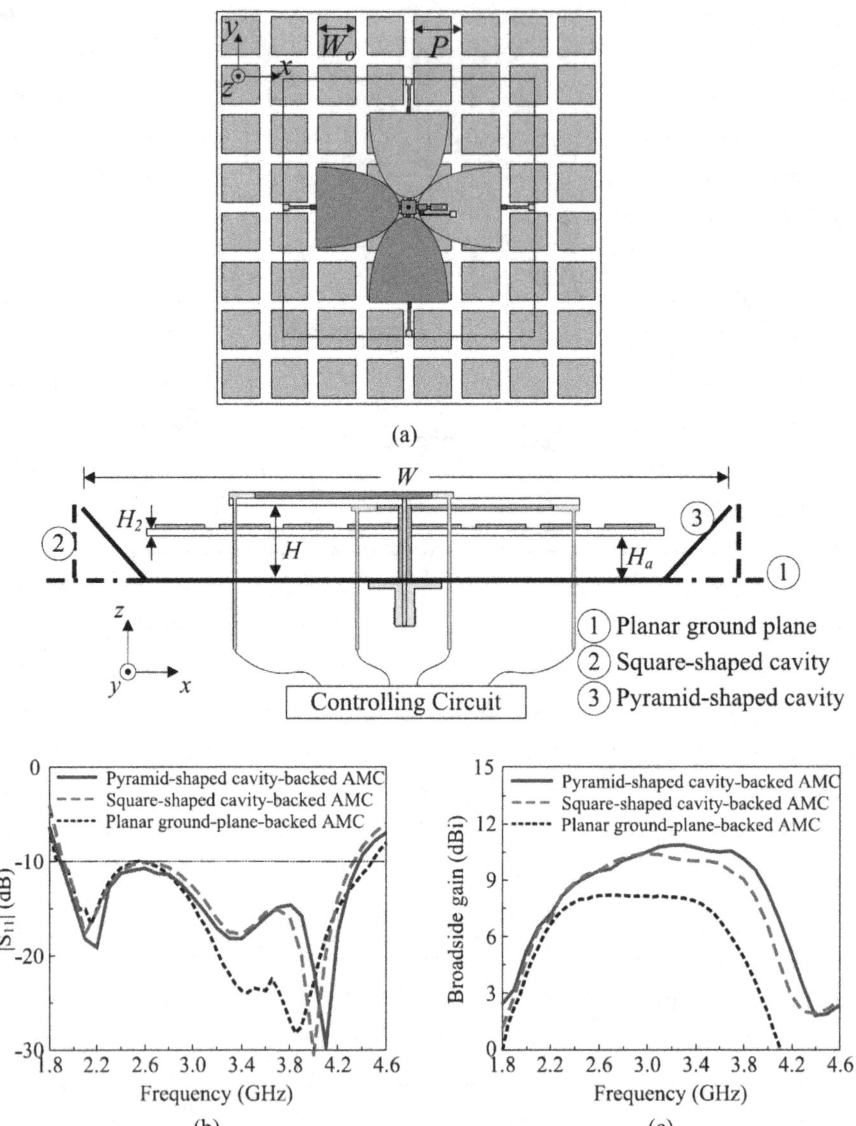

FIGURE 11.27 (a) High-gain reconfigurable antenna; (b) return loss; (c) broadside gain [96].

11.4 REAL-LIFE APPLICATIONS OF EBG PATCH ANTENNAS

11.4.1 High-Precision GPS

High-precision positioning can be achieved by combining global navigation satellite systems (GNSS) such as GPS and Galileo [97–99]. By using this system, surveyors can make measurements with sub-centimeter accuracy. In order to acquire such accurateness, some precautions are required to shield the antenna from false signals. The conventional approaches such as choke rings provide a good performance [100]. But, choke rings are generally very huge and expensive. Nowadays, EBG structures are used while maintaining good antenna performance.

A Galileo antenna on an EBG substrate is illustrated in Figure 11.28a. To achieve best gain for GPS applications, the dimension of the EBG structures is optimized. It is observed that with the combination of EBG substrate and patch antenna, an axial ratio of 2 dB is achieved. The axial ratio value for choke rings is 1 dB, which implies that the EBG patch has a better performance. Similarly, the patch and Koch fractal EBG structure [98] are shown Figure 11.28b. The tested results show the significant improvement in axial ratio bandwidth, which fits the requirement of GPS applications (Figure 11.29).

11.4.2 Wearable Electronics

Wearable electronic systems are technology that finds applications in many fields that include military, telemedicine, sports, and tracking. In [101], the jean fabric was used as a substrate having a dielectric constant of 1.7 with a thickness of 1 mm. As shown in Figure 11.29, the design consisted of a fractal monopole patch antenna with an EBG surface. This fractal antenna resonates at 1,800 MHz and 2.45 GHz for GSM and ISM applications, respectively. The antenna is placed over a 3×3 EBG array of size 150×150 mm^2. Figure 11.27 illustrates the fractal-based monopole patch antenna with an EBG surface. Another researcher used a Pellon fabric substrate [102] with a dielectric constant of 1. The designed antenna consisted of a coplanar waveguide (CPW)-fed monopole antenna with an AMC array of 4×6 units, with the overall size of 102×68 mm^2.

TABLE 11.6
Low-Profile Patch Antennas by Using Different EBG Structures

S. No.	Type of EBG Structure	Dimension (cm^3)	Height (mm)	Resonating Frequency Band	Reference
1.	EBG-CS	1.65	3	Two bands (3.5 and 4.5)	[86]
2.	Miniaturized EBG structure	4.2	0.7	One band (2.4)	[88]
3.	Square Sierpinski fractal EBG structure	10.24	1.6	Three narrow bands at 2.4, 3.5, and 4.6 GHz	[87]
4.	Uniplanar EBG structure	46.4	8	One band (1.25–29)	[95]
5.	Modified EBG structure	94.8	7.9	Two bands (2–3 and 3.8–6.3)	[93]

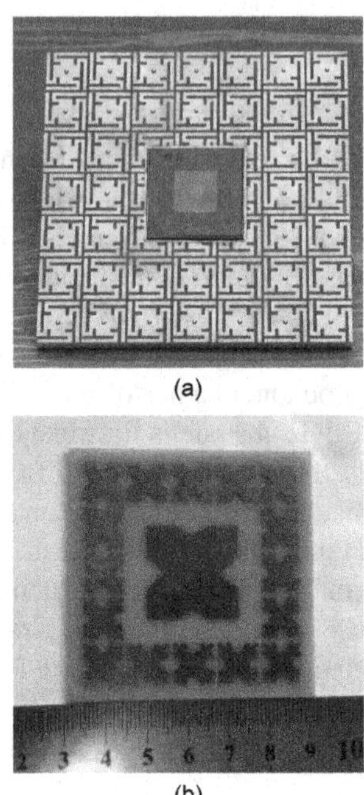

(a)

(b)

FIGURE 11.28 (a) Galileo antenna on an EBG ground plane [100]; (b) circularly polarized patch antenna with fractal HIS [102].

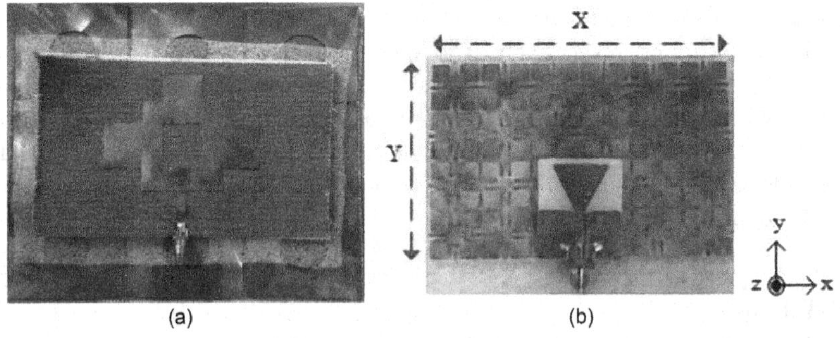

(a) (b)

FIGURE 11.29 (a) Dual-band wearable fractal-based monopole patch antenna [101]; (b) the fabricated monopole antenna with an artificial magnetic conductor as a ground plane [102].

11.4.3 Radio Frequency Identification (RFID) Systems

RFID systems have been used since World War II, and the demand for RFID systems is rapidly growing in both business and daily life. Currently, the RFID system is used in many applications such as hospitals and healthcare, passports, stores, and people identification. One of the biggest challenges associated with RFID systems is the long-range operation capability [103–106]. In [107], a CPW-fed bow tie antenna was mounted over an AMC. The AMCs reduce the backward radiation that results in

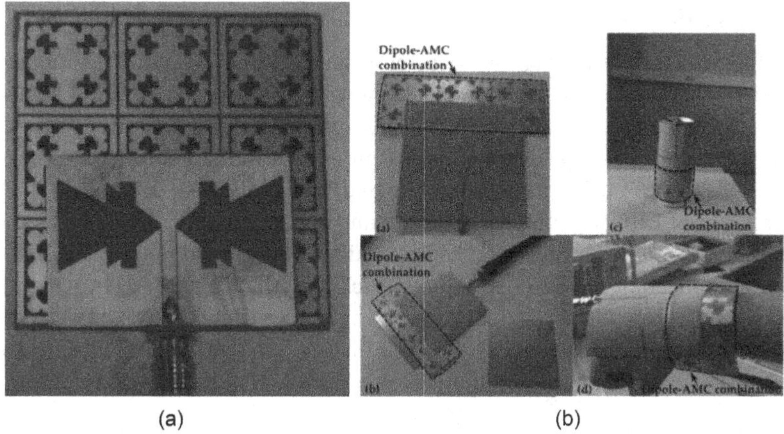

FIGURE 11.30 (a) CPW-fed bow tie antenna mounted over an AMC [107]; (b) dipole AMC [108].

FIGURE 11.31 EBG checkerboard ground plane [109].

improved gain and directivity of the overall system. The antenna was fabricated using an ARLON 25 N substrate with a thickness of 0.7 mm and a relative permittivity of 3.3. Figure 11.30a illustrates the bow tie antenna over an AMC. This structure helps in the enhancement of gain by 2.53 and 1.86 dB at 5.8 and 6.4 GHz, respectively. In [108], a dipole antenna over an AMC was designed to improve the overall performance. From the tested results, it was observed that the dipole with a balun AMC achieved an improvement of 2.9 dBi as compared to the dipole without balun AMC.

11.4.4 Radar Systems

To reduce the radar cross section, the reflection phase property of the EBG structures is used to change the direction of the fields that are scattered by a radar target. This change in direction of the scattered fields is achieved by using checkerboard EBG structures and results in a wider frequency band RCS reduction. Figure 11.31 shows the checkerboard EBG surface which is a combination of two EBG structures.

11.5 CONCLUSION

The EBG structures have greatly attracted researchers because of their unique and desirable properties. This chapter stated how the integration of EBG structures and patch antennas improves the overall performance of the antenna systems. The recent advancements of patch antenna design using EBG structures were included. Different EBG approaches to improve gain and bandwidth were discussed. The band gap property of EBG structures has been found useful to eliminate the surface wave propagation to reduce the mutual coupling and to achieve band-notch characteristics. Real-life applications of EBG structures, such as RFID, wearable electronics, and radar systems, were also included. A number of recent publications proved that the EBG technology eliminates the drawbacks of patch antenna and is most preferable for the modern-day wireless communication systems.

REFERENCES

1. Prakash, P., Abegaonkar, M. P., Kurra, L. et al. 2015. Compact electromagnetic band-gap (EBG) structure with defected ground. *IETE Journal of Research*. 62:120–126.
2. Yang, F.-R. 1999. A uniplanar compact photonic-bandgap (UC-PBG) structure and its applications for microwave circuits. *IEEE Transactions on Microwave Theory and Techniques*. 47:1–6.
3. Yang, F. and Samii, Y. R. 2003. Microstrip antennas integrated with electromagnetic band-gap (EBG) structures: A low mutual coupling design for array applications. *IEEE Transactions on Antennas and Propagation*. 51:2936–2946.
4. Shaban, H. F., Elmikaty, H. A. and Shaalan, A. A. 2008. Study the effects of electromagnetic band-gap (EBG) substrate on two patches microstrip antenna. *Progress in Electromagnetics Research B*. 10:1–20.
5. Alshamaileh, K. A., Almalkawi, M. J. and Devabhaktuni, V. K. 2015. Dual band-notched microstrip-fed vivaldi antenna utilizing compact EBG structures. *International Journal of Antennas and Propagation*. 2015:1–7.
6. Loinaz, A. I. and Bocio, C. R. 2004. EBG at microwave frequency range: Bragg and/or resonant effect. *Microwave and Optical Technology Letters*. 42:383–385.
7. Jaglan, N. and Gupta, S. D. 2015. Reflection phase characteristics of EBG structures and WLAN band notched circular monopole antenna design. *International Journal on Communications Antenna and Propagation (IRECAP)*. 5:201–233.
8. Kumar, A., Kumar, D., Mohan, J. et al. 2015. Investigation of grid metamaterial and EBG structures and its application to patch antenna. *International Journal of Microwave and Wireless Technologies*. 7:705–712.
9. Kaabal, A. M., Halaoui, E., Ahyoud, S. et al. 2016. Dual band-notched WIMAX/WLAN of a compact UltraWide Band antenna with spectral and time domains analysis for breast cancer detection. *Progress in Electromagnetics Research C*. 65:163–173.
10. Palreddy, S. 2015. Wideband electromagnetic band gap (EBG) structures, analysis and applications to antennas. Virginia Polytechnic Institute 4:1–131.
11. Choi, J. H., Hon, P. W. C. and Itoh, T. 2014. Dispersion analysis and design of planar electromagnetic bandgap ground plane for broadband common-mode suppression. *IEEE Microwave and Wireless Components Letters*. 24:772–774.
12. Bhavarthe, P. P., Rathod, S. S. and Reddy, K. T. V. 2017. A compact two via slot-type electromagnetic bandgap structure. *IEEE Microwave and Wireless Components Letters*. 27:446–448.

13. Huang, C., Ji, C., Wu, X. et al. 2018. Combining FSS and EBG surfaces for high-efficiency transmission and low-scattering properties. *IEEE Transactions on Antennas and Propagation*. 66:1628–1632.
14. Alam, M. S., Misran, N. and Yatim, B. 2013. Development of electromagnetic band gap structures in the perspective of microstrip antenna design. *International Journal of Antennas and Propagation*. 1–22. Article ID 507158.
15. Zhou, Z., Wei, Z., Tang, Z., and Yin Y. 2019. Design and analysis of a wideband multiple microstrip dipole antenna with high isolation. *IEEE Antennas and Wireless Propagation Letters*. 18: 722–726.
16. Guo, Z., Tian, H., Wang, X. et al. 2013. Bandwidth enhancement of monopole UWB antenna with new slots and EBG structures. *IEEE Antennas and Wireless Propagation Letters*. 12:1550–1553.
17. Baudha, S. and Vishwakarma, D. K. 2016. Bandwidth enhancement of a planar monopole microstrip patch antenna. *International Journal of Microwave and Wireless Technologies*. 8:237–242.
18. Alkhatib, R. and Drissi, M. 2007. Improvement of bandwidth and efficiency for directive superstrate EBG antenna. *Electronics Letters*. 43:694–702.
19. Denidni, T. A. Coulibaly, Y. and Boutayeb, H. 2009. Hybrid dielectric resonator antenna with circular mushroom-like structure for gain improvement. *IEEE Transactions on Antennas and Propagation*. 57:1043–1049.
20. Alrabadi, O., Perruisseau-Carrier, N. J. and Kalis, A. 2012. MIMO transmission using a single RF source: Theory and antenna design. *IEEE Transactions on Antennas and Propagation*. 60:654–664.
21. Braaten, B. D., Iftikhar, A., Capobianco, A. D. et al. 2015. Compact 4×4 UWB-MIMO antenna with WLAN band rejected operation. *Electronics Letters*. 51:1048–1050.
22. Kiem, N. K., Phuong, H. N. B. and Chien, D. N. 2014. Design of compact 4×4 UWB-MIMO antenna with WLAN band rejection. *International Journal of Antennas and Propagation*. 1–11. Article ID 539094.
23. Zhang, S. and Pedersen, G. F. 2016. Mutual coupling reduction for UWB MIMO antennas with a wideband neutralization line. *IEEE Antennas and Wireless Propagation Letters*. 15:166–169.
24. Elsheakh, D. N., Elsadek, H. A., Abdallah, E. et al. 2009. Enhancement of microstrip monopole antenna bandwidth by using EBG structures. *IEEE Antennas and Wireless Propagation Letters*. 8:959–962.
25. Gujral, M. J., Li, L.-W., Yuan, W. et al. 2012. Bandwidth improvement of microstrip antenna array using dummy EBG pattern on feedline. *Progress in Electromagnetics Research*. 127:79–92.
26. Wen, B.-J., Peng, L., Li, X.-F. et al. 2019. A low-profile and wideband unidirectional antenna using bandwidth enhanced resonance-based reflector for fifth generation (5G) systems applications. *IEEE Access*. 7:27352–27361.
27. Madhav, B. T. P., Sanikommu, M. M., Pranoop, N. V. et al. 2015. CPW fed antenna for wideband applications based on tapered step ground and EBG structure. *Indian Journal of Science and Technology*. 8:101–119.
28. Kim, M. 2015. A compact EBG structure with wideband power/ground noise suppression using meander-perforated plane. *IEEE Transactions on Electromagnetic Compatibility*. 57:595–598.
29. Shen, C.-K., Chen, C.-H., Han, D.-H. et al. 2015. Modeling and analysis of bandwidth-enhanced multilayer 1-D EBG with bandgap aggregation for power noise suppression. *IEEE Transactions on Electromagnetic Compatibility*. 57:858–867.
30. Yang, Q., Wang, X., Wang, J. J. et al. 2011. Reducing SAR and enhancing cerebral signal-to-noise ratio with high permittivity padding at 3 T. *Magnetic Resonance in Medicine*. 65:358–362.

31. Wang, C.-D. 2012. Bandwidth enhancement based on optimized via location for multiple vias EBG power/ground planes. *IEEE Transactions on Components, Packaging and Manufacturing Technology*. 2:332–341.
32. Taha, E. A., Ahmed, I. I. and Yahiea, A. 2015. A miniaturized lotus shaped microstrip antenna loaded with EBG structures for high gain-bandwidth product applications. *Progress in Electromagnetics Research C*. 60:157–167.
33. Cos, M. E. and Heras, F. L. 2012. Dual-band uniplanar CPW-fed monopole/EBG combination with bandwidth enhancement. *IEEE Antennas and Wireless Propagation Letters*. 11:365–368.
34. Jam, S. and Simruni, M. 2018. Performance enhancement of a compact wideband patch antenna array using EBG structures. *AEU – International Journal of Electronics and Communications*. 89:42–55.
35. Shoaib, N., Shoaib, S. R., Khattak, Y. et al. 2018. MIMO antennas for smart 5G devices. *IEEE Access*. 67:7014–77021.
36. Peddakrishna, S., Khan, T. and De, A. 2017. Electromagnetic band-gap structured printed antennas: A feature oriented survey. *International Journal RF Microwave Computational Aided Engineering*. 27:1–16.
37. Hadarig, R. C., de Cos, M. E. and Las-Heras, F. 2013. Novel miniaturized artificial magnetic conductor. *IEEE Antennas and Wireless Propagation Letters*. 12:174–177.
38. Khan, M. S., Tahir, F. A., Meredov, A. et al. 2019. A W-band EBG-backed double-rhomboid bowtie-slot on-chip antenna. *IEEE Antennas and Wireless Propagation Letters*. 18:1046–1050.
39. McKinzie, W. E., Nair, D. M., Thrasher, B. A. et al. 2016. 60-GHz LTCC patch antenna array with an integrated EBG structure for gain enhancement. *IEEE Antennas and Wireless Propagation Letters*. 15:1522–1525.
40. Chen, D., Yang, W. and Che, W. 2018. High-gain patch antenna based on cylindrically projected EBG planes. *IEEE Antennas and Wireless Propagation Letters*. 17:2374–2378.
41. Bhavarthe, P. P., Rathod, S. S. and Reddy, K. T. V. 2018. A compact two-via hammer spanner-type polarization-dependent electromagnetic bandgap structure. *IEEE Microwave and Wireless Components Letters*. 28:284–286.
42. Ketkuntod, P., Hongnara, T. Thaiwirot, W. et al. 2017. Gain enhancement of microstrip patch antenna using I-shaped mushroom-like EBG structure for WLAN application. In *2017 International Symposium on Antennas and Propagation (ISAP)*. 1–2.
43. Bharathi, M. and Phavithra, P. J. 2018. Gain enhancement of a square patch antenna using EBG structure. *International Journal of Innovative Technology and Exploring Engineering*. 8:1–4.
44. Boutayeb, H. and Denidni, T. A. 2007. Gain enhancement of a microstrip patch antenna using a cylindrical electromagnetic crystal substrate. *IEEE Transactions on Antennas and Propagation*. 55:3140–3145.
45. Mondal, K. and Sarkar, P. P. 2019. Gain and bandwidth enhancement of microstrip patch antenna for WiMAX and WLAN applications. *IETE Journal of Research*. 1–9.
46. Mark, R., Rajak, N. Mandal, K. et al. 2019. Metamaterial based superstrate towards the isolation and gain enhancement of MIMO antenna for WLAN application. *AEU – International Journal of Electronics and Communications*. 100:144–152.
47. Han, Z.-J., Song, W. and Sheng, X.-Q. 2017. Gain enhancement and RCS reduction for patch antenna by using polarization-dependent EBG surface. *IEEE Antennas and Wireless Propagation Letters*. 16:1631–1634.
48. Liu, Z., Liu, Y. and Gong, S. 2018. Gain enhanced circularly polarized antenna with RCS reduction based on metasurface. *IEEE Access*. 6:46856–46862.

49. Roseline, A., Malathi, K. and Shrivastav, A. K. 2011. Enhanced performance of a patch antenna using spiral-shaped electromagnetic bandgap structures for high-speed wireless networks. *IET Microwaves, Antennas & Propagation.* 5:1733–1750.
50. Lee, Y. J., Yeo, J., Mittra, R. et al. 2005. Application of electromagnetic bandgap (EBG) superstrates with controllable defects for a class of patch antennas as spatial angular filters. *IEEE Transactions on Antennas and Propagation.* 53: 224–235.
51. Haraz, O., Elboushi, M., Alshebeili, A. et al. 2014. Dense dielectric patch array antenna with improved radiation characteristics using EBG ground structure and dielectric superstrate for future 5G cellular networks. *IEEE Access.* 2:909–913.
52. Yang, X., Liu, Y., Xu, Y.-X. et al. 2017. Isolation enhancement in patch antenna array with fractal UC-EBG structure and cross slot. *IEEE Antennas and Wireless Propagation Letters.* 16:2175–2178.
53. Mohamed, I., Abdalla, M. and Mitkees, A. E. 2019. Perfect isolation performance among two-element MIMO antennas. *AEU – International Journal of Electronics and Communications.* 107:21–31.
54. Mohamadzade, B. and Afsahi, M. 2017. Mutual coupling reduction and gain enhancement in patch array antenna using a planar compact electromagnetic bandgap structure. *IET Microwaves, Antennas & Propagation.* 11:1719–1725.
55. Li, Q., Feresidis, A. P., Mavridou, M. et al. 2015. Miniaturized double-layer EBG structures for broadband mutual coupling reduction between UWB monopoles. *IEEE Transactions on Antennas and Propagation.* 63:1168–1171.
56. Tan, X., Wang, W., Wu, Y. Y. et al. 2019. Enhancing isolation in dual-band meander-line multiple antenna by employing split EBG structure. *IEEE Transactions on Antennas and Propagation.* 67:2769–2774.
57. Arshed, T. and Tahir, F. A. 2017. A miniaturized triple band-notched UWB antenna. *Microwave and Optical Technology Letters.* 59:2581–2586.
58. Toyota, Y. A., Engin, E. T., Kim, H. et al. 2006. Stopband analysis using dispersion diagram for two-dimensional electromagnetic bandgap structures in printed circuit boards. *IEEE Microwave and Wireless Components Letters.* 16:645–647.
59. Farahani, H., Veysi, S., Kamyab, M. et al. 2010. Mutual Coupling reduction in patch antenna arrays using a UC-EBG superstrate. *IEEE Antennas and Wireless Propagation Letters.* 9:57–59.
60. Alibakhshikenari, M., Khalily, M. B., Virdee, B. S. et al. 2019. Mutual coupling suppression between two closely placed microstrip patches using EM-bandgap metamaterial fractal loading. *IEEE Access.* 7:23606–23614.
61. Kumar, N. and Kommuri, U. K. 2019. MIMO antenna H-plane isolation enhancement using UC-EBG structure and metal line strip for WLAN applications. *Radioengineering.* 27:399–406.
62. Mavridou, M., Feresidis, A. P. and Gardner, P. 2016. Tunable double-layer EBG structures and application to antenna isolation. *IEEE Transactions on Antennas and Propagation.* 64:70–79.
63. Kushwaha, N. and Kumar, R. 2013. An UWB fractal antenna with defected ground structure and swastika shape electromagnetic band gap. *Progress in Electromagnetics Research B.* 52:383–403.
64. Jaglan, N., Kanaujia, B. K., Gupta, S. D. et al. 2016. Triple band notched UWB antenna design using electromagnetic band gap structures. *Progress in Electromagnetics Research C.* 66:139–147.
65. Pandey, G. K., Singh, H. S., Bharti, P. K. et al. 2013. Design of WLAN band notched UWB monopole antenna with stepped geometry using modified EBG structure. *Progress in Electromagnetics Research B.* 50:201–217.

66. Mouhouche, F., Azrar, A., Dehmas, M. et al. 2018. Design a compact UWB monopole antenna with triple band-notched characteristics using EBG structures. *Frequenz.* 11:479–487.
67. Peng, L. and Ruan, C.-L. 2011. UWB band-notched monopole antenna design using electromagnetic-bandgap structures. *IEEE Transactions on Microwave Theory and Techniques.* 59:1074–1081.
68. Mandal, T. and Das, S. 2014. Design of dual notch band UWB printed monopole antenna using electromagnetic-bandgap structure. *Microwave and Optical Technology Letters.* 56:2195–2199.
69. Jaglan, N., Gupta, S. D., Thakur, E. et al. 2018. Triple band notched mushroom and uniplanar EBG structures based UWB MIMO/Diversity antenna with enhanced wide band isolation. *AEU – International Journal of Electronics and Communications.* 90:36–44.
70. Jaglan, N., Kanaujia, B. K., Gupta, S. D. et al. 2017. Dual band notched EBG structure based UWB MIMO/diversity antenna with reduced wide band electromagnetic coupling. *Frequenz.* 71:11–12.
71. Liu, H. and Xu, Z. 2013. Design of UWB monopole antenna with dual notched bands using one modified electromagnetic-bandgap structure. *The Scientific World Journal.* 2013:1–9.
72. Wang, J. H., Yin, Y.-Z. and Liu, X. L. 2013. Triple band-notched ultra wideband (UWB) antenna using a novel modified capacitively loaded loop (CLL) resonator. *Progress in Electromagnetics Research Letters.* 42:55–64.
73. Li, T., Zhai, H.-Q., Li, G.-H. et al. 2012. Design of compact UWB band-notched antenna by means of electromagnetic-bandgap structures. *Electronics Letters.* 48:601–608.
74. Ghosh, A., Mandal, T. and Das, S. 2019. Design and analysis of triple notch ultrawideband antenna using single slotted electromagnetic bandgap inspired structure. *Journal of Electromagnetic Waves and Applications.* 33:1391–1405.
75. Jaglan, N., Gupta, S. D., Kanaujia, B. K. et al. 2018. Band notched UWB circular monopole antenna with inductance enhanced modified mushroom EBG structures. *Wireless Networks.* 24:383–393.
76. Jaglan, N., Kanaujia, B. K., Gupta, S. D. et al. 2018. Design of band-notched antenna with DG-CEBG. *International Journal of Electronics.* 105:58–72.
77. Peddakrishna, S. and Khan, T. 2018. Design of UWB monopole antenna with dual notched band characteristics by using π-shaped slot and EBG resonator. *AEU – International Journal of Electronics and Communications.* 96:107–112.
78. Zhang, X. 2018. Design of defective electromagnetic band-gap structures for use in dual-band patch antennas. *International Journal of RF and Microwave Computer-Aided Engineering.* 28:22126–21287.
79. Yi, H. and Qu, S.-W. 2013. A novel dual-band circularly polarized antenna based on electromagnetic band-gap structure. *IEEE Antennas and Wireless Propagation Letters.* 12:1149–1152.
80. Saurav, K., Sarkar, D. and Srivastava, K. V. 2014. Dual-polarized dual-band patch antenna loaded with modified mushroom unit cell. *IEEE Antennas and Wireless Propagation Letters.* 13:1357–1360.
81. Emadian, S. R. and Shokouh, J. A. 2015. Very small dual band-notched rectangular slot antenna with enhanced impedance bandwidth. *IEEE Transactions on Antennas and Propagation.* 63:4529–4534.
82. Liu, T. H., Zhang, W. X., Zhang, M. et al. 2000. Low profile spiral antenna with PBG substrate. *Electronics Letters.* 36:773–779.
83. Yousefi, L., Mohajer-Iravani, B. and Ramahi, O. M. 2007. Low profile wide band antennas using electromagnetic bandgap structures with magneto-dielectric materials. *International workshop on Antenna Technology: Small and Smart Antennas Metamaterials and Applications.* 431–434.

84. Li, L., Li, B., Liu, H.-X. et al. 2006. Locally resonant cavity cell model for electromagnetic band gap structures. *IEEE Transactions on Antennas and Propagation.* 54:90–100.
85. Ghabzouri, M., El Salhi, A. E., Anacleto, P. et al. 2017. Enhanced low profile, dual-band antenna via novel electromagnetic band gap structure. *Progress in Electromagnetics Research C.* 71:79–89.
86. Ma, X., Mi, S. and Lee, Y. H. 2015. Design of a microstrip antenna using square Sierpinski fractal EBG structure. *IEEE 4th Asia-Pacific Conference on Antenna and Propagation (APCAP).*
87. Ashyap, A. Y. I. 2017. Compact and low-profile textile EBG-based antenna for wearable medical applications. *IEEE Antennas and Wireless Propagation Letters.* 16:2550–2553.
88. Li, Z. and Samii, Y. R. 2000. PBG, PMC and PEC ground planes: a case study of dipole antennas. *IEEE Antennas and Propagation Society International Symposium. Transmitting Waves of Progress to the Next Millennium. 2000 Digest. Held in conjunction with: USNC/URSI National Radio Science Meeting (Cat. No.00CH37118).* 2: 674–677.
89. Cao, W., Zhang, B., Liu, A. et al. 2012. Multi-frequency and dual-mode patch antenna based on electromagnetic band-gap (EBG) structure. *IEEE Transactions on Antennas and Propagation.* 60:6007–6012.
90. Kasahara, Y., Toyao, H. and Hankui, E. 2017. Compact and multiband electromagnetic bandgap structures with adjustable bandgaps derived from branched open-circuit lines. *IEEE Transactions on Microwave Theory and Techniques.* 65:2330–2340.
91. Misran, N., Islam, M. T. and Alam, M. S. 2013. Inverse triangular-shape CPW-fed antenna loaded with EBG reflector. *Electronics Letters.* 49:86–88.
92. Tran, H. H. and Park, I. 2016. A dual-wideband circularly polarized antenna using an artificial magnetic conductor. *IEEE Antennas and Wireless Propagation Letters.* 15:950–953.
93. Chamani, Z. and Jahanbakht, S. 2017. Improved performance of double-t monopole antenna for 2.4/5.6 GHz dual-band WLAN operation using artificial magnetic conductors. *Progress in Electromagnetics Research M.* 61:205–213.
94. Lin, W., Chen, S.-L., Ziolkowski, R. W. et al. 2018. Reconfigurable, wideband, low-profile, circularly polarized antenna and array enabled by an artificial magnetic conductor ground. *IEEE Transactions on Antennas and Propagation.* 66:1564–1569.
95. Tran, H. H., Trong, N. N., Le, T. T. et al. 2018. Low-profile wideband high-gain reconfigurable antenna with quad-polarization diversity. *IEEE Transactions on Antennas and Propagation.* 66:3741–3746.
96. Bao, X. L., Ruvio, G. and Ammann, M. J. 2007. Low-profile dual-frequency GPS patch antenna enhanced with dual-band EBG structure. *Microwave and Optical Technology Letters.* 49:2630–2634.
97. Wang, E. and Liu, Q. 2016. GPS patch antenna loaded with fractal EBG structure using organic magnetic substrate. *Progress in Electromagnetics Research Letters.* 58:23–28.
98. Srivastava, R. 2014. Dual band rectangular and circular slot loaded microstrip antenna for WLAN/GPS/WiMax applications. *2014 Fourth International Conference on Communication Systems and Network Technologies.* 45–48.
99. Baggen, R. M., Vazquez, M. J., Leiss, J. et al. 2008. Low profile GALILEO antenna using EBG technology. *IEEE Transactions on Antennas and Propagation.* 56:667–674.
100. Velan, S. 2015. Dual-band EBG integrated monopole antenna deploying fractal geometry for wearable applications. *IEEE Antennas and Wireless Propagation Letters.* 14:249–252.
101. Alemaryeen, A. and Noghanian, S. 2019. On-body low-profile textile antenna with artificial magnetic conductor. *IEEE Transactions on Antennas and Propagation.* 67:3649–3656.

102. Phatarachaisakul, T., Pumpoung, T. and Phongcharoenpanich, C. 2005. Dual-band RFID tag antenna with EBG for glass objects. *IEEE 4th Asia-Pacific Conference on Antennas and Propagation (APCAP)*. 199–200.
103. Ukkonen, L., Sydanheimo, L. and Kivikoski, M. 2004. Patch antenna with EBG ground plane and two-layer substrate for passive RFID of metallic objects. *IEEE Antennas and Propagation Society Symposium*. 1:93–96.
104. Phatra, C., Krachodnok, P. and Wongsan, R. 2009. Design of a RFID tag using dipole antenna with Electromagnetic Band Gap. *International Conference on Electrical Engineering/Electronics, Computer, Telecommunications and Information Technology*. 1:1–4.
105. Konishi, T., Miura, T., Numata, Y. et al. 2009. An impedance matching technique of a UHF-band RFID tag on a high-impedance surface with parasite elements. *IEEE Radio and Wireless Symposium*. 67–70.
106. Cos, M. E. and Las-Heras, F. 2012. Dual-band antenna/AMC combination for RFID. *International Journal of Antennas and Propagation*. 2012:1–7.
107. Hadarig, R. C., de Cos, M. E. and Las-Heras, F. 2013. UHF dipole-AMC combination for RFID applications. *IEEE Antennas and Wireless Propagation Letters*. 12:1041–1044.
108. Simruni, M. and Jam, S. 2019. Radiation performance improvement of wideband microstrip antenna array using wideband AMC structure. *International Journal of Communication Systems*. 32:3954–3962.
109. Hossein, M. and Shahrokh, J. 2016. Improved radiation performance of low profile printed slot antenna using wideband planar AMC surface. *IEEE Transactions on Antennas and Propagation*. 64:4626–4638.

12 Design of Frequency Selective Surface (FSS) Printed Antennas

Kanishka Katoch, Dr. Naveen Jaglan, and Prof. Samir Dev Gupta
Jaypee University of Information Technology, Solan

Prof. Binod Kumar Kanaujia
Jawaharlal Nehru University, New delhi

CONTENTS

12.1 INTRODUCTION

One of the most desired features in microwave and optical range signal processing systems is spatial filtering. Frequency selective surface (FSS) is also a type of spatial filter, which offer transmission and reflection characteristics by modifying the electromagnetic incident wave striking its surface. FSS are two-dimensional planar structures arranged in a periodic manner. Metallic arrays (apertures or patch), as shown in Figure 12.1, are etched over a dielectric substrate, exhibiting partial or full transmission or reflection of the incident wave at a particular frequency [1]. The amplitude and phase of the transmitted wave vary after striking the FSS when compared to the incident wave. This occurs when the resonance frequency of the FSS matches the plane wave frequency. Therefore, in free space, FSS can either block or pass the incident wave at a particular frequency.

Traditionally, FSS is used as filter. However, the conventional FSS do not provide adequate spatial filtering response and have a narrow bandwidth. Therefore, from last few decades, miniaturization of the FSS and improvement of frequency response for a wider bandwidth have intensively been investigated by the researchers. It is observed that single-layered FSS are inadequate due to the unstable performance when the incident angle of the EM wave varies. To eliminate this problem, multi-layer FSS have been implemented, which provides more flexibility of varying parameters for preferred performance [2–7]. For the compact structures, miniaturized arrays and fractal elements are used these days [8–10]. Embedded FSS [11], metamaterial FSS [12,13], integration of FSS with electromagnetic band gap (EBG) structures [14], reconfigurable FSS [15], and three-dimensional FSS structures [16] are some recent advancements in FSS technology.

Apart from filtering, FSS serve a wide variety of applications according to one's requirements. These variations depend on the way the incident wave is modified. Some of the other major properties of FSS are multi-pole frequency response with

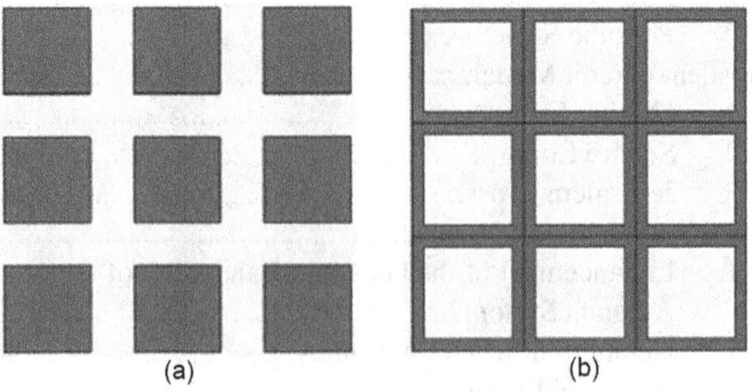

FIGURE 12.1 (a) Patch-type FSS, (b) aperture-type FSS.

huge out-of-band rejections, polarization, absorption, reduced periodicity, angular stability, low profile, low cost, ease of fabrication, and many more. However, attaining all the above features for a modified design is a difficult task for the designers [2]. To utilize the maximum EM properties of FSS, various design modification techniques are used. By varying the geometric parameters, such as dielectric substrate, element shape, and inter-element spacing, and adjusting the size of the unit cell of the FSS, the desired properties of the FSS can be achieved [17,18].

Recently, multiple features of antennas, such as low cost, compact structure, high gain with wide impedance bandwidth, lightweight, and directional beamforming, have attracted the researchers. For conventional antenna structures, it is a challenging task to achieve all the above parameters for an optimal design. FSS help in enhancing most of the aforementioned features of the antennas [19–21]. In the past, FSS were mostly used in reflector antennas [22], including resonant beam splitters, and antenna radomes [23,24]. However, these days, FSS is employed in lens antennas [25], radio frequency identification (RFID) [26], and electromagnetic shielding [27]. The most famous applications of the FSS are reconfigurable antennas, isolation in MIMO antennas, controlling radar cross section (RCS), and antenna radomes.

In this chapter, we present the basic concepts and principle of operation of the FSS. The equivalent circuit analysis is also included. Various types and techniques are discussed on the current state-of-art in the field of FSS. The applications of the FSS related to the antennas are discussed later in this chapter. Finally, the advantages and disadvantages of the FSS with future scope are summarized.

12.2 TYPES OF FSS

In this section, the classification of FSS on the basis of FSS elements, design structure, and applications is discussed, as shown in Figure 12.2.

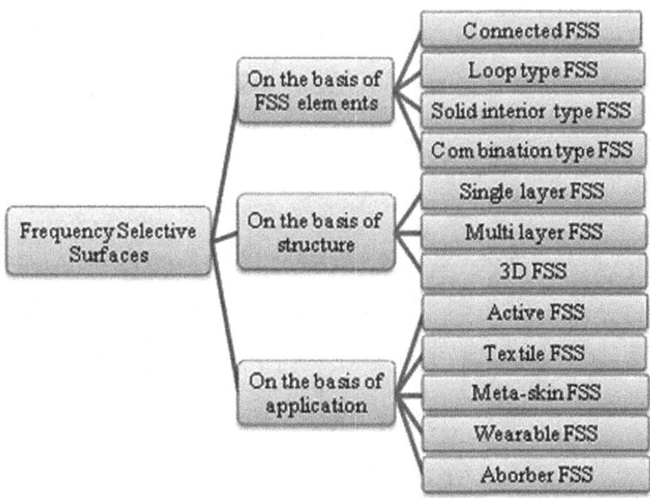

FIGURE 12.2 Classifications of the FSS.

12.2.1 On the Basis of FSS Elements

Four basic types have been classified in this category by Munk [1]. They are center-connected FSS (Group 1), which include dipoles, cross-dipoles, and Jerusalem cross; loop-type FSS (Group 2), which include square-, hexagonal-, and circular-shaped loops; solid-interior-type FSS (Group 3), which are referred to as patch-type elements; and combination-type FSS (Group 4), which are a combination of the aforementioned types. These are shown in Figure 12.3.

Thus, depending upon one's requirements, the elements are selected from these groups. From every FSS element, it is desired to have a stable frequency response at oblique angles. Group 1 consists of dipoles, cross dipoles, etc. However, this group suffers from narrow impedance bandwidth. By proper designing of the elements, the bandwidth can be enhanced with reduced inter-element spacing. Mostly, it is beneficial to select Group 2 (loop family) because it enhances the bandwidth. Resonance is achieved by loop-structured FSS when the perimeter of the loop equals the full wavelength.

Resonance also depends upon the relevant inter-element spacing and effective permittivity of the substrate [1]. By decreasing the inter-element spacing,

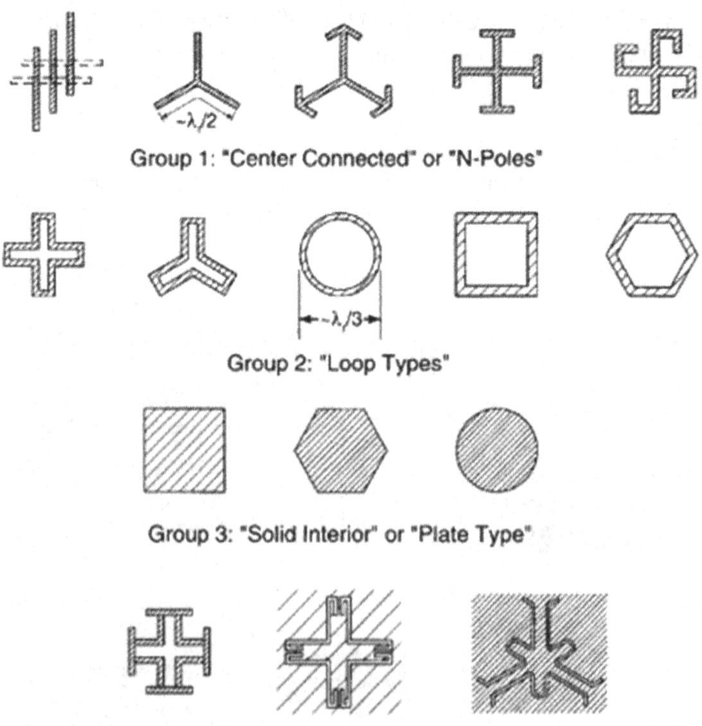

FIGURE 12.3 Basic structures of FSS [1].

bandwidth can be enhanced. However, with a certain decreased distance, inter-element capacitance builds up, dragging the resonance frequency toward lower side. By changing the shape of these structures, a wide variety in the range of bandwidth can be achieved. Therefore, for a desired resonant frequency, the perimeter of the FSS element should be minimized. Solid interior elements of Group 3 are mostly used in combination with other elements. Group 4 contains the combination of all. Generally, when the size of the patch or aperture is in the order of half the wavelength, the desired resonance is achieved. However, this large size leads to an unstable frequency response at different incident angles. To avoid this, miniaturization of the FSS elements is necessary and, for this, fractal FSS or convoluted meander FSS can be used.

12.2.2 On the Basis of Structure

On the basis of structure, FSS is classified into three categories, i.e., single-layer FSS, multilayer FSS, and three-dimensional FSS.

12.2.2.1 Single-Layer FSS

A planar periodic two-dimensional array structure is a single-layer FSS. Single-layer FSS with filtering characteristics can be used in a wide variety of applications. But due to the restricted space in a unit cell, their potential use is limited. Large element size and wide inter-element spacing result in degraded performance and realizing these surfaces is a difficult task [28]. Therefore, the role of miniaturized FSS is growing for a stable frequency response at different incident angles. These miniaturized elements are in high demand due to its ability to separate grating lobes from the desired band [29]. FSS elements, such as loops, patches, and fractal, meander, multipole and convoluted structures, help in the achievement of miniaturized FSS. Single-layer FSS are easy to design and fabricate, have low cost, and can easily be integrated with other structures, such as antennas and radars.

12.2.2.2 Multilayer FSS

Narrow operating bandwidth is one of the main issues faced by FSS, which is challenging to improve with a single-layer FSS. It is observed that even with the implementation of fractal elements and convoluted meander structures, the bandwidth performance of the FSS is narrow. Apparently, these techniques are best suitable for reducing the size and profile of FSS. The separation of individual FSS' frequency bands is possible with wider transmission/reflection bands and also faster roll-off in some of the applications. However, frequency response declines eventually on both sides of the resonant frequency in the case of single-layer FSS, making them inappropriate for such critical applications. This issue can be resolved either by making use of complex geometry for resonation or by constructing multiple layers of FSS [30]. Complex geometries typically are not easy to design and produce multiple bands instead and, moreover, widen the overall bandwidth without improving the bandwidth of the individual resonant band. Consequently, for enhancing the bandwidth of individual resonance, multilayer FSS become a better option. Due to the discontinuity in between the FSS layers, multilayer FSS achieve a wideband response at a certain frequency. Better frequency response and faster roll-off can be achieved by higher-order passband FSS when compared to single-layer FSS. However, it requires the thickness of the substrate to be approximately of the order $\lambda/4$, which is not appreciable these days. Moreover, non-resonant array elements largely increase the FSS size; consequently, a trade-off is certain. Therefore, the more are the FSS layers, the better is the roll-off and the broader is the bandwidth, but this results in a bulky and complex system.

12.2.2.3 3-Dimensional FSS

Applications, such as terahertz sensing, communication system, and radar cross section reduction, demand a broader bandwidth with a fast roll-off. Conventional FSS structures are easy to design and fabricate; however, these single-layer or multilayer FSS are unable to provide the aforementioned characteristics efficiently and therefore are not potentially suitable for various applications [31]. To overcome these limitations, a new and unique class of FSS has been introduced. These are the three-dimensional FSS [32]. For the realization of the 3D FSS, the same methods that are used by 2D FSS are applied. The only difference is that for the deduction of 3D structure, additional cavities or structures are etched over a 2D FSS. Vias, metallic lines, metal plate bits, etc., can be used in the inter-structure additions. These 3D FSS structures provide better roll-off, broad bandwidth, flexibility to the design, and angular stability due to surplus choices in their designing. For unit cell miniaturization and better angular stability, 3D FSS was implemented and then fabricated [33], as shown in Figure 12.4a. Four layers of substrate were stacked together, over which four-legged loop elements were printed, and via holes were created on this multilayer PCB board. It was noticed that from 0° to 60°, this FSS structure had a stable response at 6.4 GHz and shown a good agreement in between the simulated result and measured result.

In the literature, various methods have been proposed to describe the behavior of the FSS. A 3D band-pass filter had been implemented with high frequency selectivity [34], as shown in Figure 12.4b. Each individual unit cell contained multiple resonators, which permits multiple number of transmission poles and zeros. Open-circuited and short-circuited resonators help in the production of transmission poles in the passband, whereas transmission zeros were produced by the resonant frequencies of the

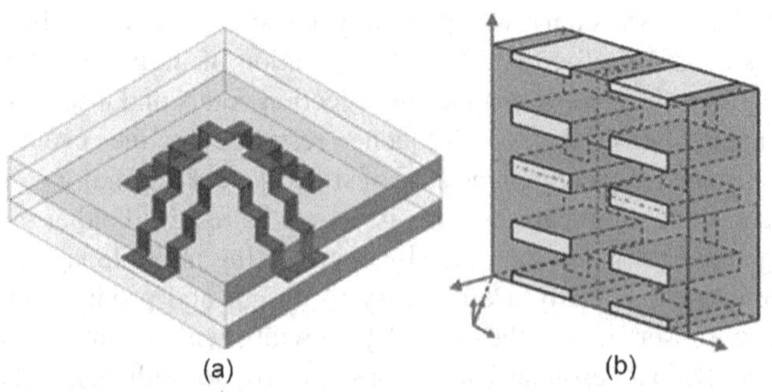

FIGURE 12.4 Unit cell structure of 3D FSS [33,34].

other short-circuited resonator. Here, a three-layered FSS structure was made, which gives an angular stability from 0° to 50°.

12.2.3 On the Basis of Application

12.2.3.1 Active FSS

Conventional FSS elements are passive in nature and, once fabricated, become inflexible. In order to achieve the design flexibility for better performance, a new class of FSS is introduced, which is known as active frequency selective surfaces (AFSS) [35]. Active elements such as PIN diodes, Schottky diodes, and varactor diodes are used for the electromagnetic beam switching and work in two states, i.e., ON or OFF. These are mostly used in reconfigurable antennas for frequency or pattern reconfiguration, and a detailed discussion on this is performed later in this chapter.

12.2.3.2 Textile FSS

Textile FSS are the new emerging class of FSS. Having a unique feature of integrating with clothes for off-body and on-body communications [36], these structures are printed over a paper or cloth by lithographic processes, screen printing, stamping, and inkjet printing. Smoothness, flexibility, higher strength-to-weight ratio, and lightweight are some benefits offered by textile structures.

12.2.3.3 Meta-Skin FSS

Recent trends in flexible and stretchable electronic devices have made it feasible to create liquid metal FSS. With the introduction of this technology, the shape, size, and characteristics of a material changes with the injection of liquid metal into an elastomeric substrate, making it an active element, which has been used in the reconfigurable antenna applications lately. Other applications of these meta-skin FSS are artificial skin sensors, probes, electrical interconnects, etc. Inside an elastomer, an array of liquid metal split ring resonators are enclosed [37]. With the injection of the liquid, wide tuning range and high FSS performance were obtained.

12.2.3.4 Wearable FSS

Owing to the features such as wide impedance bandwidth and flexibility, wearable FSS are manufactured for the applications such as heart rate monitoring and communication. As per the requirement, these devices can be placed at any part of the body. In [38], a heart rate monitor based on FSS was designed with attached photodiodes and a LED. This device was made to wear on the finger.

12.2.3.5 Absorber FSS

Devices such as missiles, airplane, radars, and airplanes have a high tendency to get detected. To eliminate the possibility of being caught, highly efficient broadband absorbers are required. Generally, FSS provide good reflection characteristics in its stopband. However, with the addition of some absorbing elements, these strong backward waves can be reduced and get absorbed in the stopband. These types of FSS are absorber FSS and are especially designed to be undetected by radars [39].

12.3 PRINCIPAL OF OPERATION

12.3.1 FSS Operational Theory

The operational theory of FSS unit cells is explained as follows [1]: When the EM waves are incident on the FSS, the current inside the FSS unit cells gets excited. The amount of coupling decides the amplitude and phase of the current generated. These produced currents serve as EM source and hence produce added scattered fields. These fields are then combined with the incident fields, thereby making the resultant fields. By proper designing of the FSS elements, the required field and current characteristics can be attained, as shown in Figure 12.5 [40]. Therefore, for example, elements such as patch type exhibit stopband characteristics and slot type exhibit passband characteristics.

12.3.2 Periodic Structure (FSS)

Periodic FSS structures are designed on the basis of Floquet theorem. It says that when a plane wave of infinite length hits the surface of an infinite planar periodic structure, then the same amount of current distribution and field is attained in every individual element present in that periodic structure. However, due to the inter-element spacing and phase of the incident EM wave, phase shift occurs.

A planar periodic infinite FSS configuration is shown in Figure 12.6. Here, identical inter-element spacing is considered in x-axis and z-axis ($D_x = D_z$). Assuming that the incoming incident wave is moving in a particular direction of â, the equation is given by [41] eq. (12.1):

$$\hat{a} = \hat{x} a_x + \hat{y} a_y + \hat{z} a_z \quad (12.1)$$

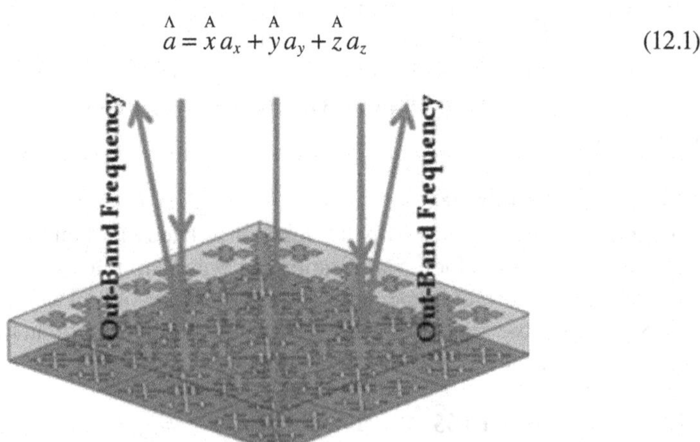

FIGURE 12.5 Operation of the FSS [40],

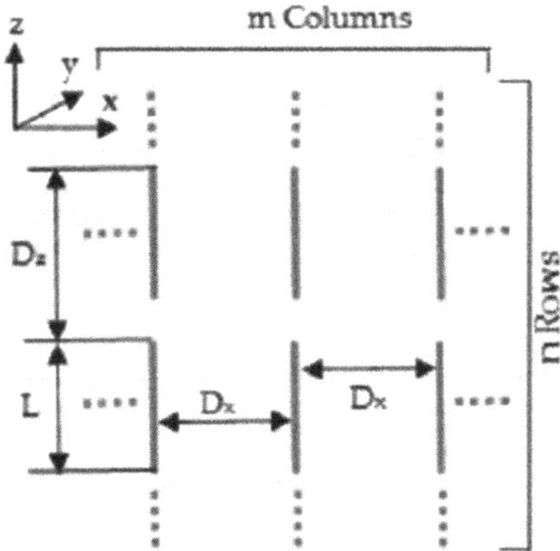

FIGURE 12.6 Periodic structure with identical inter-element spacing [41],

The current distribution inside all FSS unit cells remains identical; however, the phase varies with the phase of the currents [29], as shown in eq. (12.2).

$$I_{mn} = I_{oo} e^{-j\beta m D_x a_x} e^{-j\beta n D_z a_z} \tag{12.2}$$

By Ohm's law, the voltage of a reference unit cell is given in eq. (12.3).

$$V_{oo} = \left[Z_L + \sum_{m=-\infty}^{\infty} \sum_{n=-\infty}^{\infty} Z_{0,mn} e^{-j\beta m D_x a_x} e^{-j\beta n D_z a_z} \right] I_{oo} \tag{12.3}$$

The impedance of the scanning array is given by eq. (12.4).

$$Z_L = \sum_{m=-\infty}^{\infty} \sum_{n=-\infty}^{\infty} Z_{0,mn} e^{-j\beta m D_x a_x} e^{-j\beta n D_z a_z} \tag{12.4}$$

12.4 EQUIVALENT CIRCUIT MODEL

To understand the physical background of the designed FSS geometry, equivalent circuits are created. There are various conventional structures for which the equivalent circuit model is studied.

12.4.1 Grating Strip

The basic FSS structures are the metallic strip structures. The analysis of the metallic strip FSS is shown in Figure 12.7. If the E-field is aligned in the direction of the metallic strip, inductive strip grating filter is deduced, and if the E-field is perpendicular to the strips, then capacitive strip grating filter is deduced. For modeling the arrays at oblique angles, there is a requirement to model in transverse electric (TE) and transverse magnetic (TM) incidence.

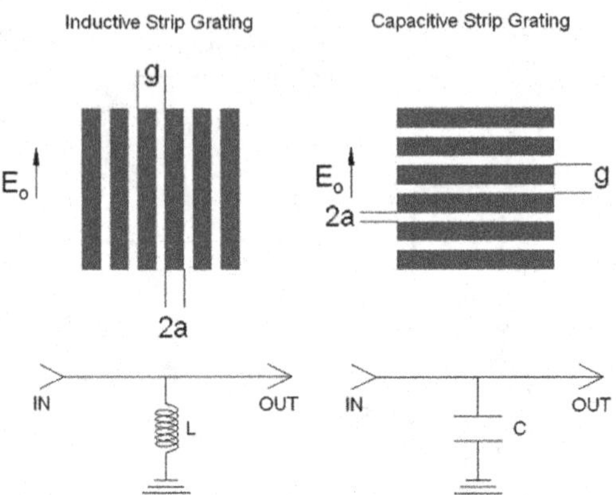

FIGURE 12.7 Strip grating filter with equivalent circuit [42].

When the E-field is polarized parallel to the incident plane, TM incidence occurs, i.e., $\theta=0°$, and when the E-field is perpendicular to the incident plane, TE incidence occurs, i.e., $\Phi=0°$. The resonance frequency is given by eq. (12.5):

$$w_r^2 LC = 1 \tag{12.5}$$

The normalized shunt inductance of the inductive strip filter [43] is given by:

$$\frac{w_r C}{Y_o} = 4\frac{d}{p}\sec\theta\left[\frac{p}{\lambda}\ln\left(\csc\left(\frac{\pi g}{2p}\right)\right) + G(p,w,\lambda,\theta)\right](\varepsilon_{\text{eff}}) \tag{12.6}$$

The normalized shunt susceptance of the capacitive strip filter [43] is given by:

$$\frac{w_r L}{Z_o} = \frac{d}{p}\cos\theta\left[\frac{p}{\lambda}\ln\left(\cos\text{ec}\left(\frac{\pi w}{2p}\right)\right) + G(p,w,\lambda,\theta)\right] \tag{12.7}$$

To avoid grating lobes,

$$p(1+\sin\theta) < \lambda \tag{12.8}$$

The correction factor is given by eq. (12.9):

$$G(p,w,\lambda,\theta) = \frac{(1-\beta^2)\left[\left(1-\frac{\beta^2}{4}\right)(A_+ + A_-) + 4\beta^2 A_+ A_-\right]}{\left(1-\frac{\beta^2}{4}\right) + \beta^2\left(1+\frac{\beta^2}{2}-\frac{\beta^4}{8}\right)(A_+ + A_-) + 2\beta^6 A_+ A_-} \tag{12.9}$$

where

$$A_{\pm} = \frac{1}{\left[1 \pm \frac{2p\sin\theta}{\lambda} - \left(\frac{p\cos\theta}{\lambda}\right)^2\right]^{1/2}} - 1 \tag{12.10}$$

$$\beta = \frac{\sin \pi w}{2p} \tag{12.11}$$

Here, d, g, p, and w are the length of the unit cell, gap in between two FSS unit cells, periodicity of the FSS, and angular resonant frequency of the unit cell, respectively. θ is the angle of incidence of the EM wave with which it strikes the surface of FSS, and λ is the resonant wavelength.

12.4.2 Square Loop

The frequency characteristics of square loop FSS have a lower-frequency transmission band with a single reflection band which can be presented in the form of capacitance and inductance using an equivalent circuit. In the FSS model, the array is represented by a single series LC circuit shunted across a transmission line of impedance Z_o as shown in Figure 12.8b, where Z_o is the characteristic impedance of free space. The normalized shunt inductive reactance expression of the square-loop FSS is given by [44] eq. (12.12):

$$X_{\mathrm{TE}} = F(p, 2w, \lambda) \tag{12.12}$$

$$X_{\mathrm{TE}} = \frac{w_r L}{Z_o} = p\frac{\cos\theta}{\lambda}\left[\ln\left(\operatorname{cosec}\left(\frac{2\pi w}{2p}\right)\right) + G(p, w, \lambda, \theta)\right] \tag{12.13}$$

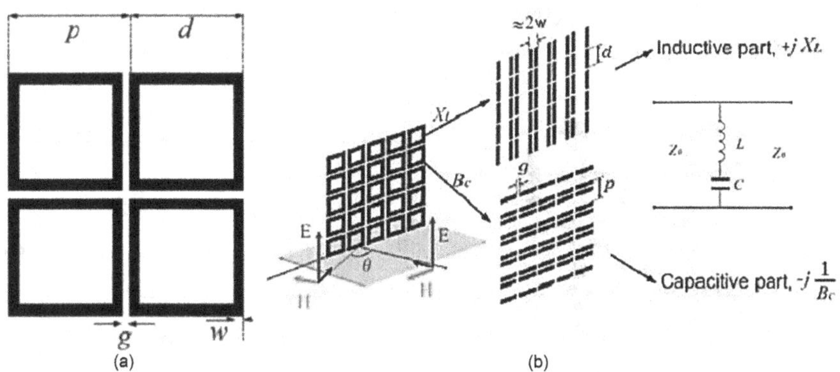

FIGURE 12.8 Square-loop FSS and its equivalent circuit [42].

The normalized shunt susceptance expression of the capacitive strip grating is given by

$$B_{\mathrm{TE}} = \frac{w_r c}{Y_o} = 4p\frac{\sec\theta}{\lambda}\left[\ln\left(\operatorname{cosec}\left(\frac{\pi g}{2p}\right)\right) + G(p,w,\lambda,\theta)\right] \tag{12.14}$$

Therefore, the inductive reactance is given by

$$\frac{X_f}{Z_o} = \frac{1}{\sqrt{\varepsilon_e}}\frac{d}{p}F(p,2w,\lambda) \tag{12.15}$$

The capacitive reactance is given by

$$B_f Z_o = \frac{4d}{p}\sqrt{\varepsilon_e}F(p,g,\lambda) \tag{12.16}$$

The net permittivity is given by

$$\varepsilon_e = \frac{\varepsilon_r + 1}{2} + \frac{\varepsilon_r - 1}{2}\cdot\frac{1}{\sqrt{1+\frac{12t}{w}}} \tag{12.17}$$

where t and ε_r are the thickness and relative permittivity of the substrate.

12.4.3 Jerusalem Cross

The structure of a J-cross is shown in Figure 12.9.

The inductive reactance XL of width w is given by eq. (12.18):

$$\frac{w_r L}{Z_o} = F(p,w,\lambda,\varphi) = p\frac{\cos\varphi}{\lambda}\left[\ln\left(\operatorname{cosec}\left(\frac{\pi w}{2p}\right)\right) + G(p,w,\lambda,\theta)\right] \tag{12.18}$$

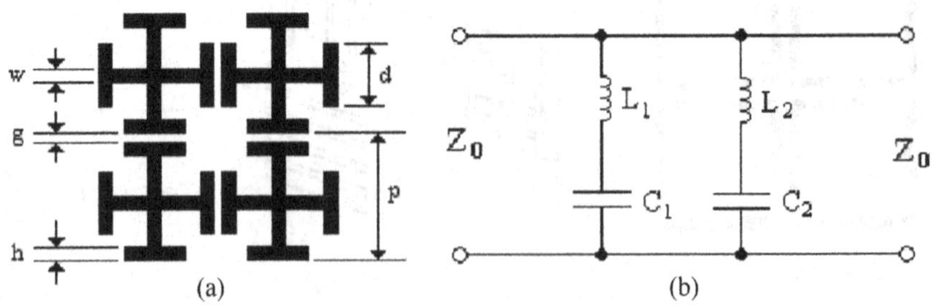

FIGURE 12.9 Jerusalem cross and its equivalent circuit [42].

The susceptance $B = B_d + B_g$, where B_g is the capacitance between the horizontal arms, with a spacing g, and is reduced by a factor of d/p

$$B_g = \frac{4d}{p} F(p, g, \lambda) \tag{12.19}$$

The susceptance B_g is between the vertical capacitors spaced by $(p - d)$, and it is given as

$$B_d = \frac{4(2h + g)}{p} F(p, p - d, \lambda) \tag{12.20}$$

The inductive reactance XL is given as:

$$\frac{X_{L2}}{Z_o} = \frac{d}{p}.F(p, 2w, \lambda, \theta) \tag{12.21}$$

12.5 APPLICATIONS OF FSS

There is a wide range of applications that FSS can provide in the field of antennas. They are as follows.

12.5.1 Enhancement of the Gain and Bandwidth of the Antenna Systems

In the fields such as communication, high-resolution medical imaging, military, and ground-penetrating radars (GPR) [45], antennas with high gain and narrow beam are required. Conventional antennas provide narrow bandwidth and low gain. Therefore, intensive research has been done by the researchers on the improvement of the antenna design. FSS, when integrated with antennas, not only provide filtering characteristics but also help in the enhancement of gain and impedance bandwidth [46–49]. Due to the reflections in between the FSS layer and the antenna, the gain gets enhanced. Moreover, FSS show very less variation in the impedance bandwidth of the antenna and provide its own impedance bandwidth; therefore, the impedance bandwidth gets enhanced.

One way to position a FSS is below the antenna as a back-reflector. The FSS is used as a back-reflector in order to extend the usable frequency range. The FSS provides additional impedance bandwidth, resulting in multi-band characteristics [50]. The back-reflector helps in the enhancement of the gain of the antenna as well, without disturbing the impedance bandwidth of the antennas. These reflectors have very less effect on the impedance bandwidth of the antennas.

In Figure 12.10, the antenna is placed at a distance L from the FSS layers. The FSS layer provide reflection at a particular frequency. ϕT is the reflection phase from the reference plane T; ϕS is the back and forth reflections in between the antenna and FSS; and ϕR is the overall reflected phase from the FSS layer.

$$\phi_T = \phi_R + \phi_S \tag{12.22}$$

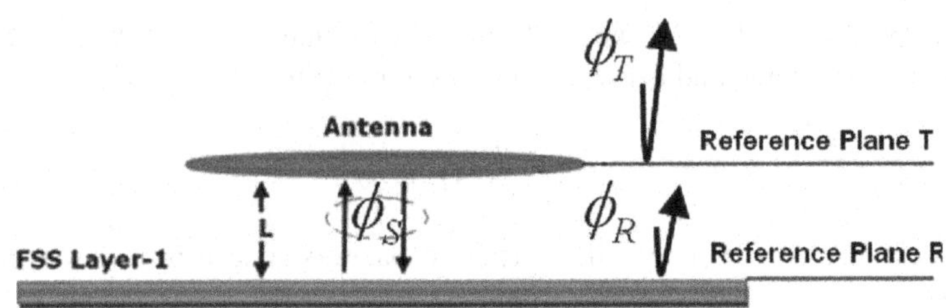

FIGURE 12.10 FSS as a back-reflector.

When an incidence wave travels toward the FSS layer, it is reflected back and gets added to the wave radiating directly from the antenna. If both the waves are in phase, they create a constructive inference, resulting in enhanced gain, and hence provide the maximum gain. However, these are suitable for small frequency bands. To enhance the front-to-back ratio with enhanced gain, the FSS was sandwiched in between the connected dipole antenna array and the ground plane in [51]. A dual-polarized radiator was mounted over a FSS back-reflector surface for UWB, and it exhibited an enhanced average gain of 9.5 dBi [52].

To obtain flattened gain over a given impedance bandwidth, the multilayer FSS are useful. A two-layer FSS was used below the antenna [53], as shown in Figure 12.11. Of the two layers of FSS used, FSS layer 1 reflects at lower frequency band and FSS layer 2 reflects at higher frequency band for a wide impedance bandwidth with stable average gain over the whole UWB frequency range. This antenna system provides an impedance bandwidth of 122% over the entire UWB frequency range with an average gain of 7.8 dBi, as shown in Figure 12.13a. Similarly, to obtain more gain, a four-layer FSS was used below the antenna in [54], as shown in Figure 12.12. It was observed that due to more reflection in between the FSS layers and the antenna, the gain gets more enhanced. The gain gets enhanced from 4 to 9.3 dBi due to the FSS layer underneath. A stable gain variation is obtained in between 3 and 15GHz in the limits of ±0.5 dBi. The bandwidth of the UWB antenna was 145% with FSS and 149% without FSS, which is well under the limits, as shown in Figure 12.13b.

Conformal FSS is a trending topic for the researchers these days, as it can be molded into any shape and size [55,56]. To enhance the radiation diversity of hybrid monopole dielectric resonator antennas, a conformal FSS was used in [57], as shown in Figure 12.14a. FSS as a parabolic reflector was placed at half-wavelength distance. It was observed that the impedance bandwidth gets enhanced by approximately 27% in 4–6GHz band, with a gain enhancement of 5–6 dBi over this band. A stable gain was obtained around 9.5 dBi with a variation of ±1.5dB. Moreover, an omnidirectional radiation was maintained in this FSS at 7–9GHz.

The classical way to position the FSS is as a superstrate. To enhance the gain, several techniques have been proposed. An X-slot was etched in a radiating dielectric resonator with FSS as a superstrate layer to resonate at MMW range [58], as shown in Figure 12.14b, and in [59], a FSS superstrate was embedded over an electromagnetic band gap (EBG) antenna. The impedance bandwidth gets enhanced by 145% with the

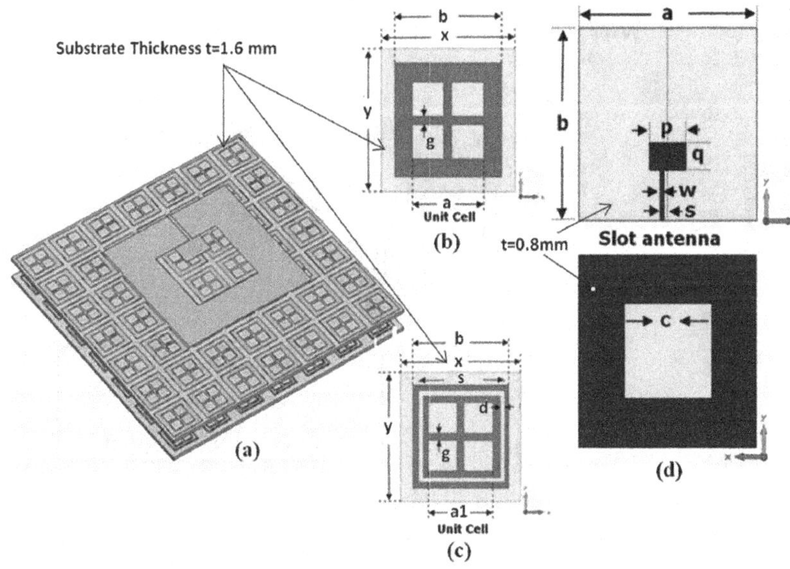

FIGURE 12.11 (a) Slot antenna with a FSS reflector: (b) FSS layer 1. (c) FSS layer 2. (d) UWB slot antenna [53].

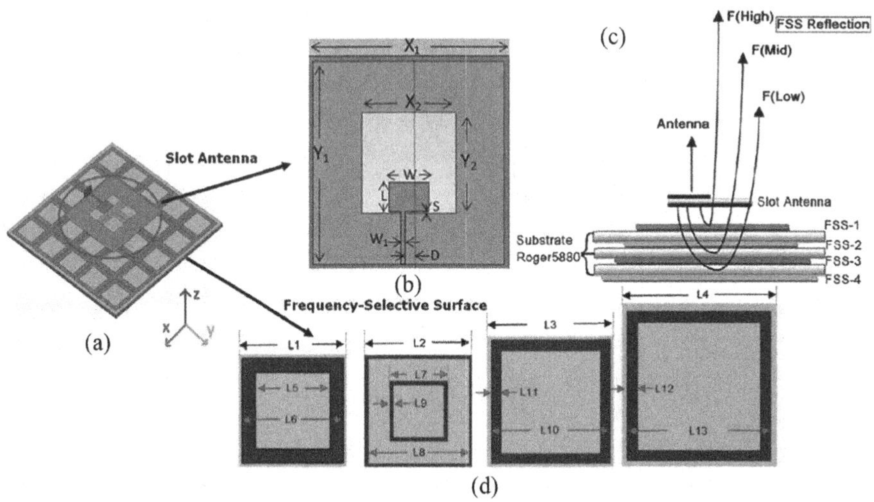

FIGURE 12.12 (a) UWB slot antenna with FSS: (b) slot antenna; (c) reflection mechanism; (d) different layers of FSS [54].

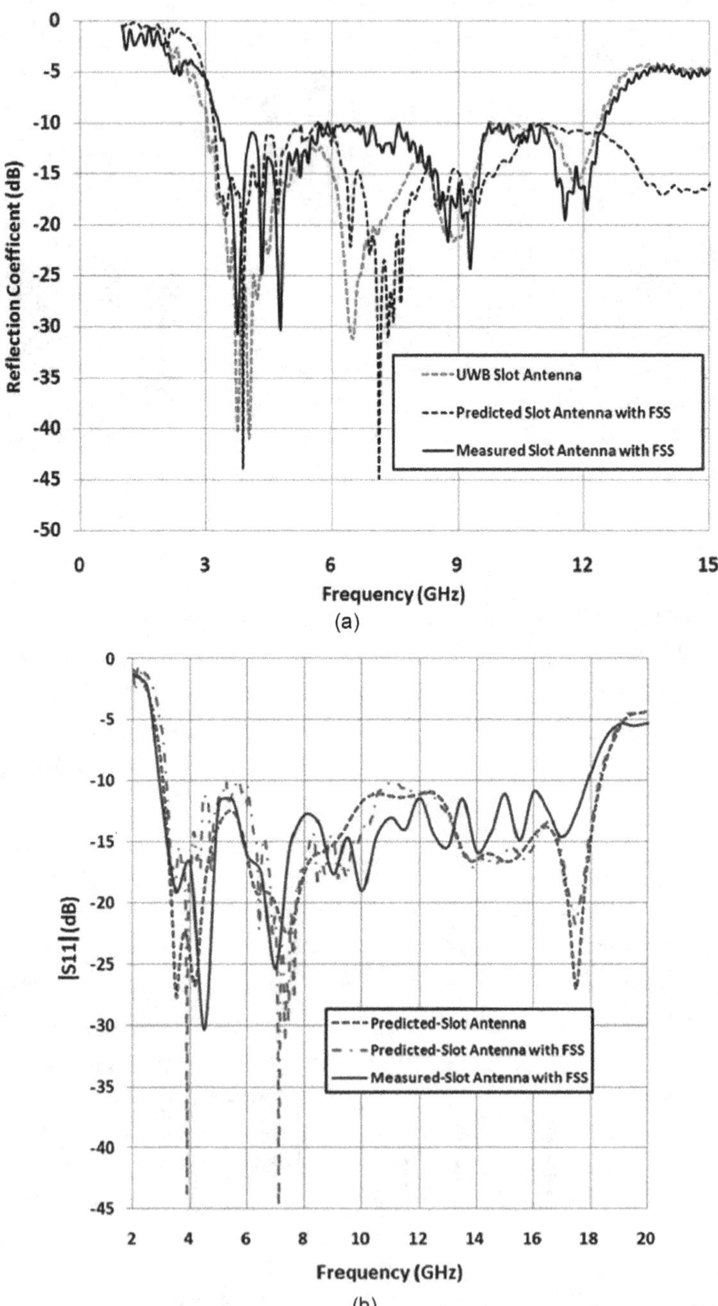

FIGURE 12.13 Impedance bandwidth with and without the use of FSS, [53] and [54].

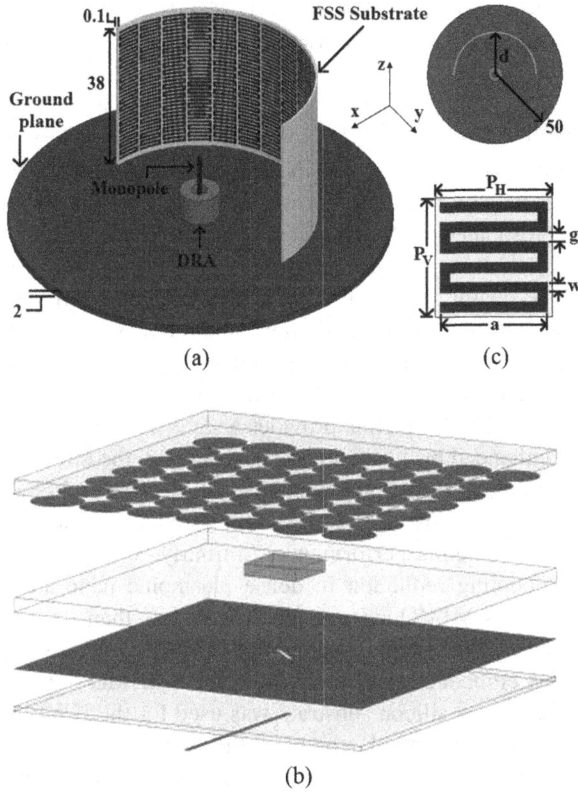

FIGURE 12.14 (a) Conformal FSS integrated with monopole DRA; (b) FSS layer as a superstrate over a DRA [57,58].

peak gain of 22.5. Apart from enhancing the gain, FSS help in the elimination of the unwanted radiation emitted from the feed line [60]. Therefore, it is evident that FSS help in the enhancement of the gain and bandwidth of the antenna.

12.5.2 Isolation in MIMO Antennas

For a high data rate with fixed power and bandwidth, multiple-input multiple-output (MIMO) technology has been chosen in the fourth-generation communication systems over its conventional counterpart, i.e., single-input single-output system. MIMO system solves the problem of multi-path fading, which tends to degrade the performance of wireless communication links. By sending multiple data streams, the data rate and the reliability of the system increase rapidly. Multiple antenna elements are integrated together in a MIMO system to provide a higher data throughput at the same operating band. The designing of an MIMO antenna is a challenging task, as the antenna elements are placed in very close proximity to each other. Due to this, the port coupling and field coupling increase. This affects the efficiency and channel capacity of the antenna system. Therefore, there is a requirement of high port isolation and low correlation level [61–64]. It is desirable to have TARC <0 dB, ECC <0.5, and CCL <0.5 bits/s/Hz for the acceptable performance of a MIMO system [65].

To obtain a wideband structure in U- and V-bands, the metal was printed on both sides of the RO4003 substrate [66]. On top of the substrate, a Jerusalem cross was imprinted, whereas at the bottom, a FAN shape was imprinted. Every element of the substrate gives its own resonance. It was observed that the J-cross resonates at 48 GHz and FAN resonates at 60 GHz, when simulated separately. Due to its better performance, this structure was used for mutual coupling reduction in a 60-GHz MIMO dielectric resonator antenna [67], as shown in Figure 12.15. Two slots of different size were etched over the ground plane, and a FSS wall was built in between the MIMO antenna elements. The slots minimize the surface current, and the wall provides the isolation in between two MIMO antenna elements by decreasing the free space radiation. This MIMO antenna provides a stable response from 57 to 63 GHz, and the gain was enhanced by 1.5 dB in comparison with the reference antenna.

It was observed that with the implementation of the slots and FSS wall, an isolation of about −30 dB was achieved. An ECC of less than 5e−6 was obtained, which implies good MIMO antenna performance. Similarly, a graphene-based FSS was used for mutual coupling reduction in dense plasmonic nano-antenna arrays for multi-band [68]. In this MIMO system, an ECC of less than 0.01 was obtained at 1.1–1.7 THz frequency with a high isolation of −25 dB. In [69], a FSS structure with Y-shaped slots was embedded over the four-element MIMO antenna to achieve a good isolation of 20 dB. A silicon substrate was used for the UWB MIMO antenna [70]. Six FSS elements were placed in between the antennas for obtaining a good isolation of about −16 dB. In [71], a dual-passband FSS was designed at 2–3.4 GHz and 5.5–6.8 GHz frequency bands. The FSS had two layers and was designed to improve the scattering performance and radiation of the antenna. This structure helps in the enhancement of the bandwidth by 31.4% and 50% at lower frequency band and higher frequency band, respectively. The gain also gets enhanced by 2.53 and 1.86 dB at 5.8 and 6.4 GHz, respectively.

The correlation coefficient deals with the individual radiation pattern of the MIMO antenna elements and measures the degree of correlation that all the radiation patterns will add along the propagation channel in a given environment. When the

FIGURE 12.15 FSS is placed in between the two elements of DRA MIMO antenna [67].

antenna elements are placed in close proximity to each other, the radiation patterns overlap; to avoid this, a practical reflective surface (PRS), which is a FSS, was used in a two-element Fabry–Perot cavity MIMO antenna [72], as shown Figure 12.16. To reduce the height of the cavity to $\lambda/4$, the reflection phase of the wave should be 0 (or 2Π). To achieve this, the PRS was made of a composite structure. On one side of the substrate, an inductive layer was embedded, and on the other side, a capacitive layer was embedded [73]. At the top layer, unit cells are identical and provide propagation in boresight direction as the wave combines with the same phase; however, at the bottom, there is a gradual variation in the dimension of the FSS unit cells. Therefore, when the wave combines with a different phase, the wave tilts in the same axis. Therefore, when the wave propagates through this MIMO antenna, due to the unevenly distributed FSS unit cell structure, the wave tilts and the correlation between the two field patterns reduces, leading to a rapid decrease in the correlation coefficient. It was observed that a change in the inductive surface gave a more tilt to the wave. It was observed that more than 95% of the correlation value was improved at 5.25 GHz.

Further, a four-element DRA MIMO antenna also used this technique for the decorrelation of the fields [74], as shown in Figure 12.17a and b. Two DR elements were placed above the substrate, and the other two elements were placed below the

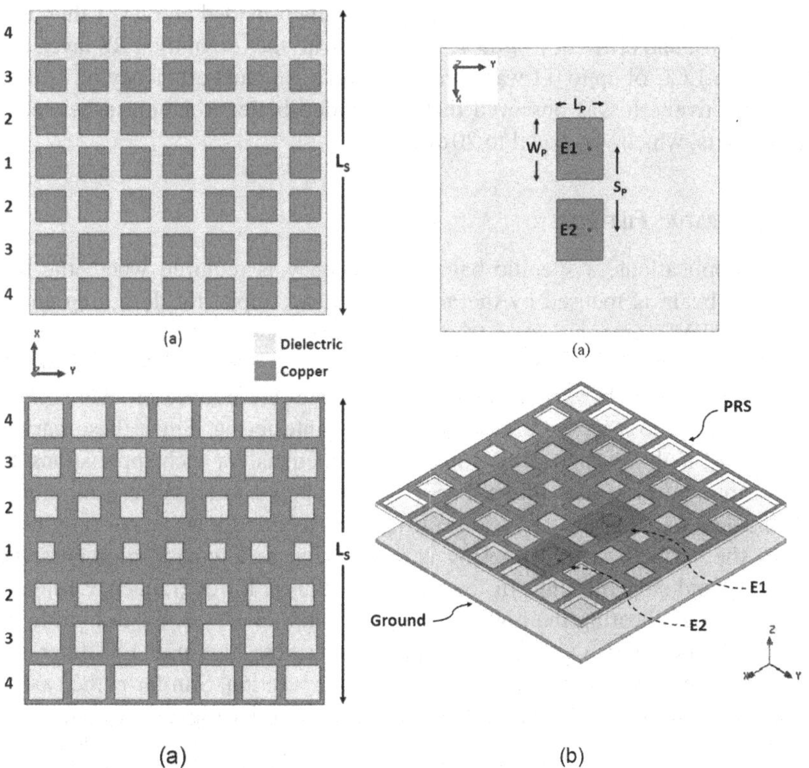

FIGURE 12.16 (a) FSS structure; (b) FSS with two-element MIMO antenna [72].

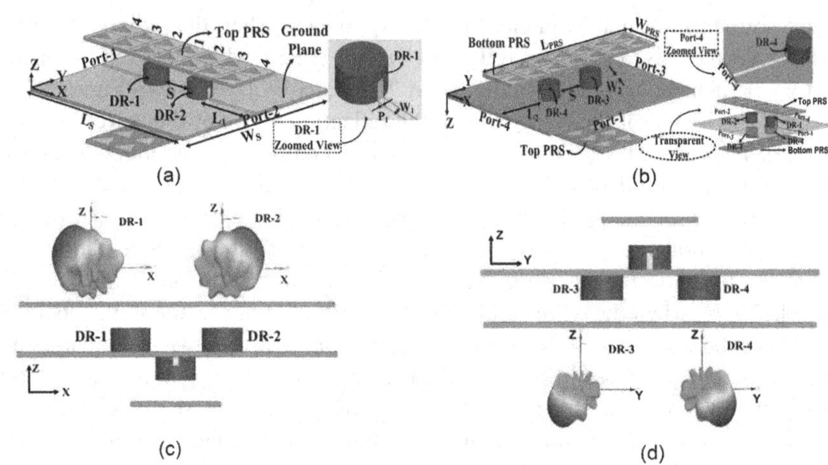

FIGURE 12.17 DRA MIMO antenna with two-layer FSS: (a) top view, (b) bottom view, (c), (d) radiation patterns at 5.25 GHz [74].

substrate, orthogonally. Two triangular-shaped FSS phase gradient layers were positioned as the superstrate to all the four DRA elements. It was observed that without using a PRS layer, the four-element MIMO antenna provided an ECC value of 0.31, gain of 4.9 dB, and efficiency of 89%. However, with the use of the PRS layer, a very low value of ECC of up to 0.1 was obtained, with a gain and efficiency of 7.2 dB and 81%, respectively. It was observed that it provided better isolation in between the antenna ports, which was equal to 20 dB.

12.5.3 Spatial Filtering

In some applications, a specific band of frequency is required where the highly directional beam is focused by the antennas. For example, for designing an ultra-wideband (UWB) antenna, some frequency ranges such as WiMAX (3.3–3.6 GHz), WLAN (5–6 GHz), and X-band satellite communication band (7.2–8.4 GHz), which are extensively used in various applications, create interference problems [75–77]. There is a need to stop these frequencies from interfering. Since these particular bands are difficult to suppress by conventional antennas, for such applications, there is a requirement of spatial filters. FSS is called spatial filter, since it modifies the wave incident on its surface by transmitting or reflecting the wave partially or fully through the surface. Spatial filtering is categorized as low-pass, high-pass, band-pass, and band-stop filtering. In low-pass FSS filters, lower frequency ranges get passed while eradicating the higher frequency band. However, for designing high-pass FSS filters, Babinet's principle is applied, since the high-pass FSS filter operation is complementary to the low-pass FSS filter operation. Similar is the case with band-pass and band-stop filtering. Band-pass filter permits a specific frequency band to pass, while band-stop filter rejects one. Size, shape, and material of the substrate and array elements are very important parameters in the designing of the FSS filters. For a required operating band, metallic patches or apertures are embedded on the

FSS substrates. Metallic patch elements act as a band-stop filter [75], whereas aperture elements act as a band-pass filter [78].

In the literature, a wide range of band-pass and band-stop structures have been presented. Single, dual, and multiple frequency bands are obtained using different shapes and sizes of the FSS. The main motive of all the researchers is to make a miniaturized structure for stable frequency response and better angular stability. To stop WLAN frequency band to pass through the FSS, a modified swastika-shaped unit cell was designed [75], as shown in Figure 12.18a and b. The structure was etched over a FR4 substrate with 35° rotation of the arms. This FSS structure provided an impedance bandwidth of 400 MHz with a stable TE and TM frequency response. For dual-band-stop characteristics [76], a modified structure was made. In this, a metallic square patch was etched at the center of the FSS structure and L-shaped arms were attached to it with a separation of 45°.

The L-shaped arms permit dual-band-stop characteristics; i.e., vertical arms have a control on lower resonating frequency (8.47 GHz), while horizontal arms have a grip over upper resonating frequency (10.45 GHz), as shown in Figure 12.18c and d.

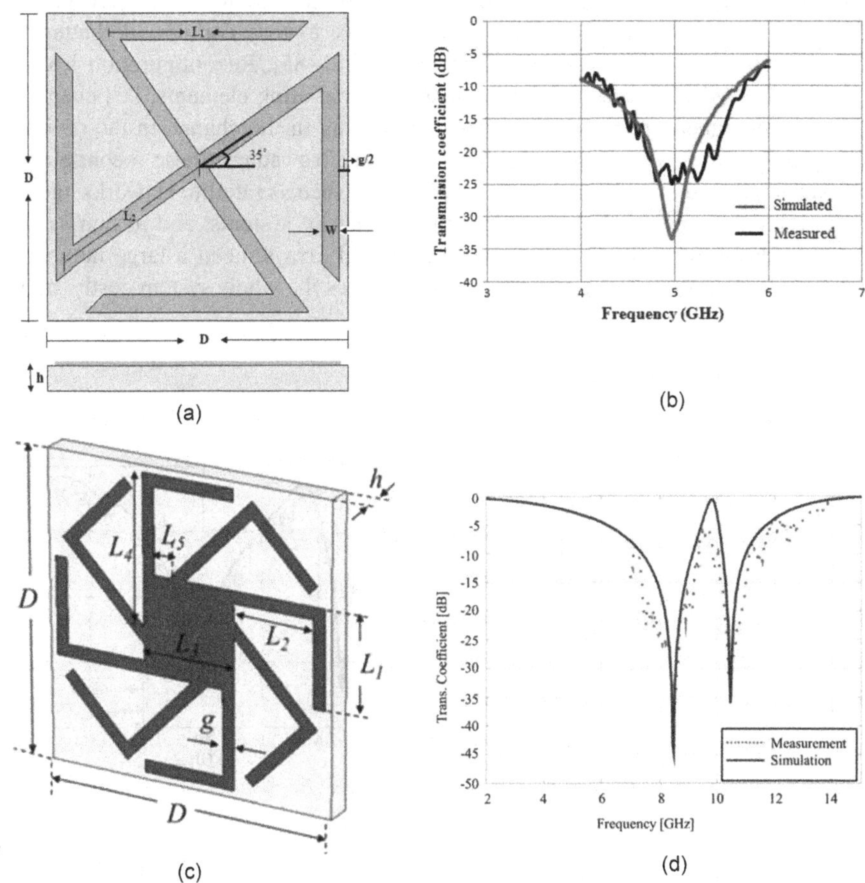

FIGURE 12.18 (a, c) Unit cell of FSS structure and (b, d) its transmission characteristics, respectively [75,76].

Similarly, for multiple bands, structure modifications are required. To obtain triple-band-stop characteristics, two square loops were embedded on both sides of a FR4 substrate; moreover, at the backside of the substrate, two folded metallic arms were etched with the square loop to obtain the triple-notch band at WiMAX, WLAN, and X-band [77], as shown in Figure 12.19a and b. A comparison of different types of FSS is made in Table 12.1. It is observed from this table that with the miniaturization of the size of FSS, the angular stability increases, which results in better frequency response.

Therefore, these structures can be implemented with the antennas for filtering purposes.

12.5.4 FSS for Reconfiguration of the Antennas

The wireless communication system has become a very fast and dynamic growing sector in the past few years. There is a requirement to make new antennas that can adapt themselves to the increasing demands of the new systems easily. Therefore, reconfigurable antennas (RA) are the smart choice as they can provide better performance to support the emerging applications [81–85]. Reconfiguration leads to the change in the current distribution inside the radiating elements, i.e., change in the electromagnetic field distribution, which results in the change in the radiation properties of the antenna and impedance as well. Nowadays, these reconfigurable antennas are used in wide varieties of applications such as satellite networks, tactical radio relay systems, cordless systems, public security systems, and military applications. However, traditional RA, such as phased arrays, need a large number of feedback networks and phase shifters, which makes the whole system costly, bulky,

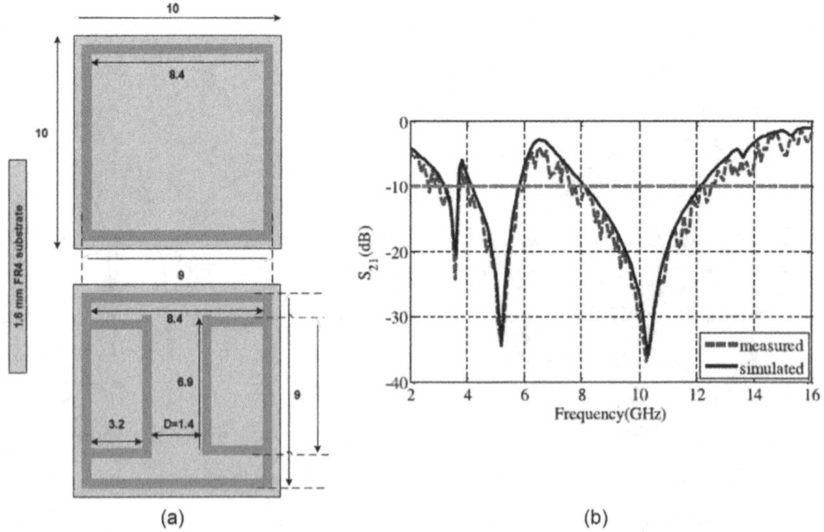

FIGURE 12.19 (a) Unit cell of FSS structure and (b) its transmission characteristics, respectively [77].

TABLE 12.1
Comparison of Various FSS Techniques

Reference	Unit Cell Dimension (mm)	Resonant Frequency (GHz)	No. of Operating Bands	Frequency Bands	Angular Stability	Rejection Bandwidth
[75]	7×7×1.6	5	1	WLAN	0°–60°	400 MHz
[76]	8.8×8.8×0.762	8.47, 10.45	2	Wideband X-band	0°–60°	–
[77]	10×10×1.6	3.5, 5.2, 10.2	3	WLAN, WiMAX, X-band	0°–30°	0.5 GHz (3.1–3.7), 1.9 GHz (4.1–6), 4.1 GHz (8–12.1)
[79]	56×56×40	1.86, 3.10	2	L- and S-bands	0°–45°	660 MHz
[80]	10×10×1.6	3.5, 4.5, 8.4	2	Wideband operation	0°–30°	1 GHz (3, 4) 11.4 GHz (4.6–16)

complex, and more power-consuming [86]. Therefore, recently, a new technique of beam-switching antennas has been introduced, which incorporates reconfigurable FSS [87]. Reconfigurable FSS are achieved by tuning and switching the response of the FSS by integrating active controlling devices. Reconfiguration can be performed on the basis of frequency response [88], radiation pattern [89], and polarization [90].

Active elements such as PIN diodes, Schottky diodes, and varactor diodes are used for the electromagnetic beam switching. These allow the rapid switching in between the ON and OFF states of the FSS depending upon the external DC bias source. When the PIN diode is in ON state, the FSS becomes a source of transmission for the incident waves generated from the dipole antenna, whereas when PIN diode is in OFF state, a high reflection of the incident waves occurs, as FSS unit cells have a high reflection coefficient [91–93]. In [94], a dipole antenna was located at the center of a cylindrical AFSS. The AFSS comprised of ten circular metallic loops per column. A metallic strip was imprinted along the ring loops. PIN diodes were positioned on this metallic line. When the diode was in OFF state, a passband was generated at 2.8 GHz, while two stopbands were also generated due to the working of both the metallic strips and ring loops. However, when the bias was applied, the stopband remained due to the ring loops. The parasitic elements of diode caused the first stopband to disappear and generated a passband at 1.8 GHz. This is shown in Figure 12.20. The maximum gain of 9.2 dB was obtained at 1.8 GHz. In [95], an omnidirectional antenna was surrounded by a cylindrical AFSS. Each AFSS unit cell comprises of discontinuous split-ring resonator (SRR) cross-shape structure with four active elements. The maximum gain of 8.1 dB was obtained at 2.1 GHz. Similarly, for dual-band frequency reconfiguration, a cylindrical AFSS was proposed in [96]. PIN diodes were used for switching the AFSS between ON and

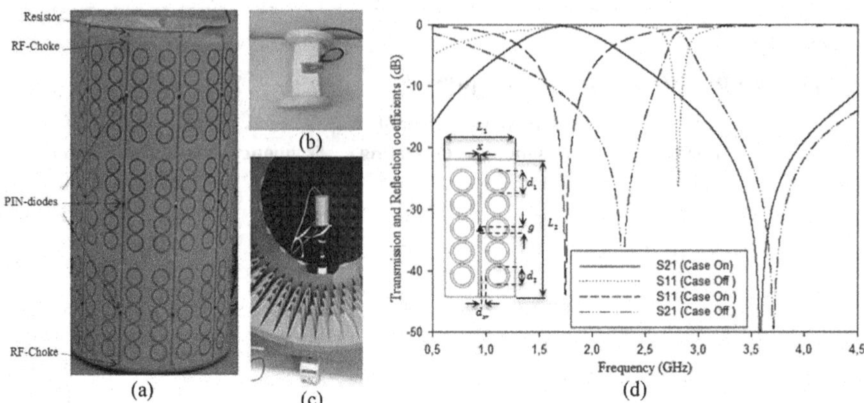

FIGURE 12.20 (a–c) AFSS embedded over antenna; (d) frequency response of AFSS [94].

OFF states. By controlling the DC bias voltage applied to different PIN diodes, the beam sweeps in the whole azimuthal plane. This antenna operated at 2.5 and 5.3 GHz when all the diodes were OFF, whereas when the diodes were ON, a transmission level below −15 dB between 2 and 6 GHz was achieved.

PIN diodes, metallic switches, etc., are readily available; however, they are costly, have a complicated bias network, and increase the system complexity. Therefore, in [97], a hexagonal-shaped cantilever-enabled FSS was mounted over a cylindrical DRA for dual-plane beam sweep. The antenna was designed to operate at a resonating frequency of 30 GHz, as shown in Figure 12.21a–c. It was observed that beam sweeping occurred in both the azimuthal and elevation planes. In six 60° steps, the whole azimuthal plane was covered. The gain of the antenna was measured as 8.1 dB.

Due to the design flexibility, the use of microfluidic technology for reconfiguration is in trend. A liquid metal is poured into the microchannels that are carved inside an elastomeric substrate. The amount of liquid injected signifies the new resonant bandwidth. In [98], a fluid-reconfigurable FSS was presented with multi-functionality, as illustrated in Figure 12.21d. On both sides of the elastomeric substrate, orthogonally polarized meandered patterns were placed. The substrate used was made of polydimethylsiloxane (PDMS), which was laser-etched to mold into a microfluidic channel of any arbitrary shape. To achieve reconfiguration, eutectic gallium–indium (EGaIn) was poured inside the meander lines, mechanically. This liquid-reconfigurable FSS provided four different pattern reconfiguration states (dual-polarized all-pass, dual-polarized band-pass, single-polarized band-pass, and single-polarized low-pass) and a stable frequency response up till 60°.

12.5.5 Electromagnetic Shielding

To avoid the interference created by any other device which can degrade the performance of the antennas, electromagnetic shielding is necessary. FSS due to its good transmission and reflection properties are a suitable candidate for this application [99]. In [100], a FSS consisting of cross-dipoles and rings imprinted on the opposite

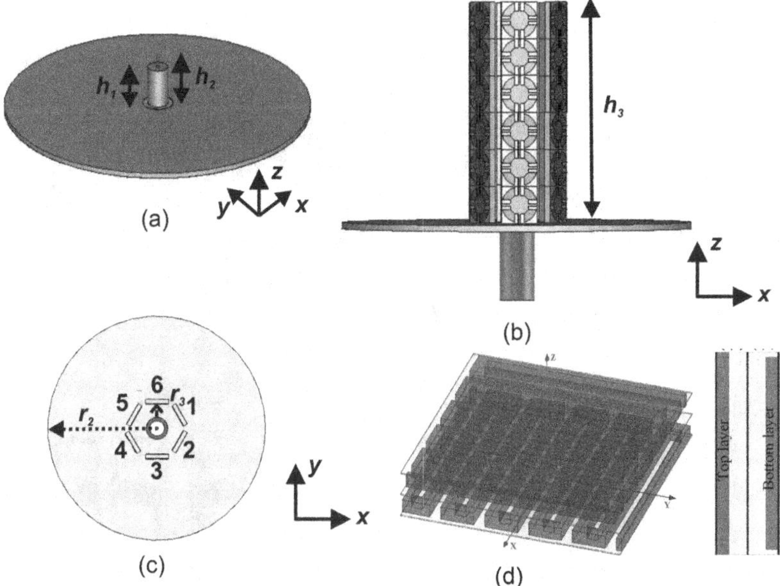

FIGURE 12.21 Structure of a beam-sweeping antenna using (a-c) cantilever-enabled FSS and (d) fluidic FSS [97,98].

sides of the substrate was presented and it exhibited a stopband of 7.5 GHz ranging from 6.5 to 14 GHz, as shown in Figure 12.22a. This FSS structure provided effective shielding in X-band and Ka-band with minimum attenuation of 20–35 dB. Electromagnetic shielding can also be achieved using reconfigurable frequency-selective surfaces (RFSS) [101], as shown in Figure 12.22b. In this, two pairs of squares were connected together to create an individual unit cell. Frequency tuning was obtained by varying the distance between the unit cells, mechanically. The frequency shifted toward the left with the increase in the distance between the FSS, and vice versa. This structure provided electromagnetic shielding for WiMAX (3.5 GHz), WLAN (5 GHz), and ISM/WiMAX bands (5.8 GHz). The reconfiguration range of 3.5–8.2 GHz was obtained by varying the space between the unit cells.

12.5.6 FSS Radomes for Antenna Protection

Antennas or radar systems are vulnerable to the physical environment; consequently, the performance gets affected when exposed to the outer world. Therefore, radomes came into existence. A radome is a protective layer that is positioned around an antenna or radar to protect it from physical wear and tear with least impact on the performance [102]. FSS due to its unique property of manipulating the EM waves is a good option to make hybrid radomes. FSS radomes can work in absorption band [103], reflection band [104], and passband [105] for various military and commercial requirements. With the development of stealth technology and detection, a low radar

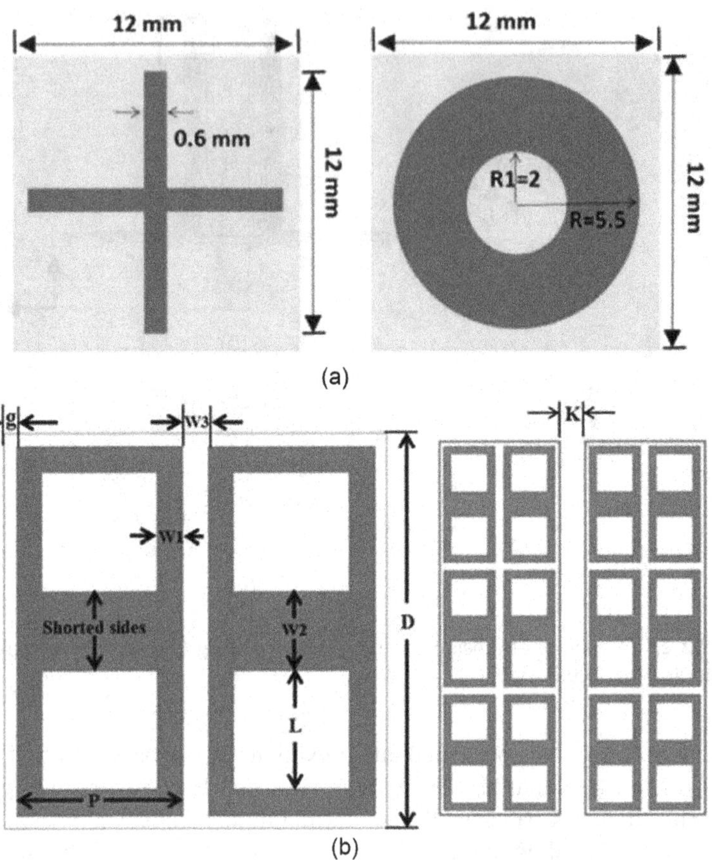

FIGURE 12.22 Unit cell dimension for electromagnetic shielding FSS [100,101].

cross section is of utmost importance. If the FSS radomes are transparent to EM waves, the transmission of the signal is smooth. However, if they are opaque, the entire incoming signal gets reflected. Due to the dome shape of the radomes, the energy is reflected and backscatters in bistatic direction; therefore, the back-reflected signals become weak in strength in different directions. Hence, the RCS is reduced [1]. Radomes are suitable for other applications as well, such as weather broadcast, telemetry, satellite, radio astronomy, and surveillance. The comparison of monostatic RCS of FSS radomes and dielectric radomes in [106] makes it clear that FSS radomes control the RCS of the antenna better than dielectric radomes. It is shown in Figure 12.23.

In [107], a dual-layer four-legged loop FSS radome was integrated over a slotted waveguide antenna to reduce the RCS. This antenna system provided a wideband of 7.8–11.2 GHz operating at 9.5 GHz. It was observed that the return loss of the FSS was greater than 9 dB over the entire passband range; however, this FSS radome had some effect on the electrical performance of the antenna in terms of reduced polarization with increased side lobe levels and more transmission losses.

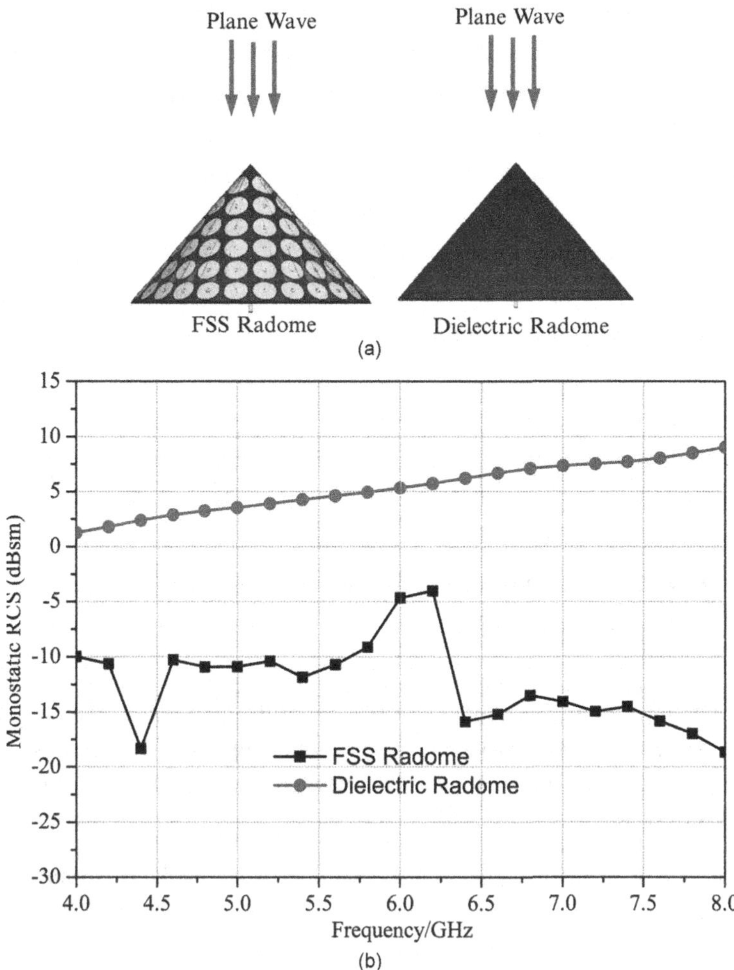

FIGURE 12.23 Comparison of monostatic RCS of FSS radome and simple dielectric radome [106].

The monostatic RCS with and without a FSS radome is shown in Figure 12.24. It was observed that the RCS was reduced effectively with the integration of the FSS at 9.5 GHz. The RCS value at 0° is high. It is assumed that at 0° the antenna behaves as a perfect electromagnetic hard surface.

A conical thick-screen FSS was integrated over a monopole antenna to obtain a narrow passband response and control reflected band RCS [108]. In [109], a folded substrate band-pass 3D FSS was implemented. Compared to its conventional 3D FSS counterpart, the thickness of this FSS unit cell was reduced by 79% (five-layer folded substrate). It was observed that this semicylindrical conformal FSS radome integrated with horn antenna provided a transmission bandwidth of 26.9% operating at 3.57 GHz and a stable performance was observed at oblique angles.

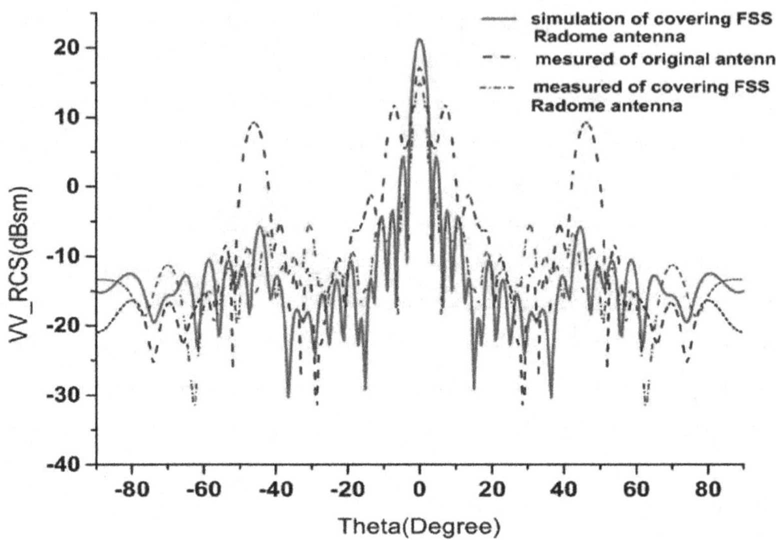

FIGURE 12.24 Monostatic RCS of the covering symmetric dual-layer FSS radome antenna at 9.5 GHz [107].

A reconfigurable AFSS reflector was presented for RCS reduction in [110], as shown in Figure 12.25a. Around 3.8 GHz, the FSS reflector was able to switch between ON state (band-pass) and OFF state (band-stop) using PIN diodes, as illustrated in Figure 12.25b. It resulted in a switchable RCS reduction in a dipole antenna with and without the bias to the FSS.

It is evident from the above discussion that hybrid FSS radomes suits better in RCS reduction in an antenna performance, without affecting the performance of the antenna.

12.6 CONCLUSION

This chapter gave an idea of what frequency selective surface is and how important it is in modern electromagnetism. Basic concepts and principle of operation were explained briefly. With rising trends in various fields such as communication, a vast amount of research on FSS and its characteristics and applications has been conducted. Different types of FSS have been studied, such as single-layer FSS, 3D FSS, multilayer FSS, wearable FSS, and active FSS. Various equivalent circuits have been explained with the LC responses. This includes simple 2D structure to complex 3D and active FSS structures. Wide bandwidth, high gain/directivity, good radiation characteristics, and better efficiency are some of the salient features of an antenna. However, a conventional antenna system is insufficient to provide all the aforementioned characteristics. So, later in this chapter, all the applications of FSS based on antennas were explained. It was studied that FSS not only work as a filter, but also helps in the enhancement of gain, bandwidth, reconfigurability, isolation between MIMO antennas, and RCS reduction.

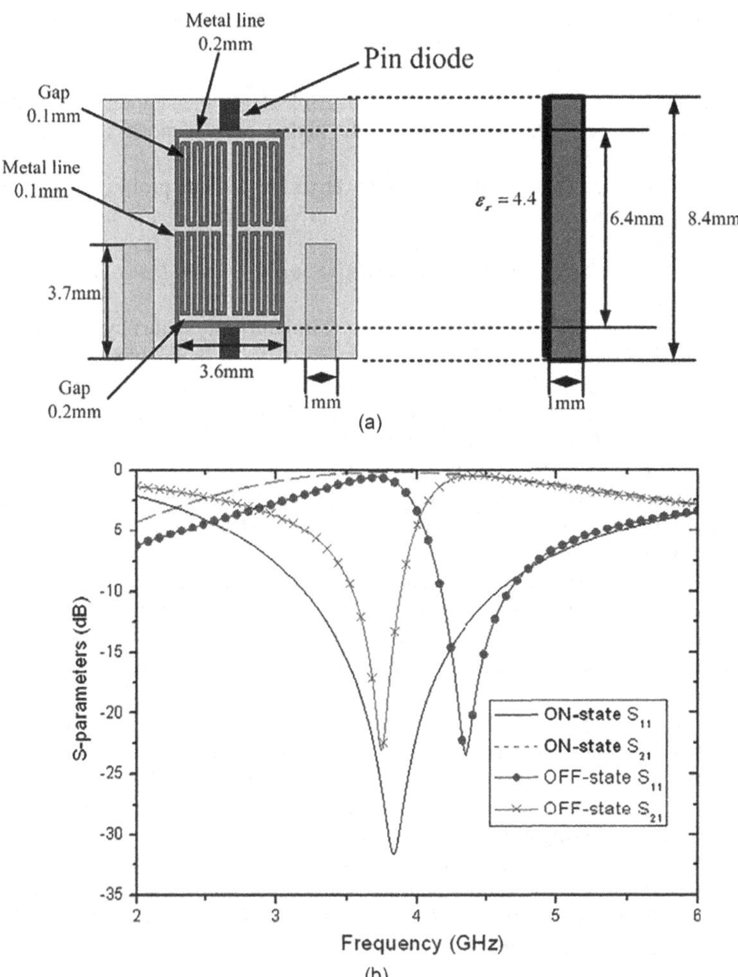

FIGURE 12.25 Configuration of reconfigurable AFSS [110].

REFERENCE

1. Munk, B.A. 2000. *Frequency Selective Surfaces: Theory and Design*. Wiley Online Library: Hoboken, NJ.
2. Abdelrahman, A.H., A.Z. Elsherbeni and F. Yang. 2014. Transmission phase limit of multilayer frequency-selective surfaces for transmitarray designs. *IEEE Trans. Antennas Propag.* 62:690–697.
3. Chiu, C.N. and K.-P. Chang. 2009. A novel miniaturized-element frequency selective surface having a stable resonance. *IEEE Antennas Wireless Propag. Lett.* 8:1175–1177.
4. Abadi, S.M.A.M.H. and N. Behdad. 2014. Design of wideband, FSS-based multibeam antennas using the effective medium approach. *IEEE Trans. Antennas Propag.* 62:5557–5564.

5. Martinez-Lopez, L., J. Rodriguez-Cuevas, J.I. Martinez-Lopez, et al. 2014. A multi-layer circular polarizer based on bisected split-ring frequency selective surfaces. *IEEE Antennas Wireless Propag. Lett.* 13:153–156.
6. Abadi, S.M.A.M.H., M. Li, and N. Behdad. 2014. Harmonic-suppressed miniaturized-element frequency selective surfaces with higher order bandpass responses. *IEEE Trans. Antennas Propag.* 62:2562–2571.
7. Xu, N., J. Gao, J. Zhao, et al. 2015. A novel wideband, low-profile and second-order miniaturized band-pass frequency selective surfaces. *AIP Adv.* 5:1–6.
8. Yu, Y.-M., C.-N. Chiu, Y.-P. Chiou, et al. 2014. A novel 2.5-dimensional ultraminiaturized-element frequency selective surface. *IEEE Trans. Antennas Propag.* 62:3657–3663.
9. Shi, Y., W. Tang, W. Zhuang, et al. 2014. Miniaturised frequency selective surface based on 2.5-dimensional closed loop. *Electron. Lett.* 50:1656–1658.
10. Campos, A.L.P., E.E.C. de Oliveira and P.H. da Fonseca Silva. 2010. Design of miniaturized frequency selective surfaces using Minkowski island fractal. *J. Microw. Optoelectron. Electromagn. Appl. (JMOe)* 9:43–49.
11. Li, B. and Z. Shen. 2014. Bandpass frequency selective structure with wideband spurious rejection. *IEEE Antennas Wireless Propag. Lett.* 13:145–148.
12. Song, K. and P. Mazumder. 2013. Design of highly selective metamaterials for sensing platforms. *IEEE Sens. J.* 13:3377–3385.
13. Sun, L., H. Cheng, Y. Zhou, et al. 2012. Broadband metamaterial absorber based on coupling resistive frequency selective surface. *Opt. Express* 20:4675–4680.
14. Huang, C., C. Ji, X. Wu, et al. 2018. Combining FSS and EBG surfaces for high-efficiency transmission and low-scattering properties. *IEEE Trans. Antennas Propag.* 66:1628–1632.
15. Li, L., Y. Li, Z. Wu, et al. 2015. Novel polarization-reconfigurable converter based on multilayer frequency-selective surfaces. *Proc. IEEE.* 103:1057–1070.
16. Harnois, M., M. Himdi, W.Y. Yong, S.K.A. Rahim, K. Tekkouk and N. Cheval. 2020. An improved fabrication technique for the 3-D frequency selective surface based on water transfer printing technology. *Sci. Rep.* 10:1–8.
17. Raiva, A.P., F.J. Harackiewicz and J. Lindsey. 2003. *Frequency Selective Surfaces: Design of Broadband Elements and New Frequency Stabilization Techniques.* Southern Illinois University at Carbondale School of Technology: Carbondale, IL.
18. Costa, F., S. Genovesi and A. Monorchio. 2009. On the bandwidth of high-impedance frequency selective surfaces. *IEEE Antennas Wireless Propag. Lett.* 8:1341–1344.
19. Moharamzadeh, E. and A.M. Javan. 2018. Triple-band frequency-selective surfaces to enhance gain of x-band triangle slot antenna. *IEEE Antennas Wireless Propag. Lett.* 12:1145–114.
20. Chatterjee, A. and S.K. Parui. 2016. Performance enhancement of a dual-band monopole antenna by using a frequency-selective surface-based corner reflector. *IEEE Trans. Antennas Propag.* 64:2165–2171.
21. Li, J., Q. Zeng, R. Liu, et al. 2018. A gain enhancement and flexible control of beam numbers antenna based on frequency selective surfaces. *IEEE Access* 6:6082–6091.
22. Arnaud, J. and F. Pelow. 1975. Resonant-grid quasi-optical diplexers. *Bell Syst. Tech. J.* 54:263–283.
23. Pelton, E. and B. Munk. 1974. A streamlined metallic radome. *IEEE Trans. Antennas Propag.* 22:799–803.
24. Lee, S.-W. 1971. Scattering by dielectric-loaded screen. *IEEE Trans. Antennas Propag.* 19:656–665.
25. Pozar, D. 1996. Flat lens antenna concept using aperture coupled microstrip patches. *Electron. Lett.* 32:2109–2111.

26. Costa, F., S. Genovesi, A. Monorchio, et al. 2015. A robust differential-amplitude codification for chipless RFID. *IEEE Microw. Wireless Compon. Lett.* 25:832–834.
27. Li, D., T.-W. Li, E.-P. Li, et al. 2018. A 2.5-D angularly stable frequency selective surface using via-based structure for 5G EMI shielding. *IEEE Trans. Electromagn. Compat.* 60:768–775.
28. Yan, M., S. Qu, J. Wang, et al. 2014. A novel miniaturized frequency selective surface with stable resonance. *IEEE Antennas Wireless Propag. Lett.* 13:639–641.
29. Sanz-Izquierdo, B., E.A. Parker, J.-B. Robertson, et al. 2010. Singly and dual polarized convoluted frequency selective structures. *IEEE Trans. Antennas Propag.* 58:690–696.
30. Zhang, J., Y. Yin and S. Zheng. 2009. Double screen FSSs with multi-resonant elements for multiband, broadband applications. *J. Electromagn. Waves Appl.* 23:2209–2218.
31. Rashid, A.K., B. Li and Z. Shen. 2014. An overview of three-dimensional frequency-selective structures. IEEE Antennas Propag. Mag. 56:43–67.
32. Azemi, S.N., K. Ghorbani and W.S. Rowe. 2013. A reconfigurable FSS using a spring resonator element. *IEEE Antennas Wireless Propag. Lett.* 12:781–784.
33. Lee, I.G. and I.P. Hong. 2014. 3D frequency selective surface for stable angle of incidence. *Electron. Lett.* 50:423–424.
34. Tao, K., B. Li, Y. Tang, et al. 2017. Analysis and implementation of 3D bandpass frequency selective structure with high frequency selectivity. *Electron. Lett.* 53:324–326.
35. Deng, F., X. Xi, J. Li, et al. 2015. A method of designing a field-controlled active frequency selective surface. *IEEE Antennas Wireless Propag. Lett.* 14:630–633.
36. Tennant, A., W. Hurley and T. Dias. 2012. Experimental knitted, textile frequency selective surfaces. *Electron. Lett.* 48:1386–1388.
37. Yang, S., P. Liu, M. Yang, et al. 2016. From flexible and stretchable meta-atom to metamaterial: A wearable microwave meta-skin with tunable frequency selective and cloaking effects. *Sci. Rep.* 6:1–8.
38. Lorenzo, J., A. Lazaro, R. Villarino, et al. 2017. Diversity study of a frequency selective surface transponder for wearable applications. *IEEE Trans. Antennas Propag.* 65:2701–2706.
39. Zhang, K., W. Jiang, J. Ren, et al. 2018. Design of frequency selective absorber based on parallel LC resonators. *Prog. Electromagn. Res.* 65:91–100.
40. Anwar, R.S., L. Mao and H. Ning. 2018. Frequency selective surfaces: A review. *Appl. Sci.* 8:1–46.
41. Yang, F. and Y. Rahmat-Samii. 2009. *Electromagnetic Band Gap Structures in Antenna Engineering*. Cambridge University Press: Cambridge, UK.
42. Haddad, M.T.A. 2016. Design of frequency selective surface (FSS) for mobile signal shielding. Phd thesis, The Islamic University of Gaza, Palestine.
43. Marcuvitz, N. 1951. *Waveguide Handbook*. New York: McGraw-Hill.
44. Langley, R. and E. Parker. 1982. Equivalent circuit model for arrays of square loops. *Electron. Lett.* 18:294–296.
45. Singh, D., A. Kumar, S. Meena, et al. 2012. Analysis of frequency selective surfaces for radar absorbing materials. *Prog. Electromagnet. Res. B* 38:297–314.
46. Foroozesh, A. and L. Shafai. 2010. Investigation into the effects of the patch-type FSS superstrate on the high-gain cavity resonance antenna design. *IEEE Trans. Antennas Propag.* 58:258–270.
47. Gangwar, D., S. Das, R.L. Yadava, et al. 2017. Frequency selective surface as superstrate on wideband dielectric resonator antenna for circular polarization and gain enhancement. *Wireless Pers.* Commun. 97:3149–3163.
48. Mondal, K., D.C. Sarkar and P.P. Sarkar. 2019. 5×5 Matrix patch type frequency selective surface based miniaturized enhanced gain broadband microstrip antenna for WlAN/WiMAX/ISM band applications. *Prog. Electromagnet. Res. C* 89:207–219.

49. Pirhadi, A., H. Bahrami and J. Nasri. 2012. Wideband high directive aperture coupled microstrip antenna design by using a FSS superstrate layer. *IEEE Trans. Antennas Propag.* 60:2101–2106.
50. Akbari, M., S. Gupta, M. Farahani, et al. 2016. Gain enhancement of circularly polarized dielectric resonator antenna based on FSS superstrate for MMW applications. *IEEE Trans. Antennas Propag.* 64:5542–5546.
51. Yahya, R., A. Nakamura, M. Itami, et al. 2017. A novel UWB FSS based polarization diversity antenna. *IEEE Antennas Wireless Propag. Lett.* 16:2525–2528.
52. Moustafa, L. and B. Jecko. 2010. Design of a wideband highly directive EBG antenna using double-layer frequency selective surfaces and multifeed technique for application in the Ku-band. *IEEE Antennas Wireless Propag. Lett.* 9:342–346.
53. Munk, B.A. 2003. *Finite Antenna Arrays and FSS.* Wiley: New York.
54. Pasian, M., S. Monni, A. Neto, et al. 2010. Frequency selective surfaces for extended bandwidth backing reflector functions. *IEEE Trans. Antennas Propag.* 58:43–50.
55. Ranga, Y., L. Matekovits, K.P. Esselle, et al. 2011. Multioctave frequency selective surface reflector for ultrawideband antennas. *IEEE Antennas Wireless Propag. Lett.* 10:219–222.
56. Ranga, Y., L. Matekovits, A.R. Weily, et al. 2013. A constant gain Ultra-Wideband antenna with a multi-layer Frequency Selective Surface. *Prog. Electromagnet. Res. Lett.* 38:119–125.
57. Edalati, A. and T.A. Denidni. 2011. High-gain reconfigurable sectoral antenna using an active cylindrical FSS structure. *IEEE Trans. Antennas Propag.* 59:2464–2472.
58. Chatterjee, A. and S.K. Parui. 2018. Beamwidth control of omnidirectional antenna using conformal frequency selective surface of different curvatures. *IEEE Trans. Antennas Propag.* 66:3225–3230.
59. Chatterjee, A. and S.K. Parui. 2017. Frequency-dependent directive radiation of monopole-dielectric resonator antenna using a conformal frequency selective surface. *IEEE Trans. Antennas Propag.* 65:2233–2239.
60. Lee, C., R. Sainati and R.R. Franklin. 2018. Frequency selective surface effects on a coplanar waveguide feedline in Fabry–Perot cavity antenna systems. *IEEE Antennas Wireless Propag. Lett.* 17:768–810.
61. Goldsmith, A. 2005. *Wireless Communications.* Cambridge University Press: Cambridge, UK.
62. Sharawi, M.S. 2014. *Printed MIMO Antenna Engineering.* Artech House: Norwood, MA.
63. Karimian, R. and H. Tadayon. 2013. Multiband MIMO antenna system with parasitic elements for WLAN and WiMAX application. *Int. J. Antenna Propag.* 2013:1–9.
64. Mikki, S.M. and Y.M. Antar. 2015. On cross correlation in antenna arrays with applications to spatial diversity and MIMO systems. IEEE Transactions on Antennas and Propagation. 63:1798–1810.
65. Chae, S.H., S.K. Oh and S.O. Park. 2007. Analysis of mutual coupling, correlations, and TARC in WiBro MIMO array antenna. *IEEE Antennas Wireless Propag. Lett.* 6:122–125.
66. Kesavan, A., R. Karimian and T.A. Denidni. 2016. A novel wideband frequency selective surface for millimeter-wave applications. *IEEE Antennas Wireless Propag. Lett.* 15:1711–1713.
67. Karimian, R., A Kesavan, M. Nedil, et al. 2017. Low-mutual-coupling 60-GHz MIMO antenna system with frequency selective surface wall. *IEEE Antennas Wireless Propag. Lett.* 16:373–376.
68. Zhang, B., J.M. Jornet, I.F. Akyildiz, et al. 2019. Mutual coupling reduction for ultra-dense multi-band plasmonic nano-antenna arrays using graphene-based frequency selective surface. *IEEE Access.* 7:33214–33225.

69. Bilal, M., R. Saleem, H. Abbasi, et al. 2017. An FSS-based nonplanar quad-element UWB-MIMO antenna system. *IEEE Antennas Wireless Propag. Lett.* 16:987–990.
70. Zhu, X., X. Yang, Q. Song, et al. 2017. Compact UWB-MIMO antenna with metamaterial FSS decoupling structure. *EURASIP J. Wireless Commun. Netw.* 2017:1–6.
71. Liu, Z., S. Jie, H. Ma, et al. 2019. A novel dual-passband net-shaped FSS structure used for MIMO antennas. *Prog. Electromagnet. Res. C* 90:29–39.
72. Feresidis, A.P. and J. Vardaxoglou. 2001. High gain planar antenna using optimised partially reflective surfaces. *IEE Proc. Microwaves Antennas Propag.* 148:345–350.
73. Hassan, T., M.U. Khan, H. Attia, et al. 2018. An FSS based correlation reduction technique for MIMO antennas. *IEEE Trans. Antennas Propag.* 66:4900–4905.
74. Das, G., N. Kumar Sahu, A. Sharma, et al. 2019. FSS based spatially decoupled back to back four port MIMO DRA with multi-directional pattern diversity. *IEEE Antennas Wireless Propag. Lett.* 18:1552–1556.
75. Natarajan, R., M. Kanagasabai, S. Baisakhiya, et al. 2013. A compact frequency selective surface with stable response for WLAN applications. *IEEE Antennas Wireless Propag. Lett.* 12:718–720.
76. Ünaldı, S., S. Cimen, G. Çakır, et al. 2017. A novel dual-band ultrathin FSS with closely settled frequency response. *IEEE Antennas Wireless Propag. Lett.* 16:1381–1384.
77. Bashiri, M., C. Ghobadi, J. Nourinia, et al. 2017. WiMAX, WLAN, and X-band filtering mechanism: Simple-structured triple-band frequency selective surface. *IEEE Antennas Wireless Propag. Lett.* 16:3245–3248.
78. Li, D., T.-W. Li, R. Hao, et al. 2017. A low-profile broadband bandpass frequency selective surface with two rapid band edges for 5G near-field applications. *IEEE Trans. Electromagnet. Comp.* 59:670–677.
79. Sivasamy, R. and M. Kanagasabai. 2017. Novel reconfigurable 3-D frequency selective surface. *IEEE Trans. Compon., Packag. Manuf. Technol.* 7:1678–1682.
80. Majidzadeh, M., C. Ghobadi and J. Nourinia. 2017. Ultrawide band electromagnetic shielding through a simple single layer frequency selective surface. *Wireless Pers. Commun.* 95:2769–2783.
81. Haupt, R.L. and M. Lanagan. 2013. Reconfigurable antennas. *IEEE Antennas Propag. Mazagine* 55:49–61.
82. Zhu, H.L., X.H. Liu, S.W. Cheung, et al. 2014. Frequency reconfigurable antenna using metasurfaces. *IEEE Trans. Antennas Propag.* 62:80–85.
83. Bakshi, S.C., D. Mitra and S. Ghosh. 2019. A frequency selective surface based reconfigurable rasorber with switchable transmission/reflection band. *IEEE Antennas Wireless Propag. Lett.* 18:29–33.
84. Li, L., J. Wang, J. Wang, H. Ma, H. Du, J. Zhang, S. Qu and Z. Xu. 2016. Reconfigurable all-dielectric metamaterial frequency selective surface based on high-permittivity ceramics. Sci. Rep. 6:1–8.
85. Niroo-Jazi, M. and T.A. Denidni. 2013. Electronically sweeping-beam antenna using a new cylindrical frequency-selective surface. *IEEE Trans. Antennas Propag.* 61:666–676.
86. Karmakar, N.C. and M.E Bialkowski. 2001. A beam-forming network for a circular switched-beam phased array antenna. *IEEE Microwave Wireless Compon. Lett.* 11:7–9.
87. Edalati, A. and T.A. Denidni. 2009. Reconfigurable beamwidth antenna based on active partially reflective surfaces. *IEEE Antennas Wireless Propag. Lett.* 8:1087–1090.
88. Ge, L. and K.M. Luk. 2016. Band-reconfigurable unidirectional antenna. *IEEE Antennas Propag. Magazine* 52:18–27.
89. Saleem, M.K., M.A.S. Alkanhal and A.F. Sheta. 2014. Dual strip-excited dielectric resonator antenna with parasitic strips for radiation pattern reconfigurability. *Int. J. Antennas Propag.* 2014:1–7.

90. Zou, L. and C. Fumeaux. 2011. A cross-shaped dielectric resonator antenna for multifunction and polarization diversity applications. *IEEE Antennas Wireless Propag. Lett.* 10:742–745.
91. Mahmood, S.M. and T.A. Denidni. 2016. Pattern-reconfigurable antenna using a switchable frequency selective surface with improved bandwidth. *IEEE Antennas Wireless Propag. Lett.* 15:1148–1151.
92. Ji, L., Z. Zhang, and N.-W. Liu. 2019. A two-dimensional beam steering partially reflective surface (PRS) antenna using a reconfigurable FSS structure. *IEEE Antennas Wireless Propag. Lett.* 18:1076–1080.
93. Jazi, M.N. and T.A. Denidni. 2010. Agile radiation-pattern antenna based on active cylindrical frequency selective surfaces. *IEEE Antennas Wireless Propag. Lett.* 9:387–388.
94. Bouslama, M., M. Traii, T.A. Denidni, et al. 2016. Beam-switching antenna with a new reconfigurable frequency selective surface. *IEEE Antennas Wireless Propag. Lett.* 15:1159–1162.
95. Bouslama, M., M. Traii, T.A. Denidni, et al. 2017. Reconfigurable frequency selective surface for beam-switching applications. *IET Microwaves Antennas Propag.* 11:69–74.
96. Gu, C., B.S. Izquierdo, S. Gao, et al. 2017. Dual-band electronically beam-switched antenna using slot active frequency selective surface. *IEEE Trans. Antennas Propag.* 65:1393–1398.
97. Kesavan, A., M. Mantash, J. Zaid, et al. 2018. A dual-plane beam-sweeping millimeter-wave antenna using reconfigurable frequency selective surfaces. *IEEE Antennas Wireless Propag. Lett.* 17:1832–1836.
98. Ghosh, S. and S. Lim. 2018. Fluidically reconfigurable multifunctional frequency-selective surface with miniaturization characteristic. *IEEE Trans. Microwave Theory Tech.* 66:3857–3865.
99. Farooq, U., M.F. Shafique and M.J. Mughal. 2019. Polarization insensitive dual band frequency selective surface for RF shielding through glass windows. *IEEE Trans. Electromagnet. Comp.* 62:93–100.
100. Syed, I.S., Y. Ranga, L. Matekovits, et al. 2014. A single-layer frequency-selective surface for ultrawideband electromagnetic shielding. *IEEE Trans. Electromagnet. Comp.* 56:1404–1411.
101. Sivasamy, R., B. Moorthy, M. Kanagasabai, et al. 2018. A wideband frequency tunable FSS for electromagnetic shielding applications. *IEEE Trans. Electromagnet. Comp.* 60:280–283.
102. Duan, Z., G. Abomakhleb and G. Lu. 2019. Perforated medium applied in frequency selective surfaces and curved antenna radome. *Appl. Sci.* 9:1–12.
103. Zhou, Q., P. Liu, K. Wang, et al. 2007. Absorptive frequency selective surface with switchable passband. *AEU – Int. J. Electron. Commun.* 89:160–166.
104. Abbasi, S., J. Nourinia, C. Ghobadi, et al. 2018. A sub-wavelength polarization sensitive band-stop FSS with wide angular response for X- and Ku-bands. *AEU – Int. J. Electron. Commun.* 89:85–91.
105. J. Huang, T. Wu and S. Lee. 1994. Tri-band frequency selective surface with circular ring elements. *IEEE Trans. Antennas Propag.* 42:166–175.
106. Zhou, H., S. Qu, B. Lin, et al. 2012. Filter-antenna consisting of conical FSS radome and monopole antenna. *IEEE Trans. Antennas Propag.* 60:3040–3045.
107. Chen, H., X. Hou and L. Deng. 2009. Design of frequency-selective surfaces radome for a planar slotted waveguide antenna. *IEEE Antennas Wireless Propag. Lett.* 8:1231–1233.
108. Lin, B.-Q., F. Li, Q.R. Zheng, et al. 2009. Design and simulation of a miniature thick-screen frequency selective surface radome. *IEEE Antennas Wireless Propag. Lett.* 8:1065–1068.

109. Omar, A.A. and Z. Shen. 2018. Thin 3-D bandpass frequency-selective structure based on folded substrate for conformal radome applications. *IEEE Trans. Antennas Propag.* 67:1–10.
110. Wang, F., K. Li, Y. Ren, et al. 2019. A novel reconfigurable FSS applied to the antenna radar cross section reduction. *Int. J. RF Microwave Comp. Eng.* 29:1–8.

Index

Note: Bold page numbers refer to tables; *italic* page numbers refer to figures.

Printed in the United States
By Bookmasters